NUCLEAR QUADRUPOLE RESONANCE IN CHEMISTRY

G. K. Semin, T. A. Babushkina,
and G. G. Yakobson

NUCLEAR QUADRUPOLE RESONANCE IN CHEMISTRY

Translated from Russian by A. Barouch
Translation edited by A. J. Pick

A HALSTED PRESS BOOK

JOHN WILEY & SONS
New York · Toronto

ISRAEL PROGRAM FOR SCIENTIFIC TRANSLATIONS
Jerusalem · London

Sole distributors for the Western Hemisphere
HALSTED PRESS, a division of
JOHN WILEY & SONS, INC., NEW YORK

Library of Congress Cataloging in Publication Data

Semin, Granit Konstantinovich.
Nuclear quadrupole resonance in chemistry.

Translation of *Primenenie îadernogo kvadrupol'nogo rezonansa v khimii.*
"A Halsted Press book."
1. Nuclear quadrupole resonance spectroscopy.
I. Babushkina, Tat'îana Aleksandrovna, joint author.
II. ÎAkobson, Georgii Gustavovich, joint author.
III. Title.
QD95.S4713 544'.6 74-8776
ISBN 0-470-77580-7

Distributors for the U.K., Europe, Africa and the Middle East
JOHN WILEY & SONS, LTD., CHICHESTER

Distributors for Japan, Southeast Asia and India
TOPPAN COMPANY, LTD., TOKYO AND SINGAPORE

Distributed in the rest of the world by
KETER PUBLISHING HOUSE JERUSALEM LTD.
ISBN 0 7065 1382 7
IPST cat. no. 22093 3

This book is a translation from Russian of
PRIMENENIE YADERNOGO KVADRUPOL'NOGO
REZONANSA V KHIMII
Izdatel'stvo "Khimiya"
Leningrad, 1972

Printed and bound by Keterpress Enterprises, Jerusalem
Printed in Israel

CONTENTS

AUTHORS' FOREWORD

Nuclear quadrupole resonance (NQR) is an important method of rf spectroscopy. NQR is observed in nuclei of elements with an electric quadrupole moment, such as chlorine, bromine, iodine, boron, aluminum, nitrogen, antimony, and others. The nuclei of over 60 isotopes of various elements have nuclear quadrupole moments.

Broadening of the spectral lines makes the observation of nuclear magnetic resonance for such nuclei difficult or sometimes even impossible. However, when similar problems can be studied by both NMR and NQR, these methods supplement each other. In addition, NQR helps to solve a number of specific problems that often cannot be solved by other methods.

In contrast to NMR, frequency shifts in nuclear quadrupole resonance depend on the electric interactions only, and thus NQR is one of the few methods by which the variations in electric fields at atomic nuclei can be studied directly. This provides a method for the direct experimental verification of quantum-chemical calculations. Since exact calculations of polyelectronic systems of complex molecules are extremely difficult, NQR can serve as an empirical basis for studying the distribution patterns of electron density. The exceptional sensitivity of NQR to the slightest variations in the electric field favors the application of this method to the study of numerous problems relating to intra- and intermolecular forces.

NQR is generally applied to solids (crystals, glass, amorphous substances); as a rule, molecules are studied by NQR in their ground state.

Due to the increasing amount of literature on the application of NQR to various fields of solid-state physics and chemistry, it is sometimes difficult to locate data required. NQR also has an exceedingly diverse and wide range of application, which will doubtlessly be further expanded in the near future.

NQR should be used to study conductivity in various organic, inorganic and hetero-organic compounds, intra- and intermolecular coordination forces, and complexing. Studies on macromolecular chemistry, particularly determinations of structure, mobility and ordered arrangement of macromolecules, may be extremely promising. It is now possible to study, by NQR, problems of prototropy and metallotropy, the importance of which can hardly be overestimated, especially for biological substances. NQR can be used to study substances with special electromagnetic properties: ferroelectrics, antiferromagnets, etc.

Known NQR reviews are either outdated or cover a comparatively restricted field. It has to be stressed that, for some reason, they scarcely report the work done by Soviet scientists.

This book is one of the first attempts at systematizing the different ways of applying NQR to chemistry. The authors concentrated mainly on chemical problems, which can be solved more easily by NQR.

Since the book is intended for chemists interested in the application of physical methods to chemical problems, it only discusses the fundamental theory of nuclear quadrupole resonance, and scarcely mentions the quantum-chemical calculations of quadrupole coupling constants. These are dealt with competently by E. A. C. Lucken in his recently published book "Nuclear Quadrupole Coupling Constants." A very brief description is given of the necessary equipment and the methods for determining the NQR parameters. Application of NQR to solid-state physics is not discussed. However, the application of NQR for solving problems of crystal structure is described in some detail (Chapters 3 and 9). Another aspect discussed in detail are molecular interactions, which are of importance in solid-state chemistry (Chapter 7).

A catalog of spectral data (spectra of some 3,000 compounds) and tables of quadrupole coupling energy eigenvalues are presented in the appendixes at the end of the book. The references cover papers published up to late 1969.

The authors wish to express their sincere gratitude to T. L. Khotsyanova and Yu. K. Maksyutin, who helped to prepare Section 1 of Chapter 3 and Sections 1—4 of Chapter 7, and to E. V. Bryukhova for her contribution to the preparation of the manuscript and valuable criticism.

The authors hope that their work will arouse the interest of scientists who have never before applied NQR to their research. The authors realize the imperfections of the book. They followed the saying that it is better to write an imperfect book than never to write a perfect one.

Any suggestions and comments will be gratefully received.

Chapter 1

ELEMENTS OF NQR THEORY

Nuclear quadrupole resonance arises due to interaction of the atomic nucleus having nonspherical symmetry with the nonuniform field of electrons surrounding it [1, 2]. The nuclear electric quadrupole moment is the deviation of the nuclear charge distribution from spherical symmetry. The magnitude of the nonuniformity of the electric field is given by the electric field gradient.

1. NUCLEAR ELECTRIC QUADRUPOLE MOMENT

There exist a large number of nuclei in which the charge distribution is nonspherically symmetric. In general, the charge distribution in nuclear spheroids can be either elongated or contracted in the direction of the nuclear spin I (Figure 1.1).

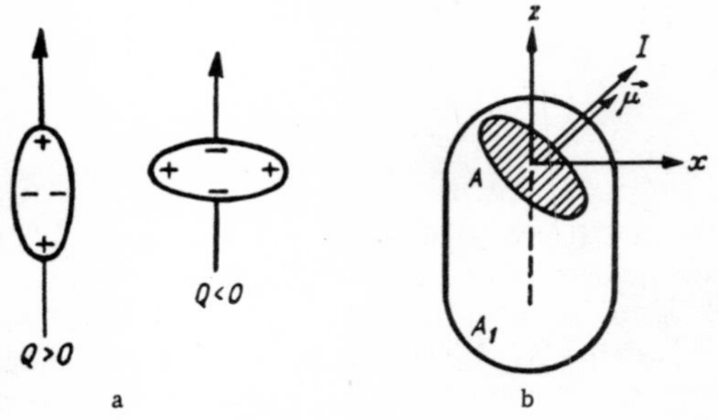

FIGURE 1.1. Nuclear charge distribution:

a) quadrupole nuclei of the atoms; b) diagram of the arrangement of the nuclear moments in the nonuniform electric field of valence electrons (the z axis is directed along the coupling axis).

In electrostatics such a charge distribution is called axial quadrupole. Accordingly, we can represent the nonspherical distribution of the charge as a sum of a monopole (the charge) and a quadrupole. It follows from general considerations that only nuclei with $I \geqslant 1$ may have a nuclear electric quadrupole moment Q (see Appendix I). The quadrupole moments of the nuclei describing the asphericity of the charge distribution are not too large, and all the known values of Q for atomic nuclei lie within the range -2 barn $< Q <$ 10 barn (1 barn $= 10^{-24}$ cm^2).

2. ELECTRIC FIELD GRADIENT AT THE NUCLEUS

Formation of a chemical bond is accompanied by a strong nonuniform electric field along the direction of the bond (the z axis). The magnitude of this nonuniformity is the electric field gradient:

$$q_{zz} = \frac{\partial^2 V}{\partial z^2}$$

where V is the electrostatic potential at the atomic nucleus created by all the surrounding charges.

In general, q is a tensor having nine components in the cartesian coordinate system (x, y, z) with

$$q_{xx} + q_{yy} + q_{zz} = 0.$$

At the nucleus the charge vanishes (the Laplace condition is fulfilled). Usually the electric field gradient is expressed in the coordinate system of principal axes where the nondiagonal terms of the tensor vanish. For $q_{xx} = q_{yy}$ there is axial symmetry, whereas for $q_{xx} \neq q_{yy}$ (i. e. when additional bonds appear), an asymmetry parameter is introduced. This parameter describes the deviation of the electron distribution in the bond from axial symmetry:

$$\eta = \frac{|q_{xx} - q_{yy}|}{q_{zz}}.$$

We shall examine the factors causing the appearance of a nonuniform electric field on the atomic nucleus.

1. The value of eq is mainly determined by the valence electrons. The largest contribution is that of the p electrons (the three fully occupied p orbitals do not contribute to the electric field gradient, and the same holds for the s electrons).

2. The polarization of the atomic core by the valence electrons can also contribute to the value of eq. For elements of atomic number larger than 40, this contribution can be very significant, for example in ^{79}Br it can be ~30%.

3. It is necessary to allow for the effect on the electric field gradient of distortions of the inner electron shells under the influence of the quadrupole moment of the nucleus and the external charge. The field gradient eq is determined by the expression:

$$eq = eq_0 [1 - \gamma (r)],$$

where $\gamma(r)$ is a parameter called the Sternheimer antiscreening factor (cf. Chapter 5).

4. The influence of charges close to the atom considered (atoms, ions, molecules), i. e. charges lying above the range of the atomic radius, are significant.

However, under normal conditions we treat atoms bound in molecules, where the state of the valence electrons (and the degree of polarization of the atomic core) can suffer considerable disturbances. These are caused by the influence of the valence electrons of a nearby atom, and the distribution of charges on adjacent atoms or ions, i. e. the charges lying far beyond the limits of the atomic core.

So far there is no rigorous theory linking the value of the electric field gradient eq with the distribution of the electron density about the nucleus. The semiempirical relationships derived by Townes and Daily [3, 4] can give only a tentative evaluation of the characteristics of the chemical bond. A rigorously substantiated selection of a given variant of distribution of the contributions introduced in eq by ionization or hybridization is impossible within the framework of these relationships. The experiment can yield only the values of e^2Qq and η, and as long as no exact theory of many-electron systems exists, the data obtained by NQR will not have a complete theoretical background.*

However, by studying the variation of the NQR frequencies or of e^2Qq in series of similar compounds we are able to make interesting comparisons, from which definite conclusions concerning the change in character of the chemical bonds can be drawn..

3. ENERGY LEVELS AND NQR FREQUENCIES

If the atomic nucleus possesses a quadrupole moment, its interaction energy with the nonuniform electric field of the charges surrounding it must depend on the mutual orientation of the nucleus and the field gradient.

Experiments show that the preferred direction (the axis of quantization of the nuclear spin I) coincides for monovalent atoms with the direction of the bond axis, i. e. with the direction of the maximum field gradient. It follows from quantum-mechanical treatment [1] that the interaction energy of a nucleus having an electric quadrupole moment in a nonuniform electric field with axial symmetry in the direction of the chemical bond has the form:

$$E_m = \frac{e^2Qq\,[3m^2 - I\,(I + 1)]}{4I\,(2I - 1)}, \qquad (1.1)$$

where m is the magnetic quantum number for the given nuclear spin I, and takes values I to $-I$, having $2I + 1$ values in all.

* This applies to the other spectroscopic methods as well. Nevertheless, the NQR data were the first to enable the chemist to establish the spatial distribution of electrons about certain atoms and provided him with a unique method to quantitatively test the predictions of theoretical chemistry.

It is seen from formula (1.1) that the quadrupole energy levels are twofold degenerate with respect to m, since E_m depends only on the absolute quantity $|m|$. Thus, for ^{35}Cl and ^{79}Br, $I = 3/2$, there are two twofold degenerate energy levels. Between these energy levels one transition occurs with frequency:

$$\nu = \frac{E_m - E_{m-1}}{h}, \tag{1.2}$$

where h is Planck's constant.*

For ^{127}I, ^{121}Sb there are three twofold degenerate levels, and therefore two transitions are permitted.

If the electric field gradient does not possess axial symmetry, the asymmetry parameter η must be introduced. For spin $I = 3/2$ the frequency ν is expressed as follows in terms of e^2Qq and η [5]:

$$\nu = \frac{e^2Qq}{2}\left(1 + \frac{\eta^2}{3}\right)^{\frac{1}{2}}. \tag{1.3}$$

It is obvious that in this case the quadrupole coupling constant e^2Qq and the asymmetry parameter η cannot be determined simultaneously by measuring a single NQR frequency of an unknown compound.

However, for organic compounds the function $\left(1 + \frac{\eta^2}{3}\right)^{1/2}$ is not too sensitive to variation in η, since in most cases $\eta \leqslant 0.2$. This causes the NQR resonance to be determined mainly by the largest field gradient eq_{zz}, i. e. by the gradient along the bond axis.

If the spin $I = 1, 2, 5/2$ and higher, e^2Qq and η are determined from the frequencies of the quadrupole spectra. The energy levels for the different cases can be obtained by solving the corresponding secular equations [5]:

$$\begin{aligned} &I = 5/2 \quad E^3 - 7(3 + \eta^2)E^2 - 20(1 - \eta^2) = 0 \\ &I = 7/2 \quad E^4 - 42\left(1 + \frac{\eta^2}{3}\right)E^2 - 64(1 - \eta^2)E + 105\left(1 + \frac{\eta^2}{3}\right)^2 = 0 \\ &I = 9/2 \quad E^5 - 11(3 + \eta^2)E^3 - 44(1 - \eta^2)E^2 + \\ &\qquad + \frac{44}{3}(3 + \eta^2)^2 E + 48(3 + \eta^2)(1 - \eta^2) = 0. \end{aligned} \tag{1.4}$$

A numerical solution of these equations for $I = 5/2$, $7/2$, and $9/2$, and for $\eta = 0.1$–1.0 with intervals of 0.1 was given by Cohen [6]. Livingston and Zeldes [7] tabulated the eigenvalues for $I = 5/2$ and for $\eta = 0$–1.0 with intervals of 0.001.

Tables of the eigenvalues of the energy which are a solution of equations (1.4) for the range $\Delta\eta = 0.01$, the differences $\Delta E_{ij} = E_m - E_{m-1}$ and

* Note that the existence of energy levels of quadrupole interaction is due to interaction of the nonuniform electric field with the electric quadrupole moment of the nucleus. The induced transitions are associated with the interaction of the magnetic moment μ of the nucleus with the magnetic vector of the rf field. The interaction energy of the electric quadrupole moment with the electric component of the electromagnetic wave is 10^{14} times smaller than the energy of the magnetic dipole interaction.

the ratios $\Delta E_{ij}/\Delta E_{i'j'}$, equal to the corresponding frequency ratios, are given in Appendix II. The behavior of the difference ΔE_{ij} and the ratio $\Delta E_{ij}/\Delta E_{i'j'}$, with an increase in the asymmetry parameter η from 0 to 1, for spins $I = {}^5/_2$, ${}^7/_2$, and ${}^9/_2$, is illustrated in Figure 1.2 (p. 5–7).

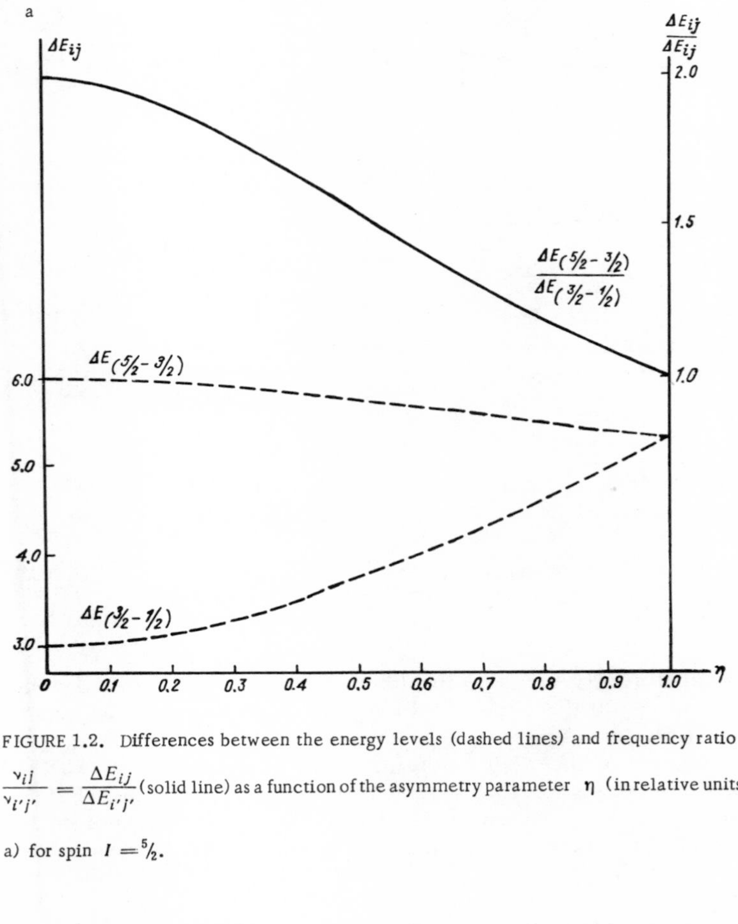

FIGURE 1.2. Differences between the energy levels (dashed lines) and frequency ratio $\frac{\nu_{ij}}{\nu_{i'j'}} = \frac{\Delta E_{ij}}{\Delta E_{i'j'}}$ (solid line) as a function of the asymmetry parameter η (in relative units):

a) for spin $I = {}^5/_2$.

For spin $I = 1$ when $\eta \neq 0$ there exist three energy levels:

$$E_{\pm} = \frac{e^2Qq}{4}(1 \pm \eta)$$

$$E_0 = -\frac{e^2Qq}{2} \tag{1.5}$$

between which two transitions are possible:

$$\nu_{\pm} = \frac{3}{4}e^2Qq\left(1 \pm \frac{1}{3}\eta\right). \tag{1.6}$$

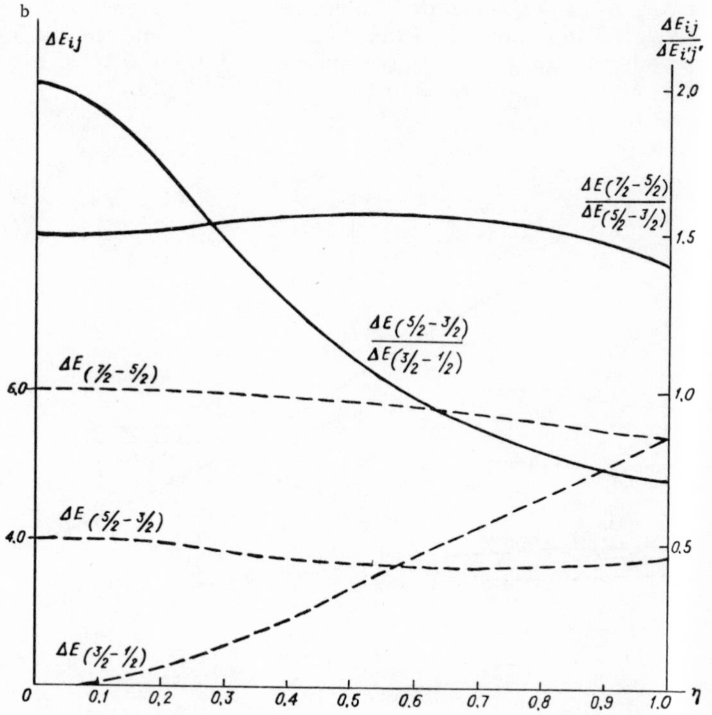

FIGURE 1.2. b) for spin $I = {}^7\!/_2$.

For a sufficiently large η, the transition $E_+ \longleftrightarrow E_-$ can also be observed. For $\eta = 0$ the upper level $E_\pm$ becomes twofold degenerate and only one transition is observed.

The intensity of the NQR lines [5, 6] for the single crystal depends on the crystal orientation with respect to the direction of the rf field. It is maximum when the principal axis of the electric field gradient is perpendicular to the rf field, and vanishes when these two directions coincide. For polycrystalline samples there is a whole set of possible directions of the field gradient with respect to the rf field, and thus the intensity is $\sim {}^2\!/_3$ of the maximum for the case when spin $I = {}^3\!/_2$.

For large half-integer spins the probability of transitions with different Δm depends strongly on the asymmetry parameter of the electric field gradient. The values of the transition probabilities for $I = {}^5\!/_2$, ${}^7\!/_2$ and ${}^9\!/_2$ as a function of η are given in Figure 1.3 and Appendix III.

In addition to the frequency, other characteristic parameters of NQR spectra are the line width $\Delta\nu$, the relaxation times T_1 (spin-lattice), T_2 (spin-spin) and T_2^* (which is inversely proportional to the line width $\Delta\nu$) [15]. As a rule the NQR line width is determined by the random scattering of the gradient values due to defects and stresses appearing in nonideal crystals containing impurities. The temperature considerably affects the frequency, width and intensity of the NQR lines. Accordingly, determinations of NQR frequencies in chemical investigations are usually performed at liquid nitrogen temperature (77°K).

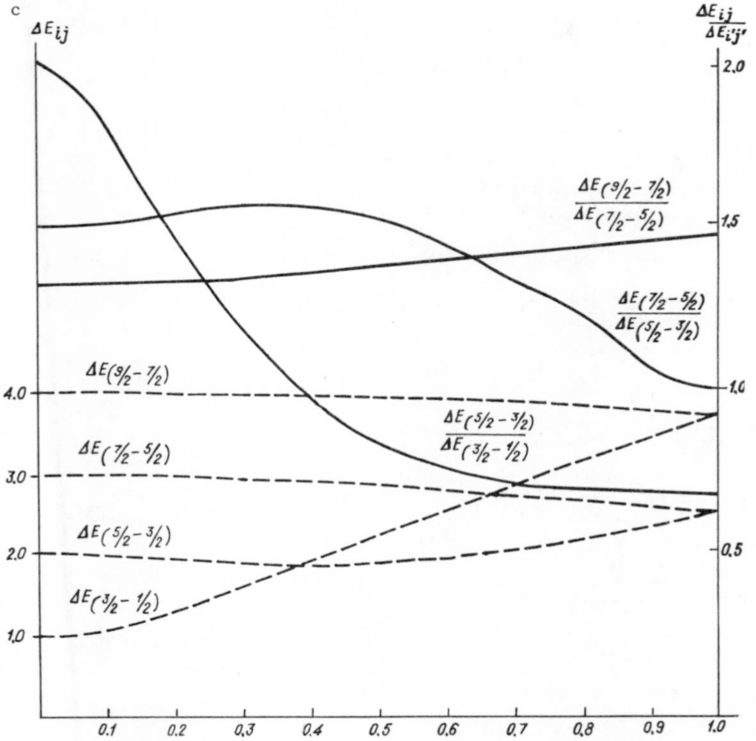

FIGURE 1.2. c) for spin $I = {}^9/_2$.

4. INFLUENCE OF A WEAK MAGNETIC FIELD

The parameters e^2Qq and η for spin $I = {}^3/_2$ can be determined from the splittings of the NQR lines in a weak magnetic field (the Zeeman effe .). Experiments of this type, which use large single crystals, are no longer widely employed.

In weak magnetic fields the degeneracy of the energy levels is removed and line splitting occurs (see Figure 1.4 [5]). The resulting interaction energy due to interaction between the magnetic dipole moment of the nucleus and the applied constant magnetic field H_0 will have different values for nuclear spin components "along" and "against" the field H_0. The total interaction energy can be described by the expression:

$$E_{\pm m} = \frac{e^2Qq}{4I(2I-1)}[3m^2 - I(I+1)] \mp mh\gamma H_0 \cos\theta, \qquad (1.7)$$

where γ is the gyromagnetic ratio of the given nucleus, and θ is the angle between the field H_0 and the direction of the maximum electric field gradient.

The spin components with $m = \pm 1/2$ are a special case. In the presence of a magnetic field with pure states, $\psi_{+1/2}$ and $\psi_{-1/2}$ do not exist, but mixed states of ψ_+ and ψ_- are formed. Thus, for spin $I = 3/2$, when a magnetic field H_0 is applied, the line of quadrupole resonance is split into four lines (see Figure 1–4). The magnitude of this splitting depends on the angle θ. For a certain orientation of the magnetic field with respect to the z axis of the electric field gradient (when $\sin^2\theta = \frac{2}{3-\eta\cos 2\varphi}$, where φ is the azimuthal angle), the inner components α of the quadruplet coincide. The geometrical locus of such points is called the cone of zero splitting.

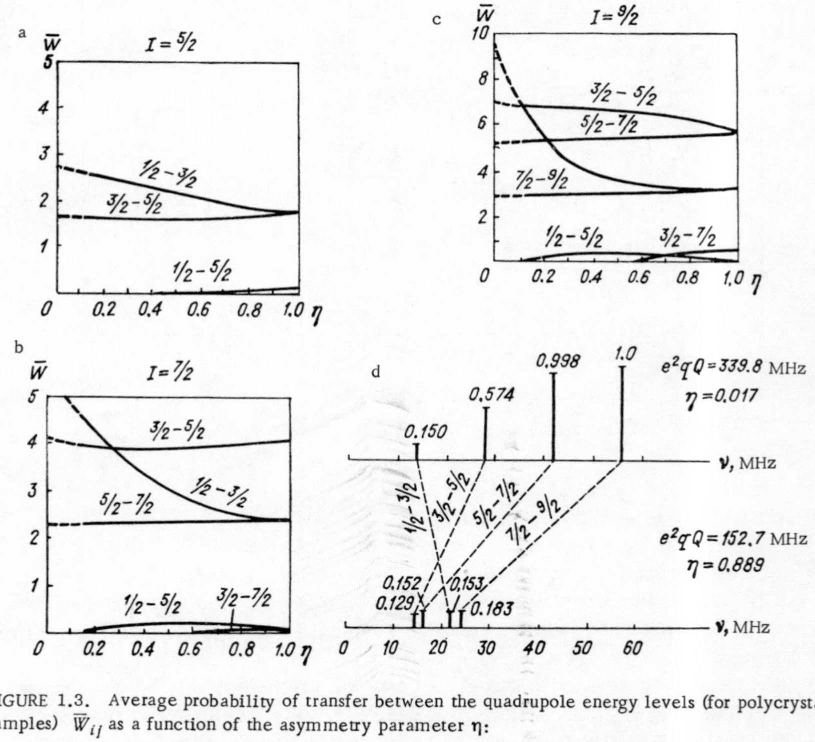

FIGURE 1.3. Average probability of transfer between the quadrupole energy levels (for polycrystalline samples) $\overline{W}_{ij}$ as a function of the asymmetry parameter η:

a) for spin $I = 5/2$; b) for spin $I = 7/2$; c) for spin $I = 9/2$; d) relative intensities of NQR lines for ^{209}Bi $\overline{W}_{ij}\nu_{ij}^2$ calculated from the spectrum of Bi_2S_3.

For spin $I = {}^3/_2$, $\theta = 54°44'$, for $I = {}^5/_2$, $\theta = 43°19'$. For $\eta \neq 0$ the section of the cone for zero splitting is an ellipse. For spin $I = {}^3/_2$, η is determined from the ratio between the long and short axes. In addition, if the value of θ_0 for the azimuthal angles $\varphi = 0$ and $\varphi = 90°$ is known, it is also possible to determine the asymmetry parameter of the electric field gradient:

$$I = {}^3/_2; \quad \eta = 3\,\frac{\sin^2\theta_0\,(0°) - \sin^2\theta_0\,(90°)}{\sin^2\theta_0\,(0°) + \sin^2\theta_0\,(90°)}. \tag{1.8}$$

The Zeeman effect on single crystals is used widely in structural analysis by NQR (see Chapter 3).

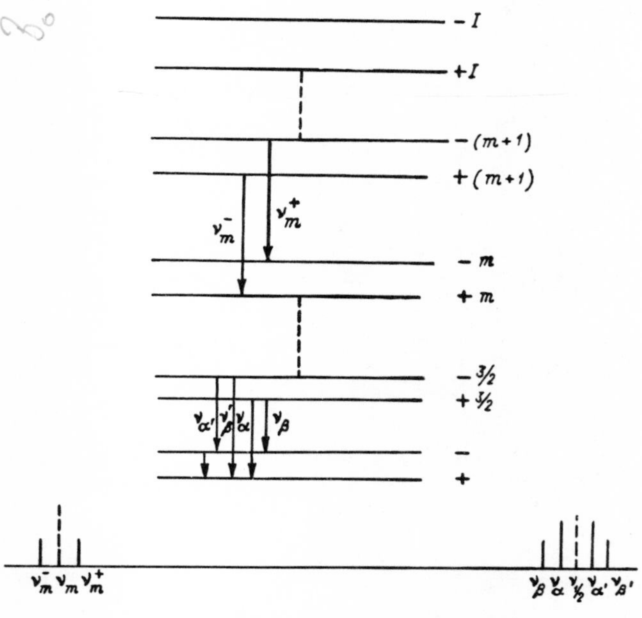

FIGURE 1.4. Zeeman splitting of the quadrupole energy levels [5]

For spin $I = 1$, when $\eta = 0$, the magnetic field removes the degeneracy and two lines appear with splitting $\Delta\nu = \gamma H_0 \cos\theta$. When $\eta \neq 0$ the energy levels $E_\pm$ are shifted in opposite directions. For $I = 1$, this is frequently used to identify the origin of the two lines in the spectrum. These two lines may appear either when $\eta \neq 0$ or when $\eta = 0$ if there are two crystallographically nonequivalent positions of the corresponding atoms.

It was attempted recently [8, 9] to use the Zeeman effect in powder samples to determine the value of η, but this method is not so accurate.

5. INFLUENCE OF EXTERNAL ELECTRIC FIELDS

When an external electric field is applied to atoms with quadrupole nuclei, the direct effect of the field on the value of the gradient is small. This effect (called the Stark effect) takes place by additional perturbation of the electron environment of the quadrupole nucleus. For NQR it was predicted by Bloembergen [10] and observed in [11–15].

The spectral manifestations of the Stark effect in NQR differ strongly for half-integer and integer spins.

For a half-integer spin ($I = {}^3/_2$) when a field E is applied we easily obtain:

$$\frac{\delta\nu}{\nu_0} = \frac{\sum_i E_i \left[R_{izz} + \frac{1}{3}\eta\,(R_{ixx} - R_{iyy})\right]}{eq_{zz}\left(1 + \frac{1}{3}\eta^2\right)}, \tag{1.9}$$

taking the asymmetry parameter η into account [10].

The expression is greatly simplified for $\eta = 0$:

$$\frac{\delta\nu}{\nu_0} = \frac{E_z \cdot R_{zzz}}{eq_{zz}}. \tag{1.10}$$

The expression for the Stark shift of the NQR frequency for integer spins ($I = 1$) and $\eta \neq 0$ is derived in a similar manner [10]:

$$\frac{\delta\nu}{\nu} = \frac{\sum_i E_i \left[R_{izz} \pm \frac{1}{3}(R_{ixx} - R_{iyy})\right]}{eq_{zz}\left(1 \pm \frac{1}{3}\eta\right)}. \tag{1.11}$$

In equations (1.9)–(1.11), R_{zzz} is the (z) component of the polarizability tensor of the atom in the given bond; E_i (or E_z) are the corresponding components of the external field strength E.

6. EQUIPMENT FOR THE OBSERVATION OF NQR

The frequencies of the nuclear quadrupole interaction range from tens of kilohertz to ~1000 MHz. It is very difficult to observe pure quadrupole transitions in the low-frequency range from 0.01 to ~2 or 3 MHz. Accordingly, indirect methods are used here, e.g., the quadrupole splittings in the NQR spectra of solids are studied. The quadrupole interaction energy is considerably lower than that of the magnetic interaction ($H_Q \ll H_M$). Indirect methods are used to study the quadrupole interaction of nuclei having a small quadrupole moment (such as ^{2}D, ^{7}Li, ^{23}Na, ^{39}K, ^{41}K, ^{133}Cs, ^{27}Al, ^{10}B, ^{9}Be, ^{14}N, etc.).

A detailed treatment of the quadrupole effects in NMR was given in the brilliant review of Cohen and Reif [16]. We should note, however, that the study of quadrupole splittings in powder substances is very difficult due to the low intensity of the quadrupole satellites, and their line width depends not only on the orientation, but also on the perfection of the crystals studied. Experiments with single crystals provide a much higher accuracy, but these experiments are used less due to the difficulty in obtaining large single crystals. For many solids the NQR signals have very long spin-lattice relaxation times. To observe satellites caused by quadrupole interaction it is necessary either to use a very low power level (which leads to signal attenuation) or to artificially shorten the spin-lattice relaxation time by irradiation.

Direct observation methods are used for large values of quadrupole interaction ($\nu > 2$ MHz). The equipment used can be divided into two basic subgroups: stationary and pulsed.

Stationary methods of observing NQR signals. Regenerative and superregenerative circuits are used as oscillator-detectors. The processes taking place in the regenerative oscillator-detector can be described briefly as follows.

The sample of the substance is placed inside the coil of the oscillator (Figure 1.5). The circuit of the oscillator sends out excited vibrations and their frequency is varied smoothly by varying the capacitance of the circuit. When the frequency satisfies the condition $\Delta E = h\nu$ (where ΔE is the energy between the quadrupole levels and ν the frequency of the resonance circuit), the substance starts to absorb rf energy, altering the active component of the conductance of the LC circuit, i. e. its Q factor. The resulting change in voltage on the LC circuit is detected and amplified.

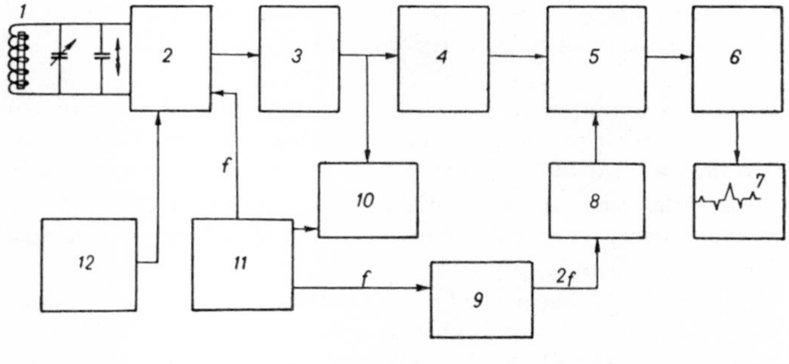

FIGURE 1.5. Block diagram of the superregenerative NQR spectrometer:

1) sample studied; 2) oscillator-detector; 3) low-frequency amplifier; 4) narrow-band low-frequency amplifier; 5) phase-sensitive detector; 6) RC-circuit, DC amplifier; 7) recorder; 8) phase inverter of the reference voltage; 9) modulation frequency doubler; 10) oscilloscope; 11) modulation frequency oscillator; 12) quenching oscillator.

In superregenerative circuits the mode of oscillations is discontinuous, i. e. the oscillations of the oscillator-detector are periodically "quenched" by an external or internal device. The detected signal is proportional to the mean rectified current. The spectrum of the superregenerator contains, in addition to the ground frequency of oscillations ν_0, a number of side bands $\nu_0 \pm n\nu_q$ (where $n = 1, 2, 3, \ldots$; ν_q is the quenching frequency). Therefore, when passing through a single quadrupole resonance line the superregenerator records $2n + 1$ "signals," of which the central one is the actual NQR signal, and the other $2n$ are superregenerative satellites. Due to the discontinuous mode of operation, a mixed absorption-dispersion signal is recorded having a ratio which is determined by the characteristics of the sample and the oscillator mode.

The NQR signals can be observed by stationary methods where the high frequency close to the absorption resonance is modulated by a lower one (of the order of tens and hundreds of hertz). At resonance, the change in the amplitude of the high-frequency oscillations gives rise, at the output, to a low-frequency voltage which is reflected by an absorption line shape. Figure 1.5 shows the block diagram of the stationary radiospectrometer.

Recently, multichannel accumulation has been used at the output of stationary NQR radiospectrometers, and this increases the sensitivity of these instruments [19].

Detailed diagrams of the regenerative and superregenerative oscillators at different frequencies from 2 to 1000 MHz can be found in [17, 18].

However, we should note that the stationary method of detecting NQR signals has a very low probability for signal detection [27]. This is attributed to the strong dependence of the sensitivity on the equivalent Q factor (i. e. the width) of the NQR line $\left(Q_{eq} = \frac{\nu}{\Delta\nu} \right)$.

The actual width of the NQR line is mainly determined by the crystal imperfection, and since the overwhelming majority of substances are not ideal crystals, NQR spectra are not detected in many samples. In a large number of substances it is even impossible to detect them by stationary methods.

In most cases the frequency is modulated to find the broad lines because this is a relatively simple operation. However, when the variations in amplitude are large, the null point of the instrument unavoidably fluctuates due to the amplitude modulation noise. The signal in the oscillator-detector may be simultaneously quenched due to the shift of the working point of the parasitic signal detector.

The same is true of magnetic modulation. Modulation methods and integrating devices have been used for many years in NMR, EPR and NQR, and were found to be capable of improving the signal to noise ratio by no more than a factor of 100. An additional increase in the accumulation time does not have any noticeable effect since the signal is quenched in the oscillator-detector. Besides, the NQR experiment, in which a large frequency range must be covered, would then become too time-consuming. In addition, the detection circuits used in the stationary detection method are very unreliable despite their apparent simplicity.

Another shortcoming of the stationary methods may be attributed to the use of superregenerative oscillator-detector circuits. The appearance of side superregenerative satellites considerably complicates the spectrum and leads to a possible error in the measured frequency equal to some multiple of the superregeneration frequency. This shortcoming is particularly conspicuous in multiplet NQR spectra where analysis of the spectrum without special methods (such as modulation of the stripping frequency) is often quite impossible. The use of such circuits for removing the side signals involves a sharp decrease in the sensitivity and is efficient only for very narrow intense NQR lines.

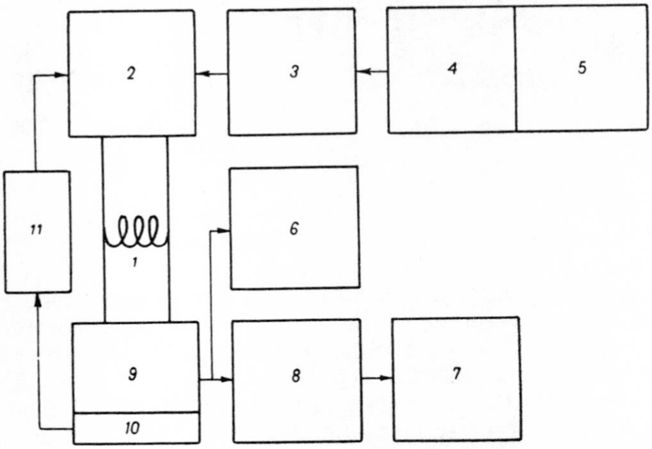

FIGURE 1.6. Block diagram of the pulsed NQR spectrometer-relaxometer:

1) sample; 2) pulsed high-frequency oscillator; 3) modulator; 4) pulser; 5) programming block; 6) oscilloscope; 7) recorder; 8) pulse accumulator; 9) receiver; 10) heterodyne; 11) automatic frequency tuner.

Therefore, the stationary method of detecting NQR signals is applicable only when the equivalent Q value of the line $Q_{eq} = \frac{\nu}{\Delta\nu}$ is not smaller than $5 \cdot 10^3$.

Pulsed methods. The method of pulsed NQR was proposed in [20, 21, 29]. The rf field is applied to the sample in the form of powerful square-wave pulses. If the amplitude and duration of the first pulse are selected on the basis of the condition $\gamma H_1 t_p^{(1)} = \frac{\pi}{2}$ (where γ is the gyromagnetic ratio of the nucleus, H_1 the amplitude of the rf pulse, and $t_p^{(1)}$ its duration), the pulse causes the total magnetic moment of the sample to rotate through 90° and become perpendicular to the z axis. On terminating the pulse, a signal of the free nuclear induction is induced in the circuit of the coil, decreasing with a time constant T_2^*.

After time τ, a second pulse, satisfying the condition $\gamma H_1 t_p^{(2)} = \pi$, is applied. This pulse causes the diverging vectors of the magnetic moments of the nuclei to rotate through 180°, so that they converge. This leads to the appearance, after time 2τ, of an echo signal decreasing with a time constant T_2. The block diagram of the pulsed NQR radiospectrometer is shown in Figure 1.6. The oscillator circuits which are programming devices for measuring the relaxation time by the spin echo method have been given in many papers [22–26].

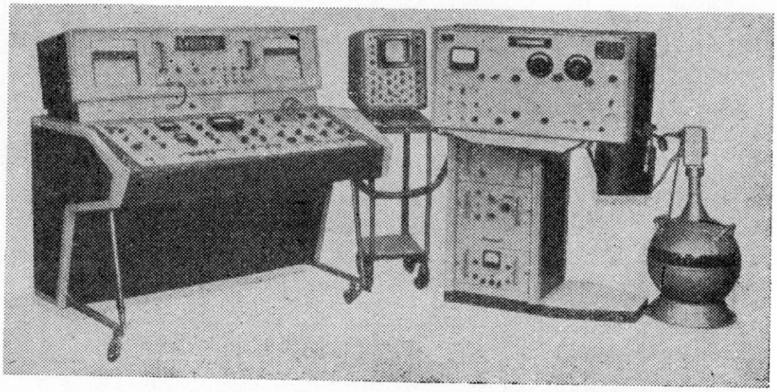

FIGURE 1.7. Pulsed NQR spectrometer-relaxometer ISSh-1 (range 2—1000 MHz) of SKB IRE AN SSSR

The main difficulty in radiospectrometers of the quadrupole spin echo type is to coordinate the tuning of the oscillator and receiver (usually of the superheterodyne type). This difficulty has been successfully overcome in the instruments developed in SKB IRE AN SSSR [27] (Figure 1.7).

The general opinion was that the sensitivity of pulsed methods was not higher than that of stationary methods. Since Hahn [5] (who helped develop radiospectrometric pulsed methods) also held this view, the introduction of pulsed methods to NQR research was considerably slowed down. This erroneous opinion evolved in the period when very little was known about the crystalline effects which lead to strong broadening of the lines. Thus, it was impossible to know exactly the frequency of samples with narrow and broad NQR lines.

Investigators were attracted by the pulsed method [27, 28] since the initial signal amplitude of the quadrupole spin echo depends little on Q_{eq} when certain conditions concerning the length of the exciting pulses are fulfilled. As a result, the pulsed method ensures a very high gain in sensitivity even for $Q_{eq} < 5 \cdot 10^3$.

The resolving capacity of the pulsed method is determined completely by the state of the sample (the line width) and depends relatively little on instrumentation factors.

BIBLIOGRAPHY

1. Casimir, H.B.R. Interaction between Atomic Nuclei and Electrons. — Teyler's Tweede Genootschap Haarlem 37 (1936), 36.
2. Dehmelt, H.G. and H. Gruger. — Naturwissenschaften 37 (1950), 111.
3. Townes, C.H. and B.P. Dailey. — J. Chem. Phys. 20 (1952), 35.
4. O'Konski, C.T. Nuclear Quadrupole Resonance Spectroscopy. — In: "Determination of organic structure by physical methods," Vol. 2, Part 11. New York, 1962.
5. Das, T.P. and E.L. Hahn. Nuclear Quadrupole Resonance Spectroscopy. — In: "Solid state physics," Suppl. 1, New York—London, 1958.
6. Cohen, M.H. — Phys. Rev. 96 (1954), 1278.
7. Livingston, R. and H. Zeldes. Tables of Eigenvalues for Pure Quadrupole Spectra, Spin $^5/_2$. — Oak Ridge National Laboratory Report ORNL-1913, 1955.

8. Toyama, M. — J. Phys. Soc. Japan 14 (1959), 1727.
9. Morino, Y. and M. Toyama. — J. Chem. Phys. 35 (1961), 1289.
10. Bloembergen, N. — Science 133 (1961), 1363.
11. Kushida, T. and K. Saika. — Phys. Rev. Lett. 7 (1961), 9.
12. Armstrong, J., N. Bloembergen, and D. Gill. — J. Chem. Phys. 35 (1961), 1132.
13. Duchesne, J., M. Read, and P. Cornil. — Physics Chem. Solids 24 (1963), 1333.
14. Read, M., P. Cornil, and J. Duchesne. — Compt. rend. 256 (1963), 5331.
15. Boguslavskii, A. A. and G. K. Semin. — Fiz. Tverd. Tela 11 (1969), 3617.
16. Cohen, M. H. and F. Reif. Nuclear Quadrupole Effects in Nuclear Magnetic Resonance. — In: "Solid state physics," Vol. 5, p. 321, 1957.
17. Semin, G. K. and E. I. Fedin. — Zh. Strukt. Khim. 1 (1960), 252, 464.
18. Grechishkin, V. S. and G. B. Soifer. — Pribory Tekhn. Eksperim., No. 5, 1964.
19. Zelenin, V. P., G. G. Kudymov, Yu. G. Svetlov, and V. A. Troshev. — Trudy ENI, Permsk. State Univ. 11, No. 4 (1966), 71.
20. Bloom, M., E. L. Hahn, and B. Herzog. — Phys. Rev. 97 (1955), 1699.
21. Herzog, B. and E. L. Hahn. — Phys. Rev. 103 (1956), 148.
22. Hahn, E. L. and B. Herzog. — Phys. Rev. 93 (1954), 639.
23. Gutowsky, H. S., J. C. Buchta, and D. E. Woessner. — Rev. Scient. Instrum. 29 (1958), 55.
24. Safin, I. A. — Pribory Tekhn. Eksperim., p. 98, 1962.
25. Safin, I. A., B. N. Pavlov, and D. Ya. Shtern. — Zavod. Lab. 30 (1964), 676.
26. Borsutskii, Z. R., A. D. Gordeev, V. S. Grechishkin, and I. V. Izmest'ev. — Trudy ENI Permsk. State Univ. 12, No. 1 (1966), 67.
27. Pavlov, B. N., I. A. Safin, G. K. Semin, E. I. Fedin, and D. Ya. Shtern. — Vestn. Akad. Nauk SSSR, No. 11 (1964), 40.
28. Shtern, D. Ya. and B. N. Pavlov. — In: "Radiospektroskopiya tverdogo tela," p. 198. Atomizdat, 1967.
29. Clark, W. — Rev. Scient. Instrum. 35 (1964), 316.

Chapter 2

CLASSIFICATION OF NQR FREQUENCY SHIFTS

As indicated in Chapter 1, the NQR frequency varies for different elements from several kHz to 1000 MHz and serves as a quantity that describes the change in e^2Qq for small η. This variation depends on the quadrupole moments of the nuclei, the states of the valence electrons and the type of chemical bond in which the particular atom participates. The factors, in decreasing order, which cause a shift in the NQR frequencies, can be classified as follows.

The NQR frequencies for different elements lie in the range 10^2—10^9 Hz depending on the value of eQ (Figure 2.1). Townes [1] gives an interesting correlation between Q/R^2 (where R is the mean value of the nucleus radius) and the number of even nucleons. The intrinsic quadrupole moments of the nuclei change in a systematic manner, increasing downward and to the left in the periodic table. Exceptions to this rule are ^{133}Cs and ^{209}Bi which have "magic" numbers of neutrons, i. e. closed nucleon shells in the nucleus. The nuclear quadrupole moments for the elements on the left-hand side of the half-groups are always larger than for those on the right-hand side. However, the opposite is true for the values of e^2Qq. This is due to the difference in the electron states of the valence electrons of the corresponding elements. In addition, most of the quadrupole elements are positive due to the tendency of the nucleus to be in an elongated form relative to the quantization axis. This shape has a lower electrostatic energy than that of a contracted nucleus having a negative value for eQ_0.

The classification of NQR frequency shifts for monovalent atoms* as a function of their positions can be represented in the following form [2].

1. The largest shifts of NQR frequencies are caused by the state of the valence electrons of the bonding atoms (i. e. the nature of the bonding atom). They can be as much as 1200—1500%.

Figure 2.2 shows the changes in the NQR frequencies of ^{35}Cl as a function of the nature of the M—Cl bond. These frequencies change very evenly, the lowest frequency corresponding to the Cl^- ion, and the highest to the chlorine atom in ClF_3. Definite conclusions concerning the structure of the molecules and the character of the bonds can be drawn from the figure.

2. The changes within one type of bond (with the same bonding atom) are only 40—50%. Figure 2—3 shows the shifts in the NQR frequencies of ^{35}Cl for different types of C—Cl bonds.

* For polyvalent atoms e^2Qq depends on the coordination number and the hybrid state.

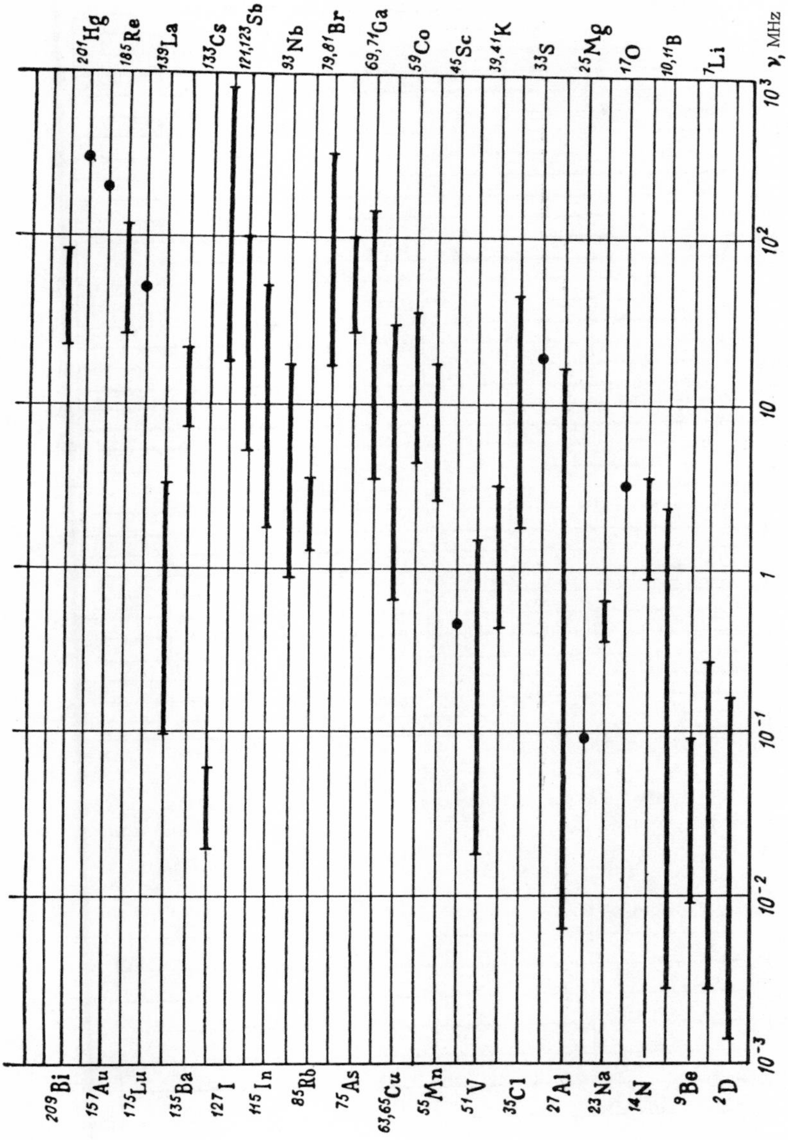

FIGURE 2.1. Experimentally found shifts in the NQR frequencies of different elements

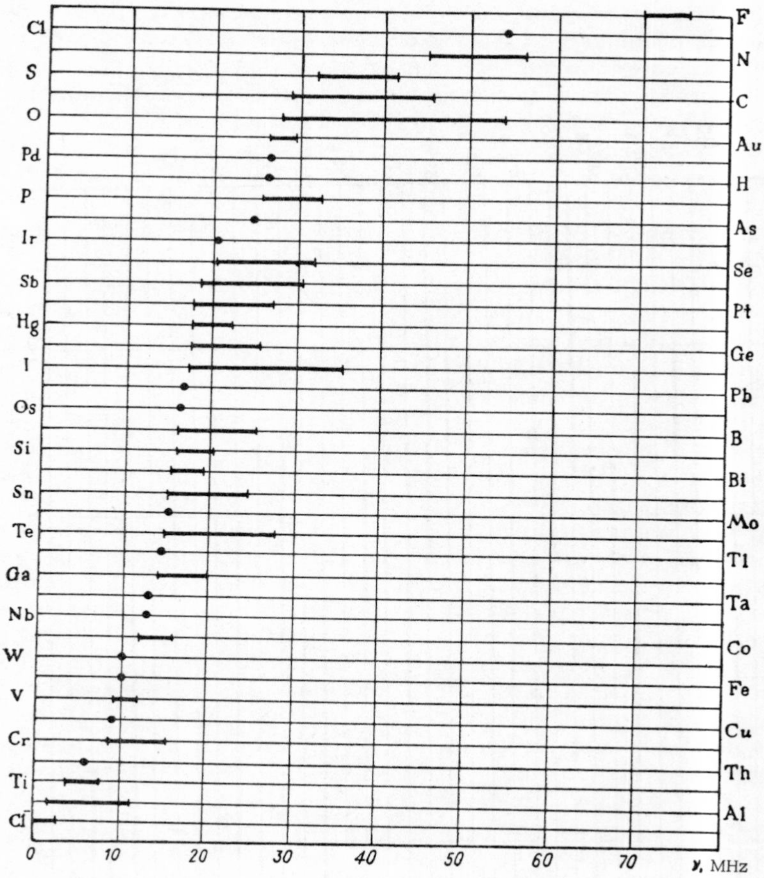

FIGURE 2.2. Variation of the NQR frequencies of ^{35}Cl as a function of the nature of the bonding atom in M—Cl

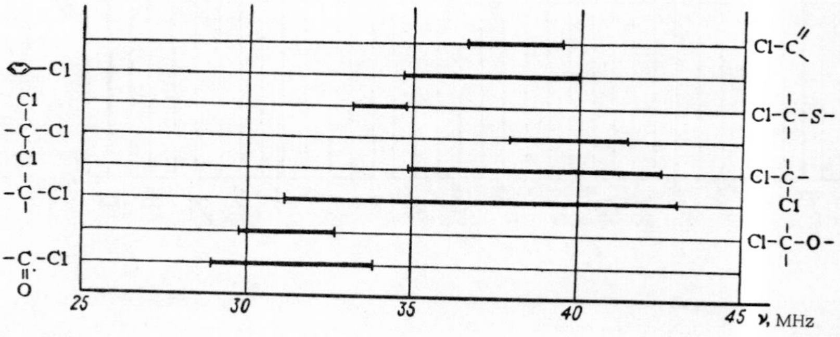

FIGURE 2.3. Variation of the NQR frequencies of ^{35}Cl in different types of C—Cl bonds

3. The range of possible shifts is reduced still further (to 10–20%) in compounds of the same class. For such molecules a shift to either side is explained by the position, number, and donor-acceptor properties of the substituent. Thus, in the series of benzenes substituted by halogens the change in frequency for the C–Cl bond is 9%, for the C–Br bond 12%, and for the C–I bond 18.5%. (The increase in the change in frequency is due to the increased mobility of the electron shell of the investigated halogen atom.)

4. Changes in the NQR frequencies caused by associative inter- and intramolecular bonds are about 3–40%.

5. Shifts of NQR frequencies in molecular crystals caused by crystal field effects are at most about 1.5–2%.* However, on studying series of similar compounds this scatter, as a rule, does not exceed about 0.3% [7].

In experiments involving chemical applications of NQR we must distinguish between the concepts of measurement accuracy, maximum possible deviations caused by the effect of the crystal, and spectral uncertainty due to the influence of the crystal field.

The accuracy of measuring NQR frequencies depends on the line width and fluctuations within 10^{-3}–10^{-7} of the measured frequency.

The maximum possible deviations (~1.5–2%) caused by the crystalline nonequivalence of chemically equivalent atoms are determined by the finite values of the dipole moments of the bonds and the intermolecular distances in the organic crystals. The order of magnitude of these contributions was estimated by examining the entire set of values of crystalline splittings known and the frequency shifts during phase transitions, as well as by analyzing the line width in statistically unordered crystals and in amorphous polymers with intermolecular interactions of the Van der Waals type.

The spectral uncertainty introduced by the crystal effect can be defined as the mean standard deviation between the calculated and measured NQR frequencies (see below). It agrees with the mean standard deviation obtained by processing the entire set of crystal splittings in molecular crystals [7]. A similar value (~0.3%) is obtained by analyzing the line shape in amorphous solids with chemically equivalent resonating atoms and intermolecular interactions of the Van der Waals type. Thus, the fact that the least squares method gave this degree of spectral uncertainty (~0.3%) means that all the frequency deviations of a chemical nature which exceed the crystal effect or are systematic have been taken into account.

The above classification of the frequency shifts in NQR enables us to define the concept of structural nonequivalence in NQR spectra. Structural nonequivalence must be understood as differences in the gradients (and frequencies) of resonating atoms. These are caused by chemical nonequivalence and are determined by the differences in distribution of the electron density in the free molecule, and also by the crystallographic nonequivalence. (The crystallographic nonequivalence arises on account of differences in the additional contributions due to the crystal field in the gradients of the resonating

* By taking into account the additional contribution of all the dipoles of the bonds of the molecules in the electric field gradient on the nucleus, we are able to use NQR for conformation analysis. The line splittings can exceed 2%, which fact enables us in many cases to determine the presence of conformers and the ratio between them [3–6].

chemically equivalent and nonequivalent atoms.) The concept of crystallographic equivalence is similar to the concept of physical nonequivalence given by Das and Hahn. Obviously, dividing the structural nonequivalence into chemical and crystallographic parts is meaningless in coordination and ionic crystals where no individual molecules exist.

BIBLIOGRAPHY

1. Townes, C. H. — Handb. Phys. 38 (1958), 377.
2. Semin, G. K. — Author's Summary of Doctoral Thesis. Inst. Elemento-Organ. Soedin. AN SSSR, 1970. (Russian)
3. Semin, G. K. and T. A. Babushkina. — Teor. Eksperim. Khim. 4 (1968), 835.
4. Maksyutin, Yu. K., E. V. Bryukhova, E. N. Gur'yanova, and G. K. Semin. — Izv. AN SSSR, chem. ser., p. 2658, 1968.
5. Englin, B. A., B. N. Osipov, V. A. Valovoi, T. A. Babushkina, G. K. Semin, V. B. Bondarev, and R. Kh. Freidlina. — Izv. AN SSSR, chem. ser., p. 1251, 1968.
6. Englin, B. A., T. A. Onishchenko, V. A. Valovoi, T. A. Babushkina, G. K. Semin, A. G. Zelinskaya, and R. Kh. Freidlina. — Izv. AN SSSR, chem. ser., p. 332, 1969.
7. Babushkina, T. A., V. I. Robas, and G. K. Semin. — In- "Radiospektroskopiya tverdogo tela," p. 221. Atomizdat, 1967.

Chapter 3

THE "CRYSTAL EFFECT" AND STRUCTURAL APPLICATIONS OF NQR

1. MULTIPLICITY OF NQR SPECTRA

We showed in the preceding chapter that the atoms of a given element which are in crystallographically equivalent positions have the same value of e^2Qq.* On the other hand, atoms of the same kind (chemically equivalent atoms) in chemically nonequivalent positions have different frequencies (crystalline shifts caused by the difference in the crystal field at the atomic nucleus).

Thus, the number of resonance lines in the multiplet for a given transition corresponds to the number of nonequivalent positions occupied by the resonating atoms in the crystal.** Before proceeding to a more detailed discussion on the interrelationship between the number and intensity of the lines in the multiplet, we shall consider some concepts of structural crystallography which are necessary for a clear understanding of the material.

We start by considering some of the relationships between the symmetry of the crystal field (internal symmetry of the crystal), the intrinsic symmetry of the molecule (or particle) and its symmetry in the crystal.

The internal symmetry of the crystal is described completely by the spatial symmetry group,† i. e. the group of symmetry operations that transform it into itself over the whole crystal space.

A full description of all the symmetry elements of the crystal space and their notation is given in the "International Tables" [1] and some other handbooks.

There are 32 types (or classes) of external crystal symmetry, among which are distributed all the known 230 space groups. This external symmetry of the crystalline polyhedrons is described by 32 point groups.

The point group represents a group of symmetry operations which leave at least one point in the same place. Each point group is characterized by a certain set of symmetry elements, multiplicity and singular point. The multiplicity of the group is determined by the number of equivalent points (elements or faces), i. e. points connected by the action of all the

* This is true since chemically equivalent atoms are usually located in crystallographically equivalent positions. In crystals with a statistically random orientation of molecules, this is not true.

** Exceptions from this rule are observed very rarely, but one of them is given in Table 3.1.

† Another term for the spatial symmetry group, namely the fedor group, is also found.

symmetrical transformations of the group. A singular point is a point which is transformed only into itself under all the symmetry operations of the group.*

On passing from the external symmetry of the crystals to the internal symmetry, we meet analogous concepts: the multiplicity of the regular system of points and the symmetry of position of the points (particles, molecules).

Consider, for instance, the molecule of tetrachloro-p-phenylenediamine with symmetry $D_{2h}(mmm)$:

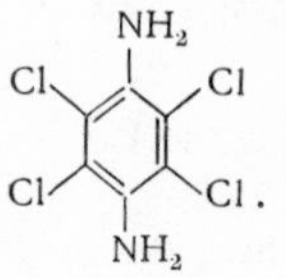

In a crystal lattice of arbitrary symmetry such a molecule can occupy an arbitrary site, i. e. not correspond to any of the symmetry elements of the lattice. It can implement an intrinsic symmetry in the crystal, corresponding to one of the symmetry elements or group of elements of the lattice, i. e. the so-called lattice complex.

Thus, the molecule in the crystal field can lie on the center of symmetry, on the twofold axis, at the intersection of three planes, two planes or the twofold axis and the mirror plane.

The position in which the molecule lies on one or several symmetry elements of the crystal lattice is called particular. In this case the molecule is transformed by the symmetry elements of the lattice complex into itself and not into other crystallographically equivalent ones. It follows that the multiplicity of the molecule (particle or point) in a particular position is smaller than its multiplicity when it is in a general position in the same space group. We should recall that not one but several molecules can be located in some general position. The number of molecules in such a position is smaller than the multiplicity of this position by a factor n (where n is the multiplicity of the lattice complex to which the molecule belongs). The multiplicity of the atoms of each of the crystallographically different molecules may differ. The difference depends on whether the crystallographically nonequivalent atom lies on an element of symmetry or not. In spectral analysis it is frequently convenient to use the number and multiplicity of the crystallographically independent atoms.

In the example, the symmetry of the molecule 1,4-$(NH_2)_2C_6Cl_4$ in the crystal (C_i — center of symmetry) is considerably lower than its intrinsic symmetry (D_{2h}) [2]. This is not surprising since the rules for the formation of molecular crystals are governed by the close-packing law [3] where the number of frequently encountered space groups is limited. Moreover, these groups, which themselves possess a high packing density of molecules of arbitrary shape, have low symmetry. Therefore, only in

* The number of singular points of the point group can be infinite. For instance, the whole set of points of plane m.

exceptional cases does the symmetry of the molecule in the crystal coincide with its intrinsic symmetry in the free state.*

We consider finally the question of the number and intensity of the multiplet lines. As was shown, the number of lines of the multiplet is equal to the number of crystallographically nonequivalent atoms. The intensity of the resonance line must be proportional to the number of resonating atoms, and therefore the intensity ratio of the lines in the multiplet must be equal to the ratio of the number of resonating atoms in each nonequivalent position or to the ratio of their multiplicities in the given space group. Obviously, such an evaluation of the intensity does not allow for line broadening caused by the spin-spin and spin-lattice interactions, and also by differences in the thermal vibrations.

If we denote the number of kinds of crystallographically nonequivalent molecules in the unit cell by N, and the number of crystallographically nonequivalent atoms in each molecule by n_j, then the number of multiplet lines will be equal to:

$$M = \sum_{j=1}^{N} n_j, \qquad (3.1)$$

where $j = 1, 2, \ldots$ etc.

The intensity ratio of the lines is then written in the form:

$$m_1 : m_2 : m_3 : \ldots : m_k = q_1 : q_2 : q_3 : \ldots : q_k,$$

where the right-hand side represents the ratio between the multiplicities of the crystallographically nonequivalent atoms in the given space group.

If the crystal structure of the substance is known, or if at least the number of molecules in the unit cell (z) and the space group have been determined, then the total number of resonating atoms z' will be equal to $z' = p_n z$, where p_n is the number of resonating atoms in the molecule $(p_n \geqslant n_j)$, or $z' = \sum_{k=1}^{K} q_k$.

In this case it is not difficult to form all the possible combinations of the distribution of atoms in nonequivalent positions (or possibly a unique combination if the space group is determined uniquely and the intrinsic symmetry of the molecule is known). The number of lines of the multiplet and the ratio of their intensities can then be specified a priori. However, knowledge of the crystal structure is not obligatory since for a given shape and symmetry of the molecule it is always possible to predetermine several admissible variants of the spectrum. It is sufficient to imagine the possible ways of arranging molecules of given symmetry in the lattice. In order to estimate the intensity of the lines in the multiplet it is sometimes convenient to use formulas for calculating the multiplicity of the molecules, namely (3.1) and

* Cases in which the molecule in the crystal has a higher symmetry than its intrinsic symmetry are also known. They correspond to statistically unordered structures or coincidently rotating molecules (or groups) and present a special problem which will be discussed below.

$$m_1:m_2:m_3 \ldots = Q_1p_{11}:Q_1p_{12}:\ldots Q_2p_{ij}:\ldots Q_ip_{ij} \tag{3.2}$$

$$p_{11} + p_{12} + \cdots + p_{1n} = p_{i1} + p_{i2} + \cdots + p_{ij}. \tag{3.3}$$

In formula (3.2), $Q_1, Q_2, \ldots Q_i$, etc., denote the multiplicity of molecules of each of the N kinds,* and p_{ij} the number of equivalent atoms of the i-th molecule belonging to the j-th position. Formula (3.3) expresses the additional trivial condition that the number of atoms N in different equivalent positions of the molecule should be equal to the total number of atoms in the molecule. Similar expressions are given in [4].

Therefore, formulas (3.1) and (3.2) determine only the number and the relative intensity of the lines in the spectrum. In general, it is impossible to split the spectrum into groups belonging to each of the N kinds of crystallographically nonequivalent molecules.

The frequency and intensity of the lines can be associated with the position of the crystallographically nonequivalent nuclei in the unit cell only by means of Zeeman-analysis (application of a constant magnetic field). These questions will be discussed in Section 2.

It follows that from the NQR spectra the configuration of the molecule in the crystal can be determined, and therefore also its intrinsic symmetry. The fact that the number of lines in the multiplet determines the number of crystallographically nonequivalent atoms can be used to solve crystallographic problems.

We shall give several examples of the detailed analysis of NQR spectra (Table 3.1) and the conclusions drawn from it. Consider the NQR spectra of ^{35}Cl in four trihalogen substituted benzenes: 1,2,3-trichlorobenzene, 1,3-dichloro-2-iodobenzene, 1,3-dichloro-2-bromobenzene and 1,3-dichloro-2-fluorobenzene. The spectrum of the first compound consists of six lines of equal intensity, which corresponds to two crystallographically independent molecules in a general position ($N = 2$):

$$m_1:m_2:m_3:\ldots:m_6 = Q_1p_{11}:Q_1p_{12}:Q_1p_{13}:Q_2p_{21}:Q_2p_{22}:Q_2p_{23}$$
$$Q_1 = Q_2 = 1; \qquad p_{11} = p_{12} = \cdots = p_{23} = 1.$$

The spectrum of the second compound corresponds to the same symmetry. It consists of four lines of equal intensity:

$$m_1:m_2:m_3:m_4 = Q_1p_{11}:Q_1p_{12}:Q_2p_{21}:Q_2p_{22}$$
$$Q_1 = Q_2 = 1; \qquad p_{11} = p_{12} = p_{21} = p_{22} = 1$$

and corresponds to two crystallographically nonequivalent molecules and four chlorine atoms in a general position.

The spectrum of the third compound consists of two lines of equal intensity:

$$m_1:m_2 = Q_1p_{11}:Q_1p_{12}; \qquad p_{11} = p_{12}; \qquad N = 1$$

and corresponds to a crystal with one crystallographically independent molecule in a general position or two crystallographically independent

* Since the true values $Q_1, Q_2, \ldots Q_i$ can be found only after a preliminary X-ray investigation, numbers proportional to them are frequently used instead.

molecules in a particular position. In the fourth compound the molecule occupies a particular position in the crystal (on axis 2 or intersected by a plane perpendicular to the plane of the molecule). Here the spectrum consists of one line which corresponds to one crystallographically independent chlorine atom in the molecule (i.e. to a molecule in a particular position). It may be seen that for a molecule of the first compound combinations 3 or 4 would have been possible in principle or a situation in which one crystallographically independent molecule would occupy a general position and the second one a particular one. These cases would correspond to the spectra:

1) three lines of equal intensity ($N = 1$)

$$m_1:m_2:m_3 = Q_1p_{11}:Q_1p_{12}:Q_1p_{13}; \qquad p_{11}:p_{12}:p_{13} = 1:1:1$$

2) two lines, with one twice as intense as the other ($N = 1$)

$$m_1:m_2 = Q_1p_{11}:Q_1p_{12}; \qquad p_{11}:p_{12} = 2:1$$

3) five lines with an intensity ratio 2 : 2 : 2 : 2 : 1 ($N = 2$)

$$m_1:m_2:m_3:m_4:m_5 = Q_1p_{11}:Q_1p_{12}:Q_1p_{13}:Q_2p_{21}:Q_2p_{22}$$
$$Q_1:Q_2 = 2:1; \qquad p_{11}:p_{12}:p_{13}:p_{21}:p_{22} = 1:1:1:2:1.$$

Cases of crystallization with a large number of crystallographically independent molecules ($N > 2$) are not met very often. Nevertheless, it is necessary to take them into account (see for example compound 6 in Table 3.1).

We shall give examples where the apparent symmetry of the molecule in the crystal, and also its intrinsic symmetry, may be determined from the number of quadrupole spectra alone. Obviously, in these cases the number of lines in the spectrum and the magnitude of the splittings are taken into account. Such a procedure may be illustrated by the spectra of the three phosphorus compounds: PCl_4F, PCl_3F_2 and PCl_2F_3 [6] (Table 3.1). The NQR spectrum of ^{35}Cl in PCl_4F consists of four lines, three of which are a very narrow triplet centered at 32.54 MHz, whereas the frequency of the fourth corresponds to 28.99 MHz. Such a pattern determines the configuration of the molecule, namely trigonal bipyramid, with three chlorine atoms in an equatorial position, and one located at a distance in an axial position. It follows from spectrum 8, which consists of two lines with intensity ratio 2 : 1 (with frequencies 31.26 and 31.89 MHz), that one of the two crystallographically nonequivalent chlorine atoms is in a general position and the second is in a particular position. Spectrum 9 (one line with $\nu = 31.49$ MHz) corresponds to a molecule in a particular position with two crystallographically equivalent chlorine atoms in an equivalent position, bonded to the plane (or axis) of symmetry in the crystal (one crystallographically nonequivalent chlorine atom in a general position). Although the determination of the symmetry of the molecule in the crystal is not unique (due to the absence of the necessary structural data), the configuration of the molecule is determined uniquely.

TABLE 3.1. Comparison of NQR spectra and crystallographic parameters

No.	Compound	Symmetry of the molecules in the crystal	Resonating atom	Crystallographic parameters	Intensity ratio of the multiplet lines
1	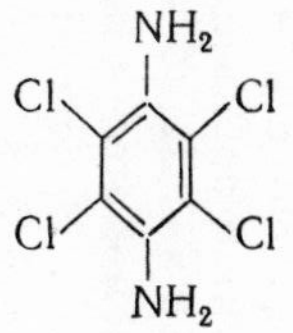	$\bar{1}\ (C_i)$	^{35}Cl	$N = 1$ $Q = 2;\ p_{11} = p_{12} = 2$ $n_1 = 2$	$m_1 : m_2 = Q_1p_{11} : Q_2p_{12} = 1:1$
2	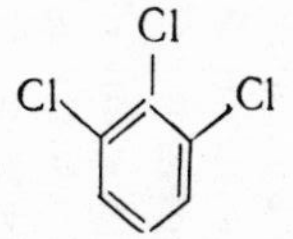	$1\ (C_1)$	^{35}Cl	$N = 2$ $Q'_1 = Q'_2 = 1;\ n_1 = n_2 = 3$ $p_{11} = p_{12} = p_{13} = p_{21} =$ $= p_{22} = p_{23} = 1$	$m_1 : m_2 : m_3 : m_4 : m_5 : m_6 =$ $= Q_1p_{11} : Q_1p_{12} : Q_1p_{13} : Q_2p_{21} :$ $: Q_2p_{22} : Q_2p_{23} = 1:1:1:1:1:1$
3	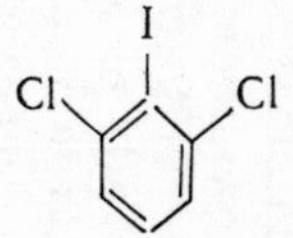	$1\ (C_1)$	^{35}Cl	$N = 2$ $Q'_1 = Q'_2 = 1;\ n_1 = n_2 = 2$ $p_{11} = p_{12} = p_{21} = p_{22} = 1$	$m_1 : m_2 : m_3 : m_4 =$ $= Q_1p_{11} : Q_1p_{12} : Q_2p_{21} : Q_2p_{22} =$ $= 1:1:1:1$
4	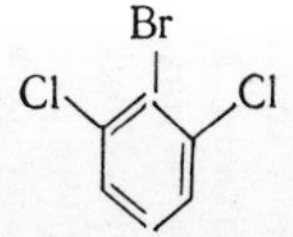	$1\ (C_1)$	^{35}Cl	$N = 1$ $Q'_1 = 1;\ n_1 = 2$ $p_{11} = p_{12} = 1$	$m_1 : m_2 = Q_1p_{11} : Q_1p_{12} = 1:1$

				or $N = 2$ $Q'_1 = Q'_2 = 1;\ n_1 = n_2 = 1$ $p_{11} = p_{12} = 2$	or $m_1 : m_2 = Q_1p_{11} : Q_2p_{21} = 1:1$
5	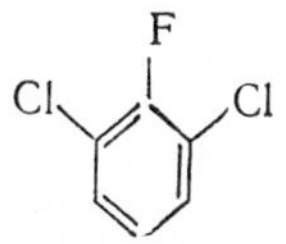	$2\,(C_2)$ $m\,(C_3)$	^{35}Cl	$N = 1$ $Q'_1 = 1;\ n_1 = 1$ $p_{11} = 2$	Singlet
6	$(C_6H_5)_3CCl$	$1\,(C_1)$	^{35}Cl	$N = 5$ $Q'_1 = Q'_2 = Q'_3 = Q'_4 = Q'_5 = 1$ $p_{11} = p_{21} = p_{31} = p_{41} = p_{51}$ $n_1 = n_2 = n_3 = n_4 = n_5 = 1$	$m_1 : m_2 : m_3 : m_4 : m_5 =$ $= Q_1p_{11} : Q_2p_{21} : Q_3p_{31} : Q_4p_{41} :$ $: Q_5p_{51} = 1:1:1:1:1$
7	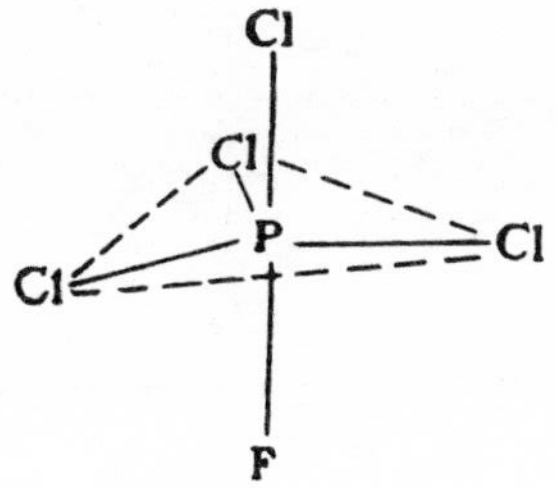	$1\,(C_1)$	^{35}Cl	$N = 1$ $Q'_1 = 1;\ n_1 = 4$ $p_{11} = p_{12} = p_{13} = p_{14} = 1$	$m_1 : m_2 : m_3 : m_4 =$ $= Q_1p_{11} : Q_1p_{12} : Q_1p_{13} : Q_1p_{14} =$ $= 1:1:1:1$

TABLE 3. 1 (continued)

No.	Compound	Symmetry of the mole-cules in the crystal	Resonat-ing atom	Crystallographic parameters	Intensity ratio of the multiplet lines
8	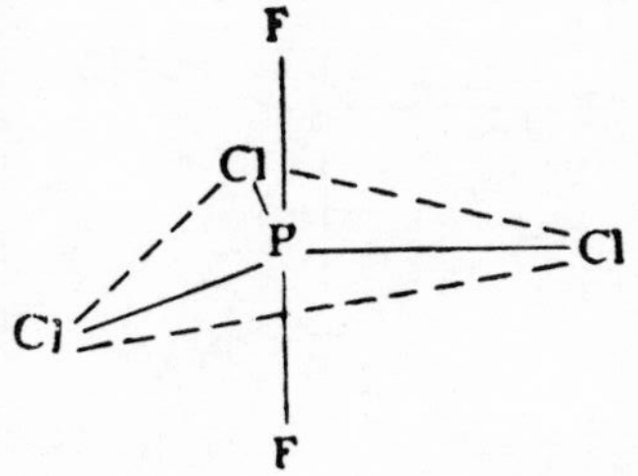	$C2\ (C_2)$ $m\ (C_s)$	^{35}Cl	$N = 1;\ Q'_1 = \frac{1}{2}$ $n = 2;\ p_{11} = 1;\ p_{12} = 2$	$m_1 : m_2 = Q_1 p_{11} : Q_1 p_{12} = 1:2$
9	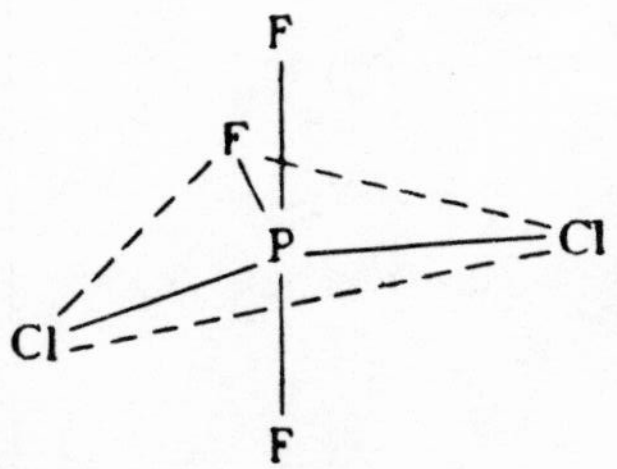	$2\ (C_2)$ $m\ (C_s)$	^{35}Cl	$N = 1;\ n = 1$ $Q'_1 = \frac{1}{2};\ p_{11} = 2$	Singlet
10	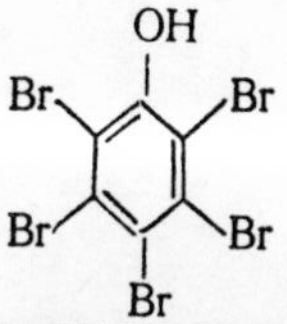	$2\ (C_s)$	^{79}Br	$N = 1;\ Q'_1 = \frac{1}{2}$ $n = 3$ $p_{11} = p_{12} = 2;\ p_{13} = 1$	$m_1 : m_2 : m_3 : m_4 = 2:1:1:1$

Note. The quantity Q' is proportional to the multiplicity Q for those cases where no structural data on Q are available.

Number 10 gives the NQR spectrum of ^{79}Br for pentabromophenol [7]. It consists of four lines with approximate intensity ratio 1 : 1 : 2 :1. Apparently the molecule in the crystal occupies a particular position (it may be intersected by the symmetry plane *m*), so that the bromine atoms in ortho and meta positions are symmetrically bonded in pairs, and the atom in the para position is on a symmetry element. However, the line from the ortho atoms of bromine is split into two due to the asymmetrical intramolecular interaction OH . . . Br. This example is an exception to the rule that the number of lines in the spectrum corresponds to the number of crystallographically nonequivalent atoms.

2. STRUCTURAL APPLICATIONS OF THE ZEEMAN EFFECT

Questions associated with the general theory of the Zeeman phenomenon in NQR spectra were treated in the corresponding section of Chapter 1. We shall discuss here only the application of Zeeman NQR spectra to purely structural, or to be more precise, to crystal structure problems. Problems associated with the determination of the asymmetry parameters and the character of the bond are dealt with in Chapter 5.

The "zero splitting cone" method is used widely to solve crystal structure problems of two types: 1) determination of the relative orientation of the crystallographic axes of the single crystal and the bond axes in the molecule; 2) determination of the number and relative orientation of the nonequivalent directions of the axes of the electric field gradient tensor.

The first type of problem may be solved for purely structural purposes or if the analysis of purely quadrupole spectra does not permit assignment of frequencies. In the first case the Zeeman-analysis is usually combined with a full X-ray study. Indeed, if a molecule of known structure contains, for example, the C—Cl bond, then by finding the mutual orientation of the crystallographic axes and these bonds, and by conducting an elementary geometrical analysis, it is possible to approximately determine the structure of the crystal, i. e. overcome one of the most difficult stages of structural analysis.

A problem of this type was formulated in the most general form by Sacurai [8]. For both molecules with a center of symmetry (1,2,4,5-tetrachlorobenzene [9], tetrachlorohydroquinone [10]) and those without (pentachlorophenol [11], 2,5-dichloroaniline [12]) X-ray analysis is simplified appreciably by NQR data on the orientation of the molecules in the unit cell. For molecules without a center of symmetry [11, 12] the directions of the C—Cl bonds (i. e. the directions of the z-axes) are determined with an accuracy of up to 1°. Thus, for molecules having two chlorine atoms in ortho or meta positions, the orientation of the benzene ring can be determined with an accuracy consistent with the assumption of coplanarity of the substituent and the ring.

The determination of the direction of the principal axes of the electric field gradient tensor gives more detailed information about the orientation of the benzene ring. The deviation of the field gradient from axial symmetry indicates that the C—Cl bond has, to some extent, double bond character. The p_x orbital of the chlorine atoms participating in the formation of the π bond is one of the principal axes of the field gradient tensor. However, the accuracy of determining the direction of the normal to the plane of the benzene ring is of the order of $\pm 5°$.

The coplanarity of the polyhalogenobenzenes has been established quite accurately [13—15]. Even for a compressed molecule like hexaiodobenzene, the X-ray structural analysis established, allowing for experimental error, that the benzene ring is planar, whereas the iodine atoms are characterized by a deviation of 0.04 Å from the plane of the ring.

A similar conclusion is reached from the NQR data for 1,2,3,4-tetrachlorobenzene [16] and 1,3,5-trichlorobenzene [17], which indicate that nonvalent interactions between the chlorine atoms may cause the chlorine atoms to deviate from the plane of the benzene ring by no more than ~1°. Investigations of this type are based on the assumption that the direction of the principal axes of the electric field gradient tensor coincides with the direction of the C—Cl bonds. The data given in Table 3.2 support this assumption.

TABLE 3.2. Comparison of NQR and X-ray structural data on the orientation of the bonds

Substance	Determined angle	Value of the angle, deg	
		NQR	X-ray analysis
Chloranil	Axis *b* and Cl_1 σ bond	63.6 [18]	63.9 [18]
	Axis *b* and Cl_2 σ bond	65.4	65.1
	Cl_1 σ bond and Cl_2 σ bond	64.8	64.9
1,3,5-Trichlorobenzene	Axis *a* and Cl_1 σ bond	119.1 [17]	118.8 [19]
	Axis *b* and Cl_1 σ bond	149.5	150.0
	Axis *c* and Cl_1 σ bond	81.7	80.9
	Axis *a* and Cl_2 σ bond	64.1	65.0
	Axis *b* and Cl_2 σ bond	31.3	30.2
	Axis *c* and Cl_2 σ bond	73.7	74.3
	Axis *a* and Cl_3 σ bond	24.8	24.0
	Axis *b* and Cl_2 σ bond	90.4	90.5
	Axis *c* and Cl_2 σ bond	114.8	114.0

If the atoms which are investigated by NQR participate in additional coordination bonds, the asymmetry parameters (determined for spin $3/2$ by the Zeeman method alone) can be used for characterizing such interactions, and also for calculating the angles between the interacting orbitals. In such cases the difference in the bond orientation values obtained by the NQR and X-ray methods is usually fairly large (see Chapter 7). The substantial difference in the values of the angles obtained from NQR and X-ray data for pentachlorophenol is explained by Sacurai [11]. The perturbing influence of the protons in the adjacent hydroxyl groups form a system of hydrogen bonds in the crystal.

The Zeeman-analysis of the NQR spectra for $SbCl_3$ crystals [20, 21] shows that the $SbCl_3$ molecule loses its intrinsic symmetry C_{3v} in the crystal, and shows that additional intermolecular bonds exist. This is also supported by the study of the temperature dependence (in the low temperature range [22]) of the NQR frequencies of ^{35}Cl in $SbCl_3$. (For more detail see Section 3 of this chapter.)

To determine the relative orientation of the nonequivalent directions z of the electric field gradient tensor, it is necessary to determine the Laue symmetry class of the crystal. Indeed, since the character of the Zeeman splitting of the NQR lines depends on the angle θ between the z-axis and the direction of the magnetic field H_0 (see equation (1.8)), the Zeeman method enables us to determine all the nonequivalent directions (nonequivalent orientations) of the z-axis of the gradient in the unit cell of the crystal. However, the positive and negative z-axis of the field gradient do not differ as regards the Zeeman splitting, and therefore the parallel and antiparallel orientations of the z-axes coalesce into one nonequivalent direction. Thus, the vectors of the crystalline space (for instance, the X—Y bonds, where X is a quadrupole atom), linked by the operation of a center of inversion or a parallel shift in the given space group, represent only one nonequivalent direction (Figure 3.1, a). It follows that the number of nonequivalent directions of the gradient axes determines the crystal symmetry only up to a Laue class. Moreover, if the X—Y bond in the crystal is accidentally parallel to a plane or axis of symmetry, the number of nonequivalent directions is halved (Figure 3.1, b). This leads to a nonunique determination of the crystal form. However, where there is an arbitrary orientation of the bonds with respect to the crystal elements of symmetry, this orientation can be determined. Thus, the possible elements of symmetry and the crystallographic axes can be determined [23].

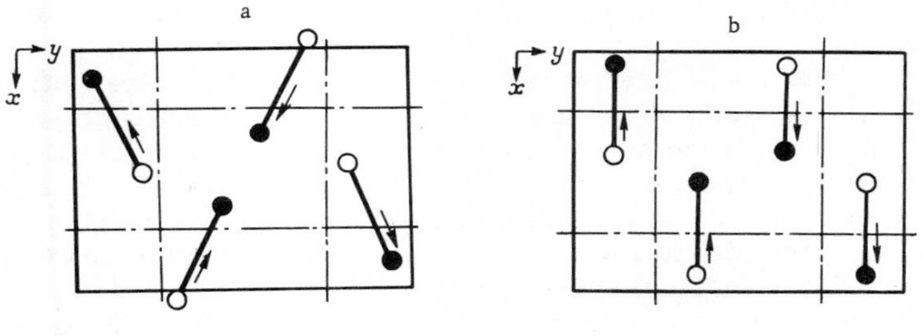

FIGURE 3.1. xy projection of a crystal consisting of XY molecules in the space group $p2_1/a$ (crystal class $2/m$ — C_{2h}):

a) molecule located arbitrarily with respect to the elements of symmetry of the space group; number of nonequivalent directions of the z-axis of the gradient (nonparallel vectors) equal to 2; b) molecule located parallel to the x-axis of symmetry of the crystal; number of nonequivalent positions of the z-axis of the gradients equal to 1.

Table 3.3 gives the number of nonequivalent orientations of the z-axis of the electric field gradient tensor for different crystal classes [23].

TABLE 3.3. Number of nonequivalent orientations of the z-axis for different crystal classes

Crystal form	Laue symmetry	Number of nonequivalent orientations of the axis for each frequency
Triclinic	$\bar{1}$ (C_i)	1
Monoclinic	$2/m$ (C_{2h})	2
Orthorhombic	mmm (D_{2h})	4
Tetragonal	$4/m$ (C_{4h})	4
	$4/mmm$ (D_{4h})	8
Cubic	$m\,3$ (T_h)	12
	$m\,3\,m$ (O_h)	24
Rhombohedral	$\bar{3}$ (C_{3i})	3
	$\bar{3}\,m$ (D_{3d})	6
Hexagonal	$6/m$ (C_{6h})	6
	$6/mmm$ (D_{6h})	12

As already mentioned, research on both the first and second types is carried out using single crystals. However, the Zeeman effect can also be studied on polycrystalline samples (see [24—26]). Such investigations are usually carried out to determine the asymmetry parameter.

3. INFLUENCE OF THE THERMAL MOTIONS OF THE MOLECULES ON THE NQR SPECTRAL PARAMETERS

It was mentioned in Chapter I that the temperature has an appreciable influence on the NQR frequency. We shall now discuss this phenomenon in greater detail.

Due to the thermal vibrations of the atoms and molecules in the crystal, the electric field gradient fluctuates in time with respect to the resonating nucleus. Since the frequency of thermal vibrations is considerably higher than the NQR frequency, it turns out that the nucleus is in a time-averaged electric field gradient. It follows that the quadrupole resonance cannot be observed, in principle, in liquids or solids where isotropic random reorientations of the molecules occur.

At small amplitudes of the thermal vibrations the NQR frequency is expressed as follows [27]:

$$\nu(T) = \nu_0\left(1 - \frac{3}{2} < \theta^2(T) > \right), \tag{3.4}$$

where ν_0 is the NQR frequency of the molecule at rest, and $<\theta^2>$ is the mean square of the amplitude of the vibrations.

The amplitude of the thermal vibrations increases with an increase in temperature, and therefore the NQR frequency decreases.

The torsional vibrations of the molecules have the greatest effect in molecular crystals. Thus, $<\theta^2>$ can be expressed in terms of the frequency of thermal vibrations ν_i and the moment of inertia J_i for that motion:

$$\nu = \nu_0 \left[1 - \frac{3h}{2} \sum_i \frac{1}{4\pi^2 J_i \nu_i} \left(\frac{1}{2} + \frac{1}{\exp(h\nu_i/kT) - 1}\right)\right]. \tag{3.5}$$

Equations (3.4) and (3.5) are called Bayer equations [27]. The predicted temperature dependence of the NQR frequency was found correct for most of the substances which were investigated (see Figure 3.2).

At temperatures above the Debye temperature T_D (for molecular crystals $T_D \approx 100°K$) the NQR frequency changes practically linearly with the temperature.

In general, when all types of thermal vibrations in the crystal are examined, the NQR frequency is equal to [28]:

$$\nu = \nu_0 \left(1 + aT + \frac{b}{T}\right) \tag{3.6}$$

where

$$a = -\frac{3}{2} k \sum_i \frac{A_i}{\omega_i^2} \tag{3.7}$$

$$b = -\frac{h^2}{8k} \sum_i A_i. \tag{3.8}$$

Here ω_i is the angular frequency of the i-th normal coordinate of the lattice vibration; and A_i has the dimensions of moment of inertia and has approximately the same order of magnitude as the moment of inertia of the given type of vibrations. At room temperature the last term in equation (3.6) is small compared with the second term.

However, equations (3.4)–(3.6) are true at constant volume if the experiment is conducted at constant pressure. The variations of the NQR frequency at constant pressure and volume are governed by the following thermodynamic relationship [29]:

$$\left(\frac{\partial \nu}{\partial T}\right)_P = -\frac{\alpha}{\chi}\left(\frac{\partial \nu}{\partial P}\right)_T + \left(\frac{\partial \nu}{\partial T}\right)_V, \tag{3.9}$$

where α is the coefficient of thermal expansion, χ is the compressibility, and $\left(\frac{\partial \nu}{\partial T}\right)_V$ is the Bayer term.

Experiments conducted on a number of crystals established the importance of the first term of the equation for ionic crystals in which the electric field gradient is proportional to V^{-1}.

In molecular crystals, a change in volume particularly affects intermolecular distances, without altering the intramolecular ones. Therefore, to a first approximation the field gradient for molecular crystals can be considered independent of the volume. The small changes in frequency with a change in volume at superhigh pressures observed in p-dichlorobenzene [28] were attributed to a change in the bond hybridization. For molecular crystals it is possible to use a simpler model suggested by Brown [32], in which the quadrupole bond constant is independent of the volume, and the frequencies of the lattice vibrations are linear functions of the temperature.

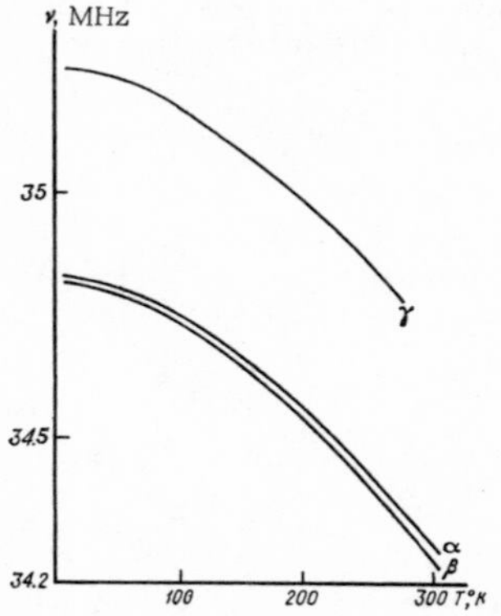

FIGURE 3.2. Typical temperature dependence of the NQR frequencies $\nu(T)$ for molecular crystals. NQR of ^{35}Cl for the three phases of p-dichlorobenzene.

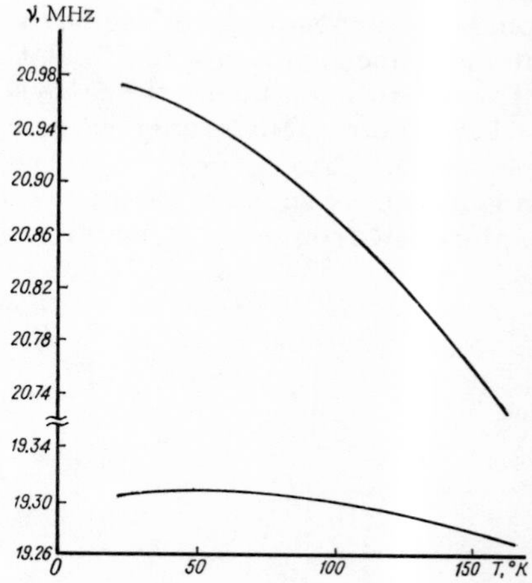

FIGURE 3.3. Temperature dependence of the NQR frequencies of ^{35}Cl in $SbCl_3$ [22]

There exist, however, a number of molecular crystals which have a positive temperature coefficient for the NQR frequency of the halogens. Such anomalies were observed for $TiBr_4$ [29], WCl_6 [33], $K_2[MHal_6]$ (where M = Pt, Ir, Os, Re, W) [34, 35]. The most consistent explanation of this phenomenon for complexes of the type K_2MHal_6 was given by Haas and Marram [36]. They showed that deformation vibrations lead to a decrease in the mean value of the π bonding of the vacant d orbitals of the transition metal with the p_π orbitals of the halogen atom. Therefore, the mean population of the p_π orbitals of the halogen atom will be proportional to the mean square of the amplitude of the deformation vibration. It follows from the theory of Townes and Dailey [37] (see Chapter 5) that e^2Qq for the halogen is proportional to the population of the p_π orbitals:

$$e^2Qq_{\mathrm{mol}} = e^2Qq_{\mathrm{at}}\left(\frac{N_x + N_y}{2} - N_z\right),$$

where N_i is the population of the corresponding p_i orbital.

Therefore, the NQR frequency increases with an increase in temperature and an increase in the amplitude of the deformation vibrations. Obviously, the overall temperature dependence of the NQR frequency is determined not only by this mechanism, but also by the Bayer mechanism.

Thus, the characteristics of the temperature dependence of the NQR frequency are frequently determined by the characteristics of the chemical bond of the atom being studied. We shall discuss one more example in which an unusual temperature relationship confirms the presence of an intermolecular coordination interaction (Figure 3.3). Whereas the high-frequency NQR line of ^{35}Cl in $SbCl_3$ of twice the intensity follows Bayer's law, the low-frequency line has two particular features: 1) the existence of a maximum at about 55°K; and 2) the low temperature coefficient of the frequency above this temperature. This behavior is due to the coordination interaction of the chlorine atom, corresponding to the low-frequency NQR line, of one $SbCl_3$ molecule with the antimony atom of the other molecule [22]. Electron density is transferred from the p_π orbital of the chlorine atom to the free d orbital of the antimony atom. This is confirmed by the drop in NQR frequency for this chlorine atom compared with the other two and the large asymmetry parameter of the field gradient for this atom ($\eta = 15.7\%$) [21]. With an increase in temperature two competing mechanisms act on the NQR frequency: deformation vibrations which weaken the coordination bond and raise the NQR frequency, and the usual Bayer averaging of the electric field gradient which lowers the NQR frequency.

In addition to the usual lattice vibrations in molecular crystals a relatively slow rotation of the molecules or different groups in the molecules (for example, CH_3, CCl_3, etc.) is often observed. If the molecules rotate about one axis, the electric field gradient is averaged, and the NQR frequency depends on the angle θ between the axis of the maximum field gradient and the axis of rotation z'. If the rotation frequency is higher than the NQR frequency, the latter is determined by the expression:

$$\nu = \frac{\nu_0}{2}(3\cos^2\theta - 1). \qquad (3.10)$$

One of the most interesting examples of such a rotation is in 1,2-dichloroethane, where the change in frequency was investigated by Ragle [38–40] and Tokuhiro [41]. In this case, $\theta = 19°23'$ (Figure 3.4). The temperature dependence of the NQR frequencies of the ^{35}Cl is presented in Figure 3.5.

The dependence of the NQR frequency in reorientation motions on the mutual orientation of the electric field gradient and the axis of rotation was examined most extensively in [42]. A molecular reorientation-rotation frequency lower than the NQR frequency results in broadening of the NQR line and shortening of the spin-lattice relaxation time T_1, i. e. of the time necessary to establish a thermal equilibrium between the spin system and the lattice. The spin-lattice relaxation time T_1 is determined mainly by interaction of the magnetic spin system with the thermal vibrations (the phonon spectrum) of the crystal lattice. Therefore, T_1 characterizes the thermal motion of the molecules in the crystal: the smaller its value, the larger the mean square of the amplitude of the thermal vibrations [43]. Thus, in a crystal of $ClH_2CCH_2CCl_2CH_2CH_2Cl$, two NQR lines of ^{35}Cl are observed with frequencies $\nu_1 = 34.18$ MHz and $T_1^{(1)} = 30$ msec for the CH_2Cl group and $\nu_2 = 35.46$ MHz and $T_1^{(2)} = 70$ msec for the CCl_2 group. It is obvious that the amplitude of oscillations about the axis perpendicular to the plane of the molecule is larger for chlorine atoms in the CH_2Cl group (T_1 is shorter) than for chlorine atoms in the CCl_2 group.

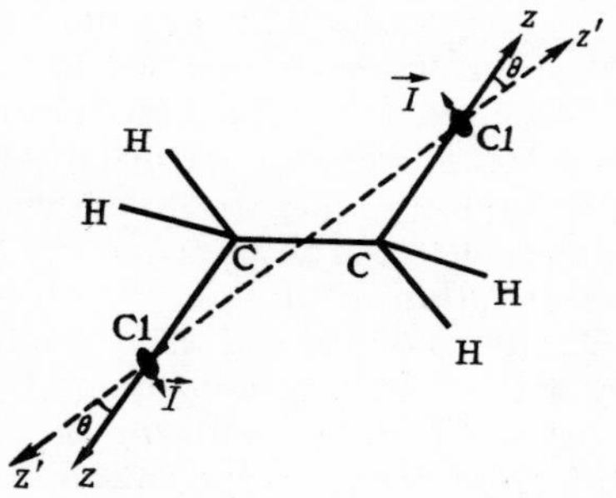

FIGURE 3.4. Diagram of the reorientating molecule 1,2-dichloroethane (rotation of the molecule about the z'-axis)

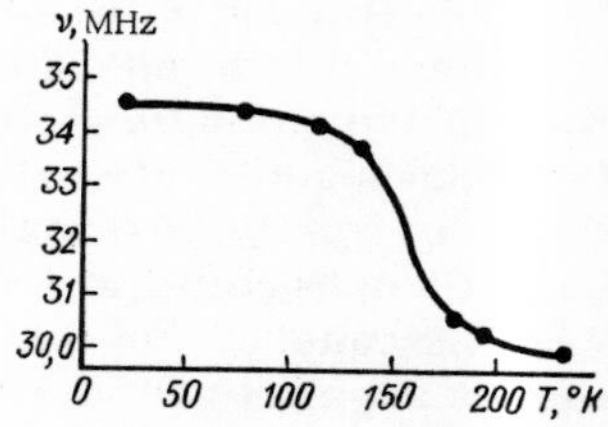

FIGURE 3.5. Temperature dependence of the NQR frequency of ^{35}Cl in 1,2-dichloroethane [38]

During reorientation T_1 decreases rapidly due to the appearance of an additional efficient mechanism of energy exchange.

It was shown [44—46] that with the appearance of hindered rotation T_1 starts to decrease exponentially when the temperature is increased. The width of the NQR line starts to increase in accordance with the following law:

$$\Delta\omega \sim \exp\left(-\frac{W}{RT}\right), \qquad (3.11)$$

where $\Delta\omega$ is the change in line width caused by the reorientation which has a potential barrier W.

Such a behavior of T_1 was observed for hexamethylenetetramine [44] and piperazine [47]. The line widths for 1,2,3-trichlorobenzene and hexachlorobenzene [48] also behaved in this manner (Figure 3.6).

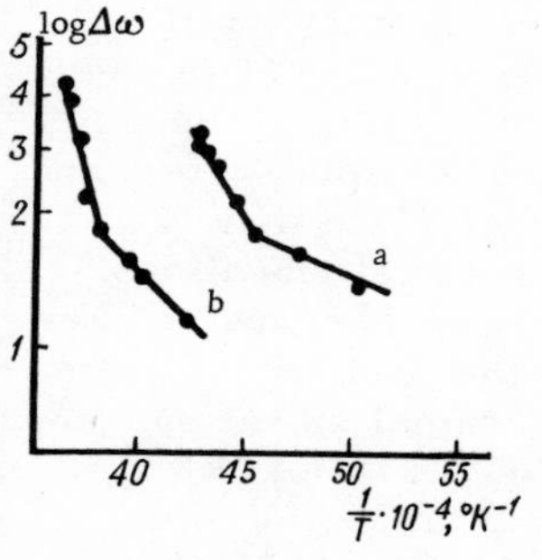

FIGURE 3.6. Temperature dependence of the line width for 1,2,3-trichlorobenzene (a) and hexachlorobenzene (b) [48]

On increasing the rotation so that the frequency becomes comparable to the quadrupole frequency, and the NQR line becomes very broad:

$$(\Delta\omega)^{-1} = \frac{2\lambda}{(2\lambda^2) + \omega_0^2}. \qquad (3.12)$$

λ is the mean velocity of the discontinuous movement of the molecules between the potential minima during rotation.

If the rotation is increased, the molecule in the solid lattice starts to rotate without "feeling" the potential wells, i. e. to participate in a motion similar to Brownian motion.

Such collective movement further narrows the NQR line and raises T_1.

4. STUDY OF PHASE TRANSITIONS BY NQR

As already mentioned in Chapter 2, the contribution of the crystal field to the electric field gradient $\frac{\partial^2 V}{\partial z^2}$ at the nucleus is not large and in frequency units does not exceed 1.5–2% of the value of ν_{NQR} for molecular crystals. However, since the experiment is accurate to within 10^{-5} on average, the spectra of nuclear quadrupole resonance are used to study phase transitions in crystalline substances. Phase transitions of the first kind, which are associated with a discontinuous rearrangement of the crystal lattice and often also with a considerable change in its symmetry, result in not only a discontinuous change of ν_{NQR} and of the spin-lattice relaxation time T_1, but also in a considerable transformation of the temperature dependence of the NQR frequencies. Very frequently (but not always) the multiplicity of the spectrum is not preserved. A phase transition of the first kind is characterized by the equality of the isobaric-isothermal potentials of the two phases at the transition point $\varphi_1(P, T) = \varphi_2(P, T)_x$ and by the discontinuous change in the first derivatives of these potentials:

$$\left(\frac{\partial\varphi_1}{\partial P}\right)_T = v_1; \quad \left(\frac{\partial\varphi_2}{\partial P}\right)_T = v_2; \quad v_1 - v_2 \neq 0$$

$$\left(\frac{\partial\varphi_1}{\partial T}\right)_P = S_1; \quad \left(\frac{\partial\varphi_2}{\partial T}\right)_P = S_2; \quad S_1 - S_2 \neq 0,$$

where v is the volume, S the entropy, and φ the Gibbs thermodynamic potential.

Each of the functions φ_1 and φ_2 on both sides of the transition point corresponds to the state of a body with strictly determined symmetry.

A phase transition of the first kind (transition from one crystal structure into another) does not necessarily lead to a considerable change in the symmetry of the crystal lattice or in the symmetry of the molecule in the crystal. It may cause a slight rearrangement of the lattice by which a more or less high packing density of the molecules is achieved. In any case, the mutual arrangement of the molecules in the crystal changes, and thus the contribution of the nearest order of crystal lattice to the quadrupole bond constant of the given atom also changes. Therefore, sharp shifts in the NQR frequencies caused by phase transitions of this type are of the order of magnitude of the usual crystal splittings [49]. Therefore, the NQR spectra observed in a sufficiently wide temperature range enable us [50]:

1) to detect phase transitions of the first kind from the observed frequency jump which is sometimes accompanied by a change in the multiplicity;

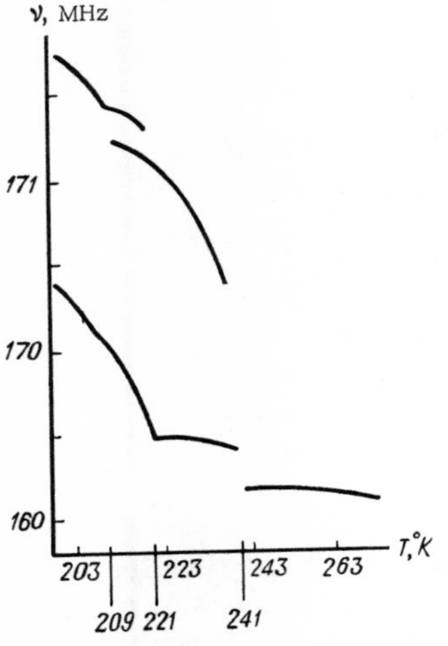

FIGURE 3.7. Temperature dependence of the NQR frequencies of ^{79}Br during phase transitions in $K_2[SeBr_6]$ [51]

2) to draw a conclusion about the character of the motion of the molecules in each of the investigated phases by following the values of the temperature coefficients of the NQR frequencies;

3) to separate the spectra of these phases when they appear simultaneously and to analyze the character of the transition (from the point of view of the change in character of the thermal motion and the packing density of the considered phases) by comparing the spin-lattice relaxation times T_1 in the different phases. If the phase transition leads to a change in the multiplicity of the spectrum, we are sometimes able to predict the symmetry of the molecule in the crystal and its change during the transition. As a rule, low-temperature phases possess a lower symmetry. This loss of symmetry elements is obviously justified by the increase in the packing density, as indicated by the longer relaxation time T_1.

Phase transitions of the second kind can also be observed in NQR spectra. However, since they are not accompanied by a discontinuous change in the state of the phases on both sides of the transition point, there is no discontinuous change in the characteristics of the spectrum. Nevertheless, since they are associated with a continuous change in the crystal symmetry, they can be characterized by a continuous change in the multiplicity. If the transition is of the "order-disorder" type, then, in addition to the change in multiplicity, a sharp broadening of the NQR lines is observed caused by the lack of order of the system.

We shall give two characteristic examples of phase transitions of the first kind.

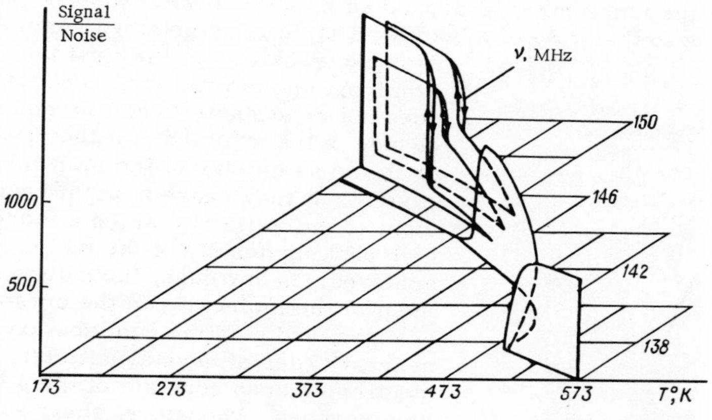

FIGURE 3.8. Temperature dependence of the NQR frequencies of ^{127}I during phase transitions in KIO_3 [52]

Figure 3.7 shows the temperature dependence of the NQR frequencies of ^{79}Br for $[K_2SeBr_6]$ [51]. Above 240°K the substance crystallizes in cubic form, which is characterized by the presence of one crystallographically nonequivalent bromine atom, i. e. by a singlet NQR spectrum of ^{79}Br.

Below 240°K the NQR spectrum consists of two lines which correspond to two crystallographically nonequivalent bromine atoms and is apparently due to the transition of the crystal into tetragonal syngony.

Below 221°K a second phase transition into a phase with three crystallographically nonequivalent atoms occurs (triplet of equal intensity; drop of the crystal symmetry to rhombic or lower). At 209°K a third phase transition is observed.

Figure 3.8 shows the phase transitions in KIO_3 in the temperature range 173 to 573°K. At 573°K the NQR spectrum of ^{127}I consists of a single line. At 483° a phase transition with no change in multiplicity occurs. This phase exists up to 363°K and then passes into a phase with three nonequivalent molecules in the cell, which corresponds to a change in the multiplicity of the spectrum (transition of the singlet into a triplet).

BIBLIOGRAPHY

1. International Tables for X-Ray Crystallography. Birmingham, 1952.
2. Khotsyanova, T.L., S.I. Kuznetsov, and T.A. Babushkina. — Zh. Strukt. Khim. 9 (1968), 713.
3. Kitaigorodskii, A.I. Organic Crystal Chemistry. — USSR Acad. Sci., 1955. (Russian)
4. Shimomura, K. — J. Sci. Hiroshima Univ., Ser. A. 21 (1958), 241.
5. Porai-Koshits, M.A. A Practical Course in X-Ray Diffraction Analysis. — Mosc. State Univ., 1960. (Russian)
6. Holms, R.R., R.P. Caster, and G.E. Peterson. — Inorg. Chem. 3 (1964), 1748.
7. Semin, G.K., V.I. Robas, L.S. Kobrina, and G.G. Yakobson. — Zh. Strukt. Khim. 5 (1964), 915.
8. Jeffrey, G.A. and T. Sacurai. Applications of Nuclear Quadrupole Resonance. — In: "Solid State Chemistry," Vol. 1. Part 10, 1964.
9. Dean, C., B. Pollak, M. Graven, and G.A. Jeffrey. — Acta crystallogr. 11 (1958), 710.
10. Sacurai, T. — Acta crystallogr. 15 (1962), 443.
11. Sacurai, T. — Acta crystallogr. 15 (1962), 1164.
12. Sacurai, T., M. Sundaralingam, and G.A. Jeffrey. — Acta crystallogr. 16 (1963), 354.
13. Strel'tsova, I.N. and Yu.T. Struchkov. — Zh. Strukt. Khim. 2 (1961), 312.
14. Gafner, G. and E. Herbstein. — Acta crystallogr. 13 (1960), 701.
15. Khotsyanova, T.L. and V.I. Smirnova. — Kristallografiya 13 (1968), 787.
16. Dean, C., C. Richardson, and T. Sacurai. — Molec. Phys. 4 (1961), 95.
17. Morino, Y. and M. Toyama. — J. Phys. Soc. Japan 15 (1960), 288.
18. Shu, S., G. Jeffrey, and T. Sacurai. — Acta crystallogr. 15 (1962), 661.
19. Milledge, J. and L.M. Pant. — Acta crystallogr. 13 (1960), 285.
20. Peterson, G., C. Dean, and G.A. Jeffrey. — Pittsburgh Diffraction Conference, Cited after [11].
21. Okuda, T., A. Nakao, M. Sciroyama, and H. Negita. — Bull. Chem. Soc. Japan 41 (1968), 61.
22. Chihara, H., N. Nakamura, and H. Okuma. — J. Phys. Soc. Japan 24 (1968), 306.
23. Shimomura, K. — J. Phys. Soc. Japan 12 (1957), 652.
24. Morino, Y. and M. Toyama. — J. Chem. Phys. 35 (1961), 1289.
25. Dinesh and P.T. Narasimhan. — J. Chem. Phys. 46 (1966), 1270.
26. Lucken, E.A.C. and C. Marzeline. — J. Chem. Soc., A, p. 153, 1968.
27. Bayer, H. — Z. Phys. 130 (1951), 227.
28. Kushida, T., G.B. Benedek, and N. Bloembergen. — Phys. Rev. 104 (1956), 1364.
29. Barnes, R.G. and R.D. Engardt. — J. Chem. Phys. 29 (1958), 248.
30. Fuke, T. — J. Phys. Soc. Japan 16 (1961), 266.
31. Matzakanin, G., T.N. O'Neel, and T.A. Scott. — J. Chem. Phys. 44 (1966), 4171.
32. Brown, R. — J. Chem. Phys. 32 (1960), 116.
33. Reddoch, A. — J. Chem. Phys. 35 (1961), 1086.
34. Ikeda, R., D. Nakamura, and M. Kubo. — Bull. Chem. Soc. Japan 36 (1963), 1056.
35. Ikeda, R., D. Nakamura, and M. Kubo. — J. Phys. Chem. 69 (1965), 2101.

36. Haas, T. E. and E. P. Marram. — J. Chem. Phys. 43 (1965), 3985.
37. Townes, C. H. and B. P. Dailey. — J. Chem. Phys. 17 (1949), 782.
38. Dodgen, H. W. and J. L. Ragle. — J. Chem. Phys. 25 (1956), 376.
39. Ragle, J. L. — Archs. Sci., Genève 12 (1959), 177.
40. Ragle, J. L. — J. Phys. Chem. 63 (1959), 1395.
41. Tokuhiro, T. — J. Chem. Phys. 41 (1964), 438.
42. Lotfullin, R. Sh. and G. K. Semin. — Zh. Strukt. Khim. 9 (1968), 813.
43. Woessner, D. and H. Gutowsky. — J. Chem. Phys. 39 (1963), 440.
44. Alexander, S. and A. Tzalmona. — Phys. Rev. Lett. 13 (1964), 546.
45. Alexander, S. and A. Tzalmona. — Phys. Rev. 138A (1965), 845.
46. Lotfullin, R. Sh. and G. K. Semin. — Kristallografiya 14 (1969), 809.
47. Tzalmona, A. — J. Chem. Phys. 50 (1969), 366.
48. Tatsuzaki, I. — J. Phys. Soc. Japan 14 (1959), 578.
49. Babushkina, T. A. and G. K. Semin. — Kristallografiya 13 (1968), 529.
50. Babushkina, T. A. — Author's Summary of Candidate Thesis. Inst. Chem. Phys. USSR Acad. Sci., 1968. (Russian)
51. Nakamura, D., K. Ito, and M. Kubo. — Inorg. Chem. 2 (1963), 61.
52. Herlach, F., H. Gramicher, and D. Itschner. — Arch. Sci. Vol. 12. Fasc. spec., p. 182, 1959.

Chapter 4

EMPIRICAL LAWS FOR THE VARIATION IN NQR FREQUENCIES OF HALOGEN ATOMS IN ORGANIC COMPOUNDS

1. TETRAHEDRAL MOLECULES WITH A CENTRAL ATOM FROM GROUPS IVA AND VA

The NQR spectra of halogen-containing compounds have been studied in greatest detail. It was found that the laws deduced for the halogen-derivatives of methane can be extended with some additions to the series of analogous derivatives of other elements of group IV (Si, Ge, Sn). Livingston [1] was the first to observe a linear dependence $\nu(n)$ of the NQR frequency on the number of halogen atoms in the chloro-derivatives of methane. Similar relationships were found [2] for a number of other bromo- and chloro-derivatives of methane of type $A_{4-n}CHal_n$ (where A are both electron-donor and electron-acceptor substituents). The character of the frequency shift (with respect to $CHal_4$) depends only on the nature of the substituent [3]. For an electron-donor substituent with respect to the halogen atom the NQR frequency of the corresponding halogen atoms is lower than the frequency of $CHal_4$, and for an electron-acceptor substituent, it is higher.

For compounds of type $A_{4-n}CHal_n$ the variation in NQR frequencies obeys the following relationship [2]:

$$\nu_{(4-n)\,A} = \frac{4-n}{3}\nu_{3A} - \frac{1-n}{3}\nu_{4Hal}, \tag{4.1}$$

where ν_{3A} is the frequency of the compound containing one halogen atom and three atoms of the substituent A, ν_{4Hal} is the frequency of the corresponding tetrahalomethane, and n is the number of halogen atoms in the molecule.

Correspondingly, the NQR frequencies of ^{79}Br in halo-derivatives of methane can be written as follows:

$$\nu_{(4-n)\,H} = \frac{4-n}{3}\nu_{3H} - \frac{1-n}{3}\nu_{4Br}. \tag{4.2}$$

A linear relationship of this type can hold only when the following conditions are fulfilled [2, 3].

1. The tetrahedral angles are almost constant for any substituent. As a rule, this is true for most of the compounds investigated, as confirmed by the extensive data from X-ray and electron diffraction investigations.

2. Spatial perturbations are absent (or are very small). This assumption can be justified by the fact that the energy of rotation of the substituent about the carbon-substituent bond axis is much smaller than the energy of deformation of the valence angle or the energy necessary to change the bond length. The deformation of the electron shell of the halogen atom itself is also not large due to the small asymmetry parameter η of the halogen atoms.

3. The effect of the substituent through the carbon atom is distributed uniformly among the three remaining bonds and is independent of the nature of the substituent.

$\nu(n)$ remains linear within 1% when the substituent A is small (CH_3, C_2H_5, CN, etc.) or planar (C_6H_5, NO_2, etc.). $\nu(n)$ is not linear only for very nonplanar substituents due to the appearance of steric hindrance (which may result from the formation of inter- or intramolecular hydrogen and similar bonds).

To compare the NQR data for the different halogens on the same scale we can plot $\nu(n)$ in relative units $\gamma = \frac{\nu_{(4-n)A}}{\nu_{4Hal}}$ [2] (Figure 4.1). For tetrahalomethanes the relative frequencies will be equal to one ($\gamma = 1$), whereas $\gamma < 1$ for donors with respect to the halogen and $\gamma > 1$ for acceptors. Comparison of the families of curves for the bromo- and chloro-derivatives of methane shows that the relative frequency shifts for bromine are always larger than the shifts in the analogous chlorine-containing compounds. This discrepancy indicates that for bromo-derivatives the contribution determined by the polarization of the atomic core by the valence electrons becomes quite high (5—7%). The initial contribution (for $CHal_4$), determined by the polarization of the atomic core, is not too large. This leads to a small difference between the relative frequency shifts for chlorine and bromine in the case of donors. However, for electron-acceptors the value of the polarization contribution in the bromine atom increases sharply due to the larger mobility of its atomic core, and this leads to substantial differences in the relative shifts of the NQR frequencies for ^{35}Cl and ^{79}Br. Thus, for $ClC(NO_2)_3$ $\gamma = 1.057$, and for $BrC(NO_2)_3$ $\gamma = 1.211$, whereas for CH_3Cl $\gamma = 0.839$ and for CH_3Br $\gamma = 0.825$.

Comparison of the relative shifts in frequency for ^{35}Cl and ^{79}Br with ^{127}I shows that the polarization of the atomic core in iodine is even larger than in bromine.

The NQR frequency of halogens in halogen-derivatives of methane with substituents of different types forms a linear combination of the frequencies $X_{4-n}CHal_n$, $Y_{4-n}CHal_n$, $Z_{4-n}CHal_n$. For compounds of the type XYZCHal, the NQR frequency of the halogen will be the arithmetic mean of the NQR frequencies of the compounds X_3CHal, Y_3CHal, Z_3CHal:

$$\nu_{XYZ} = \frac{\nu_{3X} + \nu_{3Y} + \nu_{3Z}}{3}. \quad (4.3)$$

The calculation of NQR frequencies for compounds with mixed substituents is less accurate than for compounds with identical ones, and is accurate to within about 2%. The inaccuracy in the calculation is systematic in many cases, e.g., when there is an electron-acceptor in the molecule, the calculated frequency is generally lower than the measured value.

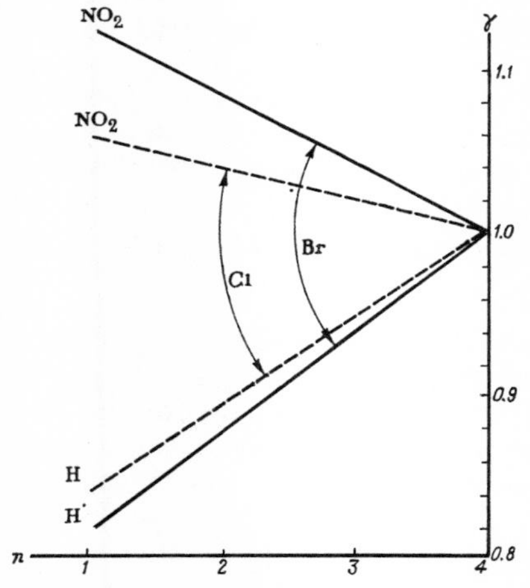

FIGURE 4.1. Ratios $\gamma = \frac{\nu_{(4-n)A}}{\nu_{4Hal}}$ as a function of the number of halogen atoms n in the molecule $CA_{4-n}Hal_n$ (where A = H, NO_2; Hal = Cl, Br)

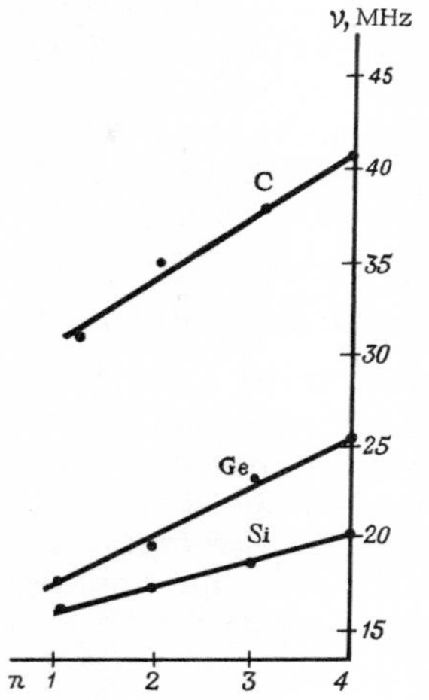

FIGURE 4.2. NQR frequency of ^{35}Cl as a function of the number of chlorine atoms n in compounds of type $(CH_3)_{4-n}MCl_n$ (where M = C, Si, Ge)

The inaccuracy in the calculation (both the scatter and the small systematic shifts) is due to the following factors: 1) the indeterminacy introduced by the crystal field; 2) the partial (very small) distortion of the tetrahedron; 3) the nonlinear change in the contribution due to the polarization of the atomic core of the halogen in molecules with mixed substituents.

Using equations (4.1)–(4.3), we can easily calculate the NQR frequencies of ^{35}Cl, ^{79}Br, ^{127}I for a large number of methane derivatives. Thus, we can predict the NQR frequencies (and, therefore, the chemical constants) and solve structural problems for compounds which are either difficult to synthesize or still unknown.

Chlorofluoro-derivatives of methane are one of the rare exceptions to the additivity rule [1, 4]. The increment for the fluorine atom, calculated for the compounds Cl_nCF_{4-n} and equal to 2.55 MHz, is not true for the compounds $CH_nCl_mF_{4-n-m}$. Moreover, substitution of chlorine by hydrogen causes a larger effect than substitution by fluorine. It was expected that the effect would be stronger and even of opposite sign since fluorine is an electron-acceptor relative to chlorine. The reasons for the lack of additivity for fluorochloromethanes are discussed in detail in Section 5 of Chapter 5.

It follows from the NQR data on halogens in halo-derivatives of silicon and germanium that the above-mentioned laws are obeyed here as well; the relationship between $\nu(n)$ and the number of halogen atoms is linear for both identical and mixed substituents [3, 5, 6] (see Figure 4.2).

The somewhat larger difference between MCl_4 and $AMCl_3$ compared to the difference between $AMCl_3$ and A_2MCl_2 or A_2MCl_2 and A_3MCl is characteristic.

A very important fact is that the ratio of the NQR frequency of ^{35}Cl in MCl_4 to that in CCl_4 is roughly constant. This is also true for other pairs of compounds with similar structures. The ratio of NQR frequencies of halo-derivatives of carbon and silicon is $k = 2.0 \pm 0.1$ [3, 5, 6], and for germanium compounds $k = \frac{\nu_{C-Cl}}{\nu_{Ge-Cl}} = 1.66$; $k = \frac{\nu_{Ge-Cl}}{\nu_{Si-Cl}} = 1.28$ [5, 6].*

The NQR frequencies of halogens in halo-derivatives of elements of group IV can also be calculated in a different way. Starting from the NQR frequency of the halogen in $MHal_4$, we can find from the frequency differences the increments in the shifts of the NQR frequencies caused by different substituents. The NQR frequency of the halogen in the compound to be calculated can then be found from the sum of the corresponding increments allowing for their signs:

$$\nu_{(4-n)\,A} = \nu_{4Hal} + \Sigma\Delta_A. \tag{4.4}$$

It is also possible to calculate in the same way the NQR frequencies in compounds of tetrahedral phosphorus, e. g., RR'P(O)Cl and RR'P(S)Cl, provided that R and R' are electron-donors (Table 4.1) [10, 11]. Lucken's tentative indications [11] that such a law does not exist were due to a number of inaccurate measurements (cf. [10] and [11]).

TABLE 4.1. Increments in the NQR frequencies of ^{35}Cl in the compounds PR'P(O)Cl and RR'P(S)Cl caused by substituents

Substituent	RR'P(O)Cl	RR'P(S)Cl	Substituent	RR'P(O)Cl	RR'P(S)Cl
CH_3	2.36	2.01	C_6H_5	2.14	1.94
C_2H_5	2.73	2.06	$N(CH_3)_2$	3.76	3.77
C_3H_7	2.68	1.88	$N(C_2H_5)_2$	3.32	—
iso-C_3H_7	2.89	—	OCH_3	1.72	—
C_4H_9	2.65	—	OC_2H_5	1.85	—
iso-C_4H_9	2.73	—			

The changes in the NQR frequencies of ^{35}Cl in the series of derivatives of trivalent phosphorus are not additive. This is due to the change in the degree of *sp* hybridization as a result of the deformation of the substantially

* For chlorine-containing lead compounds the ratio of the NQR frequency of ^{35}Cl to the frequencies of the analogous compounds of silicon, carbon and germanium is not constant [7–9]. The NQR frequency of ^{35}Cl in compounds of type $R_{4-n}SnCl_n$ decreases according to a monotonic nonlinear law with a decrease in the number of chlorine atoms. Such behavior could be due to a change in the valence angles or to the existence of coordination interactions in the solid.

less stable angles of the pyramid P$<$ by the different substituents.* However, the additivity of the NQR frequencies of ^{35}Cl in tetrahedral compounds of phosphorus is limited and can only be applied when both substituents (R and R' — alkyl and alkoxyl) are of the same type.

2. HALOGEN-DERIVATIVES OF SATURATED HYDROCARBONS

When the NQR frequencies of ^{35}Cl or ^{79}Br change in the series of compounds of type Hal$(CH_2)_n$Hal, the relative frequency shift decreases sharply with an increase in the number of methylene groups separating the halogen atoms [3, 12]. From $n = 4$ only an oscillation (from even to odd n) of the relationship $\nu(n)$ about some mean value ν_{0I} is observed (for ^{35}Cl $\nu_{0I} = 33.0$ MHz, and for ^{79}Br $\nu_{0I} = 250.0$ MHz). For sufficiently large n almost complete damping is observed due to the mutual induction of the halogens (Figure 4.3).

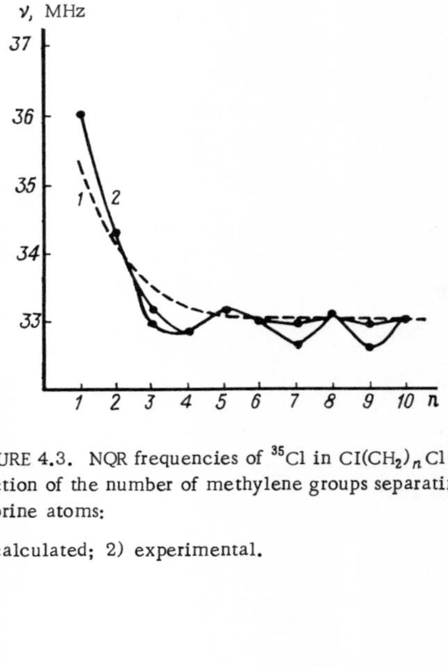

FIGURE 4.3. NQR frequencies of ^{35}Cl in $Cl(CH_2)_nCl$ as a function of the number of methylene groups separating the chlorine atoms:

1) calculated; 2) experimental.

* A direct proof of this are the changes in the angles of the pyramid As$<$: in $AsCl_3$ the angle As$<$(Cl)(Cl) is equal to 98°; in $As(CH_3)_3$ the angle As$<$(C)(C) is equal to 96°; in $As(C_6H_5)_2Cl$ the angle As$<$(C)(C) is equal to 105°, and As$<$(Cl)(C) to 96° [28, 29]. The very sharp changes in the NQR frequencies of ^{75}As for these compounds are 75, 101.5 and 163 MHz, respectively.

A similar relationship is also observed in monohalo-derivatives of the type $CH_3(CH_2)_nHal$. In this case the NQR frequencies of the halogens adjacent to the primary carbon atoms also attain the mean value ν_{0I} with an increase in n. The NQR frequencies of the halogens adjacent to the secondary and tertiary carbon atoms tend to the value ν_{0II} (32.0 MHz for ^{35}Cl, and 243.3 MHz for ^{79}Br).

The frequency shifts $\Delta_n\nu$ of a halogen atom separated from the other halogen atom by n methylene groups form a geometrical progression whose initial term is the primary shift $\Delta_1\nu$ [12]:

$$\Delta_n\nu = \xi^{n-1}\Delta_1\nu\,, \tag{4.5}$$

where ξ is the coefficient of transmission along the chain ($\xi < 1$).

The expression for the frequency of the halogen adjacent to the primary carbon atom can be written in the form:

$$\nu_{nI} = \nu_{0I} + \Delta_1\nu\xi^{n-1}. \tag{4.6}$$

Figure 4.3 shows the calculated curve $\nu_{nI} = f(n)$ for $\Delta_1\nu = 2.4$ MHz and $\xi = 0.47$. Analysis of the NQR spectra of polyhalogenated aliphatic compounds leads to the conclusion that the conditions of transmission to a given halogen atom from other atoms do not depend on the number of substituents. In other words, ν_{0I}, ν_{0II}, ξ, $\Delta_1\nu$ remain invariable and equation (4.6) can be written in the form of the following sum:

$$\nu_I = \nu_{0I} + \Delta_1\nu\sum_i \xi^{n_i-1}. \tag{4.7}$$

The values of the parameters $\Delta_1\nu$, ξ are universal and can be used with the same assumptions to calculate the NQR frequencies of the halogens adjacent to secondary or tertiary carbon atoms, i. e.:

$$\nu_{II} = \nu_{0II} + \Delta_1\nu\sum_i \xi^{n_i-1}. \tag{4.8}$$

We give as an example the calculation of the frequencies of ^{35}Cl in 1,1,1,3-tetrachloropropane. The expression for the frequency of the chlorine in the CH_2Cl group can be written in the form:

$$\nu_I^{CH_2Cl} = \nu_{0I} + 3\Delta_1\nu\xi^2\,,$$

and for the chlorine atoms in the CCl_3 group:

$$\nu_I^{CCl_3} = \nu_{0I} + \Delta_1\nu\,(2 + \xi^2)\,.$$

The corresponding frequencies will be equal to 34.59 and 38.33 compared to the experimental results 34.629 MHz and 38.43 MHz, respectively.* The agreement, within 0.1—0.3%, between the calculated and experimental

* The NQR spectrum of the trichloromethyl group is a triplet of 38.826, 38.304 and 38.150 MHz due to the nonequivalent position of each chlorine atom in the unit cell.

frequencies, averaged over the crystal splittings, can be regarded as very satisfactory since the maximum crystal splittings in molecular crystals do not exceed 1.5—2% of the measured frequency.

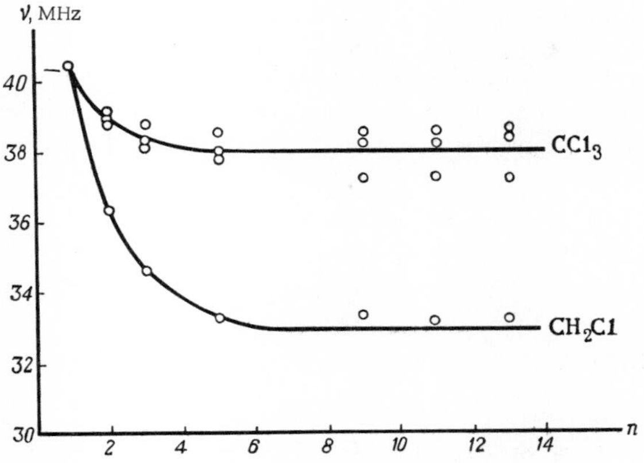

FIGURE 4.4. NQR frequencies of ^{35}Cl in $Cl(CH_2)_{n-1}CCl_3$ as a function of n [13].

The circles are the experimental results, and the lines were calculated from formulas (4.7) and (4.18).

Figure 4.4 shows the NQR frequencies of ^{35}Cl in tetrachloroalkanes $Cl_3C(CH_2)_{n-1}Cl$ as a function of the length of the carbon chain. As in $Cl(CH_2)_nCl$ and $CH_3(CH_2)_nCl$, in tetrachloroalkanes an increase in n is also accompanied by a gradual decrease in the NQR frequencies of the CCl_3 and CH_2Cl groups, and an increase in the difference between them.

From $n = 5$, the difference between the NQR frequencies of the CCl_3 and CH_2Cl groups becomes almost constant (~5 MHz), i. e. for this carbon chain length the effect of the groups on each other is almost negligible. Other changes in the NQR spectra of this series should also be noted, e. g., the triplet structure of the CCl_3 group on increasing n from $n = 5$ to $n = 9$ is changed and also a uniform increase in the splitting between the high-frequency doublet and the singlet in this triplet is observed when n is further increased. This can be associated with a change in the type of crystal lattice which accompanies the change in the molecular shape from near-globular (for $n = 5$) to clearly elongated (for $n \geqslant 9$). The large splitting in the triplet caused by the CCl_3 group is probably due to the fact that for large n the overall effect of the C—H dipoles on the two chlorine atoms lying in the same plane greatly differs from that on the third chlorine atom. By investigating a large number of chlorine- and bromine-containing aliphatic compounds we are able to estimate the degree of accuracy of the calculations (0.1—2.0%) and to predict the NQR frequencies for a large number of compounds. The differences between the calculated and measured frequencies are systematic. As a rule, in the case of a large number of substituents the calculated values are much higher than the measured ones, whereas the opposite is true for a small number of substituents.

This is apparently due to the effects of intramolecular steric hindrance and the thermal vibrations of the molecules in the crystal.

Similar relationships are also obeyed when other substituents, such as NO_2, COOH, CN, CH_3, etc., affect the halogen by transmission along the carbon chain [12]. The character of the transmission effect of these substituents along the carbon chain can be expressed by formulas similar to (4.7) and (4.8), but with other $\Delta_1\nu$ and ξ (see also Table 4.2):

$$\nu_{\mathrm{I}} = \nu_{0\mathrm{I}} + \sum_i \Delta_1\nu_i\xi_i^{n-1} \tag{4.9}$$

$$\nu_{\mathrm{II}} = \nu_{0\mathrm{II}} + \sum_i \Delta_1\nu_i\xi_i^{n-1}. \tag{4.10}$$

Thus, for $BrC(NO_2)_2CH_2(NO_2)_2CBr$ expression (4.9) is written in the form:

$$\nu_{\mathrm{I}}^{\mathrm{Br}} = \nu_{0\mathrm{I}} + 2\Delta_1\nu_{\mathrm{NO_2}}\left(1 + \xi_{\mathrm{NO_2}}^2\right) + \Delta_1\nu_{\mathrm{Br}}\,\xi_{\mathrm{Br}}^2\,.$$

The calculated frequency ν_{I} =335.52 MHz, whereas the measured value is 335.552 MHz.

Applying such a model to the polyhaloethanes $Hal_mCH_{3-m}CH_{3-n}Hal_n$ (where n =0, 1, 2, 3 and m =1, 2,3) satisfactorily describes the change in the NQR frequencies of the halogen atoms as a function of their number (Figure 4.5). When n =0 the CH_3 group can be considered an independent substituent, and the calculation of the frequencies for $Hal_mCH_{3-m}CH_3$ enables us to test the accuracy of the values of $\Delta_1\nu$ for this group.

TABLE 4.2. Empirical values of the parameters in formulas (4.9) and (4.10)

X	^{35}Cl ($\nu_{0\mathrm{I}}$ =33.0 MHz, $\nu_{0\mathrm{II}}$ =32.0 MHz)		^{79}Br ($\nu_{0\mathrm{I}}$ = 250.90 MHz, $\nu_{0\mathrm{II}}$ = 243.3 MHz)	
	$\Delta_1\nu$, MHz	ξ	$\Delta_1\nu$, MHz	ξ
Cl	2.40	0.47	—	—
Br	2.60	0.45	22.0	0.50
COOH	2.80	—	34.1	0.30
NO_2	3.50	0.34	36.5	0.31
CH_3	-0.30	—	-1.3	—

Using this model, we are able to predict the NQR frequencies of ^{35}Cl and ^{79}Br in different halogen-substituted nonbranched aliphatic hydrocarbons, including those which are difficult to obtain or those still unknown.

The results of comparing the calculated and experimental data in the series $Cl(CH_2)_{n-1}MCl_3$ (where M = C, Si, Ge) are of interest [6]. For $Cl(CH_2)_{n-1}CCl_3$ the calculated and experimental data are in almost complete

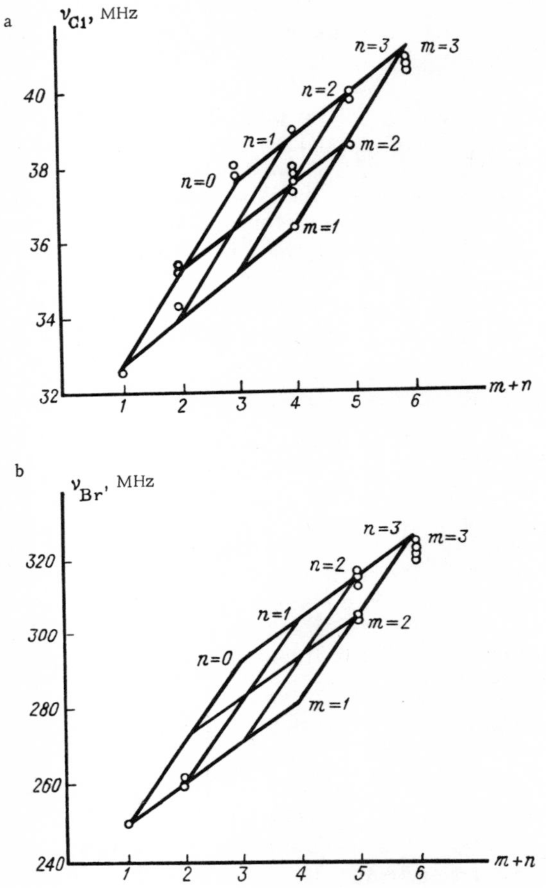

FIGURE 4.5. Comparison of the calculated (solid lines) and measured (circles) values of the NQR frequencies of the polyhalomethanes $Hal_mCH_{3-m}CH_{3-n}Hal_n$ [14]: a) Hal = ^{35}Cl; b) Hal = ^{79}Br.

agreement. The NQR frequencies of the chlorine atoms in the groups $SiCl_3$ and $GeCl_3$ can also be calculated satisfactorily using the formulas given above with an accuracy up to the factor k, i. e. the transmission effect law remains unchanged. However, the NQR frequencies of the chlorine atoms in the CH_2Cl group do not obey the same law. In silicon and germanium analogs the frequency becomes more strongly dependent on n

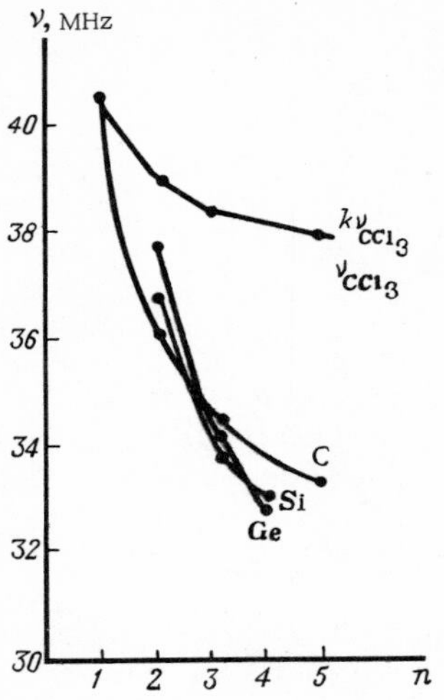

FIGURE 4.6. NQR frequencies of ^{35}Cl in $Cl(CH_2)_{n-1}MCl_3$ (where M = C, Si, Ge) as a function of the number of methylene groups n.

The frequencies of $SiCl_3$ and $GeCl_3$ were calculated from (4.7) and multiplied by k (the ratio of the frequencies of ^{35}Cl in the silicon and germanium compounds to the NQR frequencies of their carbon analogs).

and the damping does not follow a geometrical progression (Figure 4.6). The NQR frequencies of ^{35}Cl near the carbon alternate as follows: in the α position $\nu_{Ge} > \nu_{Si} > \nu_C$, in the β position $\nu_C > \nu_{Ge} > \nu_{Si}$, and in the γ position $\nu_C > \nu_{Si} > \nu_{Ge}$ (see Figure 4.6).

Thus, there exists a difference between the transmission effect along the carbon chain from element M to the terminal chlorine atom near the carbon and to the chlorine atoms in the MCl_3 group. This effect is probably due to the difference between the $\sigma - \sigma$ coupling when passing from C to Si and Ge.

3. CHLORINE-DERIVATIVES OF NONSATURATED HYDROCARBONS

Complete experimental data exist only for chlorine-derivatives of ethylene [27]. The dependence of the NQR frequencies of ^{35}Cl (averaged over the crystallographic positions) for these compounds is described by the equation:

$$\nu = \nu_0 + \Sigma\Delta_i \quad (4.11)$$

obtained by the least squares method. ν is the NQR frequency of ^{35}Cl, ν_0 is the NQR frequency of ^{35}Cl in $ClHC = CH_2$; Δ_i are the increments corresponding to the contributions of the substituents in gem (Δ_g), cis (Δ_c) and trans positions (Δ_t). The following values of the parameters were obtained: $\nu_0 = 33.84$, $\Delta_g = 2.52$, $\Delta_c = 1.48$, $\Delta_t = 1.02$ MHz.

Subsequent accumulation of data for compounds containing substituents other than chlorine will enable us to apply the formula more generally (4.11).

4. HALOGEN-DERIVATIVES OF AROMATIC COMPOUNDS

Halogen-substituted benzenes also obey the additivity law which is applied to study the influence of the substituent on the NQR frequency of the halogen in the benzene molecule. Thus, an increment α_i^X in the NQR frequency may be attributed to each substituent ortho, meta or para to the halogen atom in the benzene ring. This means that the transmission effect from the substituent to halogen through the benzene ring is characterized by three increments: ortho, meta, and para. It is assumed that the transmission effect from each substituent to the halogen is additive, and independent of the number, nature and position of the other substituents [4, 15—18, 30]. Averaging the values of the increments over a large number of NQR frequency differences due to halogens located in identical positions with respect to the same substituent enables us to obtain values of the increments very close to the real ones (Table 4.3) (see also [30]).

TABLE 4.3. Increments α_i^X (MHz) of halogen-substituted benzenes

Substituent X	^{35}Cl (ν_0=34.54 MHz)			^{79}Br (ν_0=270.29 MHz)			$^{127}I_{1/2-3/2}$ (ν_0=275.23 MHz)		
	ortho	meta	para	ortho	meta	para	ortho	meta	para
Cl	1.21	0.59	0.46	. . .	. . .	. . .	. . .	. . .	. . .
Br	. . .	. . .	. . .	11.68	5.00	2.17			
I	. . .	. . .	. . .	. . .	. . .	. . .	10.59	4.85	4.93
F	1.54	0.45	0.63	13.81	3.85	2.98	15.27	5.98	5.64
NH_2	—0.48	—0.09	—0.13	—4.91	—2.20	—2.73	—4.79	+7.12	—8.79
$NHCH_3$	+0.20	—0.06	—0.44	. . .	. . .	. . .	. . .	. . .	. . .
	—0.64								
$N(CH_3)_2$	+0.46	—0.12	—0.44	. . .	. . .	. . .	. . .	. . .	. . .
NO_2	2.56	1.03	0.65	26.30	9.73	6.60	. . .	15.93	13.18
	(1.80)			(17.7)					
CH_3	—0.29	+0.03	—0.01	—3.45	—1.35	—1.42	. . .	. . .	. . .
C_2H_5	—0.49	—0.21	—0.19	. . .	. . .	. . .	. . .	. . .	. . .
SO_2Cl	2.52	1.15	0.96	. . .	. . .	. . .	. . .	. . .	. . .
	(1.91)								
SO_2F	2.40	1.07	0.82	. . .	. . .	. . .	. . .	. . .	. . .
	(1.83)						. . .	. . .	. . .
OH	0.72	0.20	0.40	8.00	2.11	1.60	. . .	. . .	. . .
	(0.07)		(—0.12)	(0.45)		(—0.83)			
COCl	0.90	0.08	0.36	. . .	. . .	. . .	. . .	. . .	. . .
COH	0.53	—0.06	0.48	. . .	. . .	. . .	. . .	. . .	. . .
OCH_3	0.65	0.20	0.17	7.30	3.10	1.06	. . .	. . .	. . .
			(—0.12)			(—1.41)			

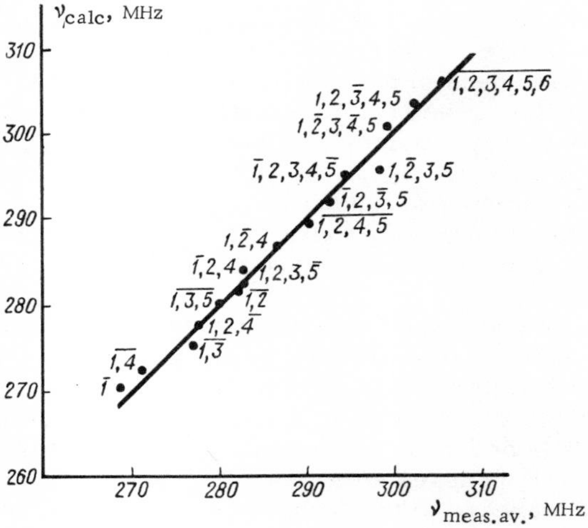

FIGURE 4.7. Relation between the calculated and measured frequencies of ^{79}Br in polybromobenzenes

The true shift is interpreted here as the shift of the NQR frequency of the halogen in the free molecule plus a contribution due to the constant mean field of the crystal. Since a large number of differences from differently substituted molecules located in differently arranged crystal lattices can be averaged, the effects due to the nearest environment of the molecule are also averaged. Only the mean values of the crystal field remain, i. e. the field determined by the averaged influence of more remote substituents. This value must be considered as roughly the same for all molecular crystals. The satisfactory agreement between the experimental data and those calculated using the additive law can be seen from Figure 4.7.

It was indicated in [17] that the sequence of increments for the different substituents agrees well with Holleman's series for both electron-donor and electron-acceptor substituents.

An exception is the increments for the OH group. This deviation is due to the fact that the phenols usually dissociate in solution and as a result the influence of this group on the halogen in the solid changes.

It was noted that the variation in the increments for the ortho, meta and para positions (α_o, α_m and α_p) of the same substituent usually obeys the rule $\alpha_o > \alpha_m > \alpha_p$ i. e. the influence of the substituent is dampened. The increments for the NH_2, $NHCH_3$, COCl and COH groups are an exception.

The characteristic sequence of frequencies in the NQR spectra of pentahalo-derivatives of benzene is [19]:

$$\nu_o < \nu_p < \nu_m \quad \text{for} \quad NH_2,\ NHCH_3,\ N(CH_3)_2$$

$$\nu_o < \nu_m < \nu_p \quad \text{for} \quad H,\ SH,\ OH,\ COOH,\ COCl$$

$$\nu_p < \nu_o < \nu_m \quad \text{for} \quad SCH_3,\ OCH_3.$$

Such a sequence agrees with the relative reactivity of these compounds with nucleophiles [20].

The shifts in the NQR frequencies of the halogen atoms meta and para* to a particular substituent are interpreted in terms of the electronic nature of the substituent. However, when the halogen atom is in an ortho position, either a sharp increase of the frequency or unequal shifts for two such atoms, or both phenomena simultaneously are observed. These effects were explained [17] by geometric considerations of the ortho substituents, namely, dimension, shape, symmetry, and the possibility of forming hydrogen bonds (see for instance α_o for the OH group). The "twisting" of the NO_2 group [21], the asymmetry of the SO_2F, SO_2Cl [22], OCH_3 and $NHCH_3$ groups [23], the anomalously large ortho effect in ortho-halogen-benzene acids and the intra- and intermolecular hydrogen bond in phenols are examples [24].

The existence of two values of α_o for the NO_2 group is due to the fact that steric hindrance between the NO_2 group and the adjacent substituent causes the NO_2 group to move out of the plane of the benzene ring and as a result the electronic environment of the resonating halogen atom is less distorted [21]. This decrease in distortion (i. e. the decrease in deviation from axial symmetry) causes a decrease in the asymmetry parameter which leads to a decrease in the frequency. Therefore, in the calculations we must allow for the type of environment adjacent to the nitro group in order to correctly select the increment.

* The expressions ortho, meta and para atom (or substituent) will now be used for brevity.

The asymmetry of the substituent leads to noticeable differences in the increments for the two halogen atoms ortho to an asymmetric substituent of the type $NHCH_3$, SO_2Cl, SO_2F (Table 4.3). The NQR frequency of an atom subjected to steric hindrance is higher since the asymmetry parameter is larger. The appearance of two increments α_p for the substituent OR is explained by the possibility of a perturbation due to coupling of OR with the benzene ring.

It follows that the general model for calculating the NQR frequencies of the halogen in polysubstituted benzene derivatives reduces to a simple addition of increments:

$$\nu = \nu_0 + \Sigma\alpha_o^X + \Sigma\alpha_m^X + \alpha_p^X, \qquad (4.12)$$

where ν_0 is the reference point allowing for the distribution of σ and π electron density in monohalogen-substituted benzene.

The additivity method has been used to study the changes in not only the NQR frequencies, but also in the number of uncompensated p electrons, determined from the NQR data both along the bond C—Hal (U_{p_z}) and perpendicular to it ($U_{p_x} - U_{p_y}$) [25] (see Chapter 5):

$$U_{p_z} = F_0 + \Sigma c_{ij} F_j; \qquad U_{p_x} - U_{p_y} = G_0 + \Sigma c_{ij} G_j, \qquad (4.13)$$

where c_{ij} is the number of given substituents (0, 1, or 2), and F_j, G_j are additive parameters listed in Table 4.4.

TABLE 4.4. Additive parameters for U_{pz} and $(U_{p_x} - U_{p_y})$

Group	G_j	F_j	Group	G_j	F_j
F_0	—	1.2890	p-Br	—0.0021	—0.0044
G_0	0.0232	—	o-NH_2	0.0027	0.0118
o-Br	0.0089	—0.0347	m-NH_2	—0.0039	0.0184
m-Br	0.0051	—0.0172	p-NH_2	—0.0120	0.0121

This examination was based on a very limited range of substances, i. e. the bromo-derivatives of benzene and aniline. In deriving the additive parameters, the known data on the asymmetry parameters of the electric field gradient for the bromine atoms were used. However, the authors of [25] cannot be accused of drawing overgeneralized conclusions since, in their opinion, the configuration of the NH_2 group is not stable.

5. CHLORINE-DERIVATIVES OF HETEROCYCLIC COMPOUNDS

The distribution of electron density in different atoms can be studied using chlorine-substituted heterocyclic compounds. In pyridine derivatives, the NQR frequencies of chlorine in the α position are much lower than in the γ position. Such a lowering (compared with chlorobenzene) is probably due to a change in the double bond [26]. Thus, for the chlorine atom the degree of double bonding increases from 3.4% in p-dichlorobenzene to 8.9% in 2-chloropyridine (see equation (5.25)).

Apparently, a similar increase in the double bonding of chlorine atoms also occurs in the derivatives of pyrimidine. The existence of a large number of chloro-derivatives of pyrimidine enabled us to develop an additive model to calculate the NQR frequencies. However, due to the initial differences in the distribution of the electron density over the pyrimidine ring we must introduce three reference points ν_{0k}, characterizing the initial distribution of the charge over the carbon atoms. The number of increments for the derivatives of pyrimidine is also much larger than for the derivatives of benzene since the transmission effect from the i-th position to the j-th is not characterized by the same increments (α_i^j) as that from the j-th to the i-th. This is understandable and agrees with the reactivity since the i-th and j-th positions have different electron density:

$$\nu_j = \nu_{0j} + \sum_i \alpha_i^j. \tag{4.14}$$

Table 4.5 gives the increments for the chloro-derivatives of pyrimidine. By accumulating experimental data on NQR we are able to establish a model of the electron density of the carbon atoms for all simple heterocyclics.

TABLE 4.5. Increments for the chloro-derivatives of pyrimidine

ν_{0k}, MHz	α_i^j, MHz		
$\nu_{02} = 34.47$	$\alpha_2^4 = 0.81$	$\alpha_2^5 = 0.22$	
$\nu_{04} = 34.65$	$\alpha_4^2 = 0.65$	$\alpha_4^5 = 1.06$	$\alpha_4^6 = 0.58$
$\nu_{05} = 36.12$	$\alpha_5^2 = 0.63$	$\alpha_5^4 = 0.95$	

BIBLIOGRAPHY

1. Livingston, R. — J. Phys. Chem. 57 (1953), 496.
2. Semin, G. K. and A. A. Fainzil'berg. — Zh. Strukt. Khim. 6 (1965), 213.
3. Semin, G. K. — In: "Radiospektroskopiya tverdogo tela," p. 205. Atomizdat, 1967.
4. Scrocco, E., P. Bucci, and M. Maestro. — J. Chim. phys. 56 (1959), 623.
5. Hooper, H. O. and P. Bray. — J. Chem. Phys. 33 (1960), 334.
6. Semin, G. K., T. A. Babushkina, V. I. Robas, G. Ya. Zueva, M. A. Kadina, and V. I. Svergun.— In: "Radiospektroskopicheskie i kvantovokhimicheskie metody v strukturnykh issledovaniyakh," p. 225. Moscow, "Nauka," 1967.
7. Swiger, E. and J. D. Graybeal. — J. Am. Chem. Soc. 87 (1965), 1461.
8. Green, P. J. and J. D. Graybeal. — J. Am. Chem. Soc. 89 (1967), 4305.
9. Semin, G. K., T. A. Babushkina, A. V. Prokof'ev, and R. G. Kostyanovskii.— Izv. AN SSSR, chem. ser., p. 1401, 1968.

10. Semin, G. K. and T. A. Babushkina. — TEKh 4 (1968), 835.
11. Lucken, E. A. C. and M. A. Whitehead. — J. Chem. Soc., p. 2459, 1961.
12. Semin, G. K. — Doklady AN SSSR 158 (1964), 1169.
13. Semin, G. K. and V. I. Robas. — Zh. Strukt. Khim. 7 (1966), 117.
14. Semin, G. K. — Zh. Strukt. Khim. 6 (1965), 916.
15. Kitaigorodskii, A. I. and G. K. Semin. — Synopses of Reports of the Conference on Physical Methods of Studying Molecules of Organic Compounds in Chemical Processes, p. 54. Frunze, 1962. (Russian)
16. Scrocco, F. — Adv. Chem. Phys. 5 (1963), 319.
17. Kitaigorodskii, A. I., G. K. Semin, and G. G. Yakobson. — In: "Radiospektroskopiya tverdogo tela," p. 202. Atomizdat, 1967.
18. Semin, G. K., L. S. Kobrina, and G. G. Yakobson. — Izv. Sib. Otdel. AN SSSR, ser. chem. sci., No. 9, Issue 4 (1968), 84.
19. Semin, G. K., V. I. Robas, L. S. Kobrina, and G. G. Yakobson. — Zh. Strukt. Khim. 9 (1964), 915.
20. Yakobson, G. G., L. S. Kobrina, and G. K. Semin. — Zh. Organ. Khim. 2 (1966), 495.
21. Semin, G. K., T. A. Babushkina, L. S. Kobrina, and G. G. Yakobson. — Izv. Sib. Otdel. AN SSSR, ser. chem. sci., No. 12, Issue 5 (1968), 69.
22. Semin, G. K., T. A. Babushkina, V. M. Vlasov, and G. G. Yakobson. — Izv. Sib. Otdel. AN SSSR, ser. chem. sci., No. 12, Issue 5 (1969), 99.
23. Semin, G. K., T. A. Babushkina, A. P. Zhukov, L. S. Kobrina, and G. G. Yakobson. — Izv. Sib. Otdel. AN SSSR, ser. chem. sci., No. 12, Issue 5 (1969), 93.
24. Semin, G. K., T. A. Babushkina, L. S. Kobrina, and G. G. Yakobson. — Izv. Sib. Otdel AN SSSR, ser. chem. sci., No. 12, Issue 5 (1968), 63.
25. Bucci, P., C. Cecchi, and A. Colligiani. — Bolletino 23 (1965), 313.
26. Lyukken, E. — In: Fizicheskie metody issledovaniya geterotsiklicheskikh soedinenii," Edited by A. K. Katritskii. "Khimiya," 1966.
27. Semin, G. K. and A. A. Boguslavskii. — Uchen. Zap. Mosc. Oblast. Pedag. Inst. im N. K. Krupskaya 222, No. 8 (1969), 74.
28. Interatomic Distances and Configurations of Molecules and Ions, London, 1958.
29. Trotter, J. — Can. J. Chem. 40 (1962), 1590.
30. Biedenkapp, D. and A. Weiss. — J. Chem. Phys. 49 (1968), 3933.

Chapter 5

INTERPRETATION OF NQR DATA

1. ELECTRIC FIELD GRADIENT IN THE ATOMS

We shall consider for simplicity an atom with one valence electron. Its wave function ψ_{nlm_l} can be represented as the product of a radial part and an angular component, the latter being a spherical harmonic function.

The electric field gradient created by this electron in the direction of its angular momentum for state $m_l = l$ will be equal to [1]:

$$eq_{nlm_l=l} = e\int |\psi_{nlm_l=l}|^2 \frac{3\cos^2\theta - 1}{r^3}\, dv = -\frac{2le}{2l+3} < \frac{1}{r^3} >_{\text{av}}. \tag{5.1}$$

For hydrogen-like wave functions:

$$< \frac{1}{r^3} >_{\text{av}} = \frac{2Z_i^3}{n^3 a_0^3\, l\,(l+1)(2l+1)}\,. \tag{5.2}$$

Here Z_i is the effective charge of the nucleus, n is the principal quantum number, and $a_0 = \frac{\hbar^2}{me^2}$ is the Bohr radius of the hydrogen atom. Thus, in the approximation of a hydrogen-like orbital we have:

$$eq_{nlm_l=l} = \frac{-4Z_i^3 e}{n^3 a_0^3 (l+1)(2l+1)(2l+3)} \quad \text{for } l \neq 0, \tag{5.3}$$

i. e. the gradient of the electric field of the atom decreases rapidly with an increase in n and l.

In what follows we will be mainly interested in the case where the valence electron participates in the chemical bond. The z-axis of the electric field gradient tensor will then lie in the direction of the chemical bond. The component of the electric field gradient created by the p electron along this axis corresponds to $m = 0$, i. e. for $l = 1$:

$$\begin{aligned} p_z &= p_0 \\ p_x &= \frac{p_{+1} + p_{-1}}{2} \\ p_y &= \frac{p_{+1} + p_{-1}}{2}. \end{aligned} \tag{5.4}$$

If q_{p_x}, q_{p_y}, q_{p_z} are the values of the field gradient created by the electrons in the p_x-, p_y- and p_z states, respectively, then $q_{p_x} = q_{p_y} = q_{1,\pm 1}$, $q_{p_z} = q_{1,0}$.

From the expression for the angular part of the atomic wave function ψ_{nlm_l} we obtain [2]:

$$eq_{at} = eq_{nl0} = -\frac{l+1}{2l-1} q_{nlm_l=l} =$$

$$= \frac{4Z_i^3 e}{n^3 a_0^3 (2l-1)(2l+1)(2l+3)} \tag{5.5}$$

$$eq_{n10} = \frac{4}{5} e < r^{-3} >_{av} . \tag{5.6}$$

For halogen atoms, where the one p electron is not in a closed shell, $eq_{at} = -eq_{n10}$. Townes and Dailey [3] examined the values of q_{nl0} using the fine structure data obtained from optical spectra for different atoms in different atomic states. They showed that the values of q_{nl0} decrease with increasing n and l more rapidly than predicted by (5.5).

TABLE 5.1. Values of q_{nl0} for the p electrons of different atoms in the ground state

Nucleus	Spin	Electron	$<r^{-3}>\cdot 10^{-24}$, cm^{-3}	$q_{at}\cdot 10^{-24}$, cm^{-3}
^{7}Li	$^3/_2$	$2p$	0.26	—0.21
^{9}Be	$^3/_2$	$2p$	1.17	—0.94
^{25}Mg	$^5/_2$	$3p$	5.2	—4.16
^{43}Ca	$^7/_2$	$4p$	7.6	—6.08
^{87}Sr	$^9/_2$	$5p$	13.5	—10.8
^{137}Ba	$^3/_2$	$6p$	19.2	—15.36
^{14}N	1	$2p$	22.5	—18.0
^{75}As	$^3/_2$	$4p$	51	—40.8
^{121}Sb	$^5/_2$	$5p$	88	—70.4
^{209}Bi	$^9/_2$	$6p$	166	—132.8
^{35}Cl	$^3/_2$	$3p$	48	—38.4

Table 5.1 gives q_{nl0} (calculated by Barnes and Smith) for the p electrons of different atoms in the ground state [4].

The methods of calculating $<r^{-3}>_{av}$ and the atomic field gradients obtained from the spectral data are discussed in detail by Townes [3] and Das and Hahn [2]. These methods include a number of corrections.

Firstly, a correction allowing for the possibility of configuration interaction was discussed by Koster [5]. Koster noted that if one assumes that the ground state has a single configuration, this is equivalent to assuming that each electron in the atom is in the field of the averaged central potential. However, as a result of the Coulomb repulsion between the electrons the different electronic configurations in the atom are mixed. Thus, for example, there is a small admixture of the configurations $4s4p5s$, $4p^3$ and $4s4p5d$ for a gallium atom in the ground $4s^2 4p$ state. Koster [5] and Lew and Wessel [6] examined the configuration interaction $nsnp(n+1)s$ (where $n=3$ and 4 for aluminum and gallium, respectively) which affects the value of eQ. Tsyunaitis and Yutsiss [7] showed that for the boron atom we must take into account the $2p^3$ state in addition to the ground $2s^2 2p$ state. The admixture alters eq_{nll}. The configuration interaction is greater for heavier atoms due to the increase in the Coulomb repulsion between the electrons compared with their attraction to the nucleus.

The second correction is due to distortion of the internal electron shells both by the nuclear quadrupole moment and by the valence electrons. This effect was detected by Pound [8] and Sternheimer [9] and studied in detail by Sternheimer [10—15].

The perturbing electrostatic potential of the electric quadrupole moment of the nucleus disturbs the spherical symmetry of the closed shells and induces in them a finite quadrupole moment. Interaction of the valence electron with this induced quadrupole moment leads to a change in the quadrupole interaction constant. The valence electron produces a similar effect, creating, as a result, a finite field gradient on the nucleus. These two additional indirect interactions can be taken into account by multiplying e^2Qq_{at} by $(1 - \gamma_\infty)$. Here, eq_{at} is given by (5.5), and γ_∞ is the so-called Sternheimer factor for the free atom. When $\gamma_\infty > 0$, it expresses the screening effect of the inner electron shell, and when $\gamma_\infty < 0$ it expresses the antiscreening effect. Appendix I lists the known values of γ_∞ for atoms and ions. It is particularly important to take the Sternheimer factor into account for ionic crystals, in which the electric field gradient is mainly determined by the charges of the adjacent ions, since the Sternheimer factor is different for the p electrons and for the charges external to the atom. In molecular crystals with covalent bonds, the influence of γ_∞ on the electric field gradient at the atomic nucleus in the molecule (created mainly by the p electrons) and in the free atom is assumed to be the same [2]. Since e^2Qq_{at} can be determined from the data obtained from atomic beam and optical spectroscopy, a special correction of $(1 - \gamma_\infty)$ in the calculations and theoretical evaluations of e^2Qq_{mol} is not necessary in these cases.

The third correction is the allowance for relativistic effects [2]. For light atoms these can be neglected.

2. TOWNES-DAILEY THEORY

We have already discussed in the introduction the main sources of the electric field gradients on the atomic nuclei in molecules. We should note again that in molecular crystals of most organic and organometallic compounds, mainly the valence electrons of that particular atom contribute to the electric field gradient.

The electric field gradient in the z direction (for example, the direction of the C—Cl bond) in the molecule eq_{mol} can be expressed in terms of the electric field gradient of the free atom:

$$eq_{mol} = U_p eq_{at}, \tag{5.7}$$

where U_p is a quantity depending on the electronic structure of the molecule and determining the number of uncompensated p electrons in the given structure.

We shall discuss in detail the method of Townes and Dailey to determine U_p to a first approximation. (This approach was described by Townes and Dailey [16], and also by O'Konski [17].)

Let atom A with a single valence electron outside the closed shell form a single covalent bond with atom B which also has one valence electron. The orbital wave function of this electron in the molecule can be written as a linear combination of the electron wave functions $\psi^{A}_{nlm_l}$ of the atom A:

$$\Psi = \Sigma a^{A}_{nlm_l}\, \psi^{A}_{nlm_l}. \tag{5.8}$$

Let the electron in each of the two bonded atoms be in the lowest energy level. The maximum component of the field gradient is directed along the bond and is equal to:

$$eq^{A}_{\mathrm{mol}} = \frac{\partial^2 V}{\partial z^2} = -e \int \Psi^* \frac{3\cos^2\theta - 1}{r^3_{A}} \Psi dv, \tag{5.9}$$

where r_A is the distance from the "point charge" $e\Psi\Psi^* dv$ to the nucleus A.

If the electron pair is equally divided between atoms A and B, the electron density in the atomic states $\psi^{A}_{nlm_l}$ will approach unity. The molecular gradient of the electric field can then be written in the form:

$$eq^{A}_{\mathrm{mol}} = e \sum_{nlm_l} \left| a_{nlm_l} \right|^2 q^{A}_{nlm_l,\, nlm_l} + e \sum_{nlm_l} a_{nlm_l} a^{*}_{n'l'm'_{l'}} q^{A}_{nlm_l,\, n'l'm'_{l'}}, \tag{5.10}$$

where

$$eq_{nlm_l,\, n'l'm'_{l'}} = e \int \psi^{*}_{nlm_l} \frac{3\cos^2\theta - 1}{r^3} \psi_{n'l'm'_{l'}} dv.$$

A large number of terms are removed in the second sum since $q_{nlm_l,\, n'l'm'_{l'}} = 0$ for $m_l \neq m'_{l'}$, and $l = l' = 0$ or $|l - l'| \neq 2$. This follows from the properties of the spherical harmonics determining the angular part of the atomic wave functions. "Integrals with cross-products" $q_{nlm_l,n'l'm'_{l'}}$ cannot be evaluated exactly, but they decrease very rapidly with increasing n, l, n' or l'. Moreover, their contribution to the total sum is small since the coefficients $a_{nlm}a_{n'l'm'_l}$ in equation (5.10) are small. For instance, if the p function is the largest component in the molecular wave function, the cross-terms including a_{n1m} will be largest, and they will contain an admixture of p states with other p and f states. However, such hybridization will be small since its contribution to the decrease of the bonding energy of the atoms A and B is small. In the hybridization of the p state with the s and d states, the cross-terms including a_{n1m} vanish since $l - l' \neq 2$.

With regard to $q_{nlm_l,\, nlm_l} = q_{nlm_l}$, for the s state ($l = 0$, $m_l = 0$) $q_{n,0,0}$ always vanishes, but when $l \neq 0$, q_{nlm_l} decreases rapidly with increasing n or l. The dominant term in the first sum corresponds to the n state of the lowest allowed n since q_{nlm_l} for this state is much larger than for states with higher energy. Moreover, in this state the electron density is larger. As a result, the amplitude q_{nlm_l} of the lowest p state is relatively large. Thus, the ratio of the molecular gradient of the electric field to the atomic gradient determining U_p will be approximately equal to the number of uncompensated p electrons in the structure.

If N_x, N_y and N_z are the populations of the p_x-, p_y- and p_z electron orbitals of the chlorine atom, the contributions of the p electrons in q_{zz} can be expressed as follows:

$$eq_{zz} = \left(\frac{N_x + N_y}{2} - N_z \right) eq_{\text{at}}. \tag{5.11}$$

We have taken into account here that the electrons contribute equally to the component of the electric field gradient along the axis of each p orbital. The factor $\frac{1}{2}$ is the contribution of the perpendicular orbital. At the nucleus where the electron density vanishes, the Laplace equation:

$$q_{xx} + q_{yy} + q_{zz} = 0$$

is fulfilled, and it follows from the condition of axial symmetry $q_{xx} = q_{yy}$ that:

$$q^z_{px} = q^z_{py} = -\frac{1}{2} q^z_{pz},$$

where $q^z_{p_i}$ is the gradient of the electric field created by one p_i electron along the z-axis.

The last expression implies that the gradient created by the p_z electron is twice as large in modulus and opposite in sign to the gradient created by the p_x or p_y electrons along the z-axis.

As an example, let us consider the chlorine atom with a $3s^2 3p_x^2 3p_y^2 3p_z$ configuration [2]. Then $U_p = -1$ and the electric field gradient at the nucleus will be negative and equal to the gradient of the field of one $3p_z$ electron. On the other hand, if the bond is 100% ionic, Cl^- will have the configuration $3s^2 3p_x^2 3p_y^2 3p_z^2$. Since the field has spherical symmetry, the gradient at the nucleus will vanish. In general, the chlorine atom, being more electronegative, usually draws off the nucleus. Then $N_z = 1 + I$, $N_x = N_y = 2$ and $eq_{zz} = \left(\frac{2+2}{2} - 1 - I \right) eq_{\text{at}} = (1 - I)\, eq_{\text{at}}$.

In the case of partial sp hybridization, the electron density of the p_z orbital decreases by the value of the s state fraction (due to the decrease in the p character by s), but the p character of the s orbital increases by $2s$ (due to the increase in the p character of the two electrons of the undivided pair by s). As a result, N_z of the atom increases by s. Moreover, since the $3d_{zz}$ orbital was vacant at first, the pd hybridization leads to a decrease in N_z by d. The participation of one of the undivided pairs $3p_x^2$ or $3p_y^2$ in the conjugation system causes N_x or N_y to decrease by the value of the degree of π double bonding. (Usually the measure of π double bonding is defined as the difference between the electron density of the p_x or p_y states of the free and bonded atom.) In general we have:

$$N_z = 1 + I + s - d$$
$$N_y = 2 - \pi$$
$$N_x = 2 - \pi,$$

and the electric field gradient on the chlorine nucleus is equal to:

$$eq_{\text{mol}} = (1 - I - s + d - \pi)\, eq_{\text{at}}. \qquad (5.12)$$

Equation (5.12) shows that the electric field gradient at the nucleus in the molecule depends on three parameters: the ionic character of the bond, the degree of hybridization, and the double bonding. This very often leads to an uncertainty in the interpretation of the quadrupole interaction data because the Townes-Dailey theory should only be applied to cases where these three factors are considered separately.

In many cases the asymmetry parameter of the electric field gradient may be of substantial help. In the same way as in equation (5.11) the number of uncompensated p electrons on the p_x and p_y orbitals can be written as:

$$eq_{xx\,\text{mol}} = \left(N_x - \frac{N_y + N_z}{2}\right) eq_{\text{at}}$$

$$eq_{yy\,\text{mol}} = \left(N_y - \frac{N_x + N_z}{2}\right) eq_{\text{at}}.$$

The asymmetry parameter of the electric field gradient will then be equal to:

$$\eta = \frac{q_{xx} - q_{yy}}{q_{zz}} = \frac{3/2\,(N_x - N_y)}{N_z - \dfrac{N_x + N_y}{2}}. \qquad (5.13)$$

For small η, equation (5.13) can be obtained in a more convenient form by neglecting terms in η^2 in the expression for the quadrupole resonance frequency:

$$I = {}^3/_2; \qquad \eta = \frac{3}{4}(N_x - N_y)\frac{e^2Qq_{\text{at}}}{\nu_{(1/2-3/2)}}$$

$$I = {}^5/_2; \qquad \eta = \frac{3}{2}(N_x - N_y)\frac{e^2Qq_{\text{at}}}{\nu_{(1/2-3/2)}}. \qquad (5.14)$$

Thus, the asymmetry parameter is a direct measure of the difference in the population of the electron states $3p_x$ and $3p_y$. For planar molecules this difference is also a measure of the double bonding [18, 19] (see Section 5 of Chapter 5).

We shall examine the application of the Townes-Dailey general formula to the iodine chloride molecule ICl, for which the quadrupole bond constants of ^{35}Cl and ^{127}I are known in the gaseous and solid states. The quadrupole interaction data for this molecule were studied by Townes and Dailey [16, 20–22], Das and Hahn [2], and Gordy [23]. We shall follow the treatment of Das and Hahn.

The iodine and chlorine atoms have the configurations $5s^25p^5$ and $3s^23p^5$, respectively. The field gradients of the free atoms are equal in magnitude and opposite in sign to the gradients from the $5p$ and $3p$ electrons. In the ICl molecule the electron orbitals overlap, leading to a decrease in energy compared with the total energy of two free atoms. The main overlapping occurs between the external p_z electrons along the internuclear axis, leading to the formation of a σ bond. The overlapping of the p_x and p_y electrons, which yields the π bond, is neglected to begin with.

The two electrons forming the σ bond are on a molecular orbital which is a linear combination of the atomic orbitals of the two atoms. This molecular orbital can be written in the following form:

$$\Psi = \frac{\psi_{\mathrm{Cl}} + i\psi_{\mathrm{I}}}{(1 + i^2 + 2iS)^{1/2}}, \tag{5.15}$$

where ψ_{Cl} and ψ_{I} are the atomic orbitals of the chlorine and iodine atoms, $S = \int \psi_{\mathrm{Cl}}\psi_{\mathrm{I}} dv$ is the overlap integral of the orbitals, and i is the measure of the relative contribution of the atomic orbitals in the molecule and depends on the ionic character of the bond I where $I = (1-i^2)/(1+i^2)$. (If $i = 1$, then $I = 0$ and the bond is purely covalent; if $i = 0$, then $I = 1$ and the bond is 100% ionic.)

The orbitals ψ_{Cl} and ψ_{I} are not in general purely p_z atomic orbitals. For the ICl molecule we should consider the hybridization of the p_z orbitals of both atoms not only with the s orbitals, but also with the d_{zz}. The normalized atomic orbitals ψ_{Cl} and ψ_{I} can then be written as follows:

$$\begin{aligned} \psi_{\mathrm{Cl}} &= \sqrt{1-s-d}\,\psi_{p_z\,\mathrm{Cl}} + \sqrt{s}\,\psi_{s\mathrm{Cl}} + \sqrt{d}\,\psi_{d_{zz}\,\mathrm{Cl}} \\ \psi_{\mathrm{I}} &= \sqrt{1-s'-d'}\,\psi_{p_z\,\mathrm{I}} + \sqrt{s'}\psi_{s_{\mathrm{I}}} + \sqrt{d'}\psi_{d'_{zz}\,\mathrm{I}}, \end{aligned} \tag{5.16}$$

where $s(s')$ and $d(d')$ are the degrees of sp and d hybridization of the chlorine (iodine) atom.

The contributions to the electric field gradient q_{zz}^{Cl} along the z-axis are: 1) two electrons on the molecular orbital Ψ (see equation (5.15)); 2) two pairs of $3p_x$ and $3p_y$ electrons of the chlorine atom; 3) the $3s^2$ electrons; 4) the electrons in the internal shells $1s^2 2s^2 2p^6$; and 5) the nucleus of the iodine atom and its electrons, which do not participate in the formation of the molecular orbital. These contributions to the electric field gradient shall be discussed separately.

1. The electrons of the molecular orbital Ψ create a field gradient eq_{MO} equal to:

$$\begin{aligned} eq_{\mathrm{MO}} &= -2e\int \Psi^2 \frac{3\cos^2\theta_{\mathrm{Cl}} - 1}{r_{\mathrm{Cl}}^3} dv = \\ &= -\frac{2e}{1+i^2+2iS}\left(\int \psi_{\mathrm{Cl}}^2 \frac{3\cos^2\theta_{\mathrm{Cl}} - 1}{r_{\mathrm{Cl}}^3} dv + \right. \\ &\left. + i^2\int \psi_{\mathrm{I}}^2 \frac{3\cos^2\theta_{\mathrm{Cl}} - 1}{r_{\mathrm{Cl}}^3} dv + 2i\int \psi_{\mathrm{I}}\,\psi_{\mathrm{Cl}} \frac{3\cos^2\theta_{\mathrm{Cl}} - 1}{r_{\mathrm{Cl}}^3} dv\right) = \\ &= -\frac{2e}{1+i^2+2iS}(q_{\mathrm{Cl}} + i^2 q_{\mathrm{I}} + 2iq_{\mathrm{ClI}}). \end{aligned} \tag{5.17}$$

Substituting expression (5.16) in (5.17), we obtain:

$$q_{\mathrm{Cl}} = (1-s-d)\,q_{p_z\,\mathrm{Cl}} + d q_{d_{zz}}\,\mathrm{Cl} + 2\sqrt{sd}q_{sd_{zz}}\,\mathrm{Cl}, \tag{5.18}$$

where q_{p_z} and $q_{d_{zz}}$ are the field gradients created by one p_z and one d_{zz} electron, respectively, and

$$q_{sd_{zz}\,\mathrm{Cl}} = \int \psi_{s\,\mathrm{Cl}} \frac{3\cos^2\theta_{\mathrm{Cl}} - 1}{r_{\mathrm{Cl}}^3} \psi_{d_{zz}\,\mathrm{Cl}}\, dv.$$

2. Each $3p_x$ and $3p_y$ electron in the chlorine atom in the z direction creates a field gradient equal to $q_{p_x\mathrm{Cl}} = q_{p_y\mathrm{Cl}} = -\frac{1}{2} q_{p_z\mathrm{Cl}}$. The total contribution to the field gradient from these π electrons is therefore equal to:

$$eq_\pi = -2e \int \psi^2_{p_x\,\mathrm{Cl}} \frac{3\cos^2\theta_{\mathrm{Cl}} - 1}{r_{\mathrm{Cl}}^3}\, dv - -2e \int \psi^2_{p_y\,\mathrm{Cl}} \frac{3\cos^2\theta_{\mathrm{Cl}} - 1}{r_{\mathrm{Cl}}^3}\, dv = +2eq_{p_z\,\mathrm{Cl}}. \quad (5.19)$$

3. The electrons of the $3s^2$ shell do not contribute to the field gradient in the free atom. In the ICl molecule the σ bond is formed by the hybridized atomic orbitals. As a result, the free pair of electrons occupy not only the orbital $3s^2$, but also the orbital $\overset{+}{\psi}_{\mathrm{Cl}}$ which is orthogonal to the sp hybridized orbital ψ_{Cl}:

$$\overset{+}{\psi}_{\mathrm{Cl}} = (1 - s)\,\psi_{s\,\mathrm{Cl}} + \sqrt{s}\,\psi_{p_z\,\mathrm{Cl}}.$$

Due to the p state fraction, these two electrons create a field gradient $\overset{+}{q}_{\mathrm{Cl}}$ equal to:

$$e\overset{+}{q}_{\mathrm{Cl}} = -2esq_{p_z\,\mathrm{Cl}}\,. \quad (5.20)$$

4. As already mentioned above, the electric field gradient created by the external charge is affected by the internal atomic shell (see Section 1 of Chapter 5). The gradients q_{Cl}, q_{I}, q_{ClI}, q_π and $\overset{+}{q}_{\mathrm{Cl}}$ change under the screening and antiscreening effects and are not equal to the amounts given by equations (5.17)–(5.20). The changes in q_{Cl}, $\overset{+}{q}_{\mathrm{Cl}}$ and q_π are taken into account by replacing q_{p_z} by q_{at} which is obtained from atomic beam experiments. Small corrections for the Sternheimer factor should be introduced in the terms $dq_{d_{zz}\mathrm{Cl}}$ and $\sqrt{sd}q_{sd_{zz}\mathrm{Cl}}$, but, as will be shown below, these terms are small. The gradient q_{I} is due to the electron on the sp orbital of the iodine atom. This electron affects both the internal shell of the chlorine atom and the other external electrons on the ψ_{Cl}, $\overset{+}{\psi}_{\mathrm{Cl}}$, $3p_x$, $3p_y$ orbitals of the chlorine atom. Thus, for q_{I} we should allow for a Sternheimer factor γ_1 which will be different from the Sternheimer factor γ_2 determined by a charge which is external to the chlorine atom and neutral due to the overlapping of the ψ_{I} and ψ_{Cl} orbitals in the molecule.

5. The effective positive charge at the iodine nucleus will create a field gradient on the chlorine nucleus equal to:

$$eq_{\mathrm{I}}^{+} = \frac{2e}{R^3}(1 - \gamma_2), \quad (5.21)$$

where R is the I—Cl bond length in ICl.

Summing equations (5.17)—(5.21), we obtain the total field gradient at the chlorine nucleus in the direction of the z-axis:

$$eq_{zz}^{\mathrm{Cl}} = \Big\{ -\frac{2e}{1 + i^2 + 2iS} \Big[q_{\mathrm{at}}^{\mathrm{Cl}} (1 - s - d) + d q_{d_{zz}\mathrm{Cl}} + 2\sqrt{sd}\, q_{s\,d_{zz}\mathrm{Cl}} \Big] + 2eq_{\mathrm{at}}^{\mathrm{Cl}} - 2esq_{\mathrm{at}}^{\mathrm{Cl}} \Big\} + \left[\frac{2e}{R^3}(1 - \gamma_2) - \frac{2i^2 eq_{\mathrm{I}}(1 - \gamma_1)}{1 + i^2 + 2iS} \right] - \frac{4eiq_{\mathrm{ClI}}}{1 + i^2 + 2iS}. \tag{5.22}$$

Townes and Dailey [20—22] introduced a number of simplifications, neglecting to begin with the terms responsible for the overlap of orbitals due to the mutual compensation of the different effects. In addition, they neglected the field gradient $q_{d_{zz}\mathrm{Cl}}$, created by the electron in the d_{zz} state, since it is much smaller than the gradient created by the $3p$ electron.

The second term in (5.22) may be neglected on examining R [20]. Das and Hahn [2] showed that even when allowing for fairly large values of $(1 - \gamma_1)$ and $(1 - \gamma_2)$ this term does not exceed 10% of $q_{\mathrm{at}}^{\mathrm{Cl}}$. Thus, for molecules in which the ionic character is not too great the approximate expression for the field gradient will be:

$$eq_{zz}^{\mathrm{Cl}} = e\left[-\frac{2(1 - s - d)}{1 + i^2} q_{\mathrm{at}}^{\mathrm{Cl}} + 2(1 - s)\, q_{\mathrm{at}}^{\mathrm{Cl}} \right] = [(1 - s + d - I) + I(s + d)]\, eq_{\mathrm{at}}^{\mathrm{Cl}}. \tag{5.23}$$

When the d hybridization is neglected, the field gradient is equal to:

$$eq_{qz}^{\mathrm{Cl}} = [(1 - s)(1 - I)]\, eq_{\mathrm{at}}^{\mathrm{Cl}}. \tag{5.23a}$$

The formation of the π bond between the iodine and chlorine atoms may lead to a decrease in the contribution of the field gradient on the chlorine nucleus from the 3_{px} and 3_{py} electrons:

$$q_{\pi} = 2\left(1 - \frac{\pi}{2}\right) q_{\mathrm{at}} = (2 - \pi)\, q_{\mathrm{at}}. \tag{5.24}$$

Except for very rare cases where I and $(s + d)$ are large, the term $I(s + d)$ can be neglected. The field gradients on the atomic nucleus will then be equal to:

$$eq_{zz}^{\mathrm{Cl}} = (1 - s + d - I - \pi)\, eq_{\mathrm{at}}^{\mathrm{Cl}} \tag{5.25}$$

$$eq_{zz}^{\mathrm{I}} = (1 - s + d + I - \pi)\, eq_{\mathrm{at}}^{\mathrm{I}}. \tag{5.26}$$

Both equations can be applied to a halogen atom bonded to any other atom. Equation (5.25) is valid when the halogen is more electronegative than the atom bonded to it, and equation (5.26) is valid when it is less electronegative.

The equations given above can be applied to covalent molecules.

3. DETERMINATION OF THE HYBRIDIZATION AND IONIC CHARACTER OF CHEMICAL BONDS FROM NQR DATA

In order to use equations (5.11), (5.12), (5.25) and (5.26) we must calculate the degree of hybridization, ionic character and double bonding of the chemical bonds by means of semiempirical procedures. Many aspects of this problem have been discussed by Dailey and Townes [20–22], O'Konski [17], and Das and Hahn [2]. We shall briefly discuss some of them here.

Hybridization. Unstrained structures with several ligands have a central atom with *sp* hybridized orbitals. The *s* character can be calculated from equation [1, 24]:

$$s = \frac{\cos\beta}{\cos\beta - 1}, \qquad (5.27)$$

where β is the angle between the bonds. It is assumed here that the orbitals are directed along the lines connecting the nuclei. This might not be true when the repulsion between the ligands is sufficiently strong. Thus, Gordy [23] attributes the observed 8° difference in valence angles in AsH_3 and AsF_3 to such a repulsion between the electronegative fluorine atoms. However, most of the angles determined by microwave radiospectroscopy are equal to the valence angles* to within the limits of experimental accuracy. There also exists a second reason for the deviation from equation (5.27), namely, the existence of partial *d* hybridization which is most important for heavy atoms.

However, the study of compounds of atoms of the first period showed that *sp* hybridization best explains the cases where the bond angles are larger than 90°. Thus, in methane the four C—H bonds have valence angles equal to 109°28' and the *s* character of each carbon atomic orbital is 0.25. In ethylene the angles between the C—H bonds are equal to 120°, which leads to a *sp* hybridization equal to 0.33, etc.

This relation is not true for atoms bonded to one atom only. Townes and Dailey [16] suggested the following rule for halogen atoms: if the halogen is bonded to an atom having an electronegativity 0.25 (or more) lower than its intrinsic value it will have 15% *sp* hybridization, and if the opposite is the case the *s* character of the hybridization will be equal to zero. It is also assumed that the *d* hybridization is always lower than 5%. Townes and Dailey believe that more accurate evaluations are not yet possible. Gordy [24] estimated that for the chlorine atom the *s* character fraction in the hybridization is 18% in FCl and 10% in CH_3Cl. He assumed that the *sp* hybridization is negligible in all bonds except those of a chlorine atom with elements of the first period.

On the basis of the hybridization rules listed, Townes and Dailey [16] and Gordy [24] used the data on the quadrupole bond constants of the different molecules to calculate the degree of ionic character and compared it with the electronegativity of the halogen and the bonded atom.

* The valence angle is defined as the angle formed by the straight lines connecting the centers of the atoms in the direction of the chemical bonds between them.

Ionic character of the bond. Gordy [23, 25] calculated the degrees of hybridization from the quadrupole interaction constants, starting from the principle that the ionic character of the bond I is related to the difference in electronegativities $(x_A - x_B)$ of atoms A and B as follows:

$$I = \frac{x_A - x_B}{2} \tag{5.28}$$

for $|x_A - x_B| < 2$ and $I = 100\%$ for $|x_A - x_B| \geqslant 2$.

Townes and Dailey [16] analyzed the quadrupole bond constants obtained by microwave spectroscopy and molecular beam methods for diatomic molecules using equations (5.25) and (5.26). This was accomplished by taking into account the hybridization rule formulated above and evaluating the degree of double bonding from the bond lengths. The results obtained are shown in Figure 5.1, and are compared with the data given by Gordy and Pauling [26] (the latter were obtained from dipole moment measurements). Pauling's curve is described by the empirical equation:

$$I = 1 - \exp\left(\frac{x_A - x_B}{2}\right)^2.$$

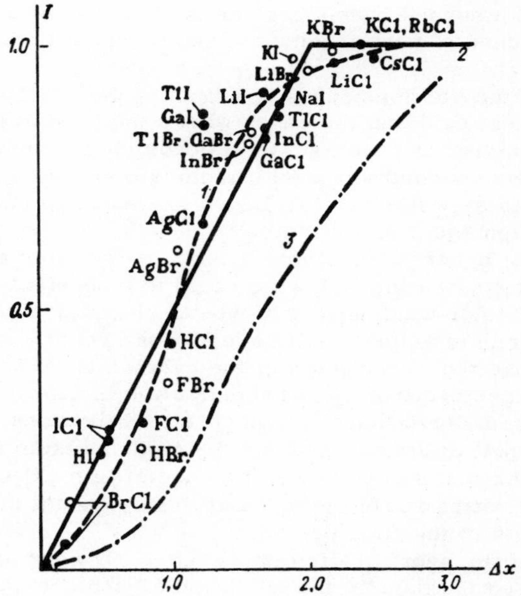

FIGURE 5.1. Ionic character I of the A— B bond as a function of the difference in electronegativities Δx:

1) Townes-Dailey curve; 2) Gordy curve; 3) Pauling curve.

Gordy's curve (2) predicts a higher ionic character than that of Townes-Dailey (1) and both are much higher than Pauling's curve (3). Thus, irrespective of the method by which the degree of *sp* hybridization was determined, the quadrupole interaction data obtained by Gordy and Townes-Dailey indicate a much higher ionic character of the bonds formed by the halogen atoms than assumed earlier.* An interesting result is the prediction of the 100% ionic bond for $|x_A - x_B| \geqslant 2$.

TABLE 5.2. Ionic character of the Hal—Cl bonds determined from the quadrupole interaction constants

Investigator	Compound	Ionic character of the bond	
		according to e^2Qq_{Cl}	according to e^2Qq_{Hal}
Townes-Daily, Gordy	BrCl	0.056	0.110
Townes-Daily	ICl	0.115	0.229
Gordy	ICl	0.248	0.229

A classic test of the validity of these two attitudes is afforded by compounds of two different halogens, in which both atoms in the molecule can be studied. The results for BrCl and ICl are shown in Table 5.2. For BrCl the Townes-Dailey theory gives the same result as that of Gordy since $\Delta x = 0.2$.

The values of the ionic character of the bond as determined from the data for the different atoms diverge noticeably. In ICI, the assumption that hybridization is absent yields a smaller discrepancy between the two values for I than the assumption that there is 15% *sp* hybridization.

Most of the data on quadrupole interactions are examined using the Townes-Dailey theory. However, the recently developed theory of orbital electronegativities shows that Gordy's theory would have been more reasonable since x_A and x_B are now regarded as the effective orbital electronegativities of atoms A and B.

The ratios between the quadrupole frequencies and the orbital electronegativities of the atoms were established by Whitehead and Jaffé [28, 29]. Coulson [30, 31] had already shown that the criterion of maximum overlapping of orbitals for the formation of stable bonds is fulfilled in the case when the energy of the molecular orbital E_{ab} is smaller than the sum of the energies of the corresponding atomic orbitals E_a and E_b. This leads to the requirement that E_a and E_b should be close in magnitude. Investigation of the energy of the *s* and *p* orbitals of halogens (Figure 5.2) shows that in diatomic halogen compounds the halogen having a higher *p* orbital energy

* Wilmshurst [27] suggested the following relation for the ionic character of the bond:

$$I = \frac{|x_A - x_B|}{x_A + x_B}.$$

On the basis of this definition Wilmshurst found much higher values of the *s* hybridization than those used by Townes and Dailey, and this resulted in a lower ionic character.

and fractional positive charge δ^+ will be hybridized [28]. This conclusion contradicts that of Townes and Dailey who held that the atom with δ^- is hybridized. If the halogen is bonded to an atom which has δ^+ and cannot be hybridized (for instance, hydrogen or deuterium), the halogen can be hybridized even though it has δ^-. Application of this principle to molecules of type XY (where X = Hal, H, D, CN, and Y = Cl, Br, I) leads to the approximate equality of the energies of the combining orbitals and satisfactorily explains the observed NQR frequency. It was shown that *d* hybridization must be taken into account in all compounds of bromine and iodine with other halogens [28].

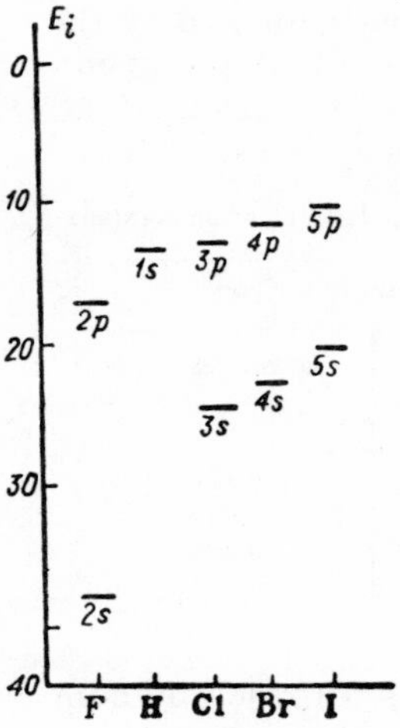

FIGURE 5.2. Energy of the *s*- and *p*-orbitals for fluorine, hydrogen, bromine, chlorine, and iodine [28]

Assuming that the orbital electronegativity χ is the measure of the energy of the valence state of the *n* electrons, Whitehead and Jaffé established the relation between the number of uncompensated *p* electrons U_p and χ_i [29].

If the energy of the valence state of the *n* electrons is equal to:

$$E(n) = k + an + bn^2 + cn^3 + \cdots, \qquad (5.29)$$

the electronegativity is defined as:

$$\chi = \frac{dE(n)}{dn} = a + 2bn, \qquad (5.30)$$

and the ionic character is then defined as the charge transfer which equalizes the electronegativity of the bond orbitals:

$$\chi_A^* = \chi_B^*.$$

The system with two valence electrons is electrically neutral:

$$n_B^* = 2 - n_A^*$$

since

$$a_A + 2b_A n_A^* = a_B + 2b_B n_B^*.$$

Since

$$I = n_A^* - 1,$$

then

$$I = \frac{1}{2} \cdot \frac{\chi_A - \chi_B}{b_A - b_B}. \qquad (5.31)$$

Thus, I is dimensionless, and if the difference $(b_A - b_B)$ is equal to one energy unit, this definition reduces to the Gordy equation (5.28).* In polyatomic molecules the orbital electronegativity of the central atom depends on the atoms and groups to which it is bonded. The effective electronegativity χ^{eff} is introduced in this case.

It was found [33] that the orbital electronegativity χ_0^{hyb} can be written in terms of the electronegativities of the pure s and p orbitals χ_0^s and χ_0^p as follows:

$$\chi_0^{hyb} = s\chi_0^s + (1 - s)\chi_0^p. \tag{5.32}$$

To calculate the bond in diatomic molecules AHal (where A = H, D, F, CN) on the basis of the Townes-Dailey theory (see equation (5.23), without allowing for the d hybridization), we write for the A atoms $\chi_0^A = \chi_0^{hyb}$ and for the halogen atoms $\chi_0 = \chi_0^{hyb}$, which leads to:

$$\left(\frac{\chi_0^s}{2} - \frac{\chi_0^p}{z}\right)s^2 - \left(\frac{\chi_0^s}{2} - \chi_0^p + 1 + \frac{\chi_0^A}{2}\right)s + \left(1 + \frac{\chi_0^A}{2} - \frac{\chi_0^p}{2}\right) - U_p = 0. \tag{5.33}$$

In BrCl and ICl NQR are observed on both halogen atoms, and

$$\begin{aligned} U_p^{Cl} &= (1 - s_{Cl})(1 - I) \\ U_p^{X} &= (1 - s_X)(1 + I), \end{aligned} \tag{5.34}$$

where χ_0^{hyb} for the chlorine and halogen atoms are given by (5.32). From equations (5.34) s_X can be expressed in terms of s_{Cl}, U_p^{Cl} and U_p^X:

$$s_X = \frac{\left(2 - U_p^X\right)\left(1 - s_{Cl}\right) - U_p^{Cl}}{2\left(1 - s_{Cl}\right) - U_p^{Cl}}. \tag{5.35}$$

Instead of equation (5.33) we obtain a cubic equation for s_{Cl}. In both equations there is only one unknown, s_{Cl}. The quantities U_p are measured by experiment, and the values of the orbital electronegativities are given by Hinze and Jaffé [33]. After the values of s_{Cl} have been calculated it is easy to determine χ_0^{hyb} and hence the ionic character of the bond I. The results of the calculation of Whitehead and Jaffé for compounds of type AHal are given in Table 5.3. The calculation for the molecules BrCl and ICl has so far not been satisfactory. The application of this method to other spectra is described in the following chapters.

This method, called the bond electronegativity equalization method (BEEM), was used by Kaplansky and Whitehead to calculate the s character and ionic character of the σ and π bonds X—Cl [34, 35].

Subsequent studies of the problem of quadrupole interaction have been carried out in greater depth.

Bassompierre [35—38] performed the exceptionally difficult and rigorous calculation of e^2Qq of nitrogen in the molecule HCN and e^2Qq of mercury in

* It was shown in [32] that $|b_A - b_B| > 1$.

the molecule $HgCl_2$, and obtained a satisfactory agreement with the experimental results. Scrocco and Tomassi [40] calculated e^2Qq of chlorine in the molecule HCl, using the MO LCAO method of a self-consistent field. A similar calculation is now known for many simple covalent molecules, e.g., $C^{17}O$, [41], N_2 [41, 42], HCN [43, 44], NH_3 [45, 46]. A large number of papers deal with the calculation of the quadrupole coupling constants of deuterium [47–53]. Table 5.4 gives some experimental and theoretical data of e^2Qq for deuterium.

TABLE 5.3. Theoretical values of s_{Hal}, χ_0^{hyb} and I for compounds of type AHal [28, 29]

Compound A	Compound Hal	U_p	χ_0^s	χ_0^p	s	χ_0^{hyb}	χ_0^{eff}	I
F		...	...	3.90	...	...	...	...
	Cl	1.33	5.69	2.95	0.048	3.08	...	0.40
F		...	...	3.90	...	...	...	...
	Br	1.41	5.93	2.61	0.067	2.82	...	0.54
H		...	...	2.21	...	...	...	...
	Cl	0.49	5.69	2.95	0.074	3.15	...	0.47
D		...	...	2.21	...	...	...	...
	Cl	0.50	5.69	2.95	0.069	3.14	...	0.47
D		...	...	2.21	...	...	...	...
	Br	0.69	5.93	2.61	0.107	2.97	...	0.38
D		...	...	2.21	...	...	...	...
	I	0.80	5.06	2.51	0.050	2.64	...	0.22
CN		...	...	...	...	...	2,6	...
	Cl	0.60	5.69	2.95	0.147	3.35	...	...
CN		...	...	...	...	...	2,6	...
	Br	0.74	5.93	2.61	0.140	3.08	...	0.24

TABLE 5.4. Theoretical and experimental values of e^2Qq for deuterium in a number of molecules

Molecule	e^2Qq, MHz calculation	e^2Qq, MHz experiment	Molecule	e^2Qq, MHz calculation	e^2Qq, MHz experiment
DC=CCl*	368	225±18	HDCO	176	170±2.5
	380			$\eta = 0.09$	
HDO	649	314.3±1.5	DC≡N	310	290±120
LiD*	45.3	33±1	DC≡CD*	311	200±10
	46.1			278	
	38.1			270	
DF*	416.2	340±40		414	
	415.9			287	
	372.5		C_6D_6	154	147
			CD_3COCD_3	121	135

* The different calculated values of the electron field e^2Qq_D correspond to different selections of the base functions.

Papers have appeared recently in which the number of uncompensated electrons is calculated using MO LCAO theory:

$$U_p = \frac{N_x + N_y}{2} N_z.$$

Thus, Sichel and Whitehead [54] calculated the quadrupole coupling constant $e^2Qq_{mol} = U_p e^2Qq_{at}$ for the halogens in a number of compounds: HHal, CH_3Hal, FHal, $(Hal)_2$, HalHal' and HalCN (Hal = Cl, Br, I) and for nitrogen in the compounds: NH_3, N_2, N_2O, HCN, CH_3CN, FCN, HalCN, by the expanded Huckel and MO theory of a self-consistent field, completely neglecting the differential overlap. The values of the quadrupole coupling constants calculated for the halogens, on the basis of the semiempirical theory neglecting the differential overlap, agree satisfactorily with the experimental results, whereas the e^2Qq calculated for nitrogen are unsatisfactory. The expanded Huckel theory (allowing for the orbital overlap term) leads to poor results, both for the halogens and for nitrogen.

4. RELATION BETWEEN THE ASYMMETRY PARAMETER OF THE ELECTRIC FIELD GRADIENT AND THE DEGREE OF DOUBLE BONDING IN PLANAR MOLECULES

The relation between the asymmetry parameter of the electric field gradient of the halogen atom in planar molecules and the degree of double bonding in which this atom participates was first established by Goldstein and Bragg [19]. They based their results on an analysis of the microwave data for vinyl chloride using equations analogous to (5.14). The equation obtained by them can be rewritten in our symbols in the form:

$$\pi = \frac{2}{3} \cdot \frac{q_{xx} - q_{yy}}{q_{at}}. \qquad (5.36)$$

Assuming that the principal axis of the electric field gradient tensor in vinyl chloride is directed along the C—Cl bond, they found $e^2Qq = -67$ MHz and $\pi = 0.06$. The value of π was estimated to within $\pm 20\%$. Similar values for vinyl iodide [55] and vinyl bromide [56] give for the C—Hal bond a degree of double bonding $\pi = 0.03$ and $\pi = 0.04$, respectively. Kivelson and Wilson [57], who studied nine isotopic isomers of vinyl chloride, showed that the principal axis of the field gradient forms an angle of 5° with the C—Cl axis, but that the degree of double bonding does not change substantially.

Bersohn [18] gave a detailed analysis of the degree of double bonding in the C—Cl bonds. Considering the planar molecules of chlorobenzene or vinyl chloride, each carbon atom has three sp^2 hybrid orbitals directed at an angle of 120° to each other and forming σ bonds with the adjacent atoms, and one $2p_\pi$ electron together with the π electron of the adjacent carbon atom participates in the formation of a double bond. When a chlorine atom substitutes the hydrogen atom, the $3s3p_z$ hybrid orbital forms a σ bond with the adjacent carbon atom. From the two pairs of electrons, $3p_x^2$ and $3p_y^2$, one orbital (e.g., $3p_x^2$) is in the plane perpendicular to the plane of the benzene ring, and participates in the conjugation with the π orbitals of the carbon atom. A similar pattern is also observed for vinyl chloride. According to MO theory a degree of double bonding in the C—Cl bond implies the loss of a π electron of the chlorine atom (cf. equation (5.24)). In this case the experimental value of the asymmetry parameter of the electric field gradient η can be expressed in terms of the degree of π double bonding as follows:

$$\eta = \frac{3}{2}\,\pi\,\frac{e^2Qq_{\text{at}}}{e^2Qq_{\text{mol}}},$$

(cf. equations (5.13) and (5.36)).

Bersohn, in his calculations of the degree of π double bonding in the C—Cl bond, used the simple molecular orbital theory of Coulson and Loguet-Higgins [58] for conjugated π electron systems. He assumed that the delocalized π electrons are in the average field of the nuclear charges and the σ electrons, so that the π electron system can be treated separately from the σ electron system. Let the total molecular wave function for the π electrons be Ψ_i. It can be expressed as a linear combination of the wave functions of the π electron p_x states φ_r of all the carbon and chlorine atoms:

$$\Psi_i = \sum_r c_{ri}\varphi_r. \tag{5.37}$$

The number of electrons of the r-th atom is equal to:

$$n_r = \sum_i |c_{ri}|^2 n_i. \tag{5.38}$$

The summation is performed over all possible molecular orbitals, and $n_i = 0$ and 2 for the unoccupied and occupied i orbitals, respectively. By definition, the degree of double bonding (see equation (5.24)) is equal to:

$$\pi = 2 - n_{\text{Cl}} = 2 - \sum_i |c_{\text{Cl}\,i}|^2 n_i. \tag{5.39}$$

Taking into account the normalization condition $\sum_i |c_{\text{Cl}_i}|^2 = 1$, we have:

$$\pi = 2\Big(1 - \sum_{i'} |c_{\text{Cl}\,i'}|^2\Big) = 2\sum_{i''} |c_{\text{Cl}\,i''}|^2, \tag{5.40}$$

where i' denotes the occupied molecular orbitals, and i'' the unoccupied ones. In order to obtain the coefficients c_{Cl_i} and $c_{\text{Cl}\,i''}$ we must solve the secular equation for the hamiltonian of the whole π electron system. Using perturbation theory, Bersohn obtained the following final expression:

$$\pi = 2\lambda^2 \sum_{i''} \frac{|c_{ri''}|^2 \beta^2}{(H_{\text{ClCl}} - \varepsilon_{i''})^2}, \tag{5.41}$$

where the summation is performed over the unoccupied states of the π electron system without allowing for the chlorine atom; $c_{ri''}$ is the amplitude of the atomic wave function φ_r on the r-th molecular orbital; $\varepsilon_{i''}$ is the energy of the i'' orbital, λ is a parameter determined from the condition $H_{ClCl} = \lambda\beta = \lambda H_{C_1C_2}$ (H_{ClCl} is the Coulomb integral). The value of the resonance integral H_{CCl} depends on the degree of overlap ($S_{CCl} = \langle\varphi_c|\varphi_{Cl}\rangle$) between the π electrons of the carbon and chlorine atoms. Bersohn [18] obtained $S_{CC} = 0.272$ and $S_{CCl} = 0.128$ for $\lambda = 1/2$. The resonance energy due to the conjugation of the π electrons of the chlorine atom with the aromatic system is equal to [18]:

$$E = -2\lambda^2 \sum_{i''} \frac{|c_{ri''}|^2 \beta^2}{H_{ClCl} + \varepsilon_{i''}}. \quad (5.42)$$

Application of this theory to the NQR data for chloro-derivatives of benzene indicates that if a chlorine atom is bonded to an adjacent carbon atom, the degree of double bonding in the C—Cl bond is altered to a small extent. Taking into account that $\lambda = 1/2$, we obtain:

$$\Delta\pi_r = -6 \sum_{i''} \frac{|c_{ri''}|^2 |c_{ti''}|^2}{(H_{ClCl} - \varepsilon_{i''})^4}. \quad (5.43)$$

This fact will be taken into account when discussing the NQR data for the halogen-derivatives of benzene.

TABLE 5.5. Measured values of the asymmetry parameters, and degrees of double bonding of the π bonds C—Cl calculated from equation (5.41)

Compound	T, °K	e^2Qq, MHz	η	π
$CH_2=CHCl$	20	33.61	0.07 ± 0.03	0.444
$CFCl=CH_2$	—	36.65	0.085 ± 0.015	0.428
1,4-$C_6H_4Cl_2$	77	34.78	0.08 ± 0.02	0.350
p-$ClC_6H_4CH_2Cl$	77	34.57	0.07 ± 0.02	0.350
p-$ClC_6H_4NH_2$	77	34.15	0.06 ± 0.03	0.264
$COCl_2$	77	35.58—36.23	0.25 ± 0.03	1.158

Table 5.5 gives the values of η and π derived by Bersohn [18]. These data confirm that η is proportional to π, and indicate that $\lambda \approx 1/3$ for most C—Cl bonds.* The best agreement between the calculation and experiment is obtained by using the values of η for free molecules obtained from microwave data. In this case η is neither affected by the intermolecular interactions nor by the averaging effect of the torsional vibrations of the molecules in the solid.

* Analysis of the experimental NQR data for the C—Br and C—I bonds indicates that in these cases $\lambda \approx 1/3$. The overlap of the $2p_z$ orbital of carbon with the $4p_x$ orbital of bromine or the $5p_x$ orbital of iodine differs negligibly from the overlap with the $3p_x$ orbital of chlorine.

5. INTERPRETATION OF THE EMPIRICAL RELATIONSHIPS BETWEEN THE NQR FREQUENCIES OF SOME HALOGEN-CONTAINING ORGANIC COMPOUNDS

Tetrahedral molecules. Schawlow [60] was the first to analyze the NQR data for tetrahalides of the IVB group of elements on the basis of the Townes-Dailey theory. Assuming that the *sp* hybridization of the halogen is 15%, he calculated the ionic character of the M—Hal bond. The degree of double bonding was calculated using Pauling's formula [26]:

$$\pi = \frac{R_1 - R}{R_1 + 2R - 3R_2}, \tag{5.44}$$

where R_1, R_2 and R are the bond lengths of the single, double and intermediate bonds, respectively. In these calculations, the Schomaker-Stevenson rule, concerning the change in bond length by $\Delta R = -0.09\,(x_A - x_B)$ as a result of the ionic character, was used. Schawlow's results (Table 5.6) indicate a decrease in the ionic character with a decrease in the difference in electronegativities. However, the numerical agreement with the Townes-Dailey curve (Figure 5.2) is unsatisfactory, almost all points lying below this curve.

TABLE 5.6. Properties of the M—Hal bonds in tetrahalides of elements of group IVA

Bond	Population U_p (experimental data)	Difference in electronegativity		Orbital hybridization of the halogen s, % [29]	Ionic character of the bond I, %		Degree of π double bonding, %	
		$x_{Hal}-x_M$ [60]	$\chi^{eff}_{Hal}-\chi^{hyb}_M$ [29]		[60]	[29]	[60]	[29]
C—Cl	0.746	0.55	0.35	9.7	19	17.4	0	—
C—Br	0.83	0.35	0.17	6.9	8	8	0	—
C—I	0.529	—	0.01	0.4	—	0.2	—	—
Si—Cl	0.372	1.25	0.82	37.1	33	40.8	45 ± 12	37
Si—Br	0.46	1.05	0.67	31.5	16	33.5	10 ± 15	37
Si—I	0.580	—	0.43	26.2	—	21.5	—	—
Ge—Cl	0.467	1.25	0.62	32.3	25	31	41 ± 16	28
Ge—Br	0.54	1.05	0.38	28.6	27	29.5	16 ± 6	29
Ge—I	0.587	0.75	0.31	29.3	19	16.9	11 ± 7	34
Sn—Cl	0.439	1.25	0.67	33.8	38	33.8	22 ± 11	31
Sn—Br	0.50	1.05	0.56	30.7	28	27.5	25 ± 8	33
Sn—I	0.594	0.75	0.35	27.8	17	17.7	21 ± 1	34

Note. The following values were used in [29]: χ^{hyb} (C) = 2.51; χ^{hyb} (Si) = 2.32; χ^{hyb} (Ge) = 2.59; χ^{hyb} (Sn) = 2.52; χ^s_0 (Cl) = 5.69; χ^s_0 (Br) = 5.93; χ^s_0 (I) = 5.06; χ^p_0 (Cl) = 2.95; χ^p_0 (Br) = 2.61; χ^p_0 (I) = 2.51.

In the gas phase these molecules have a C_3 axis and, in spite of the high degree of double bonding, the asymmetry parameter of the electric field gradient on the halogen must vanish.

In the solid state the intermolecular forces may alter the symmetry of the molecules, and as a result the values of the asymmetry parameter may be finite.

Whitehead and Jaffé [29] examined these data using electronegativities. The results of their calculations are given in Table 5.6.

For tetrahalides, χ_0^{hyb} of the central atom is the effective orbital electronegativity, which for the C—W bond in the compound CXYZW is equal to the corresponding hybrid orbital electronegativity multiplied by $1/6$ of the difference between this value and the corresponding orbital electronegativities of X, Y and Z:

$$\chi_0^{eff}(C) = \chi_0^{hyb}(C) + \frac{1}{6}\left[\chi_0^{hyb}(X) - \chi_0^{hyb}(C)\right] + \frac{1}{6}\left[\chi_0^{hyb}(Y) - \chi_0^{hyb}(C)\right] + \frac{1}{6}\left[\chi_0^{hyb}(Z) - \chi_0^{hyb}(C)\right] = \frac{1}{2}\chi_0^{hyb}(C) + \frac{1}{6}\left[\chi_0^{hyb}(X) + \chi_0^{hyb}(Y) + \chi_0^{hyb}(Z)\right]. \quad (5.45)$$

For compounds $MHal_4$:

$$\chi_0^{eff}(M) = \frac{1}{2}\chi_0^{hyb}(M) + \frac{1}{2}\chi_0^{hyb}(Hal). \quad (5.46)$$

Combining relations (5.23a), (5.31), (5.32) and (5.46), we obtain:

$$U_p = (1 - s_x)(1 - I) = (1 - s_x)\left\{1 - \frac{1}{2}\left[\chi_0^{eff}(M) - \chi_0^{hyb}(Hal)\right]\right\} = (1 - s_x)\left[1 - \frac{1}{4}\chi_0^{hyb}(M) + \frac{1}{4}s_x\chi_0^{s}(Hal) + \frac{1}{4}(1 - s_x)\chi_0^{p}(Hal)\right]. \quad (5.47)$$

Since the molecules $MHal_4$ are tetrahedral, the central atom must have sp^3 hybridization and $s_M = 0.25$.

One would expect the hybridization from chlorine to bromine and iodine, in carbon tetrahalides, to gradually decrease as the difference between $\chi_0^{eff}(C)$ and $\chi_0^{p}(Hal)$ decreases (Table 5.6).

This decrease does occur, but the *sp* hybridization for iodine sharply drops to almost zero. This sharp drop is attributed [29] to the possible *d* hybridization of the orbitals participating in the formation of the σ bond of bromine and iodine, which shields the actual *s* hybridization [62]. Actually, one would expect an increase in the *s* character of the bond from chlorine to bromine and iodine, because the *s* hybridization increases the difference between $\chi_0^{eff}(Hal)$ and $\chi_0^{eff}(M)$. Apparently, the *p* orbitals of the chlorine atom and the sp^3 orbital of the carbon atom have a similar spatial distribution. Therefore, for an increase in the overlapping, the C—Cl bond must have a considerable *s* character. In bromine and iodine, to achieve maximum overlap, the orbitals must be more directed so that the *d* hybridization compensates the discrepancy in χ, to which the increase of the *s* character of the bond would have led.

For the remaining elements of group IVA (M = Si, Ge, Sn) the behavior of the tetrahalides is similar to $CHal_4$, but the absolute value of the changes in the *sp* hybridization and the degree of the ionic character of iodine when passing from chlorine to bromine and from bromine to iodine are very different. The absolute values of these differences are difficult to explain. However, by using equation (5.23) and assuming that the values of *s* and *I* for all $MHal_4$ are the same and equal to the values for CCl_4, Whitehead and

Jaffé obtained, using the concept of $d_\pi - p_\pi$ conjugation, the values of the double bonding given in Table 5.6. Indeed, the latest experimental data on the values of the asymmetry parameters of the electric field gradient in MCl$_4$ show that the $d_\pi - p_\pi$ conjugation of the M—Cl bonds is important [63].

In order to elucidate the question of the participation of the chlorine atom in $d_\pi - p_\pi$ conjugation with the silicon atom it is possible to use Stark effect experiments. It was shown that the change in the field gradient caused by the π electrons under the influence of the electric field is opposite in sign to the change of the field gradient along the axis of the σ bond [64]. It can be concluded that the shift caused by the Stark effect is considerably smaller for a multiple bond than for a single one.

The experimental values of the NQR line splittings under the influence of the electric field $\frac{d\Delta\nu}{dE}$ for the C—Cl bond are equal to 40—70 Hz/(kV/cm), and for the Si—Cl bond to 23 Hz/(kV/cm), whereas the values calculated without allowing for the conjugation are 95 and 130 Hz/(kV/cm), respectively [64].

Another interesting series, the chlorofluoromethanes, were investigated by Livingston [65]. As shown in Chapter 4, the NQR frequencies of halogen-substituted methanes change linearly with the change in the number of halogen atoms. However, for the given series a deviation from additivity was observed, i.e. the increments derived for the compounds Cl_nCF_{4-n} were found to be unsuitable for the series $Cl_nCH_mF_{4-n-m}$.

Data on the chloro-derivatives of methane indicate that the NQR frequency of ^{35}Cl increases with an increase in the number of chlorine atoms, which according to the Townes-Dailey theory points to a decrease in the ionic character of the bond. This can be explained by the fact that the chlorine atom, being more electronegative than the carbon atom, tends to repel the electrons from the carbon and, as a result, the ionic character of the C—Cl bond is increased.

The fluorine atom is more electronegative than hydrogen and, therefore, the NQR frequencies of ^{35}Cl for CFCl$_3$ are higher than for CHCl$_3$, and those for CHFCl$_2$ are higher than for CH$_2$Cl$_2$, etc. However, in the series Cl_nCF_{4-n} the NQR frequency of ^{35}Cl ought to increase with an increase in the number of fluorine atoms since it is more electronegative than chlorine, but this does not occur. Livingston [65] suggested that this may be due to the increase in the degree of double bonding in the C—Cl bond, resulting from the contribution of structures of the type:

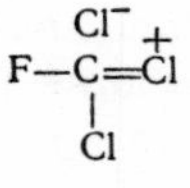

$$\begin{array}{c} Cl^- \\ \overset{|}{F-C}\overset{+}{=}Cl \\ \overset{|}{Cl} \end{array} \text{ for } CFCl_3 \text{ and } \begin{array}{c} Cl^- \\ \overset{|}{F-C}\overset{+}{=}Cl \\ \overset{|}{F} \end{array} \text{ for } CF_2Cl_2.$$

The double bonding reduces the number of uncompensated p electrons U_p in accordance with formula (5.23) and thus lowers the NQR frequency. Pauling [26] explained in a similar manner the decrease in length of the C—Cl bond in chloromethanes when the chlorine is replaced by fluorine.

Lucken's present ideas [66] concerning the influence of the $p - \sigma$ conjugation of type F—C—Cl on the NQR frequency of the chlorine atom are

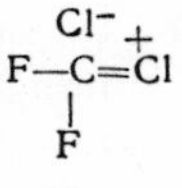

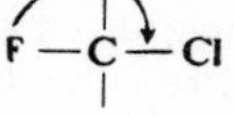

generally accepted. With such conjugation the *p* orbital of fluorine overlaps the antibonding σ^* orbital of C—Cl. The supply of electrons from the fluorine increases the ionic character of the bond and decreases the frequency. Such hyperconjugation makes the carbon more electronegative than would a simple inductive effect through the σ bond. This series was also analyzed by Whitehead and Jaffé [29]. According to their calculations (Table 5.7), the change in the effective orbital electronegativity of the carbon atom with respect to chlorine is the decisive factor. When the number of fluorine atoms in the fluorochloromethanes increases, the value of $\chi_0^{eff}(C)$ also increases, and in order to fulfill the condition of consistency of χ the *sp* hybridization of the chlorine atom must increase. A simultaneous increase in $\chi_0^{eff}(C)$ involves a decrease in magnitude of the carbon orbital such that the value of *s* necessary for maximum overlap decreases. Thus, the absolute *sp* hybridization increases, but at a decreasing rate.

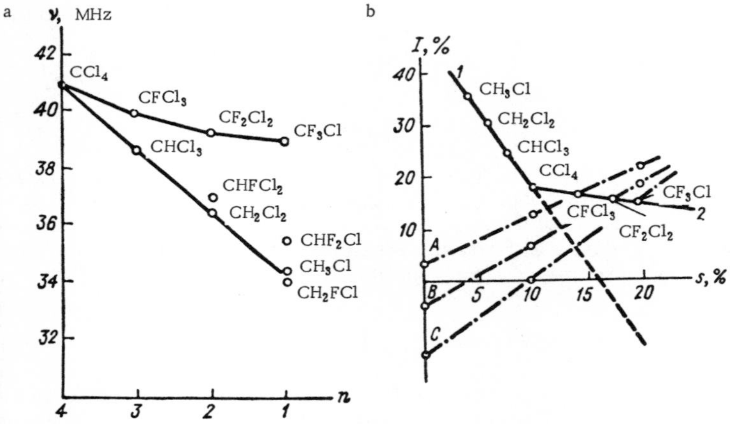

FIGURE 5.3. NQR spectra of ^{35}Cl in polyhalomethanes:

a) NQR spectra of ^{35}Cl in the series Cl_nCF_{4-n} and Cl_nCH_{4-n} as a function of the number of chlorine atoms *n* [49]; b) the ionic character *I* of the C—Cl bond as a function of the degree of *sp* hybridization *s* for polychloro- and polyfluorochloromethanes [29].

This is accompanied by the transfer of charge from the chlorine atom to the carbon atom, thus decreasing the ionic character of the bond.

In chloro-derivatives of methane $\chi_0^{eff}(C)$ decreases, but part of the hybridization is necessary for improving the overlap, which increases the difference in electronegativities. A compromise is achieved by the transfer of charge from carbon to chlorine (which increases the ionic character of the bond) and a certain reduction in the *sp* hybridization.

Whitehead and Jaffé discussed in [29] Lucken's hypothesis [65] and found it consistent with their own concepts. They calculated the value of the hyperconjugation according to Lucken. Figure 5.3,b shows the ionic character as a function of the *sp* hybridization for the series Cl_nCH_{4-n} (line 1) and Cl_nCF_{4-n} (line 2). We would expect that without $p-\sigma$ conjugation the fluorochloromethanes would lie on the continuation of line 1. The lines *A*, *B*, *C** show the change of *I* as a function of *s* for $CFCl_3$, CF_2Cl_2 and CF_3Cl. The points of intersection of these lines with the continuation of line 1 give the values of the degree of ionic character and the *sp* hybridization in the absence of hyperconjugation. The deviations of line 2 from the extrapolated line 1 give the value of the $\pi^*p-\sigma$ conjugation for $CFCl_3$, CF_2Cl_2 and CF_3Cl (their values, given in Table 5.7, lie within the range of Lucken's evaluation).

TABLE 5.7. Ionic character of the bond and hybridization in asymmetrical tetrahedral molecules Cl_nCY (where Y = F, H)*

Compound	U_p	*s* hybridization, %		Orbital electronegativity			Ionic character of the C—Cl bond *I*, %	Hyperconjugation of the π^* double bonding, %
		Cl	Y	χ_0^{hyb} (Cl)	χ_0^{hyb} (Y)	χ_0^{eff} (C)		
Cl_3CF	0.719	14.3	0.00	3.33	3.90	3.01	0.161	5
Cl_2CF_2	0.701	17.4	0.00	3.42	3.90	3.12	0.152	10
$ClCF_3$	0.694	19.3	0.00	3.47	3.90	3.19	0.140	13
Cl_3CH	0.698	7.9	1.00	3.16	2.21	2.68	0.242	
Cl_2CH_2	0.656	6.1	1.00	3.11	2.21	2.50	0.301	
$ClCH_3$	0.620	4.3	1.00	3.06	2.21	2.36	0.352	

* It must be kept in mind that the *s* hybridization of carbon is 25%. For hydrogen the pure *s* orbital χ_0^s (H) = 2.21 is used, for fluorine the pure *p* orbital χ_0^p (F) = 3.90, and for chlorine χ_0^s (Cl) = 5.69 and χ_0^p (Cl) = 2.95.

We note in conclusion that only for elements of group IVA does the NQR frequency of the halogen in tetrahedral molecules decrease when it is replaced by a donor substituent. If the central atom is an element of the transition group (such as Ti), the NQR frequency of the halogen increases when it is replaced by a donor substituent (Figure 5.4).

It is known that the metals of the transition groups are additive acceptors of the p_π electrons of halogen. The sharp drop in the NQR frequencies of the halogens in $TiHal_4$ compared with $MHal_4$ (where M = C, Si, Ge, Sn) may be attributed, to a considerable extent, to the decrease in electron density on the p_x and p_y orbitals of the halogen (see equation (5.11)). When a bromine atom is replaced by a cyclopentadienyl ring or an ethoxy group, which are π donors compared with bromine, the electron density N_x and N_y of the bromine atom increases, which in turn increases the NQR frequency of the bromine. Naturally, it is difficult to speak in this case of the value of the ionic character and hybridization of the Ti—Br σ bond.

Halogen-substituted benzenes. As shown in Chapter 4, the NQR frequency of polyhalogen-substituted benzenes can be calculated by the additive scheme:

* The lines *A*, *B*, *C* are found as follows: for a given *s*, having determined χ_0^{hyb}(Cl) from formula (5.32) and χ_0^{eff} (C) from formula (5.45), *I* is found by formula (5.31).

$$\nu = \nu_0 + m\alpha_o + n\alpha_m + p\alpha_p, \tag{4.12}$$

where m, $n = 0, 1, 2$ and $p = 0, 1$.

Table 4.2 shows that all α_o, α_m, $\alpha_p > 0$ if the substituent is a halogen atom.

The behavior of the NQR frequencies in polychloro-substituted benzenes as a function of the number of chlorine atoms was discussed by Bray et al. [67], and also by Monfils and Duchesne [68].

Figure 5.5. shows the behavior of the NQR frequencies of halogen-derivatives of benzene as a function of the number of halogen atoms in the molecule. All the frequencies split into three groups (three lines) according to the number, m, of groups or atoms ortho to the given atom.

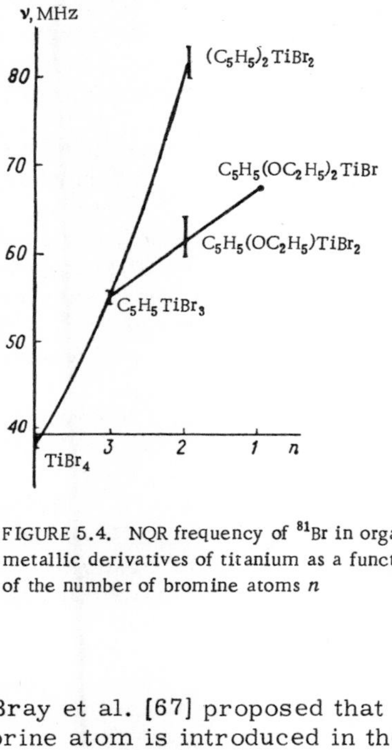

FIGURE 5.4. NQR frequency of ^{81}Br in organometallic derivatives of titanium as a function of the number of bromine atoms n

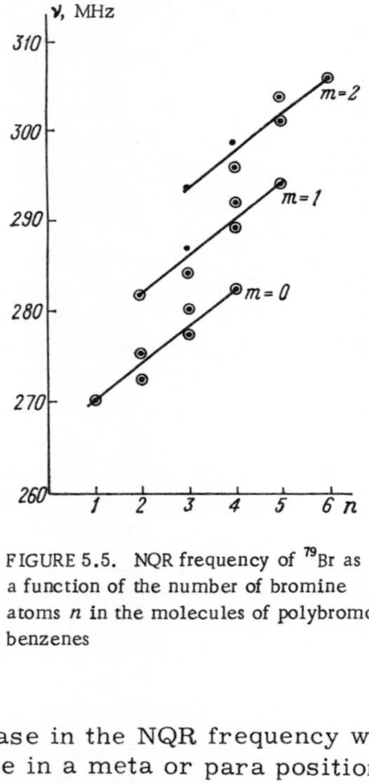

FIGURE 5.5. NQR frequency of ^{79}Br as a function of the number of bromine atoms n in the molecules of polybromobenzenes

Bray et al. [67] proposed that the increase in the NQR frequency when a chlorine atom is introduced in the molecule in a meta or para position is due to the inductive effect. As seen from equation (5.43), the introduction of a second substituent in a meta or para position does not substantially alter the original double bonding of the C—Cl bond. But the two chlorine atoms start to compete in accepting π electrons from the carbon atom and, as a result, the ionic character of the bond of the chlorine atom is lowered. According to the Townes-Dailey theory (4.23), this increases the NQR frequency.

The reason for the increase in the NQR frequency of the halogen atom when a substituent is introduced in an ortho position was the subject of a lengthy discussion. It was assumed that in the presence of an ortho group, the two chlorine atoms deviate from the plane of the benzene ring in different directions (for the free molecule this was established by electron diffraction [70]). However in the solid state even the hexahalobenzenes are planar [71].

Starting from the assumption that the chlorine atoms deviate from the plane of the ring, Duchesne and Monfils [68] proposed that the double bonding in the C—Cl bond decreases as a result of the decrease in the overlap of the p_x orbitals of the carbon and halogen. This, of course, results in an increase in the frequency. However, Bray [67] indicated that the deviation of the atoms from the plane must lead to an increase in the double bonding since, in this case, the overlap is due to both the p_π and the p_σ electrons.

Actually there are no deviations of the halogen atoms from the plane of the benzene ring and, thus, the observed increase in the asymmetry parameter of the halogens (compare p-$I_2C_6H_4$, $\eta = 4.1 \pm 0.5\%$ and o-$I_2C_6H_4$, $\eta = 8.5 \pm 1.5\%$) is probably due to intramolecular steric hindrance. The small drop in the NQR frequency caused by the increase in the double bonding is overlapped by a much stronger inductive effect, decreasing the degree of the ionic character of the C—Hal bond. This explains the presence of the three lines on Figure 5.5.

Scrocco discusses in detail the quantum-mechanical treatment of the additive character of the change in the NQR frequencies in different series of simple molecules [72, 73].

The theory is simplified by neglecting all the direct contributions to the field gradient from nonadjacent atoms. Only indirect effects through deformation of the electron shell of the resonating atom caused by adjacent nonvalence-bonded atoms or groups are allowed.

In this treatment, the theory of molecular wave functions of "separated electron pairs" is used [74, 75].

The wave function of the $2n$ electrons of the molecule is formed by means of the antisymmetrization equation:

$$A = [\Lambda_A(1,2)\,\Lambda_B(3,4)\ldots\Lambda_M(2n-1,2n)], \tag{5.48}$$

where the pair functions Λ_i consist of the pair functions of both valence and free electrons. The pair function Λ_i is, in general, a linear combination of the two-electron functions:

$$\Lambda_i = A_{i1}\Phi_{i1} + A_{i2}\Phi_{i2} + A_{i3}\Phi_{i3}, \tag{5.49}$$

where

$$\begin{gathered} \Phi_{i1} = \varphi_1(1)\,\varphi_1(2) \\ \Phi_{i2} = 2^{-1/2}\,[\varphi_1(1)\,\varphi_2(2) - \varphi_1(2)\,\varphi_2(1)] \\ \Phi_{i3} = \varphi_2(1)\,\varphi_2(2). \end{gathered} \tag{5.50}$$

These two-electron functions describe three singlet pairs obtained from the bonding and antibonding orbitals:

$$\varphi_1 = \frac{\chi_a + \chi_b}{\sqrt{2(1+S)}}; \qquad \varphi_2 = \frac{\chi_a - \chi_b}{\sqrt{2(1-S)}}; \qquad S = \int \chi_a^*\,\chi_b\,dv. \tag{5.51}$$

Since the change in the electron density of each orbital occurs as a result of the charge transfer from atom A to atom B, we introduce σ_i, the bond polarity parameter, which is a measure of the transferred positive electric charge:

$$\sigma_i = \frac{\sqrt{2}\, A_{i2}\,(A_{i1} + A_{i3})}{\sqrt{1 - S^2}}. \qquad (5.52)$$

Considering the influence of all the other substituents in the molecule on the k-th pair of electrons, the principal effect is the change in electronegativity of the orbitals of the χ_a and χ_b bonds and the resulting change in the polarity parameter σ_k. To a first approximation the change in σ_k depends linearly on the polarity parameters of all the other electron pairs of the molecule.

This result was used when examining the quadrupole coupling constants of the chlorine atom [72, 73].

In the Parks-Parr theory, the eight electrons of the chloride atoms form the following pairs:

1) one pair of electrons in the σ bond C—Cl;
2) one pair in the nonbonding σ orbital;
3) two pairs in two nonbonding π orbitals.

The quadrupole coupling constant of the chlorine atom is equal to:

$$e^2Qq_{\text{mol}} = e^2Q\,(q_{\Lambda_1} + q_{\Lambda_2} + q_{\Lambda_3} + q_{\Lambda_4}), \qquad (5.53)$$

where

$$q_{\Lambda_i} = \iint \Lambda_i(1,2)\left[\left(\frac{3\cos^2\theta - 1}{r_0^3}\right)_1 + \left(\frac{3\cos^2\theta - 1}{r_0^3}\right)_2\right]\Lambda_i(1,2)\, d\tau_1 d\tau_2. \qquad (5.54)$$

Using the specific form of the functions for the four pairs of electrons, we obtain:

$$e^2Qq_{\text{mol}} = -e^2Qq_{p_z}\left[\frac{1 + 2S + \beta}{(1 - \beta)(1 + S)} - \sigma_k\right](1 - \alpha - \beta), \qquad (5.55)$$

where and β are the *sp* and *pd* hybridization parameters of the chlorine orbitals, and σ_k is the polarity parameter.

We now separate the contributions from the atoms A and B in the k-th bond:

$$\sigma_k = \sigma_k^0 + \frac{1}{(1 - S^2)\left(H_{22}^k - H_{11}^k\right)} \sum_{i \neq k} (b_i + c_i\sigma_i). \qquad (5.56)$$

The first term is the polarity parameter in the absence of all other bonds and electrons in the molecule. The second term allows for the changes in σ_k induced by all the other electron pairs. The term $(H_{22}^k - H_{11}^k)$ is the energy difference between the two configurations described by the functions Φ_{k1} and Φ_{k2} in equation (5.50), and each term $(b_i + c_i\sigma_i)$ is the change in the difference between the potential energies of two electrons lying on the χ_{ak} and χ_{bk} orbitals of the k-th bond, i.e. the change caused by the charge distribution on the i-th bond.

Substituting of (5.56) into (5.55) yields:

$$e^2Qq_{\text{mol}} = e^2Q\overset{\circ}{q}_{\text{mol}} + \frac{(1-\alpha-\beta)\,e^2Qq_{p_z}}{(1-S^2)\left(H^k_{22}-H^k_{11}\right)} \sum_{i\neq k} (b_i + c_i\sigma_i). \qquad (5.57)$$

The second term in this expression is a function of the nature of the i-th bond and its position in the molecule. It was used successfully by Scrocco [73] to calculate the NQR constants for the homologs of dichlorobenzene.

Aliphatic chlorine-containing compounds. The periodic dependence of the NQR signal frequency on n (see Chapter 4) observed for $Cl(CH_2)_nCl$ for $n > 3$ is explained by the MO LCAO calculation [76] as being due to the partial delocalization of the C—C σ bond. In a zero approximation the MO of the model system $C-(CH_2)_n-C$ were calculated. The effect of substituting chlorine atoms onto the terminating carbon atoms was taken into account using perturbation theory. The C—H bonds were not allowed for. It follows from the calculation that, for $n > 3$, the negative charge on the chlorine atoms is somewhat larger for even n than for odd. However, for a full agreement between the experiment and the calculation it is also necessary to allow for the influence of the internal electric field of all C—H and C—Cl dipoles, since for $n > 6$ the sequence of NQR frequencies of ^{35}Cl (and ^{79}Br) changes.

BIBLIOGRAPHY

1. Townes, Ch. H. and A. L. Schawlow. Microwave Spectroscopy. — McGraw-Hill, 1959.
2. Das, T. P. and E. L. Hahn. Nuclear Quadrupole Resonance Spectroscopy. — In: "Solid state physics," Suppl. 1. New York-London, 1958.
3. Townes, C. H. — Handb. Phys. 38 (1958), 377.
4. Barnes, R. G. and W. V. Smith. — Phys. Rev. 93 (1954), 95.
5. Koster, G. — Phys. Rev. 86 (1952), 148.
6. Lew, H. and G. Wessel. — Phys. Rev. 90 (1953), 1.
7. Tsyunaitis, G. K. and A. P. Yutsiss. — Zh. Eksper. Teoret. Fiz. 28 (1955), 452.
8. Pound, R. V. — Phys. Rev. 79 (1950), 685.
9. Sternheimer, R. M. — Phys. Rev. 84 (1951), 244.
10. Sternheimer, R. M. — Phys. Rev. 86 (1952), 316.
11. Sternheimer, R. M. and H. M. Foley. — Phys. Rev. 95 (1954), 736.
12. Sternheimer, R. M. and H. M. Foley. — Phys. Rev. 92 (1953), 1460.
13. Sternheimer, R. M. and H. M. Foley. — Phys. Rev. 102 (1956), 731.
14. Sternheimer, R. M. — Phys. Rev. 159 (1967), 266.
15. Sternheimer, R. M. — Phys. Rev. 132 (1963), 1637.
16. Townes, C. H. and B. P. Dailey. — J. Chem. Phys. 17 (1949), 782.
17. O'Konski, C. T. Nuclear Quadrupole Resonance Spectroscopy. — In: "Determination of organic structure by physical methods," Vol. 2, Part 11. New York, 1962.
18. Bersohn, R. — J. Chem. Phys. 22 (1954), 2078.
19. Goldstein, J. H. and J. K. Bragg. — Phys. Rev. 75 (1949), 1453.
20. Dailey, B. P. and C. H. Townes. — J. Chem. Phys. 23 (1955), 118.
21. Townes, C. H. and B. P. Dailey. — J. Chem. Phys. 20 (1952), 35.

22. Dailey, B. P. — Discuss. Faraday Soc. 19 (1955), 255.
23. Gordy, W. — Discuss. Faraday Soc. 19 (1955), 14.
24. Gordy, W., W. V. Smith, and R. F. Trambarulo. Microwave Spectroscopy. — New York, Dover Publications Inc., 1966.
25. Gordy, W. — J. Chem. Phys. 19 (1951), 792.
26. Pauling, L. Theory of the Chemical Bond. — Goskhimizdat, 1947. (Russian)
27. Wilmshurst, J. — J. Chem. Phys. 30 (1959), 561.
28. Whitehead, M. A. and H. H. Jaffé. — Trans. Faraday Soc. 57 (1961), 1854.
29. Whitehead, M. A. and H. H. Jaffé. — Theoret. chim. Acta 1 (1963), 209.
30. Coulson, C. A. — Proc. Camb. Phil. Soc. Math. — Phys. Sci. 53 (1931), 1367.
31. Coulson, C. A. — Q. Rev. Chem. Soc. 1 (1947), 144.
32. Baird, N. C. and M. A. Whitehead. — Theoret. chim. Acta 2 (1964), 259.
33. Hinze, J. and H. H. Jaffé. — J. Am. Chem. Soc. 84 (1962), 540.
34. Kaplansky, M. A. and M. A. Whitehead. — Molec. Phys. 15 (1968), 149.
35. Kaplansky, M. A. and M. A. Whitehead. — Molec. Phys. 16 (1969), 481.
36. Bassompiere, L. — Ann. Phys. 13, No. 2 (1957), 676.
37. Bassompiere, L. — J. Chem. Phys. 21 (1954), 614.
38. Bassompiere, L. — Compt. rend. 240 (1955), 285.
39. Bassompiere, L. — Compt. rend. 237, No. 39 (1953), 1224.
40. Scrocco, E. and J. Tomasi. — Molec. Phys. 4 (1961), 193.
41. Richardson, J. W. — Rev. Mod. Phys. 32 (1960), 461.
42. Ransil, B. J. — Rev. Mod. Phys. 32 (1960), 245.
43. Foster, J. M. and S. F. Boys. — Rev. Mod. Phys. 32 (1960), 303.
44. Sholl, C. A. — Proc. Phys. Soc. 87 (1966), 827.
45. Scott, T. A. — J. Chem. Phys. 36 (1962), 1459.
46. Bassompiere, L. — Discuss. Faraday Soc. 19 (1955), 260.
47. White, R. L. — J. Chem. Phys. 23 (1955), 253.
48. Bersohn, R. — J. Chem. Phys. 32 (1960), 85.
49. Kolber, H. J. and M. Karplus. — J. Chem. Phys. 36 (1962), 960.
50. Lowe, J. T. and W. H. Flugare. — J. Chem. Phys. 41 (1964), 2153.
51. Kern, C. W. and M. Karplus. — J. Chem. Phys. 42 (1965), 1062.
52. Bonera, G. and A. Rigamonti. — J. Chem. Phys. 42 (1965), 175.
53. Olimpia, P. L., I. Y. Wei, and B. M. Fung. — J. Chem. Phys. 51 (1969), 1610.
54. Sichel, J. and M. A. Whitehead. — Theoret. chim. Acta 11 (1968), 263.
55. Goldstein, J. H. — J. Chem. Phys. 24 (1956), 100.
56. Cornwell, C. D. — J. Chem. Phys. 18 (1950), 1118.
57. Kivelson, D., E. B. Wilson, and D. R. Lide. — J. Chem. Phys. 32 (1960), 205.
58. Coulson, C. and H. C. Loquet-Higgins. — Proc. R. Soc. 192A (1947), 16.
59. Coulson, C. Valence. 2nd edition. — Oxford, Univ. Press, 1961.
60. Schawlow, A. L. — J. Chem. Phys. 22 (1954), 1211.
61. Schomaker, V. and D. P. Stevenson. — J. Am. Chem. Soc. 63 (1941), 37.
62. Whitehead, M. A. — J. Chem. Phys. 36 (1962), 3006.
63. Graybeal, J. D. and P. J. Green. — J. Chem. Phys. 73 (1969), 2948.
64. Lucken, E. A. C. Nuclear Quadrupole Coupling Constants, p. 163. London-New York, 1969.
65. Livingston, R. — J. Chem. Phys. 19 (1951), 1434.
66. Lucken, E. A. C. — J. Chem. Soc., p. 2954, 1959.
67. Bray, P. J., R. G. Barnes, and R. Bersohn. — J. Chem. Phys. 25 (1956), 813.
68. Duchesne, J. and A. Monfils. — J. Chem. Phys. 22 (1954), 562.
69. Semin, G. K., L. S. Kobrina, and G. G. Yakobson. — Izv. Sib. Otdel. AN SSSR, ser. chem. sci., No. 9, Issue 4 (1968), 84.
70. Bastiansen, O. and O. Hassel. — Acta chem. scand. 1 (1947), 489.
71. Khotsyanova, T. L. and V. I. Smirnova. — Kristallografiya 13 (1968), 787.
72. Scrocco, E., P. Bucci, and M. Maestro. — J. Chim. phys. 56 (1959), 623.
73. Scrocco, E. — Adv. Chem. Phys. 5 (1963), 319.
74. Lennard-Jones, J. E. — Proc. Camb. Phil. Soc. Math. — Phys. Sci. 27 (1931), 469.
75. Parks, J. M. and R. G. Parr. — J. Chem. Phys. 28 (1958), 335.
76. Sokolov, N. D. and S. Ya. Umanskii. — Teor. Eksperim. Khim. 2 (1966), 171.

Chapter 6

CORRELATION BETWEEN THE SPECTRAL PARAMETERS OF NQR AND THE DIFFERENT CHEMICAL AND PHYSICAL-CHEMICAL CONSTANTS

Many attempts have been made to correlate the NQR frequency shifts of the resonating atoms with their chemical properties.

The correlations between the NQR frequencies and the ionization constants pK_a of carboxylic acids, and the correlations between the spectral parameters of NQR and the Hammet-Taft σ parameters are very significant.

Such correlations, which are useful for predicting σ parameters or NQR frequencies, enable us to obtain a greater understanding of the inductive effects of substituents. Comparison of the NQR spectra of halogens with IR and NMR spectra is of great value.

1. DIFFERENCES IN THE ELECTRONEGATIVITIES AND THE DIPOLE MOMENTS OF THE BONDS

In accordance with the Townes-Dailey theory, there is a linear relation between the NQR frequency of the halogen atom and the electronegativity of the atom to which it is bonded. This follows from equation (5.23):

$$e^2Qq_{zz} = (1 - I)(1 - s)\, e^2Qq_{\text{at}},$$

where I is the ionic character of the bond and s is the degree of sp hybridization.

Since the hybridization of the halogen atom is practically constant, the quadrupole bond constant will be a linear function of the degree of the ionic character of the bond I, which is roughly proportional to the difference in electronegativity. If the electronegativity of the atom to which the halogen atom is bonded decreases, the quadrupole coupling constants likewise decrease (Figure 6.1). Thus, the NQR frequencies of the chlorine atom bonded to carbon in organic compounds is roughly double the corresponding frequencies in chlorosilanes and are close to the frequencies of a chlorine atom bonded to sulfur. The NQR frequencies of chlorine atoms bonded to nitrogen are much higher than the frequencies of chlorine atoms bonded to carbon in organic compounds and are close to the NQR frequency in the Cl_2 molecule [1]. On the basis of the sequence of frequencies of the tetrachlorides of the elements of group IVA, the following order of electronegativities is given in [2]: C > Ge > Sn > Si.

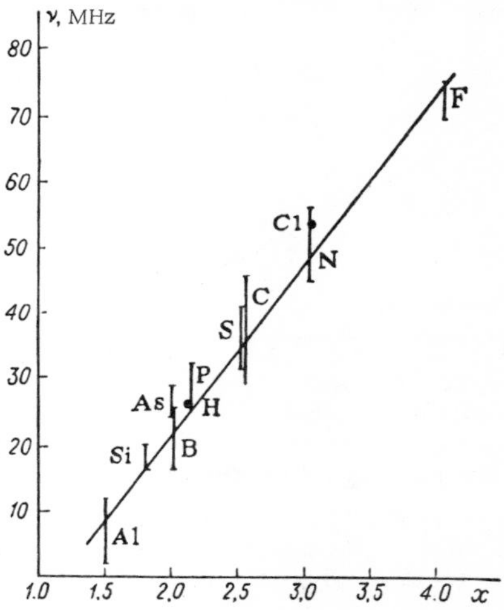

FIGURE 6.1. NQR frequencies of ^{35}Cl in the M—Cl bonds of the monovalent chlorine atom as a function of the electronegativity x of element M

Ogawa's research on the trihalides of arsenic and antimony [3] led him to propose the following linear relationship for the NQR of the halogen atoms:

$$U_p = 0.86 - 0.38\,(x_A - x_{Hal}), \tag{6.1}$$

where x_A and x_{Hal} are the electronegativities of arsenic and antimony, and the halogen atoms, respectively. For the chlorides of the transition elements ($TiCl_4$, WCl_6, $ThCl_4$, etc.), the NQR frequencies of the chlorine atoms are much lower than would be expected from the difference in the electronegativity [4]. Apparently, these deviations are due to the considerable participation of the vacant *d* orbitals of the metal in the metal-halogen bond. Thus, the Townes-Dailey equation, which assumed an *sp* hybridized σ bond, cannot be applied.

A linear relation was also found between the calculated dipole moments of the bond and the NQR frequencies of ^{35}Cl in a number of chloro-derivatives of hydrocarbons [5] (Figure 6.2). However, Hamano pointed out [6, 7] that the dipole moments of alkyl halides increase with an increase in the number of carbon atoms, and the quadrupole coupling constants in the solid state decrease in an oscillating manner (see Chapter 6). Hamano [6, 7] further pointed out that this correlation is similar to the relation between the boiling and melting points of these substances. Such oscillatory behavior was found to be due to the crystal field effects by examining the dimensions of the unit cells of compounds with an even and odd number of carbon atoms [7, 8].

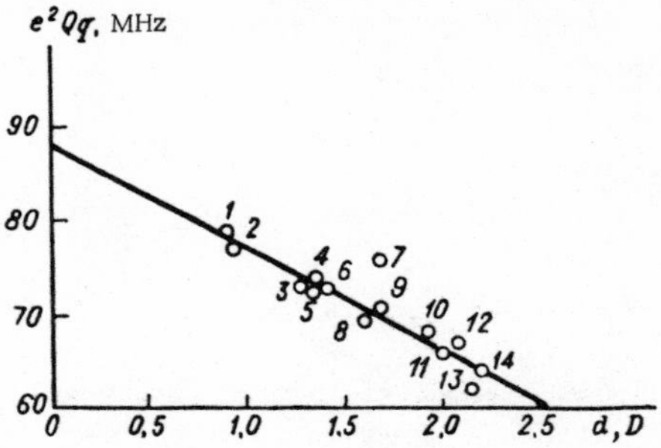

FIGURE 6.2. Quadrupole coupling constants of ^{35}Cl in aliphatic chlorine-containing compounds as a function of the calculated values of the dipole moments of the C—Cl bonds [5]:

1) CCl_3CHCl_2; 2) $CHCl_3$; 3) γ-$C_6H_6Cl_6$; 4) δ-$C_6H_6Cl_6$; 5) α-$C_6H_6Cl_6$; 6) CH_2Cl_2; 7) CH_3CHCl_2; 8) CH_2ClCH_2Cl; 9) CH_3CCl_3; 10) CH_3Cl; 11) CH_3CH_2Cl; 12) $CH_3(CH_2)_2Cl$; 13) $(CH_3)_3CCl$; 14) $(CH_3)_2CHCl$.

2. CORRELATION BETWEEN THE NQR FREQUENCIES OF THE HALOGEN ATOMS AND THE HAMMET-TAFT σ CONSTANTS

Repeated attempts have been made to correlate the NQR frequencies with the chemical reaction constants. Such an attempt was made by Meal, who plotted the NQR frequencies of ^{35}Cl in polysubstituted chlorobenzenes as a function of the corresponding Hammet σ constants [9] and established that the relationship $\nu(\sigma)$ is roughly linear (Figure 6.3). It was found that the NQR frequency increases systematically with the increase in the σ Hammet constants of the substituents, i.e. with the increase in the electron-acceptor properties. Similar correlations were found for bromo-derivatives of benzene [10, 11].

An equation of type $\nu = \nu_0 + \alpha\Sigma\sigma$ was first obtained for polychloro-substituted benzenes by Bray and Barnes [12]:

$$\nu_{^{35}Cl} = (34.826 + 1.024\Sigma\sigma) \pm 0.36 \text{ MHz}, \qquad r = 0.96 \tag{6.2}$$

where r is the correlation coefficient.

Equation (6.2) was derived from data obtained for 85 compounds, which is about 5 times the number of compounds studied by Meal. The constant σ varies from -0.6 to $+4.9$, which is also considerably greater than the range obtained by Meal. The corresponding frequencies lie in the range 34—37 MHz. For the groups COOH, $COCH_3$ and CHO in a para position, new refined values of σ were used and, thus, a better correlation was achieved as a result. The σ constants from the Bray-Barnes equation for chlorobenzene derivatives were determined to within ± 0.50.

A similar equation was given by Hooper and Bray [1] for ortho, meta and para bromobenzenes:

$$\nu_{^{81}Br} = (226.932 + 7.639\Sigma\sigma)\ \text{MHz}. \qquad (6.3)$$

The equation was derived from the results on 37 compounds. The Hammet σ constant lies in the range $-1 \leqslant \sigma \leqslant +4$; the NQR frequencies of ^{81}Br lie in the range $220 \leqslant \nu \leqslant 255$ MHz.

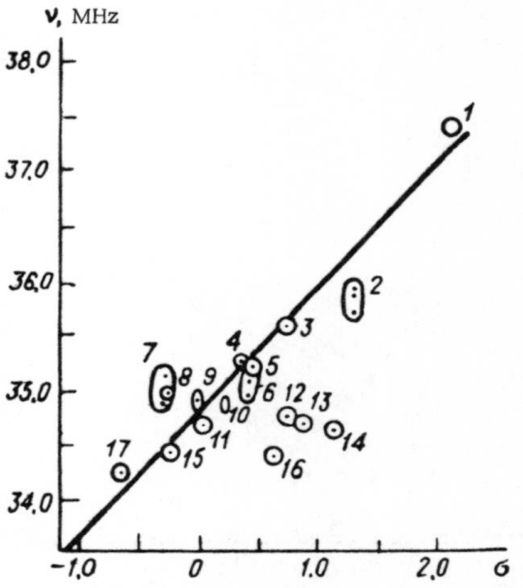

FIGURE 6.3. Meal correlation [9] of the NQR frequencies of ^{35}Cl of disubstituted chlorobenzene derivatives with the Hammet σ parameters:

1) o-NO_2; 2) o-Cl; 3) m-NO_2; 4) m-COOH; 5) m-CF_3; 6) m-Cl; 7) p-OH; 8) p-OCH_3; 9) H; 10) p-Cl; 11) p-CH_2Cl; 12) p-COOH; 13) p-$COCH_3$; 14) p-CHO; 15) p-OC_2H_5; 16) p-CH=CHCOOH; 17) p-NH_2.

In order to compare the C—Cl and C—Br bonds in a number of chloro- and bromobenzenes the correlation of the quadrupole frequencies with the Hammet σ parameters was reduced to electron charge units by dividing the two sides of the equation by $\frac{1}{2}e^2Qq_{at}$ for ^{35}Cl and ^{81}Br, respectively [13]:

$$\begin{aligned} -U_p^{Cl} &= 0.6352 + 0.0187\sigma \\ -U_p^{Br} &= 0.7066 + 0.0254\sigma. \end{aligned} \qquad (6.4)$$

The smaller value of the constant term in the equation for chlorine corresponds to its higher electronegativity, which leads to a higher degree of the ionic character of the C—Cl bond compared with the C—Br bond. The assumption that the electrons in the C—Br bond are more mobile under the influence of the substituent than the electrons of the C—Cl bond follows from the larger value of the coefficient of σ. Moreover, it is reasonable to assume that a large degree of bond ionic character leads to a lower mobility of the electrons.

In order to improve the correlation coefficients and to increase the accuracy of determining the σ constants from the NQR frequencies of ^{35}Cl, the correlation between the meta-substituted chlorobenzenes and the Hammet σ parameter was established [14]. The equation obtained has the form:

$$\nu_m = (34.55 + 1.223\sigma_m) \pm 0.09 \text{ MHz}; \quad r = 0.973$$
$$\sigma_m = (-26.71 + 0.7735\nu_m) \pm 0.07 \quad r = 0.973. \qquad (6.5)$$

The equation was derived on the basis of 10 plotted points (Figure 6.4).

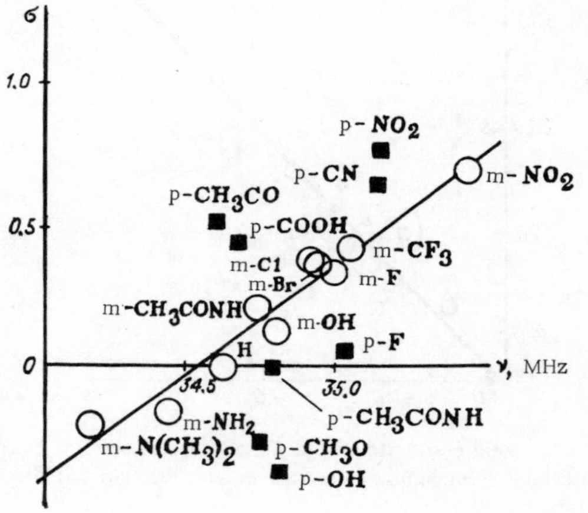

FIGURE 6.4. Correlation ν(σ) for para- and meta-substituted chlorobenzenes

It was observed that the points corresponding to para-substituted chlorobenzenes deviate from the straight line obtained for the meta-substituted ones [14]. These deviations Δν (or Δσ) were compared with σ_R, the resonance constants of the substituents (Figure 6.5). It was found that a linear relationship between Δν and σ_R exists for the para-substituted ones:

$$\Delta\nu = (0.15 + 1.466\sigma_R) \pm 0.07; \quad r = 0.99. \qquad (6.6)$$

The sum of the two equations leads to the relation [14]:

$$\nu_p = (34.40 + 1.223\sigma_p - 1.466\sigma_R) \pm 0.07; \quad r = 0.990. \qquad (6.7)$$

It is seen that the factors in front of σ_p and σ_R are very close. We can therefore assume, as a rough approximation, that:

$$\nu_p = \nu_0 + a(\sigma_p - \sigma_R).$$

But since $\sigma_p - \sigma_R = \sigma_I$, for the considered group of substituents the variation in the NQR frequency of ^{35}Cl in para-substituted benzenes is mainly determined by the induction effect. However, for polysubstituted chlorobenzene [12] a correlation with σ_p is observed and we must, therefore, assume that the conjugation effect also contributes to the change in the NQR frequencies.

This can be shown by the two-parameter correlation method $\nu = \nu_0 + \alpha\sigma_I + \beta\sigma_c$ for the NQR frequencies using the induction σ_I and conjugation σ_c parameters of the substituents. Biedenkapp and Weiss [24] discussed the NQR frequencies of polysubstituted chlorobenzenes. The equation obtained by them shows that the influence of the substituent para or meta to the chlorine atom is transmitted mainly by induction.

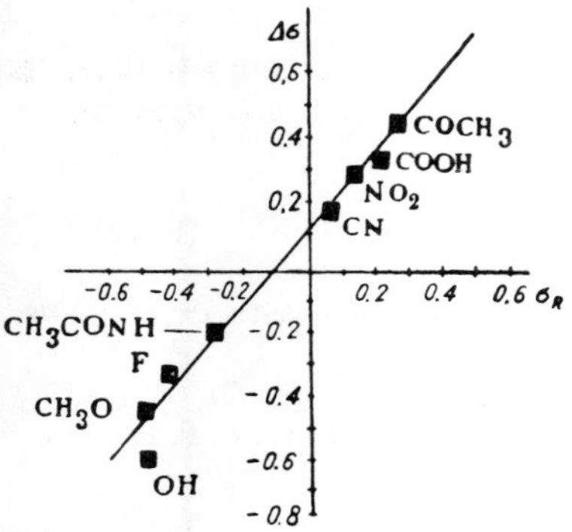

FIGURE 6.5. Correlation σ_R ($\Delta\sigma$) for the same para-substituted chlorobenzenes as in Figure 6.4

The influence of the π donors ($\sigma_c < 0$) and π acceptors ($\sigma_c > 0$) must be dealt with separately. In para-substituted chlorobenzenes with π donors and weak π acceptors, the contribution of the conjugation to the transmission effect is indeed small and is not greater than the uncertainty introduced by the crystal field. With the appearance of strong π acceptors, the contribution from the conjugation mechanism becomes comparable to the induction mechanism ($\alpha = 2.03$, $\beta = -2.49$). The negative sign of β for σ_c shows that strong π acceptors attract the π electron density, reducing, as a result, the population of the p_x orbitals of the halogen and lowering the NQR frequency.

A different approximation of (6.2) and (6.3) was proposed by Bray et al. [15, 16] and by Dewar and Lucken [17]. They extended these equations to heterocyclic compounds, namely, pyrimidines and pyridines.

The heteroatom exerts mainly an inductive effect by transmission through the σ bond, and increases the frequency of the chlorine atoms bonded to the carbon. The observed deviations from the linear relationship (6.2) were explained by the increase in the double bonding in the C—Cl bond due to the influence of the nitrogen atom. However, in many cases the value obtained for the degree of double bonding was too large. This can be due to errors in recording the spectra, or to the incorrect application of the σ values obtained for benzene derivatives to heterocyclics, or by the absence of additivity in the heterocyclics.

Numerous correlations of the NQR frequencies with the Taft induction σ constants were established to find the reason for the change in the halogen NQR frequencies in the RHal and $R_1R_2R_3MHal$ systems (where M is an element of group IVA). The first example is the correlation for the RCl series (where R = Alk) performed by Taft [18] for five compounds. He obtained the equation:

$$e^2Qq_{^{35}Cl} = (68.3 + 20.32\sigma^*) \pm 1.08 \text{ MHz}.$$

In [19] the series RCl (where R = Alk, C_6H_5, $CH_2 = CH$, CH_2X; X is the electronegative substituent) were studied with many compounds (25 points being plotted).

The equation has the form:

$$\nu_{^{35}Cl} = (33.19 + 2.99\sigma^*) \pm 0.82 \text{ MHz}; \qquad r = 0.938. \qquad (6.8)$$

Similar equations for the RCl series were obtained in [20, 21]. For the $R^{79}Br$ series the following equation was derived [20]:

$$\nu_{^{79}Br} = (253.56 + 27.148\sigma^*) \pm 0.150; \qquad r = 0.78. \qquad (6.9)$$

A comparison of the NQR frequencies of halogens in systems $R_1R_2R_3CHal$ (where Hal = Cl, Br) with the σ^* constants was made by Voronkov et al. [20–22]:

$$\nu_{^{35}Cl} = (32.05 + 1.019\Sigma\sigma^*) \pm 0.35 \text{ MHz}; \qquad r = 0.965. \qquad (6.10)$$

The sum of the σ^* constants varies roughly within the limits 0–10. The number of points plotted was 48.

However, compounds of the series Cl_nCH_{4-n} (where X = F, Cl, Br, CCl_3, CF_3, COOH) do not follow the general relationship and are described by a different equation [21, 22]:

$$\nu_{^{35}Cl} = (81.18 - 4.704\Sigma\sigma^*) \pm 0.38; \qquad r = 0.761. \qquad (6.11)$$

Eight points were plotted.

Unlike in the above equations, an increase in the number of electronegative substituents capable of conjugation does not raise the frequency of the halogen but lowers it. To explain this relationship, the authors of [19, 21] used Lucken's hypothesis on the $p - \sigma$ conjugation to explain the anomalous frequencies of fluorochloromethanes [23]. The $RCCl_3$ and $R_1R_2CCl_2$ series are described by somewhat more accurate correlation equations.

Similar effects are also observed for the $R_1R_2R_3C^{79}Br$ series [20].

Numerous papers [2, 20, 22, 24–26, 38] have been devoted to the laws of variation of the NQR frequencies of the tetrahedral molecules of group IVA $R_1R_2R_3MHal$ (where M = Si, Ge; Hal = Cl, Br, I) (Table 6.1).

TABLE 6.1. Correlations $\nu(\sigma^*)$ for the compounds $R_1R_2R_3MHal$ [38]

System	$\nu_{(1/2-3/2)}$, MHz	r	Number of points	References
$R_1R_2R_3Si^{35}Cl$	$(16.628 + 0.406\,\Sigma\sigma^*) \pm 0.12$	0.969	56	[20]
$R_1R_2R_3Si^{81}Br$	$(113.08 + 3.953\,\Sigma\sigma^*) \pm 0.81$		5	[25]
$R_1R_2R_3Si^{127}I$	$(143.09 + 7.92\,\Sigma\sigma^*) \pm 0.97$	0.998	4	[38]
$R_1R_2R_3Ge^{35}Cl$	$(17.382 + 0.941\,\Sigma\sigma^*) \pm 0.16$	0.920	19	[20, 38]
$R_1R_2R_3Ge^{81}Br$	$(120.13 + 6.68\,\Sigma\sigma^*) \pm 0.60$	0.999	5	[38]

Comparison of the data on the derivatives of group IVA elements shows that the frequencies of halogens in the molecules $R_1R_2R_3MHal$ (where M = C, Si, Ge) increase as the number of electronegative substituents increases. However, for compounds of type $R_1R_2R_3CHal$ the introduction of substituents capable of conjugation can substantially distort this pattern as a result of $p - \sigma$ conjugation of the substituent with the halogen.

An interesting example of the influence of the $p - \sigma$ conjugation on the NQR frequency is described in [27, 28]. The series XCH_2Cl (where X is an electronegative substituent which does not contain an unshared pair of electrons) is described by the linear correlation:

$$\sigma^* = (-7.27 + 0.2237\nu_{^{35}Cl}) \pm 0.007; \qquad r = 0.983. \qquad (6.12)$$

Twenty-seven points were plotted. If X is an alkyl group, the equation becomes:

$$\sigma^* = (-3.75 + 0.111\nu_{^{35}Cl}) \pm 0.05; \qquad r = 0.929. \qquad (6.13)$$

Nine points were plotted.

For substituents with unshared pairs of electrons (X = RO, F, RS, Cl, Br) noticeable deviations from equation (6.12) are observed. In accordance with Lucken [23], the authors of [27, 28] propose that this is due to the $p - \sigma$ conjugation. The values of the deviations were compared with the resonance parameters of the substituents $\overset{\circ}{\sigma}_R$. It was found that the deviations $\Delta\nu$ of the measured frequencies from the calculated ones depend linearly on $\overset{\circ}{\sigma}_R$*:

$$\sigma^* = (-7.84 + 0.2237\nu_{^{35}Cl} - 4.164\overset{\circ}{\sigma}_R) \pm 0.06; \qquad r = 0.995. \qquad (6.14)$$

The equation was derived for X = OCH_3, F, Cl, Br, I. The maximum and minimum deviations are for OCH_3 and I, respectively, i.e. the increase in σ_R is accompanied by a natural (within the framework of the $p - \sigma$ conjugation hypothesis) decrease in the NQR frequencies of ^{35}Cl.

Thus, on the basis of the available data we are able to conclude that the NQR frequencies are determined only to a first approximation by the induction effect of the substituents. Indeed, there exists a large series of haloderivatives for which satisfactory correlation with the Taft σ* constants is observed (Table 6.1). However, this law does have some limitations. Unlike the aromatic systems, in which the $p_\pi - p_\pi$ conjugation does not substantially affect the NQR frequency, the $p - \sigma$ conjugation in alkyl halides does have a noticeable effect. In addition to the indicated correlations, the decrease in the NQR frequencies when passing from β- to α-chlorosubstituted ethers is an example of the conjugation effect (33.45 MHz for $CH_3OCH_2CH_2Cl$ and 30.01 MHz for CH_3OCH_2Cl). Because of its vacant 3*d* orbital, the sulfur atom does not participate (or participates weakly) in the conjugation, and an increase in the NQR frequency when passing from $CH_3SCH_2CH_2Cl$ (32.634 MHz) to CH_3SCH_2Cl (33.104 MHz) is observed. The largest conjugation effect is shown by the dialkylamine group, which in β-halogenated amines behaves as an electron-acceptor substituent, increasing the NQR

* For the NQR frequencies of ^{79}Br the equation $\nu = (245.35 + 12.555\sigma^* + 59.36\overset{\circ}{\sigma}_R) \pm 1.39$; $r = 0.998$ can be derived.

frequency, and in α-halogenated amines as an electron-donor, inducing a partial shift of the free pair of nitrogen electrons to the C—Cl σ bond [29].

A similar variation in the NQR frequencies was noted by Lucken and Whitehead [30] for carbonic, sulfonic and phosphonic acid chlorides. They showed that the quadrupole frequencies of $RP(O)Cl_2$ and RCOCl are symbatic, and that the NQR frequencies of $RP(O)Cl_2$ and the valence vibrations of the carbonyl group in the IR spectra are also symbatic. Since the COCl group is planar, the substituent R can be conjugated to both the chlorine atom and the carbonyl group, but to different extents. The fact that $\nu_{NQR}[RP(O)Cl_2] \sim \nu_{NQR}[RCOCl]$ indicates that the phosphonic acid chlorides possibly have $p_\pi - d_\pi$ conjugation. This conclusion is verified by the two linear correlations which are obtained when the effective electronegativities of group R and the NQR frequencies of ^{35}Cl in $RPOCl_2$ are compared [30].

The question of the conjugation in these series (RCOCl, RSO_2Cl, $RPOCl_2$) has been examined in greater detail in [29, 31, 32]. Comparison of the NQR frequencies of ^{35}Cl in compounds of the type RCOCl with the Taft σ^* constants [29] shows that the variation in the NQR frequencies mainly corresponds to the inductive effect of the substituents. Moreover, the groups R = Cl, CH_3O, C_2H_5O in this case are more electronegative than would be expected from the Taft σ^* constants (and the NQR frequency is thus higher), while in alkyl halides these substituents are conjugated. This apparent contradiction is associated with the fact that mainly the carbonyl group participates in the conjugation in the acid chlorides of carbonic acid.

The fact that the P = O bond in the acid chlorides of phosphonic and phosphinic acids is less polarized than the C = O bond indicates a greater conjugation effect in the NQR frequencies of ^{35}Cl. Indeed, substitution of a chlorine atom in $POCl_3$ or CH_3POCl_2 by a fluorine atom or a dialkylamino group lowers the NQR frequency (just as in alkyl halides). Substituents which are less conjugated than the $N(Alk)_2$ group, for example AlkS and AlkO, cause an increase in the NQR frequency in accordance with their electronegativity [29].

In the series RR'POCl (where R and R' are alkyl groups) the NQR frequencies of ^{35}Cl are linearly dependent on the Taft σ^* constants [31]. However, the NQR frequency is almost the same for the ethyl and vinyl radical in the series $RPOCl_2$, despite the substantial difference in the σ^* values of these groups ($\sigma^*_{C_2H_5} = -0.100$; for $\sigma^*_{CH_2=CH}$ two values are known, namely, +0.653 and +0.40 [33]) [32]. Similarly, substitution of the methyl group by a phenyl group causes a very slight increase in the NQR frequency (0.19 MHz), although the difference in the σ^* values is large ($\sigma^*_{CH_3} = 0.000$; $\sigma^*_{C_6H_5} = +0.600$). In these cases, the difference in the inductive effect is substantially compensated by the $d_\pi - p_\pi$ conjugation. This conjugation is also shown by the two linear relationships between the NQR frequencies and the σ_Φ constants of the substituents, which characterize the effect of the electrons of the substituents at the phosphorus atom [34]. The first of these links the NQR frequency of the acid dichlorides of alkylphosphonic acids ($AlkPOCl_2$):

$$\sigma_\Phi = (-13.72 + 0.4806\nu) \pm 0.03; \qquad r = 0.985. \tag{6.15}$$

Another relationship with a similar slope describes the acid dichlorides of vinyl- and arylphosphonic acids, and also dichlorophosphates ($AlkOPOCl_2$, $PhOPOCl_2$, $PhPOCl_2$, $CH_2 = CHPOCl_2$):

$$\sigma_\Phi = (-14.43 + 0.5224\nu) \pm 0.04; \qquad r = 0.988. \qquad (6.16)$$

The conjugation is reflected less in the σ_Φ constants, and this is seen from the considerable difference in the σ_Φ values for the ethyl and vinyl groups, namely, −1.101 and −0.694, respectively.

Semin and Bruchova [35] separated the contributions of the induction and conjugation effects. The NQR frequencies were thus correlated simultaneously with the induction σ_I and conjugation σ_c characteristics of the substituents. Since σ_I and σ_c are determined on the same scale, it is possible to compare these contributions. It was assumed that the scale of variation of the NQR frequencies is constant for both electron-donor and electron-acceptor substituents and is proportional (allowing for the sign) to σ_I and σ_c of the corresponding substituents. Therefore, the polarizability of the investigated atom must be determined by the constants σ_I and σ_c only.

It should be kept in mind that, as a rule, changes in the NQR frequencies corresponding to the chemical reaction constants are observed in the absence of intra or intermolecular coordination interactions (such as hydrogen bonding, etc.). The frequency shifts caused by the coordination interactions can be much greater than the changes caused by the chemical nature of the substituent. However, the case of constant contribution of the coordination interactions (conservation of the geometry of the mutual arrangement and the intermolecular distances in the crystal) for the whole series of compounds investigated is an exception. In this case, a correlation with the chemical constants is also possible.

In the absence of coordination interactions an uncertainty in the NQR frequency is introduced by the crystal effect (see Chapter 2). The scatter in frequencies of the chemically equivalent atoms is due to their different environment in the crystal.

For series of compounds with similar structures the most probable value of the uncertainty introduced in the NQR frequency by the "crystal fluid" is ~ 0.3%; the maximum value of this uncertainty for molecular crystals with Van der Waals interaction between the molecules is 1.5—2% (see Chapter 2). The coordination interactions and the crystal effect limit the applicability of the method of correlating the NQR frequencies with the chemical reaction constants.

In all cases, averaging was performed for an expression of the form:

$$A = A_0 + \alpha\Sigma\sigma_I + \beta\Sigma\sigma_c, \qquad (6.17)$$

where A is the NQR frequency or the quadrupole coupling constant e^2Qq_{zz}, $\Sigma\sigma_I$, $\Sigma\sigma_c$ are the sums of the induction and conjugation constants of the substituents, determined from the chemical shift of the fluoro-derivatives of benzene in inert solvents [36], α and β are the coefficients characterizing the transmission properties of the system with respect to the induction and conjugation effects of the substituents, and A_0 is the frequency or the quadrupole coupling constant for compounds in which $\Sigma\sigma_I = \Sigma\sigma_c = 0$.

The following series of tetrahedral molecules of group IVA were studied:

$R_1R_2R_3CHal$ (Hal = Cl, Br, I)
$R_1R_2R_3SiHal$ (Hal = Cl, Br, I)
$R_1R_2R_3GeHal$ (Hal = Cl, Br)
$R_1R_2R_3SnHal$ (Hal = Cl, Br).

For some of these systems [37] an additive model to calculate the NQR frequencies was suggested (see Chapter 4). It was postulated when deriving the empirical relations that:

1) the angles of the tetrahedron remain practically constant for any substituent (this requirement follows from the necessity of conservation of the sp^3 hybridization of the central atom and is confirmed by numerous X-ray and electron-diffraction data);

2) spatial perturbations are absent or very small;

3) the influence of the substituent is transmitted uniformly and independently of the other substituents on the other three carbon (or Si, Ge, Sn) bonds.

These assumptions also remain valid when the correlation equations for the series of compounds under investigation are derived.

The largest amount of data are available for the series of compounds $R_1R_2R_3C^{35}Cl$ (Table 6.2). Despite the good correlation coefficient, the equation does not accurately describe the whole set of experimental data. This is possibly due to the different character of the electron-donors ($\sigma_c < 0$) and the electron-acceptors ($\sigma_c > 0$). In addition, as already mentioned, the frequencies of the series are greatly affected by the $p-\sigma$ conjugation if the substituent has a free pair of electrons. It seems, therefore, reasonable to classify the series $R_1R_2R_3CHal$ according to the sign of σ_c of the substituents, and in the case where $\sigma_c < 0$ to isolate the substituents having a free pair of electrons. For compounds where $\sigma_c > 0$ the conjugation effect was found to be small, not larger than the uncertainty introduced by the crystal field. For compounds where $\sigma_c < 0$, substituents with free pairs are best described. These equations are given in Table 6.2.

TABLE 6.2. Correlations $\nu(\sigma_I, \sigma_c)$ for the compounds $R_1R_2R_3MHal$

System	ν, MHz	Number of points	r
$R_1R_2R_3CCl$	$(34.35 + 5.03\ \Sigma\sigma_I + 2.69\ \Sigma\sigma_c) \pm 0.29$	70	0.944
$R_1R_2R_3CCl$ ($\sigma_c<0$)	$(33.90 + 10.64\ \Sigma\sigma_I + 13.32\ \Sigma\sigma_c) \pm 0.61$	16	0.946
$R_1R_2R_3C^{79}Br$ ($\sigma_c<0$)	$(264.04 + 67.89\ \Sigma\sigma_I + 73.01\ \Sigma\sigma_c) \pm 1.64$	. . .	0.996
$R_1R_2R_3CI$ ($\sigma_c<0$)	$(267.71 + 54.00\ \Sigma\sigma_I + 43.32\ \Sigma\sigma_c) \pm 2.12$	. . .	0.988
$R_1R_2R_3SiCl$	$(17.08 + 2.42\ \Sigma\sigma_I - 0.27\ \Sigma\sigma_c) \pm 0.19$	30	0.980
$R_1R_2R_3Si^{81}Br$	$(111.15 + 22.76\ \Sigma\sigma_I - 13.18\ \Sigma\sigma_c) \pm 0.33$	7	0.999
$R_1R_2R_3SiI$	$e^2Qq = (849.95 + 243,72\ \Sigma\sigma_I - 368.18\ \Sigma\sigma_c) \pm 5.32$	4	0.999
$R_1R_2R_3GeCl$	$(18.68 + 5.18\ \Sigma\sigma_I - 0.92\ \Sigma\sigma_c) \pm 0.21$	10	0.995
$R_1R_2R_3Ge^{81}Br$	$(118.75 + 35.65\ \Sigma\sigma_I - 20.97\ \Sigma\sigma_c) \pm 0.29$	5	0.999
$R_1R_2R_3SnCl$	$(9.91 + 8.38\ \Sigma\sigma_I - 7.88\ \Sigma\sigma_c) \pm 0.24$	6	0.996
$R_1R_2R_3Sn^{81}Br$	$(103.23 + 40.61\ \Sigma\sigma_I - 16.64\ \Sigma\sigma_c) \pm 0.08$	3	0.999

TABLE 6.3. Summary of the correlation equations for the halides of group IVA elements

Element	Resonance atom	$\frac{2\nu_0}{e^2Qq_{at}}$	$\frac{2\alpha}{e^2Qq_{at}}$	$\frac{2\beta}{e^2Qq_{at}}$	$\frac{\beta}{\alpha}$
C	^{35}Cl	0.625	0.196	0.245	1.25
	^{79}Br	0.690	0.177	0.191	1.08
	^{127}I	0.802	0.162	0.130	0.80
Si	^{35}Cl	0.314	+0.045	−0.04	−0.10
	^{81}Br	0.348	+0.071	−0.041	−0.58
	^{127}I	0.391	+0.112	−0.170	−1.51
Ge	^{35}Cl	0.344	+0.095	−0.17	−0.18
	^{81}Br	0.372	+0.111	−0.065	−0.59
Sn	^{35}Cl	0.183	+0.154	−0.145	−0.94
	^{81}Br	0.323	+0.127	−0.052	−0.41

To illustrate the relationships obtained for the series $R_1R_2R_3CCl$ ($\sigma_c < 0$) we present graphical solutions of the equations.

Figure 6.6 shows the NQR frequencies of ^{35}Cl as a function of the induction constants of the substituents. The line $\nu = 33.90 + 10.64\ \Sigma\sigma_I$ is the projection of the spatial line $\nu = 33.90 + 10.64\ \Sigma\sigma_I + 13.32\,\Sigma\sigma_c$ onto the plane $\nu - \sigma_I$.

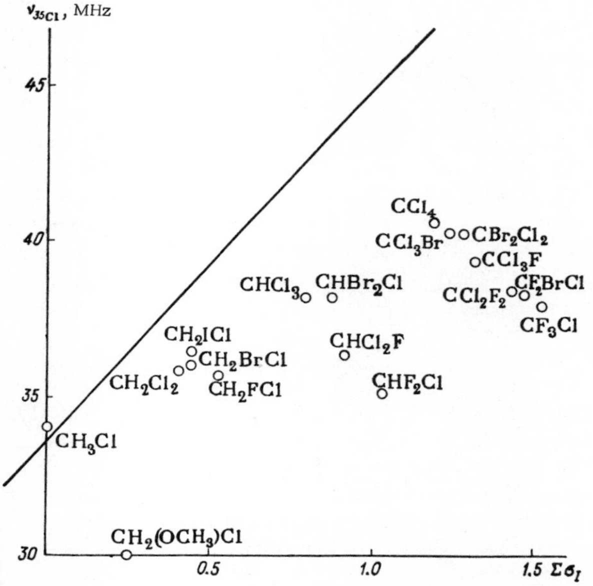

FIGURE 6.6. Correlation between the NQR frequencies of ^{35}Cl in the series $R_1R_2R_3CCl$ ($\sigma_c < 0$) and $\Sigma\sigma_I$ of the substituents R

The deviations of the experimental points from this line are proportional to the conjugation contribution. As an example, we have plotted in Figure 6.7

values of $\Delta\nu = \nu_{exp} - \nu_0 - \alpha\Sigma\sigma_I$ as a function of $\Sigma\sigma_c$. The linear character of this relationship indicates the conjugation nature of the deviations.

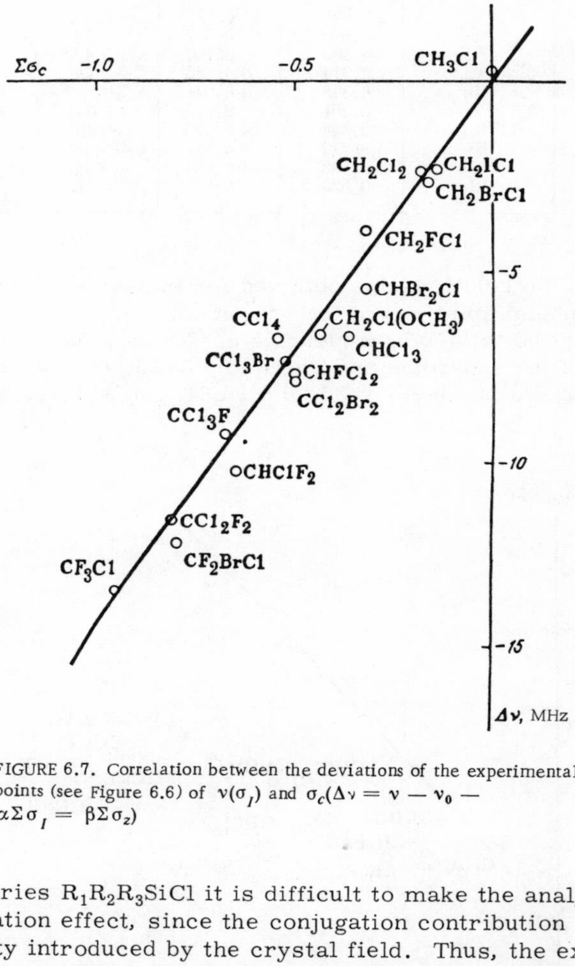

FIGURE 6.7. Correlation between the deviations of the experimental points (see Figure 6.6) of $\nu(\sigma_I)$ and $\sigma_c(\Delta\nu = \nu - \nu_0 - \alpha\Sigma\sigma_I = \beta\Sigma\sigma_z)$

For the series $R_1R_2R_3SiCl$ it is difficult to make the analysis in terms of the conjugation effect, since the conjugation contribution is smaller than the uncertainty introduced by the crystal field. Thus, the existence of satisfactory correlations between the NQR frequencies of this series and the Taft σ^* parameters is understandable (Table 6.1). To a great extent, the same is true of the conjugation contribution in the series $R_1R_2R_3GeCl$, which is somewhat larger than for $R_1R_2R_3SiCl$ but close to the uncertainty introduced by the crystal field. Bromine- and iodine-containing compounds have a very large conjugation contribution.

All the data on the halides of group IVA elements are summarized in Table 6.3. All the equations are divided by e^2Qq_{at} (or by $^1/_2e^2Qq_{at}$ for the transition frequencies $\pm^1/_2 \longleftrightarrow \pm^3/_2$).

The following values, determined at 77°K, were used to calculate $^1/_2 e^2Qq_{at}$:

for	$^{35}Cl_2$	$\nu = 54.24$ MHz
»	$^{79}Br_2$	$\nu = 382.43$ MHz
»	$^{81}Br_2$	$\nu = 319.46$ MHz
»	$^{127}I_2$	$\nu = 333.94$ MHz
		$e^2Qq = 2172.26$ MHz.

Since, according to the Townes-Dailey model [39]:

$$\frac{e^2Qq_{zz}}{e^2Qq_{at}} = U_p = \frac{N_x + N_y}{2} - N_z, \tag{6.18}$$

(where N_x, N_y, N_z are the populations of the p_x, p_y, p_z orbitals of the halogen atom, respectively), reduction of the correlation equations to electron charge units enables us to compare the M—Cl, M—Br, M—I bonds (where M is a group IVA element). It can be assumed, to a fair degree of accuracy, that the induction term $\alpha\sigma_I$ corresponds to the variation in the number N_z of electrons in the p_z orbital of the halogen, participating in the formation of the σ bond. The conjugation term $\beta\sigma_c$ corresponds to the allowance for $\frac{N_x+N_y}{2}$, i.e. the variation in the number of electrons in the p_x and p_y orbitals of the halogen participating in the formation of the π bonds. However, it must be kept in mind that such an assumption involves a big approximation.

The following rules are obtained from Table 6.3. In the series C—Hal, passing from chlorine to bromine and iodine involves an increase in $2\nu_0/e^2Qq_{at}$, i.e. a decrease in ionic character of the C—Hal bond. This agrees with the electronegativity values of the halogens, i.e. Cl > Br > I.

In addition, in the series C—Hal, the relative part of the conjugation contribution (i.e. the ratio β/α) increases in the order Cl ~ Br < I. The latter is probably due to the difference in the radial distribution of the p valence electrons of the atoms of chlorine, iodine and bromine, which for iodine and bromine can affect the conjugation effects (i.e. alter the region of overlap).

Similar relationships hold for the series Si—Hal and Ge—Hal. In the series Si, Ge, Sn there is an increase in both the induction (α) and conjugation (β) contributions (in the given case the same halogens are compared with different elements of group IVA). This trend is due to the increase in the polarizability of the halogen atoms in the M—Hal bond in the order Si < Ge < Sn. This is indicated by the NQR Stark-effect data, i.e. the investigation of the splittings in the NQR spectra under the influence of strong electrostatic fields [40, 41]. The polarizability of the halogen atom in a given type of bond is larger the lower the NQR frequency, i.e. the larger the ionic character of the bond.

Comparison of the values of $2\nu_0/e^2Qq_{at}$ for the different elements shows that the degree of covalence of the C—Hal bond is much higher than that of the Si—, Ge—, Sn—Hal bonds. For 100% covalence these ratios should be equal to unity since the comparison is carried out with the covalent molecules Cl_2, Br_2, I_2. Moreover, for the series C—Hal the signs of the induction and conjugation contributions agree, whereas for Si—, Ge—, Sn—Hal they are different. No overall explanation exists for this, but it can be assumed that for Si—, Ge—, Sn—Hal the multiplicity of the M—Hal bond increases slightly. As follows from (6.18), the changes in the gradient caused by the *p* electrons, N_z (the σ bond) and N_x, N_y (the π bond), have opposite signs. For C—Hal the signs of the two contributions are the same. Continuing with this hypothesis, we must assume the absence of multiplicity in the C—Hal bond. In the given case, the conjugation contribution may result from the $p - \sigma$ and $\pi - \sigma$ conjugation [23].

The question of the participation of the *d* orbitals in the transmission effect of the substituents through the central atom to the resonance atom still remains open. Within the framework of the concept of $p_\pi - d_\pi$ conjugation the pattern obtained can be explained as follows.

The central atom may partially accept halogen free-electron pairs on the vacant *d* orbital. In this case, a substituent which is a donor according to the the π system ($\sigma_c < 0$) must increase the population of the *d* orbitals of the central atom. This in turn must lead to a decrease in the conjugation of the p_x, p_y orbitals of the halogen with the *d* orbitals, i.e. to an increase in the populations N_x, N_y. It follows from (6.18) that such a change leads to an increase in the gradient and, therefore, in the NQR frequencies of the halogen. This increase agrees with the sign of the conjugation contribution for the series $R_1R_2R_3MHal$ (where M = Si, Ge, Sn).

The two-parameter correlation method was also used [35] to analyze compounds of type $R_1R_2P(O)Cl$ and $R_1R_2P(S)Cl$ (where R_1, R_2 = Alk, Ar, Hal).

For the series $R_1R_2P(O)^{35}Cl$ the following equation was obtained:

$$\nu = (22.89 + 5.13\Sigma\sigma_I - 5.65\Sigma\sigma_c) \pm 0.15 \text{ MHz}; \qquad r = 0.995. \qquad (6.19)$$

The system $R_1R_2P(S)Cl$ is described by the equation:

$$\nu = (23.52 + 3.43\Sigma\sigma_I - 9.59\Sigma\sigma_c) \pm 0.15 \text{ MHz}; \qquad r = 0.995. \qquad (6.20)$$

Comparing these two equations, we note the higher conjugation contribution for the system $R_1R_2P(S)Cl$, which agrees satisfactorily with the greater degree of double bonding in P = O compared to P = S.

The conclusion concerning the preferential participation of the CO and SO_2 groups in the conjugation of the carbonic and sulfonic acid chlorides was confirmed by means of two-parameter correlations [42, 43].

3. CORRELATION BETWEEN THE QUADRUPOLE COUPLING CONSTANTS FOR POLYVALENT ATOMS AND THE CHEMICAL REACTION CONSTANTS

So far we have discussed only the variation in the electric field gradient on the nucleus of a monovalent halogen atom. In order to establish a correlation with the chemical constants for polyvalent atoms we need a more rigorous conservation of the molecule's geometry on varying the substituents, and also of the coordination number and geometry of the coordination environment of the investigated atom. The change in the coordination number and the formation of additional associative bonds can substantially alter the quadrupole coupling constants.

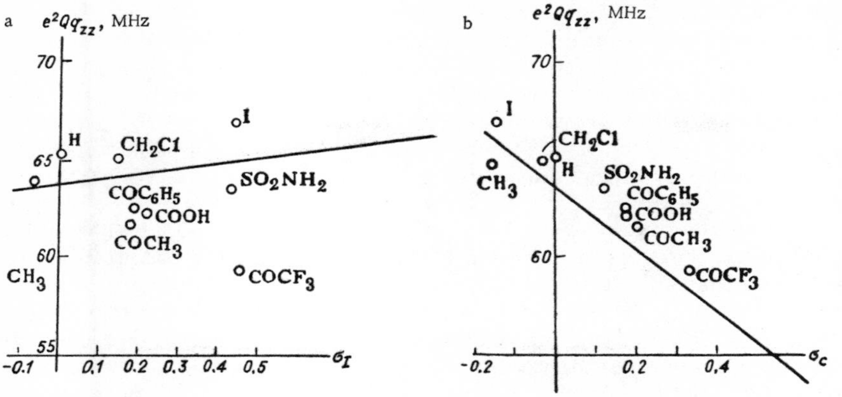

FIGURE 6.8. Correlation between e^2Qq of ^{55}Mn and the induction (a) and conjugation (b) constants of the substituents R in $RC_5H_4Mn(CO)_3$

It is therefore natural to expect that a necessary condition for the existence of such correlations is that a system of principal axes of the tensor of the electric field gradient must be rigidly fixed in the molecule on varying the substituents. The additional inter- and intramolecular interactions, which affect the asymmetry parameter, should not alter the direction of the maximum gradient of the electric field. Only then may we expect the fine variations in the quadrupole coupling constants, caused by the nature of the substituents, to be proportional to the values of its reaction constants.

The series $(CO)_3{}^{55}MnC_5H_4R$ includes a wide variety of compounds [44, 45]. Unfortunately, data on the transition frequencies $\pm{}^1/_2 - \pm{}^3/_2$ and the quadrupole coupling constants are not available for all these compounds. Accordingly, correlation equations for both the quadrupole coupling constants e^2Qq (6.21) and the transition frequencies $(\pm{}^3/_2 - \pm{}^5/_2)$ (6.22) were derived for this series. Equation (6.22) describes a wider range of substituents.

$$e^2Qq_{zz} = (63.72 + 2.89\sigma_I - 14.18\sigma_c) \pm 0.42 \text{ MHz}; \qquad r = 0{,}971 \qquad (6.21)$$

$$\nu_{(\pm 3/2 - \pm 5/2)} = (19.12 + 0.65\sigma_I - 4.32\sigma_c) \pm 0.12 \text{ MHz}; \qquad r = 0{,}973. \qquad (6.22)$$

To illustrate these equations we have plotted e^2Qq_{zz} as a function of σ_I and σ_c in Figure 6.8. The line drawn on Figure 6.8,a is described by the expression:

$$e^2Qq_{zz} = (63.72 + 2.89\sigma_I) \text{ MHz},$$

and is the projection of the spatial line of equation (6.21) on plane (e^2Qq_{zz}, σ_I). Similarly, on Figure 6.8,b the line is given by the equation:

$$e^2Qq_{zz} = (63.72 - 14.18\sigma_c) \text{ MHz},$$

and is the projection of a spatial line on plane (e^2Qq_{zz}, σ_c). The deviations of the measured points from the straight lines are caused by the induction contribution (Figure 6.8,b) and the conjugation contribution (Figure 6.8,a).

To illustrate the latter we have plotted on Figure 6.9 the deviations of the experimental points from the calculated line of Figure 6.8,a ($\Delta e^2Qq_{zz} = e^2Qq_{exp} - 63.72 - 2.89\ \sigma_I$) as a function of σ_c. The linear character of this relationship shows that the observed deviations have a conjugation nature.

We see that the measured value of e^2Qq_{zz} for $C_5H_5Mn(CO)_3$ (this point was neglected in the calculation) is ~1.5 MHz greater than the calculated value. It is assumed that such a large discrepancy is due to a fulvene-type rearrangement of the cyclopentadienyl ring when the substituent R is introduced [44, 45]. In addition, introduction of the substituent sharply changes the symmetry of the cyclopentadienyl part of the molecule.

It is important that, for the given series, the conjugation contribution be much larger than the induction contribution. This may be seen from equations (6.21) and (6.22) where the coefficient of σ_c is several times greater than the corresponding coefficient of σ_I. This means that, for the group of compounds under discussion, the transmission effect from the substituent in the ring to the metal atom takes place more by conjugation than by induction.

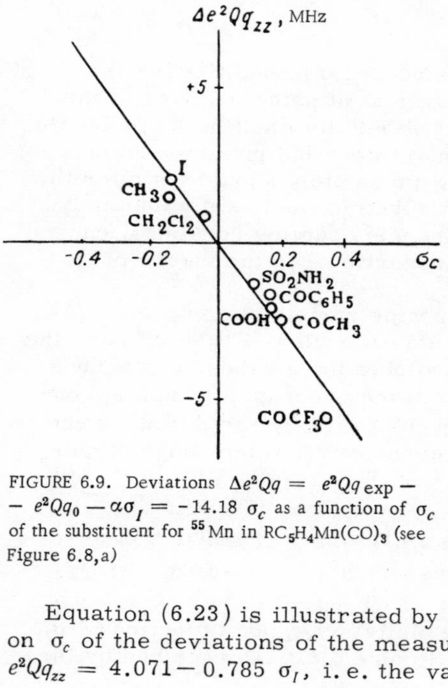

FIGURE 6.9. Deviations $\Delta e^2Qq = e^2Qq_{exp} - e^2Qq_0 - \alpha\sigma_I = -14.18\ \sigma_c$ as a function of σ_c of the substituent for ^{55}Mn in $RC_5H_4Mn(CO)_3$ (see Figure 6.8,a)

The correlation for the series $RC^{14}N$ was studied in [46]. The equation obtained for this series has the form:

$$e^2Qq_{zz} = (4.071 - 0.785\sigma_I + 2.583\sigma_c) \pm 0.051\ \text{MHz};\ r = 0.986. \quad (6.23)$$

Application of the two-parameter correlation method enables us to avoid having to split the series under investigation into subrelationships for R = Alk, R = Hal, etc., as was done in [47].

Equation (6.23) is illustrated by Figure 6.10 which shows the dependence on σ_c of the deviations of the measured points from the calculated line $e^2Qq_{zz} = 4.071 - 0.785\ \sigma_I$, i.e. the values:

$$\Delta e^2Qq_{zz} = e^2Qq_{zz} - 4.071 + 0.785\sigma_I.$$

As seen from the figure, the deviations Δe^2Qq_{zz} depend linearly on the conjugation constants σ_c of the substituents. The changes in the quadrupole coupling constants of ^{14}N in the CN group depend to a great extent on the conjugation capacity, i.e. the conjugation term in equation (6.23) is significantly larger than the induction term.

It can be assumed that the character of the transmission effect in the system under discussion is due to the C≡N bond.

Among the correlations of the quadrupole constants of polyvalent atoms with the σ constants we should also note the correlation $e^2Qq_{zz} = f(\sigma^*)$ [48] for pentavalent antimony in the series R_3SbHal_2. Since the symmetry of the central atom is rigorously preserved, we would expect the electron distribution on the antimony atom to depend on the donor-acceptor properties of the substituents. Two linear relationships were found between the quadrupole coupling constants of ^{121}Sb and the difference in the sums σ* of the equatorial and axial substituents (Figure 6.11). Since the equatorial substituents, those both capable and incapable of conjugation, yield e^2Qq_{zz} described by different equations, in this case we should use the two-parameter correlation. When the spatial symmetry of the antimony bipyramid is altered, there is a sharp change in the quadrupole interaction constant for Sb.

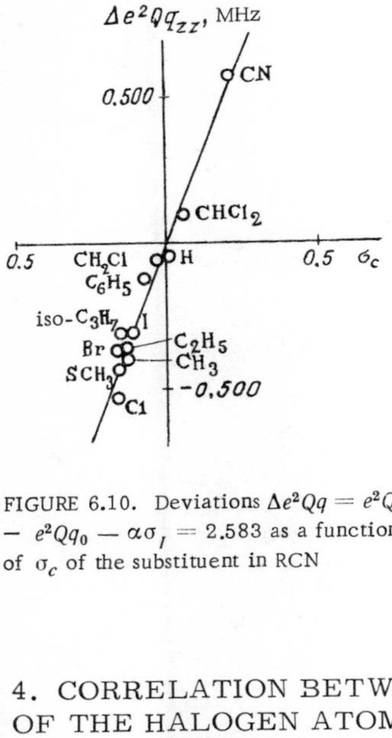

FIGURE 6.10. Deviations $\Delta e^2Qq = e^2Qq_{exp} - e^2Qq_0 - \alpha\sigma_I = 2.583$ as a function of σ_c of the substituent in RCN

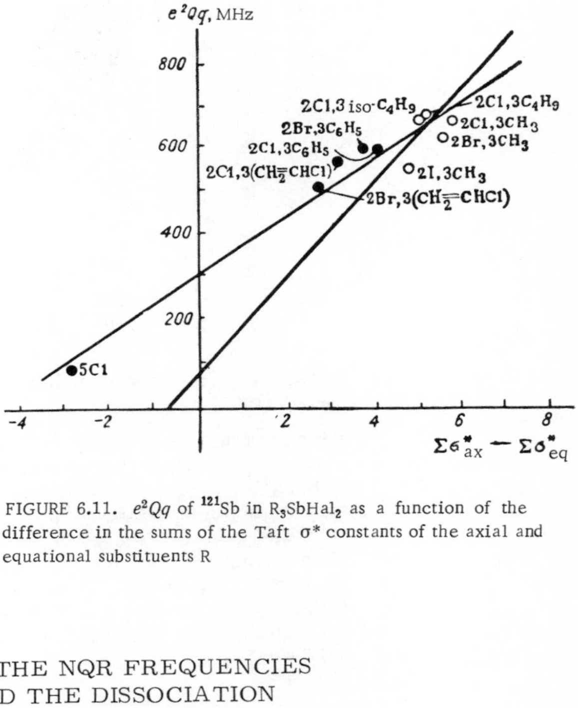

FIGURE 6.11. e^2Qq of ^{121}Sb in R_3SbHal_2 as a function of the difference in the sums of the Taft σ* constants of the axial and equational substituents R

4. CORRELATION BETWEEN THE NQR FREQUENCIES OF THE HALOGEN ATOMS AND THE DISSOCIATION CONSTANTS OF CARBOXYLIC ACIDS, AMINES AND ALCOHOLS

In [19, 28] the correlations $\nu = f(pK_a)$ were found for a number of compounds containing C—Cl and C—Br bonds (Figure 6.12).

The NQR frequencies of ^{35}Cl and ^{79}Br in halogen-containing compounds of type RX were compared with the pK_a of the corresponding carboxylic acids RCOOH [19]. The following approximate linear relationship holds for aliphatic and aromatic compounds:

$$\nu_{^{35}Cl} = (40.67 - 1.56pK_a) \pm 0.71 \text{ MHz}; \qquad r = 0.925, \qquad (6.24)$$

number of points 90;

$$\nu_{^{79}Br} = (319.31 - 12.58pK_a) \pm 6.70 \text{ MHz}; \qquad r = 0.842,$$

number of points 34.

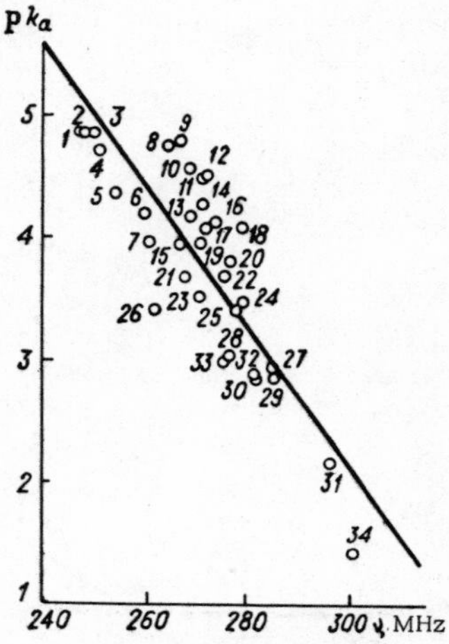

FIGURE 6.12. Correlation of $pK_a(\nu)$ with the NQR of ^{79}Br in the series RX.

X = Br, COOH; R corresponds to the points: 1) C_2H_5; 2) C_4H_9; 3) $Br(CH_2)_4$; 4) C_3H_7; 5) p-$CH_3C_6H_4CH_2$; 6) $HOOCCH_2CH_2$; 7) $BrCH_2CH_2$; 8) CH_3; 9) p-$C_2H_5OC_6H_4$; 10) p-OHC_6H_4; 11) p-$CH_3OC_6H_4$; 12) p-$C_6H_5OC_6H_4$; 13) C_6H_5; 14) p-$CH_3CONHC_6H_4$; 15) $C(C_6H_5)_3$; 16) p-FC_6H_4; 17) m-HOC_6H_4; 18) o-$CH_3OC_6H_4$; 19) p-BrC_6H_4; 20) m-BrC_6H_4; 21) α-$C_{10}H_7$; 22) p-$CH_3COC_6H_4$; 23) p-$HOOCC_6H_4$; 24) m-$NO_2C_6H_4$; 25) p-$NO_2C_6H_4$; 26) 2,4,6-$(CH_3)_3C_6H_2$; 27) o-$HOOCC_6H_4$; 28) $HOOCCHCH_3$; 29) $HOOCCH_2$; 30) o-BrC_6H_4; 31) o-$NO_2C_6H_4$; 32) $BrCH_2$; 33) o-HOC_6H_4; 34) 2,4-$(NO_2)_2C_6H_3$.

The relationship found is best satisfied by compounds in which R is a methyl group or an ortho aromatic radical, or an aromatic radical having electron-acceptor substituents.

Large deviations are observed for aromatic radicals having substituents with unshared pairs of electrons (RO, OH, CH_3, COO, $HNCOCH_3$, F, etc.).

The reason for these deviations lies in the different electronic nature of the carboxyl group and the chlorine atom and in the different character of the sensitivity of these "reaction centers" to the electron effect of the substituents. The deviations from the linear relationship in the case of aromatic radicals are due to the conjugation, which affects the NQR frequencies less than it affects the dissociation constants of the acids. Comparison* of $\Delta\sigma = \sigma - \sigma'$ with the deviations from the linear relationship $\Delta pK_a = pK_a^{calc} - pK_a^{exp}$ established that an approximate linear relationship exists between them:

$$\Delta pK_a = (-0.08 + 1.12\Delta\sigma) \pm 0.15; \qquad r = 0.865.$$

In order to establish more rigorous correlations in the series of aliphatic and aliphatic-aromatic compounds, extensive data on the dissociation constants of the corresponding carboxylic acids, amines and alcohols were used.

In the case of carboxylic acids and amines two linear relationships were obtained, one of them formed by the unsubstituted saturated aliphatic acids and primary alkylamines:

$$\begin{aligned} &\text{RCOOH } pK_a = (6.82 - 0.0592\nu) \pm 0.04; \qquad r = 0.834 \\ &\text{RNH}_2 \; pK_a = (8.96 + 0.0502\nu) \pm 0.05; \qquad r = 0.627. \end{aligned} \qquad (6.25)$$

A different relationship was obtained for substituted acetic acids of the type XCH_2COOH and primary amines XCH_2NH_2 with the frequencies of the alkyl halides XCH_2Cl:

$$\begin{aligned} &\text{XCH}_2\text{COOH } pK_a = (17.47 - 0.3930\nu) \pm 0.11; \qquad r = 0.986 \\ &\text{XCH}_2\text{NH}_2 pK_a = (39.79 - 0.9061\nu) \pm 0.19; \qquad r = 0.991. \end{aligned} \qquad (6.26)$$

The existence of two linear relationships confirms that the division of the substituents and their respective σ^* constants (see p. 91) into two groups (alkyl radicals and hydrogen on the one hand, and electronegative groups on the other) was correct [28]. Judging from the correlations with the NQR frequencies, the nature of these two groups of substituents is different.

The low correlation coefficients in equations (6.25) are due to the small slope of the lines. However, in spite of the small value of r, such correlations enable us to calculate with satisfactory accuracy the corresponding parameters.

Examination of the correlation $\nu = f(pK_a)$ for the substituted acetic acids XCH_2COOH established [28] considerable deviations from the linear relationship for substituents with unshared pairs of electrons (X = RO, F, RS, Cl, Br). In [28], just as in the case of the correlation with the Taft σ^* constants, this is explained by a p—σ conjugation of type $\overset{\frown}{X - CH_2 -} Cl$ [23], i.e. the p—σ conjugation alters the NQR frequency in a direction opposite

* According to Taft [18], an approximate measure of the conjugation is the difference between the Hammet σ constants and the σ' constants for bicyclo-(2,2,2)-octane systems, which are characterized by the inductive effect only.

to the variation of the electronegativity of the substituent. Thus, in the series XCH_2Cl (where X = F, Cl, Br, I), the NQR frequency increases instead of decreasing with a decrease in the electronegativity x. A similar effect is not observed in the acids XCH_2COOH, and the acidity increases with an increase in the electronegativity x. This causes deviations in the case of substituents with unshared pairs of electrons, and these deviations increase with an increase in the donor effect of the unshared pair.

In comparing the values of the decrease in the NQR frequencies due to the conjugation, and the resonance parameter σ_R°, used as the measure of the donor effect of the unshared pair, the following linear relationship was detected [28]:

$$\Delta\nu = \left(-2.166 - 19.24\sigma_R^\circ\right) \pm 0.16; \qquad r = 0.998. \qquad (6.27)$$

This confirms the resonance nature of the discussed phenomenon. At the same time, it opens the somewhat unexpected possibility of fairly accurately calculating the resonance parameters σ_R° from the ionization constants of the substituted acetic acids and the NQR frequencies of the halo-derivatives XCH_2Cl:

$$\sigma_R^\circ = (-2.44 + 0.0521\nu + 0.132pK_a) \pm 0.01; \qquad r = 0.999. \qquad (6.28)$$

5. CORRELATION BETWEEN THE NQR FREQUENCIES AND THE HALF-WAVE POTENTIALS IN POLAROGRAPHIC REDUCTION

The first attempts to correlate the NQR frequencies with the half-wave potential in polarographic reduction were made in 1956 by Iredale [49] on the basis of the experimental data of [50, 51] for six iodo-derivatives of benzene. The linear relationship $E_{1/2} = f(\nu)$ is easily explained, since an increase in the electron density causes an increase in the half-wave potential, whereas the NQR frequency decreases.

It was also found that the NQR frequencies of ^{35}Cl in chloronitroalkanes are linked by a linear relationship with the half-wave potentials of polarographic reduction [52] (Figure 6.13):

$$E_{1/2} = 0.092\nu - 3.69. \qquad (6.29)$$

It is characteristic that there are two values of the half-wave potential $[Cl(NO_2)_2C]_2CH_2$. One of the points is located on the correlation line, and apparently describes the ground state, i.e. the separation of one of the two chemically equivalent chlorine atoms. The second point corresponds to the separation of the second chlorine atom from the molecule in an excited state, which leads to the exclusion of this point from the general relationship. From the NQR frequency of ^{35}Cl, found from (6.29), for this value of the half-wave potential we can estimate the change in anisotropy of the electron environment of the chlorine atom in the transition complex.

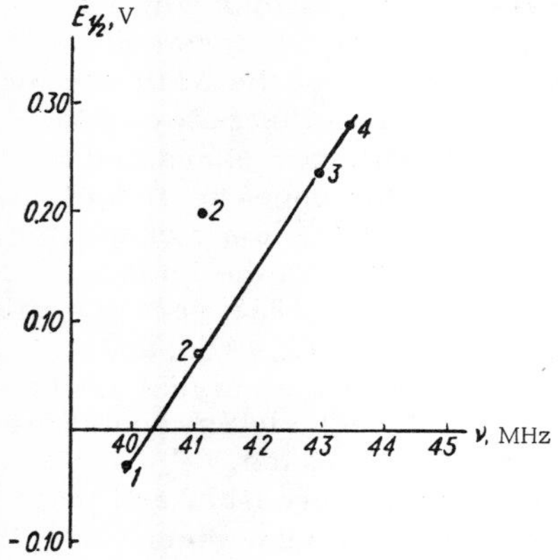

FIGURE 6.13. Half-wave potentials of polarographic reduction as a function of the NQR frequencies of ^{35}Cl in chloronitroalkanes:

1) $Cl(NO_2)_2CC_2H_5$; 2) $[Cl(NO_2)_2C]_2CH_2$; 3) $ClC(NO_2)_3$; 4) $Cl(NO_2)_2CC(NO_2)_2Cl$.

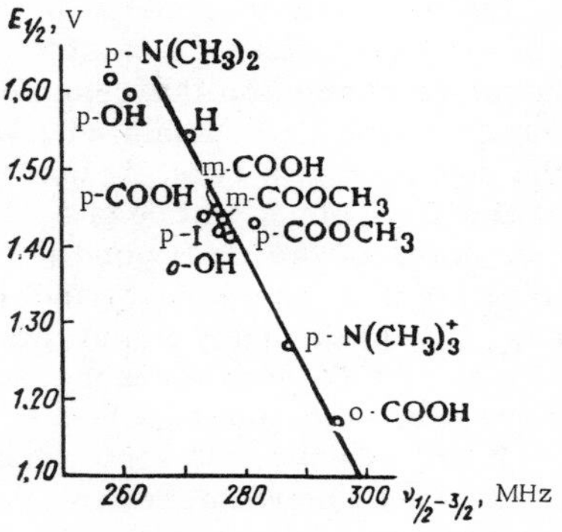

FIGURE 6.14. NQR frequencies of ^{127}I (transitions $\Delta m = {}^1/_2 - {}^3/_2$) of disubstituted iodobenzenes as a function of the half-wave potentials of polarographic reduction [53]

The correlation between the NQR frequencies and the polarographic data was studied most thoroughly by Caldwell and Hacobian [53]. They assume that for the iodine-containing compounds the polarographic reduction passes through two stages: $R-I + e^- \rightarrow T^-$ (transition state) slow stage; $T^- + e^- + H^+ \rightarrow R-H + I$ kinetic stage. Since they determine the value of the half-wave potential of the electron transfer stage and the formation of T^-, which is independent of the pH of the electrolyte, the organic iodine-containing compounds are very convenient for studying the influence of substituents on the iodine atom by the polarographic method. It was found [53] that the half-wave reduction potentials of iodobenzenes and alkyl iodides form different relationships with the NQR frequencies (Figure 6.14). This completely agrees with the fact that for iodobenzene and alkyl iodides αn_a (where α is the transmission coefficient $0 < \alpha < 1$, and n_a is the number of electrons participating in the slow stage of the electrochemical process) has different values: 0.56 ± 0.04 and 0.30 ± 0.04, respectively. In each of these relationships, the following law is obeyed: an increase or decrease in $E_{1/2}$, meaning an increase or decrease in the electron density on the iodine atom, corresponds to a decrease or increase in the NQR frequency as a result of the change of ionic character of the C—I bond.

6. CORRELATION BETWEEN THE NQR FREQUENCIES AND THE OTHER CHARACTERISTICS OF THE CHEMICAL BOND

Correlation with NMR data. The inconsistent and insufficient data on the causes of the variation in the NMR chemical shifts of ^1H and ^{19}F led to attempts to find general relationships between the NMR chemical shifts and the molecular structure on the basis of correlations with the NQR data of halogens, whose physical meaning are better understood.

On the basis of general principles [55] we would expect the NMR chemical shift of ^1H and ^{19}F in strong fields to increase with an increase in the negative charge on the resonance atom. However, the ionic character of the C—F and C—H bonds will vary differently due to the opposite signs of the dipoles: as a rule, δ_F increases with an increase in the ionic character of the C—F bond, whereas δ_H decreases with an increase in the ionic character of the C—H bond [55]. Meyer and Gutowsky in 1953 were the first to attempt to compare δ_H and δ_F with e^2Qq for the series $Cl_{4-n}CH_n$ and $Cl_{4-n}CF_n$ [54]. They found that in chloromethanes the proton signal shifts in weak fields with an increase in the number of chlorine atoms, and at the same time the quadrupole coupling constant of ^{35}Cl increases.

It follows that the ionic character of the C—H bond increases, and that of the C—Cl bond decreases, i.e. the main cause of this phenomenon is the inductive effect.

However, a direct correlation between the theoretical ionic character of the C—F bond and the value of δ_F was not detected. Moreover, in the $Hal_{4-n}CH_n$ series the change of δ_F has an opposite sign to that expected [54, 55]. Meyer and Gutowsky [54], who observed the shift of δ_F in weak fields and the small increase of e^2Qq_{35Cl} for the series of chloromethanes, proposed that the inductive effect decreases the ionic character of the two C—F and C—Cl bonds. They also suggested that the double bonding in structures of type:

$$\begin{array}{ccc}
\overset{-}{Cl} & & \overset{+}{Cl} \\
| & & \| \\
Cl-C\overset{+}{=}F & \text{and} & Cl-\overset{}{C}\ \overset{-}{F} \\
| & & | \\
Cl & & Cl
\end{array}$$

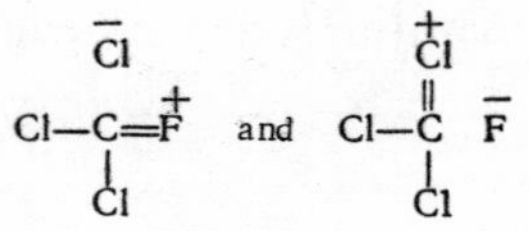

plays an important role.

A wider range of fluorine-containing compounds was studied in [56], where δ_F in the series $FCH_n(NO_2)_{3-n}$, $F_{4-n}CH_n$, FCH_nCl_{3-n}, $F_nC(NO_2)_{4-n}$ and F_nCCl_{4-n} and the NQR frequencies of ^{35}Cl and ^{79}Br in the series $Hal_{4-n}CH_n$, $Hal_nC(NO_2)_{4-n}$, Cl_nCF_{4-n} and $Hal_nCH(NO_2)_{3-n}$, respectively, are compared.

It was found that the influence of the nitro group and the chlorine atom on δ_F has different signs. At the same time, the difference in the influence of the nitro group and the fluorine atom is not large and is additive, i.e. depends linearly on the number of fluorine atoms. The presence of the nitro group makes the conjugation effect on δ_F stronger than the inductive effect. For the NMR of ^{19}F in the case of electron-donor substituents (F_nCH_{4-n}) a specific deviation from additivity is observed; the shift increases nonlinearly with a decrease in the number of fluorine atoms. The correlation of the chemical shifts of δ_F (with respect to CF_4) with the relative NQR chemical shifts* of ^{35}Cl is illustrated in Figure 6.15.

---

* The relative NQR chemical shifts Δ_{Cl} (in thousandths with respect to CCl_4) are defined as:

$$\Delta_{Cl} = \left(\frac{v_X}{v_{CCl_4}} - 1\right) \cdot 10^3.$$

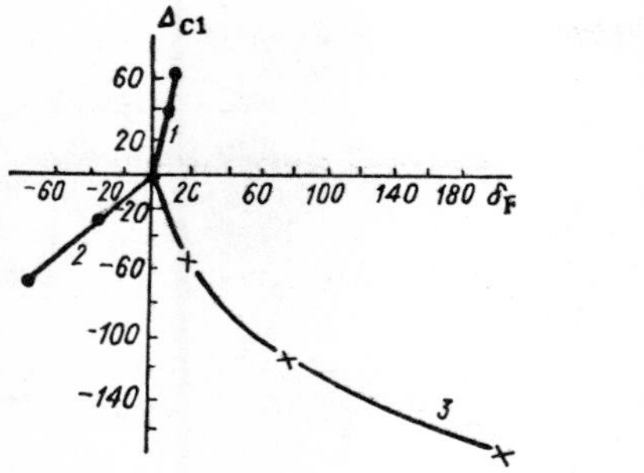

FIGURE 6.15. Correlations of Δ_{Cl} with (δ_F):
Δ_{Cl} is the relative chemical shift of the NQR frequency of ^{35}Cl $\left[\Delta=\left(1-\frac{\nu_X}{\nu_{CCl_4}}\right)\cdot 10^3\right]$;
δ_F is the NMR chemical shift of ^{19}F with respect to CF_4 in ppm.
1) $Hal_nC(NO_2)_{4-n}$; 2) $F_{4-n}CCl_n$;
3) $H_{4-n}CHal_n$.

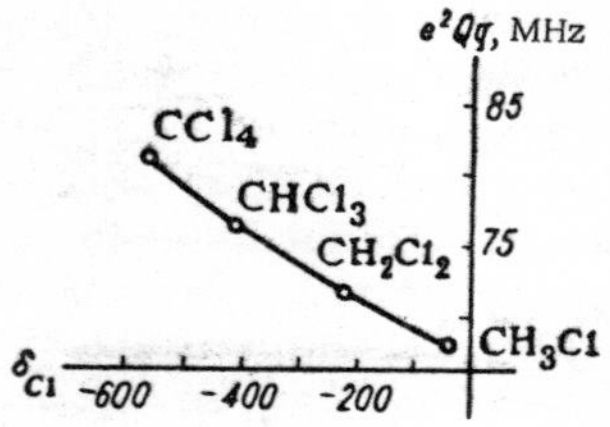

FIGURE 6.16. NMR chemical shifts of ^{35}Cl as a function of e^2Qq in chloromethanes [58]

The NQR frequencies of the halo-derivatives of methane depend linearly on the number of halogen atoms (see Chapter 4) [57]. Therefore, the existence of a linear relationship between δ_F and Δ_{Cl} indicates that the interactions result in a change in the NQR frequencies, and the NMR chemical shifts of ^{19}F are of the same type.

Since Lucken [23] suggested a mechanism of $p-\sigma$ conjugation to explain the range of NQR frequencies in the series Cl_nCF_{4-n}, it is possible that the same effects play an equally important role in the change of δ_F in the series Cl_nCF_{4-n}.

For halonitromethanes the dependence of δ_F on Δ_{Cl} can be written in the form $\delta_F = 0.16\,\Delta_{Cl}$.

A direct comparison of the NMR and NQR data was made by Saito [58], who investigated the NMR of ^{35}Cl and ^{37}Cl in the chloromethane series. Since the paramagnetic contribution to the chemical shift and the quadrupole coupling constant is determined by the same electron distribution, we must observe a linear relationship between these quantities for compounds possessing similar electron excitation energies ΔE [58]:

$$\delta = \delta_{dia} + K < r^{-3} >_{av} \Delta E^{-1}$$
$$e^2Qq = e^2Q\,\frac{l+1}{2l-1}\cdot\frac{2l}{2l+3} < r^{-3} >_{av} \qquad (6.30)$$
$$\delta = \delta_{dia} + K'\Delta E^{-1}\cdot e^2Qq.$$

Such a relationship was found for the chloromethane series (Figure 6.16). The electron excitation energies for CCl_4, $CHCl_3$ and CH_2Cl_2, evaluated from the electron absorption spectra [59], were found to be close to 5.25, 5.42, and 5.77 eV, respectively.

Another type of comparison of the NMR and NQR data was performed by Gilson [60]. He compared the number of uncompensated p electrons

$U_p = \frac{e^2Qq_{mol}}{e^2Qq_{at}}$ in halogen-substituted saturated hydrocarbons with the spin-spin coupling constant $J_{^{13}C-H}$ (Figure 6.17). Such a correlation is possible, since U_p is determined by the s hybridization of the halogen and the ionic character of the C—Hal bond (see equation (5.23a)).

The changes in the hybridization of the carbon atom can be obtained from the coupling constant $J_{^{13}C-H}$, which is determined by the Fermi contact term. If the hybridization of the carbon atom, and therefore its effective orbital electronegativity, are determined from the NMR data, it is possible to calculate by the Whitehead-Jaffé theory [61] the hybridization of the halogen atom and the degree of ionic character of the C—Hal bond. The ionic character of the bond decreases with an increase in the number of halogen atoms in the molecule since χ^{eff} of carbon increases and the difference in electronegativity decreases. The increase in the sp hybridization of the halogen is necessary for improving the overlap with the sp^3 orbital of carbon.

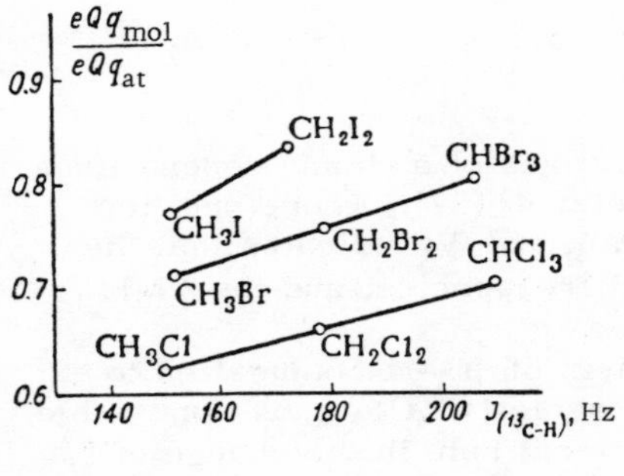

FIGURE 6.17. Relation between the number of uncompensated p electrons in halogen-substituted saturated hydrocarbons and the spin-spin interaction constants $J_{^{13}C-H}$ in these compounds [60]

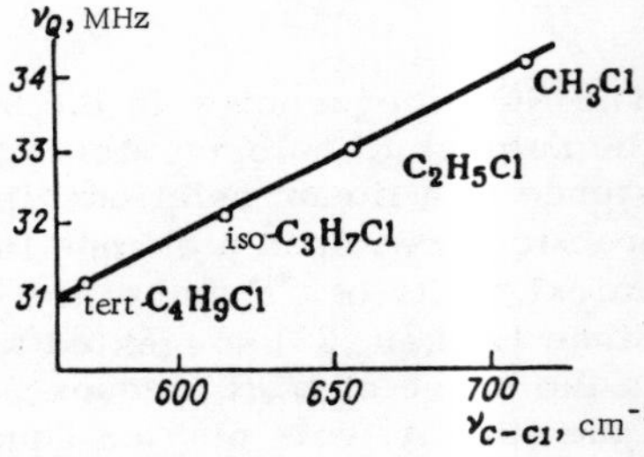

FIGURE 6.18. Variation in the NQR frequencies of ^{35}Cl as a function of the frequencies of valence oscillations of C—Cl in the alkyl chloride series [62]

C o r r e l a t i o n w i t h d a t a o f i n f r a r e d s p e c t r o s c o p y. Gerdil [62] detected a linear relationship between the NQR frequencies and the valence oscillations ν_{C-Hal} of the C—Hal bond in the alkyl halide series R—Hal (where R = CH_3, C_2H_5, iso-C_3H_7, tert-C_4H_9; Hal = Cl, Br, I) (see Figure 6.18). This means that the electric field gradient is proportional to $\sqrt{K}$ (where K is the force constant) if the effect of the mass can be neglected. Since ν_{NQR} and ν_{C-Hal} are linear functions of the Taft induction constants σ^* for the series R—Hal, it was natural that a linear relationship between these same quantities would be obtained. However, Gerdil found that the correlation coefficient of the relationship $\nu_{NQR} = f(\nu_{C-Hal})$ is higher than his correlations of these quantities with σ^*.

The linear relationship $\nu_{NQR} \approx \sqrt{K}$ also holds for simple diatomic molecules such as FCl, Cl_2, BrCl, ICl, ClCN [62], with a correction for the mass effect μ. The correlation coefficient $\nu_{NQR}\sqrt{\mu} \approx \sqrt{K}$ is equal to $r = 0.934$.

A linear relationship was found [63] between the NQR frequencies of ^{35}Cl in chlorosilanes R'R"R'''SiCl and the vibration frequency ν_{Si-H} of the corresponding silanes R'R"R'''SiH:

$$\nu_{^{35}Cl} = (-41.295 + 0.0273\nu_{Si-H}) \pm 0.11 \text{ MHz}; \qquad r = 0.93. \qquad (6.31)$$

Fourteen points were plotted.

A linear relationship was also found between the NQR frequencies and the sums of the spectroscopic (induction) Thompson constants E [21, 63, 64]. For chlorosilanes three equations for the series $RSiCl_3$, $RR'SiCl_2$, RR'R"SiCl were found [63].

One of the first correlations between the NQR frequencies and the IR frequencies was established by Lucken [30] who found a linear relationship between the NQR frequencies of ^{35}Cl in the series $RPOCl_2$ and the valence frequencies of C = O in the series RCOCl.

Correlation with Mössbauer spectra. Correlations of NQR and NMR data are interesting in that they enable us to compare the electric field gradients on different atoms in the same molecule. Such a comparison was carried out for the organotin compounds R_3SnHal [26], and a linear relationship was found for R_3SnBr:

$$\nu_{^{79}Br} = 164.24 - 19.06\Delta, \qquad (6.32)$$

where Δ is the quadrupole splitting of ^{119}Sn.

The existence of such a relationship is fairly obvious since the atoms are valence bonded, and the changes in the electric field gradients are mainly caused by redistribution of the p electrons in the two atoms.

The nature of the relationships between the NQR frequencies and the δ isomeric shifts is more complex [26].

BIBLIOGRAPHY

1. Hooper, H. and P. Bray. — J. Chem. Phys. 33 (1960), 334.
2. Semin, G. K., T. A. Babushkina, V. I. Robas, G. Ya. Zueva, M. A. Kadina, and V. I. Svergun.— In: "Radiospektroskopicheskie i kvantovokhimicheskie metody v strukturnykh issledovaniyakh," p. 225. "Nauka," 1967.
3. Ogawa, S. — J. Phys. Soc. Japan 13 (1958), 618.
4. Hamlen, R. and W. Koski. — J. Chem. Phys. 25 (1956), 360.
5. Miyagawa, I. — J. Chem. Soc. Japan, Pure Chem. Sect. 75 (1954), 1061.
6. Hamano, H. — Bull. Chem. Soc. Japan 31 (1958), 832.
7. Hamano, H. — Bull. Chem. Soc. Japan 37 (1964), L583.
8. Mueller, A. — Proc. R. Soc. A120 (1928), 437.
9. Meal, H. — J. Am. Chem. Soc. 74 (1952), 6121.
10. Bray, P. J. — J. Chem. Phys. 22 (1954), 1787.
11. Bray, P. J. — J. Chem. Phys. 22 (1954), 2023.
12. Bray, P. J. and R. G. Barnes. — J. Chem. Phys. 27 (1957), 551.
13. O'Konski, C. T. Nuclear Quadrupole Resonance Spectroscopy. — In: "Determination of organic structure by physical methods," Vol. 2, Part 11. New York, 1962.
14. Tsvetkov, E. N., G. K. Semin, D. I. Lobanov, and M. I. Kabatchnik. — Tetrahedron Lett., p. 2521, 1967.

15. Hooper, H. and P. J. Bray. — J. Chem. Phys. 30 (1959), 957.
16. Bray, P. J., S. Moskowitz, H. Hooper, R. G. Barnes, and S. Segal. — J. Chem. Phys. 28 (1958), 99.
17. Dewar, M. and E. A. C. Lucken. — J. Chem. Soc., p. 2653, 1958.
18. Taft, R. — In: "Space Effect in Organic Chemistry," p. 617. — I. L., 1960. (Russian translation)
19. Tsvetkov, E. N., G. K. Semin, D. I. Lobanov, and M. I. Kabachnik. — Doklady AN SSSR 161 (1965), 1102.
20. Biryukov, I. P. and M. G. Voronkov. — Izv. AN LatvSSR, chem. ser., No. 10 (1966), 39.
21. Biryukov, I. P. and M. G. Voronkov. — Colln. Czech. Chem. Commun. Engl. Edn. 32 (1967), 830.
22. Biryukov, I. P. and M. G. Voronkov. — Izv. AN LatvSSR, chem. ser., No. 1 (1965), 115.
23. Lucken, E. A. C. — J. Chem. Soc., p. 2954, 1959.
24. Biedenkapp, D. and A. Weiss. — J. Chem. Phys. 49 (1969), 3933.
25. Biryukov, I. P., E. Ya. Lukevits, M. G. Voronkov, and I. A. Safin. — Izv. LatvSSR, chem. ser., No. 6 (1967), 754.
26. Bryukhova, E. V., G. K. Semin, V. I. Goldanskii, and V. V. Chrapov. — Chem. Communs., p. 491, 1968.
27. Tsvetkov, E. N., G. K. Semin, D. I. Lobanov, and M. I. Kabatchnik. — Tetrahedron Lett., p. 2933, 1967.
28. Tsvetkov, E. N., G. K. Semin, D. I. Lobanov, and M. I. Kabachnik. — Teor. Eksperim. Khim. 4 (1968), 452.
29. Neimysheva, A. A., G. K. Semina, T. A. Babushkina, and I. L. Knunyants. — Doklady AN SSSR 173 (1967), 585.
30. Lucken, E. A. C. and M. A. Whitehead. — J. Chem. Soc., p. 2459, 1961.
31. Neimysheva, A. A., G. K. Semin, V. A. Pal'm, N. A. Loshadkin, and I. L. Knunyants. — Zh. Obshch. Khim. 37 (1967), 2255.
32. Tsvetkov, E. N., G. K. Semin, T. A. Babushkina, D. I. Lobanov, and M. I. Kabachnik. — Izv. AN SSSR, chem. ser., p. 2375, 1967.
33. Pal'm, V. A. — Uspekhi Khimii 30 (1961), 1069.
34. Kabachnik, M. I. — Doklady AN SSSR 110 (1956), 393.
35. Semin, G. K. and E. V. Bruchova. — Chem. Communs., p. 605, 1968.
36. Zhdanov, Yu. A. and V. I. Minkin. Correlation Analysis in Organic Chemistry. — Rostov State Univ., 1966. (Russian)
37. Semin, G. K. — In: "Radiospektroskopiya tverdogo tela," p. 205. Atomizdat, 1967.
38. Bryukhova, E. V., T. A. Babushkina, V. I. Svergun, and G. K. Semin. — Uchen. Zap. Oblast. Pedag. Inst. im. N. K. Krupskoi 222, No. 8 (1969), 64.
39. Townes, C. and B. P. Dailey. — J. Chem. Phys. 17 (1949), 782.
40. Dixon, R. and N. Bloembergen. — J. Chem. Phys. 41 (1961), 1720, 1739.
41. Cornil, P. — Dissert. Université de Liege, 1965.
42. Semin, G. K. — Author's Summary of Doctoral Thesis. Inst. Elemento-Organ. Soedinenii AN SSSR, 1970.
43. Semin, G. K., A. A. Neimysheva, and T. A. Babushkina. — Izv. AN SSSR, chem. ser., p. 486, 1970.
44. Nesmeyanov, A. N., G. K. Semin, T. A. Babushkina, K. N. Anisimov, N. E. Kolobova, and Yu. V. Makarov. — Izv. AN SSSR, chem. ser., p. 1953, 1968.
45. Nesmejanov, A. N., G. K. Semin, E. V. Bruchova, T. A. Babushkina, K. N. Anisimov, N. N. Kolobova, and Yu. V. Makarov. — Tetrahedron Lett., p. 3987, 1968.
46. Bryukhova, E. V. — Author's Summary of Candidate Thesis. Inst. Elemento-Organ. Soedinenii AN SSSR, 1969.
47. Biryukov, I. P., V. P. Feshin, and M. G. Voronkov. — Doklady AN SSSR 171 (1966), 645.
48. Svergun, V. I., A. E. Borisov, E. V. Bryukhova, N. V. Novikova, T. A. Babushkina, and G. K. Semin. — Izv. AN SSSR, chem. ser., p. 484, 1970.
49. Iredale, T. — Nature 177 (1956), 36.
50. Colichman, E. L. and S. K. Liu. — J. Am. Chem. Soc. 76 (1954), 913.
51. Hatton, J. and B. Rollin. — Trans. Faraday Soc. 50 (1954), 358.
52. Semin, G. K. and A. A. Fainzil'berg. — Zh. Strukt. Khim. 6 (1965), 213.
53. Caldwell, R. A. and S. Hacobian. — Aust. J. Chem. 21 (1968), 1.
54. Meyer, L. H. and H. S. Gutowsky. — J. Phys. Chem. 57 (1953), 481.

55. Pople, J. A. et al. High Resolution Nuclear Magnetic Resonance. — McGraw-Hill, 1959.
56. Semin, G. K., A. V. Kessenikh, L. V. Okhlobystina, A. A. Fainzil'berg, N. N. Shapet'ko, and L. A. Kurkovskaya. — Teor. Eksperim. Khim. 3 (1967), 233.
57. Kessenikh, A. V., L. V. Okhlobystina, V. M. Khutoretskii, A. A. Fainzil'berg, T. A. Babushkina, and G. K. Semin. — Teor. Eksperim. Khim., Vol. 5, 1969.
58. Saito, Y. — Can. J. Chem. 43 (1965), 2530.
59. Lacher, J. P., L. E. Hummel, E. F. Bohmfalk, and J. D. Park. — J. Am. Chem. Soc. 72 (1950), 5486.
60. Gilson, D. F. R. — J. Chem. Phys. 43 (1965), 312.
61. Whitehead, M. A. and H. H. Jaffé. — Teoret. chim. Acta 1 (1963), 209.
62. Gerdil, R. — Nature 212 (1966), 922.
63. Biryukov, I. P., M. G. Voronkov, and I. A. Safin. — In: "Radiospektroskopiya tverdogo tela," p. 252. Atomizdat, 1967.
64. Voronkov, M. G. and I. P. Biryukov. — Izv. AN LatvSSR, chem. ser., p. 170, 1968.

Chapter 7

INVESTIGATION OF COMPLEXES AND COORDINATION INTERACTIONS BY NQR

1. EFFECT OF COMPLEX FORMATION ON NQR SPECTRA

The formation of donor-acceptor complexes is due to a partial transfer of charge (or a shift of electron density) from the donor molecules D to the acceptor molecules A. Changes in the charge distribution and the accompanying changes in the electron structure of the interacting molecules cause significant changes in the electric field gradient on each atom of the complex.

On the other hand, during complex formation new spatial interactions may appear and the character of the old ones may change, which is naturally accompanied by additional changes in the electric field gradient. Crystal effects also continue to play an important role, and as a result of complex formation, very complex spectra are obtained. It is therefore expedient to start analyzing the influence of the different factors and to elucidate each one separately.

We shall discuss the part played by the charge transfer or the electronic factor.

The wave function of the ground state of the complex AD can be represented in the form [1]:

$$\psi_N = a\psi_0\,(\mathrm{DA}) + b\psi_1\,(\mathrm{D^+A^-}). \tag{7.1}$$

On the basis of this approximation, relations linking the change in the field gradient Δq with the degree of charge transfer b^2 and the character of its distribution c^2 were obtained in [2]. For the acceptor

$$\Delta q_A = \frac{b^2}{1+S^2}\sum_t c_{At}^2\, q_t \tag{7.2}$$

and for the donor

$$\Delta q_D = -\left(1 - a^2 + \frac{b^2S^2}{1+S^2}\right)\sum_r c_{Dr}^2\, q_r, \tag{7.3}$$

where S is the overlap integral of the interacting molecular orbitals (MO), and t is the atomic orbital of the atom which participates in the interaction.

We recall that $q_s = 0$, $q_{pz} = q_{at}$ (z directed along the σ bond), $q_{px} = q_{py} = -\frac{1}{2} q_{at}$, and q_d and q_f can be neglected.

We shall discuss the use of relation (7.2) for different types of acceptors (σ, π, v). For simplicity, we shall restrict our discussion at this stage to halogen-containing molecules. Firstly, relation (7.2) shows that the change in the gradient is proportional to the degree of charge transfer. However, the direction of the shift can also be predicted without calculating the MO. For σ acceptors (CCl_4, $CHCl_3$, etc.), the increase in the σ electron density of the molecule causes an increase in the population of the p_z orbitals of the halogen atom, and therefore a decrease in the electric field gradient tensor. In the simplest case of a diatomic σ acceptor (I_2, ICl, IBr, etc.), equation (7.1) leads to a simple expression for the relative change in the electric field gradient or the relative frequency shift:

$$\frac{\Delta q_{zz}}{q_{zz}} = \frac{\Delta \nu}{\nu} = -\frac{b^2}{2(1+S^2)}. \qquad (7.4)$$

For a π acceptor (chloranil, picryl chloride, etc.), the charge transfer to the π orbital increases the population of the p_x orbital of the halogen atom and, thus, the field gradient on it also increases.

The direction of the frequency shift for σ donors is opposite to that for π donors. Analysis of the shifts is more complicated when n donors (amines, ethers) or v acceptors participate in the complex formation; the shift of electron density, in an n donor molecule, to the atom with the unshared pair of electrons decreases the ionic character of the R—X bond, which increases the electric field gradient on atom X.

The complex formation of v acceptors involves, in most cases, a full rearrangement of the whole electron system of the molecule and the formation of one or several additional vacant orbitals. Naturally, the rearrangement of the geometry of the molecules causes changes in the M—X bond parameters (ionic character, double bonding, etc.). Therefore, with some exceptions, a detailed analysis of the changes in the field gradient is possible only on the basis of a detailed calculation of each specific complex. The whole process can be represented in the form of two stages: rearrangement of the acceptor molecule MX_n and charge transfer to the vacant molecular orbitals formed with corresponding distribution of this charge. The fraction of this charge in the acceptor system can be considered as the degree of charge transfer. It can also be assumed that within a given series of complexes of the same acceptor with different donors the change in the electron distribution caused by rearrangement of the electron system of the acceptor will be roughly the same for all complexes. The change within this series on passing from one complex to another can then be ascribed completely to the charge transfer. The increase in the degree of transfer will contribute to the increase in the electron density in the σ system of the acceptor and, therefore, to the decrease in the field gradient on atoms X. This is the qualitative aspect of the problem. The quantitative connection between the change in the NQR frequency and the charge transfer can be elucidated by detailed calculations only. However, this relationship can be very complex. Sometimes the valence bond method can be used to analyze the NQR spectra of such complexes, in particular when the frequencies of the central atom can also be observed. Since this is a special case, examples of such analyses are given in the following sections.

Thus, the influence of the electronic factor on the electric field gradient, and, therefore, on the direction of the frequency shift, may be different for the different types of complex. The quantitative measure of this influence is mainly determined by the degree of charge transfer.

With regard to steric factors, during complex formation the coordination centers of the interacting molecules may approach to within a distance between the sum of the Van der Waals radii (weak complexes) and the sum of the covalent radii (strong complexes). Since NQR investigations are carried out in the solid state, steric effects will be very important. Both the structure of the interacting molecules and their arrangement in the complex are also of great importance [3, 4]. The steric effects can be divided into three types. Firstly, they can prevent the approach of the donor and acceptor molecules, restricting the possibility of electron transfer. Although the dative properties of the donor may be very high in this case, the degree of charge transfer will not be large, i. e. the steric effects influence the electronic factor. Secondly, the steric interactions can distort the geometrical structure of the molecules: changes in the lengths of the bonds, deformations of the valence angles, etc. This results in changes in the ionic character of the bond, the hybridization of the central atom, etc. Finally, steric interactions can lead to deformation of the electron shells of the atoms through polarization. The last two factors are not associated with the degree of charge transfer.

Since these three factors taken together may differ greatly on passing from one complex to another, even within the same series, the degree of their influence on the magnitude and direction of the electric field gradient is unknown.

Thus, the change in the electric field gradient as a result of complex formation is determined by two basic factors: electronic and steric. These factors influence each other, with one or the other predominating.

It can be assumed that, depending on the strength of the donor-acceptor interaction, the electronic factors will dominate in the NQR spectra of strong complexes, and the geometrical factors in those of weak complexes. These two factors together lead to a very complex pattern of shifts, which must be taken into account on analysis.

An important characteristic of the NQR spectra is their multiplicity. If there are several resonance atoms of the same type in the molecule, their frequencies will be different. This may be due to electronic differences in the atoms (chemical nonequivalence), steric effects and the influence of the crystallographic environment. The crystal splittings lie, as a rule, within 2% of the measured frequency, while the splittings caused by chemical nonequivalence can be very large. For complexes of equal geometrical structure it can be assumed that, within reasonable limits, the symmetry of the electron distribution is preserved. As a result, the splittings in the NQR spectra caused by the chemical nonequivalence of the atoms will be of the same character. Therefore, knowing the electron distribution in the complex, we can predict the character of the splittings

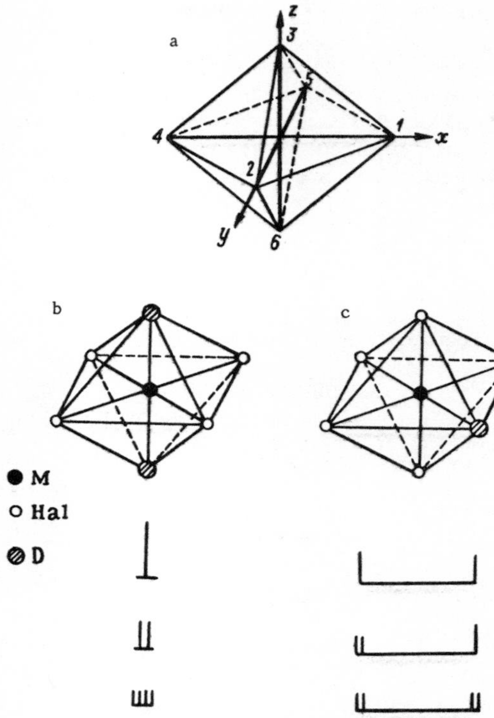

FIGURE 7.1. NQR of ^{35}Cl in the six-coordination complexes of tin tetrachloride: a) diagram of the orientation of the axes in the complexes; b) trans isomer and expected spectra; c) cis isomer and expected spectra.

in the spectrum, and on the basis of this we can draw conclusions on the geometrical and electronic structure of the complex. A tentative analysis of this type was carried out in [5] on the basis of the ligand field theory for cis and trans octahedral complexes of type $MHal_4 \cdot 2D$. Using the molecular orbitals of symmetry O_h, the authors showed that in trans isomers all four halogen atoms have the same σ electron density and are chemically equivalent [5]. Therefore, splittings in the NQR spectra of trans octahedral complexes are possible only through crystal effects. Figure 7.1,b shows the expected shape of the spectrum of such a complex. A similar calculation for the cis isomer reveals that the σ electron density is different for the two positions of the halogen atom (equatorial and axial) [5]. Therefore, large splittings in the NQR spectra of such complexes should be observed due to the chemical nonequivalence of these two positions (Figure 7.1,c). Another interesting fact follows from this calculation, namely, that for the same degree of charge transfer the magnitude of the average shift of the halogens is different for the cis and trans isomers. This fact must be taken into account when comparing the basic properties of the donors in complexes of different structure. In the simplest case, the character of the splitting due to the chemical nonequivalence can be determined from the symmetry properties of the complex. Under symmetry operations, the molecular orbitals of the complex transform according to its point group. Therefore, under symmetry operations, chemically equivalent atoms will pass into one another. It is easy to determine from this the number of nonequivalent positions of the atoms. The influence of steric interactions on the splitting of the NQR frequencies should be analyzed in great detail.

2. COMPLEXES WITH π ACCEPTORS

Complexes with polyhalogen-substituted benzenes [3, 4], chloranil [6, 7] and picryl chloride [8–10] were investigated by NQR, measuring the frequencies of ^{35}Cl.

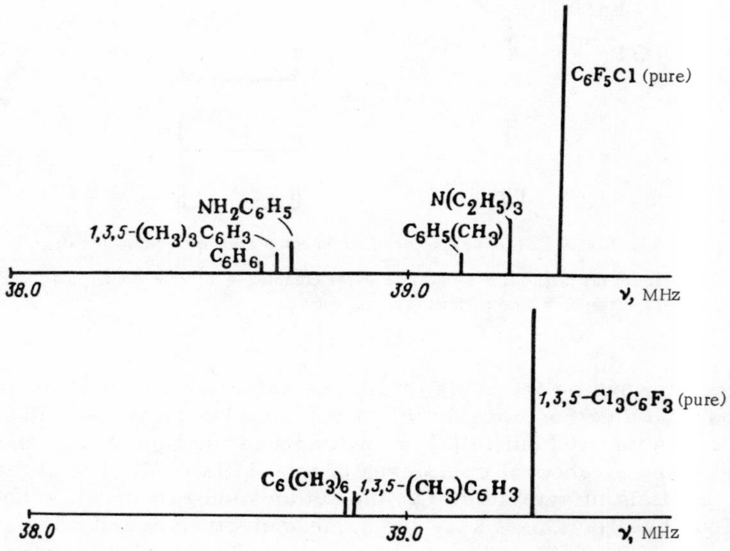

FIGURE 7.2. Diagram of the NQR frequencies of ^{35}Cl for 1 : 1 molecular compounds of type C_6F_5Cl + X and 1,3,5–$F_3C_6Cl_3$ + X

Charge transfer in complexes of this type increases the electron density in the π system of the acceptor, and this causes a shift in the spectrum to higher frequency. The degree of charge transfer is not large and usually equals about 1–6% [11]. Therefore, even when the transferred charge is localized on the halogen atom, the shift of the NQR frequency will not be greater than 5%. Therefore, it is clear that in complexes with these acceptors the steric interactions will have the greatest influence on the NQR spectra.

The influence of the geometrical structure of the molecules (or the geometrical factor) on the shift of the NQR frequencies in complex formation was shown for the first time [4] with π complexes of halogen-substituted benzene derivatives. Hückel's calculation [12] shows that all the halogen-substituted benzenes studied were weak acceptors. According to X-ray analysis [13], the distance between the components of the complex is not smaller than the sum of the Van der Waals radii. Both factors indicate that in the given case there are mainly intermolecular interactions. This point of view is also confirmed by the data in [14, 15].

In five complexes of pentafluorochlorobenzene with different bases [4] the magnitudes of the shifts of the ^{35}Cl frequencies, as seen from Figure 7.2, are independent of the dative properties of the donors (described by their ionization potentials) and are mainly determined by the geometrical structure of the molecules. Thus, the ionization potentials of mesitylene and benzene (8.39 [14] and 9.24 eV) are very different, but the two molecules have similar symmetry, and the shift of the ^{35}Cl frequency in the two complexes is practically the same (0.73 and 0.69 MHz, respectively).

On passing to other complexes, the symmetry of the donor molecule changes, and the frequency shift also undergoes a sharp change. This indicates that in complexes of pentafluorochlorobenzene the degree of charge transfer is negligible because it is limited by the geometry of the molecules which has the greatest influence on the spectrum.

The nature of the negligible low-frequency shifts for these complexes is so far not understood. We can only note that the extremely short spin-lattice relaxation time for these complexes suggests that the substituted benzene rings rotate (with frequencies lower than the observed NQR frequency of ^{35}Cl).

The spectra of complexes of chloranil with mesitylene [6] and hexamethylbenzene [7], as expected, show negligible shifts in the high-frequency range (0.075 and 0.107 MHz, respectively). The degree of charge transfer in these complexes is about 3% [11]. Therefore, the expected shift should be about 0.8 MHz. However, X-ray data show [17] that the distance between the components in the complex is somewhat smaller than the sum of the Van der Waals radii, and the planar structure of the donor and acceptor molecules is perturbed. It is, therefore, difficult at this stage to draw definite conclusions by comparing these spectra.

A more systematic investigation of the influence of the electronic and geometrical factors on the shift of the quadrupole frequencies was carried out in [9, 10] with complexes of picryl chloride with various aromatic compounds as donors.

Picryl chloride is a very convenient molecule for NQR investigations. Two nitro groups ortho to the chlorine atom deviate from the plane of the ring by 80—90° [18], and the chlorine atom is protected, to a certain extent, from the direct steric interaction of the donor molecule. If we select donors of similar symmetry, the electronic factor can be isolated in almost pure form. In complexes with monosubstituted benzenes, the symmetry of the donor molecules is the same, and we can therefore assume that the steric interactions are also roughly the same. The degree of charge transfer in a number of complexes with the same acceptor is linearly dependent on the ionization potentials of the donors. The shifts of the quadrupole frequencies and the ionization potentials were compared in [19] (Figure 7.3). The linear character of the relationship is confirmed by relation (7.4), i.e. by the linear character of the dependence of the changes in the NQR frequency shifts on the value of the charge transfer. In complexes with para derivatives of toluene the symmetry of the donors is also the same. Para substituents strengthen the influence of the steric interactions. Thus, the donor molecule can occupy two different positions in the complex with respect to the dipole of the picryl chloride molecule. The distance between the components in the complex varies as a function of the size of the substituent, and this in turn determines the degree of charge transfer. It is, therefore, obvious that when the volume of the para substituent increases, the frequency shift should decrease. This relationship is fairly approximated in Figure 7.4 by a straight line. The influence of the donor symmetry on the frequency shift is well illustrated by the spectra of the complexes of picryl chloride with polymethyl-substituted benzenes and condensed aromatic systems (Figure 7.5, curves I and II). With an increase in the number of methyl groups in the donor molecule the ionization potential of the donor decreases, but the influence of the symmetry of the donor molecule is so large that the value of the shift in different complexes becomes unpredictable. Such a trend is also observed in the spectra of complexes of picryl chloride with polycyclic aromatic compounds.

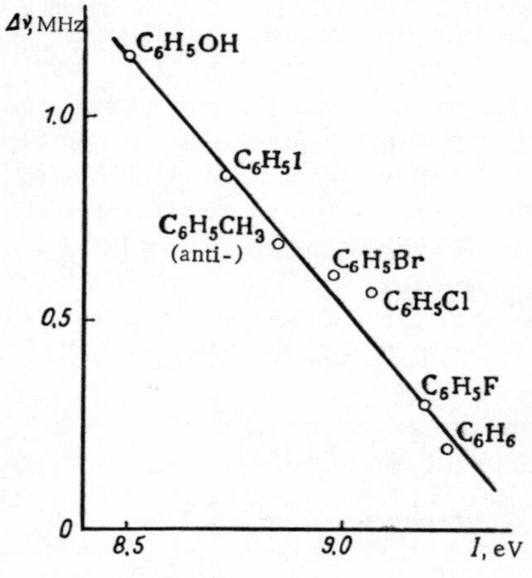

FIGURE 7.3. NQR frequency shifts of ^{35}Cl ($\Delta\nu = \nu_{comp} - \nu_{pure}$) in the complexes of picryl chloride as a function of the ionization potential I of the donor molecules C_6H_5X

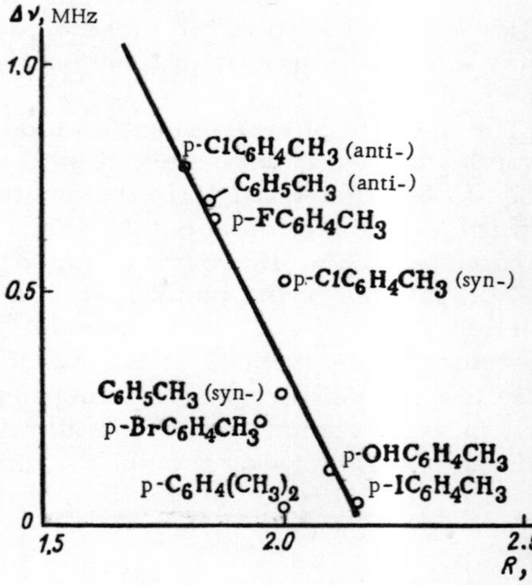

FIGURE 7.4. NQR frequency shifts of ^{35}Cl in complexes of picryl chloride with para substituted toluenes as a function of the size of the substituent R

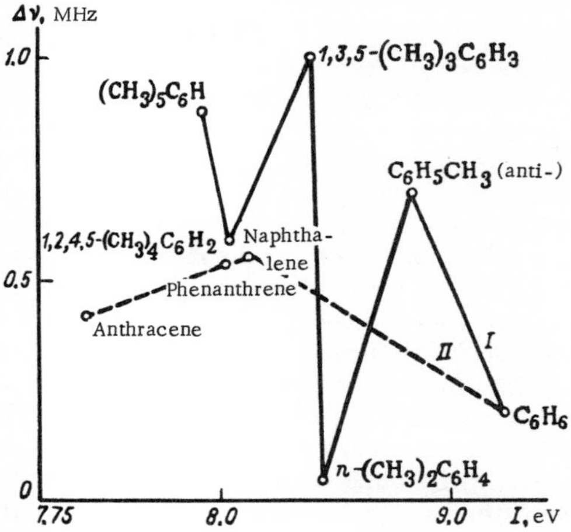

FIGURE 7.5. NQR frequency shifts of ^{35}Cl in complexes of picryl chloride as a function of the ionization potential for donor molecules with arbitrarily chosen geometry

Thus, the investigations show that: 1) steric interaction has the greatest influence on the spectra of complexes with π acceptors; 2) when the selection rules for the symmetry of the components of the complex are preserved, it is possible to obtain information on the electronic structure of the complexes; 3) the correlation between the shift and the ionization potential of the donor suggests that there is a linear relation between the shift and the magnitude of the charge transfer.

3. COMPLEXES WITH σ ACCEPTORS

As already mentioned (see Section 1), charge transfer to the antibonding σ orbital of the acceptor causes a shift to lower frequencies. The magnitude of this shift is determined by the distribution of the transferred charge. Since in most complexes of this type the degree of charge transfer is small, the frequency shifts must depend considerably on the geometrical factor.

The complexes of bromine with different π and n donors have been investigated most extensively [20—23]. The observed low-frequency shifts are usually small, about 0.01—4%. The NQR spectra of ^{81}Br in the complexes $Br_2 \cdot C_6H_6$ and $Br_2 \cdot O(C_2H_5)_2$, where very small high-frequency shifts were detected, are exceptions. The values of the shifts are determined by the dative properties of the donors. Here, just as in the case of complexes with π acceptors, the steric interactions have the greatest influence on the spectrum.

The NQR frequency of ^{81}Br was also observed for the aromatic system in the complex $Br_2 \cdot C_6H_5Br$. A low-frequency shift (equal to about 4%) was obtained as expected. In the given case, complex formation does not proceed through the π system of the ring, but through the unshared pair of $4p_x$ electrons of the bromine atom [24]. Therefore, the value of the shift directly determines the decrease in the electron density on the p_x orbital of the bromine atom caused by the charge transfer. From the relative change in the field gradient $\Delta q_{zz}/q_{zz(\mathrm{obs})}$ the degree of charge transfer was found to be 0.08e [23]. However, it would have been more correct to use q_{at} instead of $q_{zz(\mathrm{obs})}$, since the population of the $4p_x$ orbital is affected neither by the degree of sp hybridization nor by the ionic character of the C—Br bond, on which $q_{zz(\mathrm{obs})}$ depends. Using q_{at} leads to a value of ~0.05e for the charge transfer [25, 48], whereas the degree of charge transfer obtained on the basis of the shifts for the acceptor Br_2 is equal to 0.02e. This divergence of results is due to the participation of the vacant d orbitals of the bromine atom of the molecule Br_2 in the complex formation, as was assumed for the complex of carbon tetrabromide with p-xylene [20, 26]. The participation of the d orbital leads to a decrease in the p_z electron density, and this causes the high-frequency shift in the quadrupole spectrum of the bromine atom.

A curious feature of the spectra which is characteristic for all bromine complexes was also noted [23], namely, the frequencies of the two bromine atoms of the acceptor molecule coincide without any signs of band broadening. This must mean that their chemical and crystallographic environments are equivalent. X-ray investigations [27—30] showed that the complexes of bromine with benzene, acetone, and dioxane have a chain structure. It may be assumed that a chain structure also exists in all the other cases, or splittings would have been observed.

The complexes of I_2, ICl, and IBr with some nitrogen-containing donors are of considerable interest [31, 32]. Unfortunately, only the range 3000—3100 MHz, within which lie the quadrupole interaction constants of ^{127}I for nine complexes, was studied in [31]. These values are much larger than the corresponding constants of the initial compounds: I_2, ICl, IBr. Instead of the expected decrease in the quadrupole interaction constants we observe an increase. The assumption of reverse electron transfer from the IHal molecule to the donor [31] does not agree with the results obtained by other methods. It seems more correct, in accordance with [33], to assume that the iodine atom is sp^3d hybridized. We shall discuss this case in more detail. The five sp^3d hybrid functions of the iodine atom have the form [33]:

$$\begin{aligned}
\sigma_1 &= \frac{1}{\sqrt{6}} s - \frac{1}{\sqrt{6}} d_{z^2} + \frac{2}{\sqrt{6}} p_x \\
\sigma_2 &= \frac{1}{\sqrt{6}} \left(s - d_{z^2} - p_x + \sqrt{3}\, p_y\right) \\
\sigma_3 &= \frac{1}{\sqrt{6}} \left(s - d_{z^2} - p_x - \sqrt{3}\, p_y\right) \\
\sigma_4 &= \frac{1}{2} \left(s + d_{z^2} + \sqrt{2}\, p_z\right) \\
\sigma_5 &= \frac{1}{2} \left(s + d_{z^2} - \sqrt{2}\, p_z\right).
\end{aligned} \tag{7.5}$$

According to the linear structure of the complex Hal—I...D [34], the orbitals σ_1, σ_2, σ_3 are occupied by unshared pairs of electrons, σ_4 forms the covalent I—Hal bond, and σ_5 is the acceptor orbital. The ionic character of the I—Hal bond can be neglected. For an approximate evaluation of the degree of charge transfer b^2 we shall use only the population of the p orbitals. The value of b^2 obtained with these assumptions is $\sim 0.4\,e$ [25, 48]. This is very close to the values obtained from the IR spectra ($0.5\,e$) [35] and the dipole moments [36]. Thus, if the molecule IHal in weak complexes with aromatic compounds is a σ acceptor, in strong complexes with amines it is transformed into a v acceptor with the formation of an additional vacant orbital.

Investigations of the complexes of CCl_4 and CBr_4 with different n and π donors [6, 20, 26, 37, 38] did not lead to definite conclusions concerning the charge transfer. The spectra of both the acceptors themselves and the complexes are very complex, and the disjointed character of the investigations made comparison impossible. Moreover, the results obtained by different authors are contradictory [37, 38].

The complexes $AsI_3 \cdot 3S_8$, $SbI_3 \cdot 3S_8$, $SnI_4 \cdot 2S_8$ and $SnI_4 \cdot 4S_8$ [39—42] can also be considered as complexes with σ acceptors, since complex formation does not occur through the central atom of the metal but through the iodine atoms [43—45]. The large changes in the quadrupole interaction constants of ^{127}I, ^{121}Sb, ^{123}Sb and the frequencies of ^{75}As in the complexes $AsI_3 \cdot 3S_8$ and $SbI_3 \cdot 3S_8$ are not due to charge transfer, but to the different coordination numbers of arsenic (or antimony) in the individual compound and the complex. Thus, the crystal of pure AsI_3 has a layered structure with six iodine atoms around each arsenic atom [46]. However, the AsI_3 molecule has a pyramidal structure in the complex, and under such a rearrangement the changes in the electric field gradient will be considerable. Changes in the quadrupole interaction constants of ^{127}I in the other complexes are negligible, and their interpretation is difficult.

Thus, on the basis of the investigations on the NQR spectra of complexes with σ acceptors, the following can be stated: 1) the geometrical factor has the greatest influence on the form of the spectra; 2) there is a trend toward low-frequency shifts due to the charge transfer.

The lack of systematic research on compounds and their symmetry requirements prevents us from drawing more specific conclusions at this stage.

4. COORDINATION COMPLEXES (v ACCEPTORS)

Mainly the NQR frequencies of halogens bonded to the central atom in metal halides have been studied. As a rule, their shifts are in the low-frequency range and are interpreted as an increase in the ionic character of the M—Hal bond. In cases where the central atom also possesses a nucleus with a quadrupole moment, it is possible to study in more detail the changes in the electronic structure of the acceptor molecule caused by complex formation. It is convenient to discuss the data according to the number of the group of the central metal atom.

From the halides of the second group, the complexes of mercury chloride and bromide with different n donors of composition 1 : 1 were investigated [47, 48]. The low-frequency NQR shifts of ^{35}Cl of $HgCl_2$ in the complexes are fairly large and equal to 7—15%. The $HgCl_2 \cdot D$ complexes are trigonal bipyramids with the mercury atom at the center and the donor molecule in the equatorial plane [49]. The bipyramids form infinite polymeric chains by joining at the edges. The same chlorine atom is equatorial with respect to one bipyramid and axial with respect to the next. Therefore, the splittings in the NQR spectra of such complexes can be caused by crystal effects only. If we assume that the change in the parameters of the Hg—Hal bond during the change in the hybridization of the mercury atom is the same for all the complexes, then the shift of the quadrupole frequency must be proportional to the degree of charge transfer. Indeed, as with the picryl chloride complexes, the average shift of the NQR frequency of ^{35}Cl in $HgCl_2$ increases with a decrease in the ionization potentials of the donors (Figure 7.6). A strict linear relationship can hardly be expected here since the geometrical factor will play a considerable part, but the general trend is preserved [25, 48].

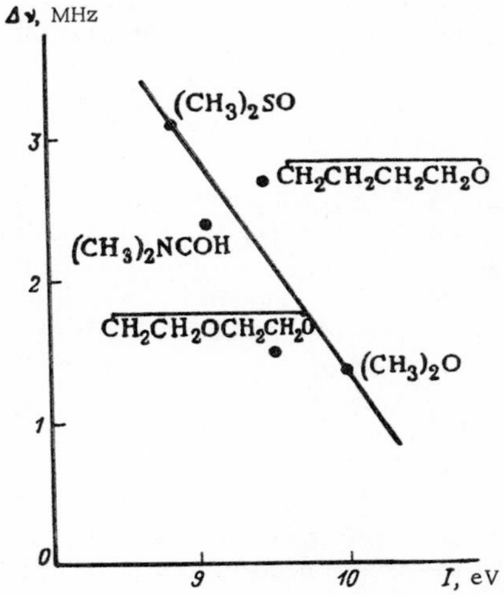

FIGURE 7.6. Shifts of the NQR frequencies of ^{35}Cl in complexes of mercury chloride as a function of the ionization potentials I of the donor molecules

For the $HgBr_2 \cdot D$ complexes both low-frequency and high-frequency shifts are observed [48]. One cannot conclude that the geometrical factor alone is responsible for such a high-frequency shift (~8%) in the complex $HgBr_2 \cdot CH_3OCH_2CH_2OCH_3$. One must assume that the change in character of the Hg—Br double bonding also plays a very important part.

From the elements of the third group the NQR spectra of complexes of aluminum chloride and bromide with some n donors were investigated. The only known spectrum of the complex of $AlCl_3$ with diethyl ether [50] does not enable us to draw any definite conclusions about its structure, since so far a NQR spectrum of ^{35}Cl in pure aluminum chloride has not been obtained. In all the studied spectra of aluminum bromide [51] considerable low-frequency shifts for the bromine atoms are observed. Because the complex of $AlBr_3$ with pyridine is stronger than the complexes of $AlBr_3$ with ethers [51], the degree of charge transfer is larger [52], and therefore the shift to low frequency is also larger. The complexes of aluminum bromide with ethers are of interest from another point of view. The splittings in the NQR spectra of $AlBr_3$ in complexes with ethers show considerable steric interactions between the radicals near the oxygen atom in the ether and the bromine atoms of the $AlBr_3$ molecule. The character of this interaction is such that the quadrupole frequency of a bromine atom nearest to a radical occupying a large volume is greater than that of a bromine atom nearest to a radical occupying a smaller volume.

From the complexes of halides of the third group, the complexes of allium chloride with a number of n donors were investigated in most detail 52a, 69]. In the NQR spectra of these complexes both the frequencies of he ^{35}Cl atoms and the frequencies of the central atom ^{69}Ga were observed. The variations of the NQR frequencies within the series of complexes in- estigated are very considerable: 16—18.5 MHz for ^{35}Cl and 0—25 MHz for ^{9}Ga. Unfortunately, the actual results of the measurements are not given n [52a], and all the results are presented in the form of the graphs: $\nu_{Cl}(\nu_{Ga})$, $\nu_{Cl}(\Delta H)$, $\nu_{Ga}(\Delta H)$, where ΔH is the energy of complex formation in he gaseous phase. In this paper there is no discussion of the results from he point of view of the chemical structure. Although the absence of specific ata restricts the possibility of interpreting the NQR spectra, a number of nteresting conclusions can be drawn from the correlations found. The elationships $\nu_{Cl}(\Delta H)$ and $\nu_{Ga}(\Delta H)$ indicate that both the frequencies of the alogens and the frequency of the central atom shift to the low-frequency ange with an increase in the strength of the donor-acceptor bond, i.e. with n increase in the degree of charge transfer. Moreover, the existence of hese relationships directly confirms the conclusion regarding the prevail- ng influence of the charge transfer on the frequency shift in the NQR pectra of the complexes with v acceptors compared with the steric factors.

It is worth noting the correlation between the frequencies of ^{35}Cl and ^{9}Ga, which can be shown graphically and also by equation [52a]:

$$\nu_{Cl} = 16.326 + 0.826\nu_{Ga} \qquad (7.6)$$

This equation enables us to estimate the character of the distribution of ransferred charge in the acceptor molecule $GaCl_3$. Assuming that the sp^3 ybridization of the gallium atom remains unchanged, the equation can be ewritten as follows:

$$\Delta e_{3Cl} = 2.5\Delta e_{Ga}, \qquad (7.7)$$

where Δe_{3Cl} is the total change in the electron density on all chlorine atoms s a result of the charge transfer, and Δe_{Ga} is the analogous change on he gallium atom. This equation describes the distribution of transferred harge in the acceptor molecule $GaCl_3$, i.e. $^2/_7$ of this charge is localized n the central gallium atom, while $^5/_7$ is distributed among the three chlorine toms. These results are of interest because they belong to one of the irst experimental estimates of the distribution of transferred charge.

In [5, 31, 47, 50, 53] the NQR spectra of six-coordi nation complexes of in chloride with oxygen-containing donors were investigated. As already oted, the character of the splitting can yield information on the geometrical tructure of the complexes. Comparison of the character of the splitting n the spectra of ^{35}Cl in $SnCl_4 \cdot 2D$ (Figure 7.7) with the proposed splitting nodel for cis and trans isomers (Figure 7.1) in [5] led to the establishment f the structure of a number of octahedral complexes. The complexes $SnCl_4 \cdot 2C_2H_5OH$, $SnCl_4 \cdot 2POCl_3$, $SnCl_4 \cdot C_6H_5NO_2$ were the cis isomers. The donors in these complexes were arranged in the following order accord- ng to their basicity, on the basis of the average value of the low-frequency hift: $C_2H_5OH > POCl_3 > C_6H_5NO_2$. The complexes $SnCl_4 \cdot 2O(C_2H_5)_2$, $SnCl_4 \cdot CH_3OCH_2CH_2OCH_3$ and $SnCl_4 \cdot CH_3O(CH_2)_4OCH_3$ were the trans isomers, the last two having a polymeric chain structure in the solid state. Since the complex $SnCl_4 \cdot RO(CH_2)_2OR$ in solution has a cis configuration, it was concluded that the cis form isomerizes into the trans form during crystallization of these complexes from solution. The existing data (IR and X-ray on the structure of these complexes [54–56] confirm the results obtained. For the complex $SiCl_4 \cdot 2POCl_3$ a high-frequency shift of 4% for the frequen of ^{35}Cl of $POCl_3$ in the complex was obtained, which corresponds to a chan of ~14% in the ionic character of the bond.

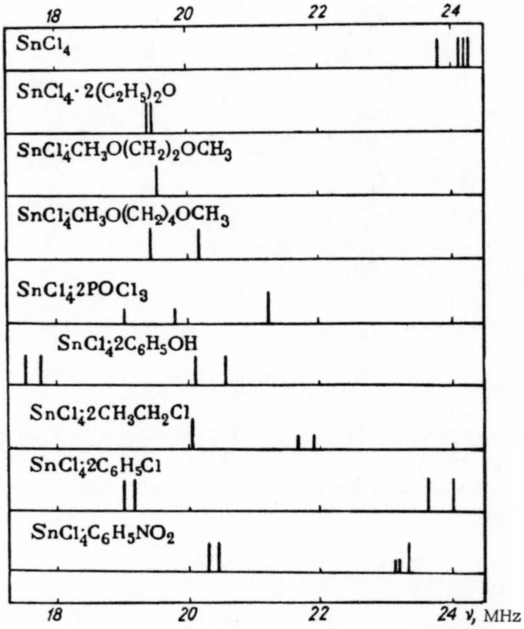

FIGURE 7.7. Diagram of the frequency distribution in the NQR spectra of six-coordination complexes of tin tetrachloride

The octahedral complexes of antimony pentachloride of type $SbCl_5 \cdot D$ ar very similar in structure to the octahedral complexes of lead tetrachlorid Their symmetry belongs to point group C_{4v}. It is easily seen that from th five chlorine atoms four must be equivalent (the equatorial atoms), and on nonequivalent (the axial atom, which is in a trans position to the donor molecule). Thus, ignoring steric factors and crystal effects the NQR spec trum of ^{35}Cl in the complex $SbCl_5 \cdot D$ must consist of two lines with an intensity ratio 4 : 1. In [53, 57] the spectra of two complexes of antimony pentachloride: $SbCl_5 \cdot CH_3CN$ and $SbCl_5 \cdot POCl_3$ were obtained. In accordan with the above arguments, the NQR spectra of ^{35}Cl in the complex of $SbCl_5$ with acetonitrile consist of two lines with an intensity ratio 4 : 1 and splitti 1.5 MHz. The assignment of the frequencies in the given cases are beyon doubt correct. However, in the complex $SbCl_5 \cdot POCl_3$, the steric influence of the donor molecule on the quadrupole frequencies of the equatorial chlorine atoms is so large that the splittings caused by it are greater than the frequency difference caused by the chemical nonequivalence. Therefore, in order to refer the frequencies the authors had to use the temperature dependence of the frequencies $\nu_{Cl^{35}}(T)$ [57]. In this paper the frequencies of the different quadrupole transitions for the isotopes ^{121}Sb and ^{123}Sb, having spins $^5/_2$ and $^7/_2$, respectively, were also measured. This enabled them to calculate the quadrupole coupling constants e^2Qq of ^{121}Sb and the asymmetry parameter η of the antimony atoms. Knowing e^2Qq of ^{121}Sb and the frequency shifts of ^{35}Cl we are able to tentatively estimate the degree of charge transfer in these complexes [25]. Assuming that the ionic character of the Sb—Cl bond does not change with the change in hybridization of the antimony atom from sp^3d to sp^3d^2, we can easily calculate using equation (7.4) with the aid of the hybrid atomic orbitals that the charge on the antimony atom increases as a result of complex formation by $0.08e$ for $SbCl_5 \cdot POCl_3$ and by $0.05e$ for $SbCl_5 \cdot CH_3CN$. Allowing for the part of the transferred charge which is distributed among the chlorine atoms, the total transferred charge is $0.46e$ and $0.41e$, respectively.

Taking into account the approximate nature of these evaluations, we may say that the degree of charge transfer in both the complexes is about 0.4–$0.5e$. The high-frequency shift of ^{35}Cl of $POCl_3$ in the complex indicates a considerable decrease in the ionic character of the P—Cl bond.

About 40 complexes of antimony trichloride and tribromide with various aromatic compounds having compositions 1 : 1 and 2 : 1 have been systematically investigated by NQR [38, 47, 58–66]. As with the complexes of antimony pentachloride, the existence of two isotopes with spins $^5/_2$ and $^7/_2$ (^{121}Sb and ^{123}Sb) enables us to determine e^2Qq_{zz} and η. The following rules governing the variation in the spectra during complex formation were noted [25, 48, 62].

1. e^2Qq_{zz} of antimony increases.
2. e^2Qq_{zz} is usually larger for complexes with composition 2 : 1 than for complexes with composition 1 : 1.
3. For complexes of composition 1 : 1, e^2Qq_{zz} of antimony increases symbatically with the ionization potential of the donors.
4. The average frequency of ^{35}Cl in the $SbCl_3$ complexes is slightly shifted to a lower frequency than that of $SbCl_3$ itself. In complexes of antimony tribromide, such a frequency shift of the halogen is rarely observed.
5. In complexes of $2SbHal_3 \cdot D$ the acceptor molecules are found to be nonequivalent in most cases.

To interpret the variations of e^2Qq_{zz} in these complexes is very difficult since in the original molecule antimony is in a sp^3 hybrid state, and during complex formation the sp^3 hybrid wave functions acquire, in addition, a partial d character. Electron transfer to the newly formed vacant sp^3 hybrid orbital complicates the analysis of the factors responsible for the change in e^2Qq_{zz} of antimony. Ignoring the d hybridization, Grechishkin and Kyuntsel' [65] proposed that a decrease in e^2Qq_{zz} is due to the acceptance of an electron on the p_x and p_y orbitals of antimony, and that an increase is due to the acceptance of an electron on p_z. However, such an interpretation does not provide consistent results since the d hybridization cannot be neglected [25, 48]. Firstly, the degree of d hybridization can be considerable for the energetically most efficient overlap between the vacant hybrid orbital of the antimony atom and the donor π orbital of the aromatic system. Secondly the sp^3d hybridization leads to a change in the distribution of electron densi between the s, p and d orbitals, which must effect the electric field gradien Direct acceptance of the electron on the p_x , p_y and p_z orbitals is impossibl since three of the sp^3 hybrid orbitals are occupied by the covalent bonds, and the fourth one by the unshared pair of electrons.

The complete change in e^2Qq_{zz} of antimony during complex formation can be considered, apparently, as the result of two basic effects (excluding the steric and crystallographic effects): the change in the sp^3 hybridization to sp^3d with a corresponding change in the population of the p orbital, and the charge transfer to the vacant orbital [25, 48].

IR spectroscopy has shown that the complex $SbCl_3 \cdot \frac{1}{2}C_6H_6$ has symmetry C_{2v} [67]. Taking this into account and also allowing for maximum repulsion between the "ligands" and the unshared pair of electrons on the antimony atom, the sp^3d_{xy} hybridization is most acceptable [68]. The increase in the degree of d hybridization leaves the p_z character of the unshared pair of electrons almost unaltered, and substantially decreases the population of the p_x and p_y orbitals. Therefore, the change in hybridization leads to an increase in e^2Qq_{zz}. The charge transfer to the vacant hybrid orbital corresponds to the increase in the electron density on the p_x and p_y orbitals and leads to a decrease in e^2Qq_{zz}. This point of view provides a more consiste explanation of the available experimental data. It can be assumed that for all complexes of the same acceptor, for instance $SbCl_3$, the degree of d hybridization is roughly the same and leads to the same increase in the value of e^2Qq_{zz}. Then, when the degree of charge transfer in the series of these complexes increases, e^2Qq_{zz} will decrease. Such a trend is observed very clearly in the $SbCl_3 \cdot D$ complexes: when the ionization potential decreases the degree of charge transfer increases and e^2Qq_{zz} decreases. In the $2SbHal_3 \cdot D$ complex the degree of charge transfer on one acceptor molecule will be smaller than in the $SbHal_3 \cdot D$ complex, and therefore e^2Qq_{zz} will be larger. An increase in the electric field gradient for most complexes and the negligible low-frequency shifts of the halogens indicate the small degree of charge transfer in these complexes. The steric factors will then play a very important part. In spite of the large amount of experimental data, a more detailed analysis of the influence of the electronic and steric factors for these complexes seems difficult at this stage. However, we must note one important characteristic of the $SbBr_3 \cdot D$ complexes The NQR spectrum of bromine is a high-frequency doublet with negligible splitting and a low-frequency line at some distance from it. The characteristic "rocking-form" structure of four-coordination complexes of trivalent antimony [67] leads us to assume that the high-frequency doublet belongs to the two bromine atoms lying closer to the donor molecule than the third one. Thus, as in the case of $AlBr_3 \cdot D$ complexes, the steric interactions lead to a high-frequency shift of the bromine frequencies. This is not observed for chlorine atoms since their volume and polarizability are lower than those of bromine atoms. In $2SbBr_3 \cdot D$ complexes such assignments are difficult due to the two nonequivalent positions of the $SbBr_3$ molecules in the complex.

The NQR spectra of ^{35}Cl in the $AsCl_3$ complexes have been studied less thoroughly [47, 70]. As in the complexes of antimony trihalides, the shifts of the NQR frequencies are negligible. From the similarity between the splitting of the ^{35}Cl frequencies and the splittings in the spectra of antimony trichloride complexes it was established that the structure of $2AsCl_3 \cdot D$ complexes is similar to that of the corresponding $2SbCl_3 \cdot D$ complexes [70].

Thus, the following are characteristic of metal halide complexes.

1. The change in the electric field gradient on the central atom is mainly determined by two factors, namely, the change in the hybridization of its valence shell and the degree of charge transfer. The electronic factors are usually very large, while the steric factors are less important.

2. The frequencies of the quadrupole spectra of the halogens shift to the low-frequency range as a result of the increase in the ionic character of the M—Hal bond. This shift is larger for larger degrees of charge transfer. Weak complexes are characterized by the considerable influence of the steric factors.

3. In addition to the steric factors, the symmetry of the complex also determines the character of the splitting, which is due to the chemical non-equivalence of the halogen atoms. The magnitude and character of the splitting can serve as an additional source of information on the electronic and steric effects in the complexes.

Thus, the shift of the NQR frequencies in the complexes depends on the degree of charge transfer. Although there are so far no direct experimental determinations of the form of this dependence, theoretical investigations and a number of indirect experimental proofs indicate that it must be linear. The character of the splittings depends on the distribution of electronic density in the molecule, and on the symmetry of the complex. The steric interactions can screen the effect of the electronic factors, but they can also serve as an additional source of information on the geometrical, spatial structure of the complexes in the crystal. For insoluble complexes, NQR is practically the only method which yields information on their electronic structure.

5. INORGANIC COMPLEXES

The first inorganic complex studied by NQR was K_2HgI_4 [71]. This method was subsequently used to investigate a large number of complexes, mainly of types $R_2[MHal_6]$ [72—83] and $R[MHal_4]$ [78, 84—86]. The electric field gradient on the nuclei of halogen atoms in R_2MHal_6 complexes is mainly determined by the electron distribution in the complex anion, since the NQR frequencies of halogens differ little for salts with identical anions. Using the Townes-Dailey theory (equation (4.23a)) it is possible to estimate the degree of ionic character I of the M—Hal bonds and the average charge on the central metal atom M. We assume that the *sp* hybridization of the halogen is equal to 15% if the electronegativity of the metal to which it is bonded is larger than the electronegativity of the halogen by 0.25. The average charge ρ on the central metal atom M is calculated from:

$$\rho = F - Z(1 - I), \qquad (7.8)$$

where F and Z are the formal charge and coordination number of the centr metal atom, respectively. For $R_2[MHal_6]$ $F = 4$ and $Z = 2$. The calculated values of the ionic character of the M—Hal bonds lie between 30 and 68%. Figure 7.8 shows the relationship between the calculated ionic character of the bond and the difference between the electronegativities (Δx) of the ligands and the central atom. Three subrelationships $I(\Delta x)$ can be clearly seen in the figure corresponding to the differences in the external electror ic configurations of the central atoms.

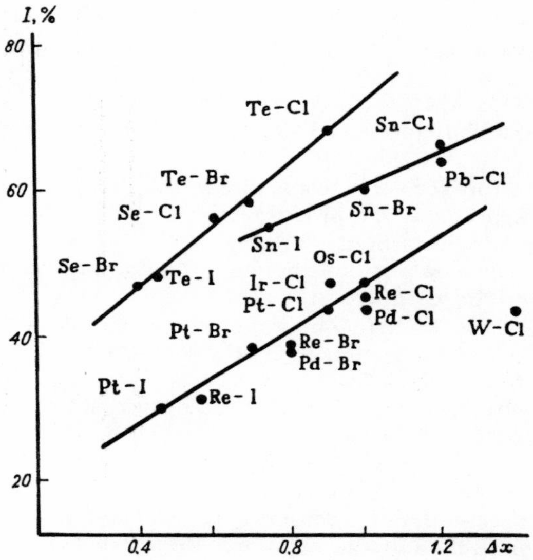

FIGURE 7.8. Ionic character of the M—Hal bond in $R_2[MHal_6]$ complexes as a function of the electronegativities of the halogen atom and the central metal atom [87]

The average charge on the central metal atom also depends on the configuration of the external shell. Thus, in the case of platinum and palladiu which have six equivalent *spd* orbitals for forming six valence bonds, the average charge can be very small: $\rho = 0.64$ for $[PtCl_6]^{2-}$ and 0.58 for $[PdCl_6]^{2-}$. For elements of groups IVA and VIA the average charge is greater than one: $\rho = 1.36$ for $[SeCl_6]^{2-}$, 2.08 for $[TeCl_6]^{2-}$ and 1.96 for $[SnCl_6]^{2-}$ [8

The lowest line on Figure 7.8 also applies to platinum and palladium complexes, and to the paramagnetic complexes of iridium, osmium, rheniu and tungsten, i.e. to atoms which have one or more vacancies in the d_ε orbitals. The existence of vacancies presupposes the formation of a partia π bond between the d_π orbitals of the metal and the p_π orbitals of the haloge In this case, however, one should not use formula (5.25), since this relatio presupposes the formation of a π bond in which the p_x and p_y orbitals of the halogen participate to different extents. In the case of complex ions with symmetry O_h [87]:

$$\pi_x = \pi_y = \frac{\pi}{2}, \tag{7.9}$$

and

$$\begin{aligned} N_x &= 2 - \pi_x \\ N_y &= 2 - \pi_y \\ N_z &= 2 - (1 - s)(1 - I - \pi).^* \end{aligned} \tag{7.10}$$

Then,

$$e^2Qq_{\text{mol}} = e^2Qq_{\text{at}}\left[(1 - s)(1 - I - \pi) - \frac{\pi}{2}\right]. \tag{7.11}$$

The degree of π bonding for paramagnetic complexes can be estimated from EPR data. Thus, for $[IrCl_6]^{2-}$ $\pi = 5.4\%$ [88]. It is reasonable to assume that the value of the π bonding is proportional to the number of vacancies. Since Os^{4+}, Re^{4+} and W^{4+} have two, three and four vacancies, respectively, and the ion Ir^{4+} has one, $\pi = 10.8\%$ for $[OsCl_6]^{2-}$, 16% for $[ReCl_6]^{2-}$ and 22% for $[WCl_6]^{2-}$. The value of the ionic character I for paramagnetic complexes calculated using equation (7.11) depends linearly on the electronegativity difference $x_{\text{Hal}} - x_{\text{M}}$ (see the lowest line in Figure 7.8). The π character of the M—Hal bond in paramagnetic complexes is confirmed by the positive temperature coefficient of the NQR frequency of the halogen. This question is discussed in detail in Section 3 of Chapter 3.

Cyanide complexes of the type $R_n[M(CN)_m]$ were investigated by ^{14}N NQR [89—91]. It was shown that for M = Zn, Cd, Hg, Cu the quadrupole coupling constant of ^{14}N is about 4.0—4.2 MHz, while for M = Pt, Co it is about 3.5—3.7 MHz. The observed difference is due to the fact that the first group of metals has a full external d shell, whereas platinum and cobalt have vacancies in the d shell, i.e. in this case $d_\pi - p_\pi$ metal-ligand acceptor properties are possible. If the d shell of the metal accepts p_π electrons from the ligand, then the number of uncompensated p electrons of nitrogen in the $-C \equiv N$ bond and, therefore, the electric field gradient will both be reduced. The probability of finding $d_\pi - p_\pi$ metal-ligand acceptor properties in complexes of metals with filled d shells is very low.

The cations in octahedral $R_2[MHal_6]$ and tetragonal $R[AuCl_4]$ complexes exert different influences.

The NQR frequency for nuclei with spin $^3/_2$ can be written in the form [87]:

$$\nu = \frac{1}{2} e^2Qq_{zz} = \frac{1}{2} e^2Q\,(q_{\text{el}} + q_{\text{a.ion}}), \tag{7.12}$$

where eq_{el} is the electric field gradient created by all the charges (except the halogen atom considered) inside the complex ion, and $eq_{\text{a.ion}}$ is the electric field gradient created by the adjacent ions of the crystal lattice.

The gradient $eq_{\text{el}} > 0$, since the halogen atom has a positive charge along the z axis (deficit on the p_z orbital). The gradient $eq_{\text{a.ion}} < 0$ and is proportional to the square of the distance M—Hal and inversely proportional to the fifth power of the distance between adjacent ions. Therefore, the total

* It is assumed that $I + \sigma + \pi = 1$, where σ is the degree of covalence of the M—Hal bond [87].

electric field gradient decreases with an increase in the size of the cation (and therefore of the lattice parameters). This agrees qualitatively with the NQR data [75].

However, the calculations made for $K_2[PtCl_6]$ and $Cs_2[PtCl_6]$ show that the frequency difference between these complexes, $\Delta\nu = 0.79$ MHz, is not due to the difference between the electric field gradients created by the adjacent ions ($\Delta\nu_{calc} = 0.01$ MHz). The direct electrostatic effect of the changes of the other ions is negligible due to the highly symmetrical structure of these crystals. The field gradient is mainly determined by the electronic structure of the complex ion which is polarized by the surrounding ions [75].

A different pattern is observed for chloroaurates. These complexes $R[AuCl_4]$ have low symmetries. In the planar ion $[AuCl_4]^-$ the distance Au—Cl may vary depending on which cation is present. The NQR spectra of ^{35}Cl are multiplets. However, as the calculations have shown [92], the multiplicity is not due to the difference in the Au—Cl distances, but is due to the contributions of the adjacent ions in the electric field gradient of the given halogen atom. The value of the mean NQR frequency is naturally determined by the electron distribution inside the complex atom, but the splittings in the spectrum are due to the influence of the charges of the adjacent ions.

6. INTERMOLECULAR COORDINATION INTERACTIONS IN NQR SPECTRA

We are able to recognize and study the inter- and intramolecular associative bonds if we know the final value of the crystal splittings of the NQR frequencies in molecular crystals. The magnitude of the splitting of the NQR frequencies of chemically equivalent atoms in those cases when they participate differently in the complex formation is greater than the final possible value for molecular crystals (maximum ~2%) [93]. This is natural, since in molecular crystals the differences are under the influence of the Van der Waals forces and do not affect the state of the electrons in the atoms studied.

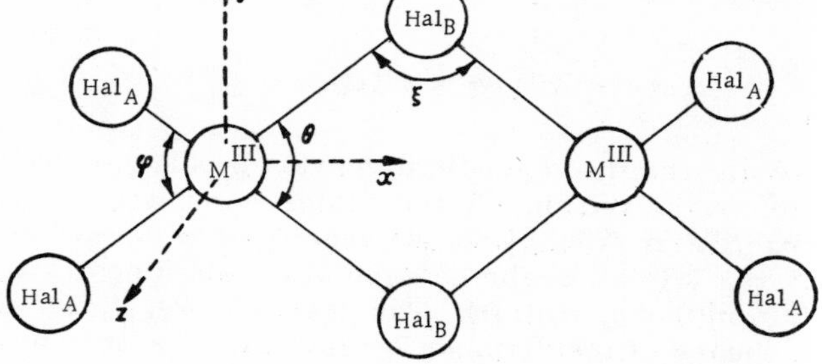

FIGURE 7.9. Diagram of the dimer molecules $M^{III}Hal_3$

The dimers of group III elements are typical examples.

The halides of group III elements ($MHal_3$) are known to form dimers of the type M_2Hal_6 [94—97] in which two halogen atoms belonging to two different monomeric molecules are located between the two central atoms of the metal and are coordinated with it to the same degree (Figure 7.9). This was shown for $AlBr_3$ by X-ray analysis [98]. Accordingly, the NQR spectrum of halogens in these compounds consists of three components: two of them are at a distance of ~1% from the measured frequency, and the third one, belonging to the "bridge" halogen atoms, is 15—20% lower (Table 7.1).*

TABLE 7.1. NQR spectral parameters of the halides of group III elements

Compound	NQR frequency ν, MHz			Asymmetry parameter η, %		
	Hal_A	Hal_B	M	Hal_A	Hal_B	M
$GaCl_3$	19.084 20.225	14.667	29.080	8.9 3.4	47.3	86.7
$GaBr_3$	140.660 140.700 140.790 140.845	100.795 100.840	26.494			
$GaI_{3(1/2-3/2)}$	173.650 174.589	133.687	21.441	0.9 2.8	23.7	
$AlBr_3$	113.790 115.450	97.945	e^2Qq=13.57	. . .	. . .	74.1
$AlI_{3(1/2-3/2)}$	129.327 129.763	111.017	. . .	0	18.1	
$InI_{3(1/2-3/2)}$	173.177 173.633	122.728	e^2Qq=321.08	1.1	29.7	68.77
$InBr_3$	128.110 128.608	104.112				

The electric field gradient of the central atom of the metal in the "bridge" halogen atoms has a large asymmetry parameter. The analysis [97] established that the asymmetry of the electric field gradient on the atoms is closely linked with the corresponding angles between the orbitals (see Figure 7.9):

$$\eta_M = -3\cos\theta$$
$$\eta_{X_B} = +3\cos\xi \qquad (7.13)$$
$$\cot^2\frac{\theta}{2} + \cot^2\frac{\varphi}{2} = 1.$$

The values of the angles obtained by this method and by X-ray analysis are compared in Table 7.2.

Complete agreement between the values of φ was obtained. The differences in the values of θ are due to the fact that NQR data give the angle between the directions of the hybrid orbitals, while X-ray data give the

* The spectrum of $AuCl_3$, whose dimeric structure has been confirmed by X-ray analysis, has a similar shape [46, 98a].

angle between the lines connecting adjacent atoms. The considerable difference between the two values of the angles indicates that the M—Hal bond is "bent." Examination of the angle ξ leads to a similar conclusion, e.g., for $(AlBr_3)_2$ $\xi = 82°$. However, it is impossible to imagine two hybrid orbitals with an angle smaller than 90° if only *sp* hybridization is used. Thus, the angle between the hybrid orbitals of bromine is larger than the angle between the lines connecting the two atoms. Moreover, the sum of the calculated angles $\theta + \xi$ for InI_3 is equal to 188°. If the connecting lines were straight, this sum would have been 180°. Therefore, the bonds in InI_3 must also be slightly bent.

TABLE 7.2. Angles (in degrees) between the bonds in the halides of group III elements

Compound	NQR			X-ray analysis		
	θ	φ	ξ	θ	φ	ξ
$AlBr_3$	104	116	—	98	114	82
InI_3	103	117	85	—	—	—
$GaCl_3$	107	—	80	89 ± 3*	—	—

* Electron diffraction data obtained from the gaseous phase.

The NQR spectra of chlorides of group IVA elements, and also the spectrum of the low-temperature phase of $SnBr_4$ have a very similar shape to the spectra of group III halides, i. e. a high-frequency triplet whose splitting is smaller than 1%, and a singlet 3—5% lower than the triplet. The similarity between these spectra leads us to assume that these structures are dimeric [99]. However, the splitting between the high-frequency triplet and the low-frequency singlet in dimers of group IVA is smaller than in group III. In addition, for the group IVA halides an essential part is played by the relation between the effective radii of the metal and the halogen atoms (the chlorides are dimers, the bromides and iodides are ordinary molecular crystals, and $SnBr_4$ has a structural phase transition from a dimer to an ordinary molecular crystal).

It is interesting to note that the NQR spectrum of ^{35}Cl in $(ICl_3)_2$ is similar, the only difference being that the splitting between the high-frequency doublet and the singlet is 60% (!) (at 24°C the NQR frequencies of ^{35}Cl are 34.918, 33.413, and 13.740 MHz) [100]. The qualitative evaluation of the formal charges in the dimer made by Evans and Lo [100] yielded the following pattern:

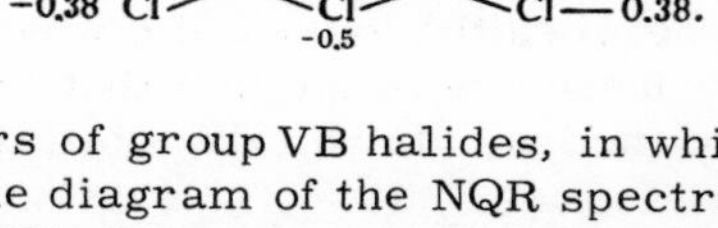

For the known dimers of group VB halides, in which the central atom is a transition element, the diagram of the NQR spectra of the halogens is different, namely, the NQR line corresponding to the coordinated halogen atom is higher than the NQR lines of the uncoordinated atoms [101] (see Table 7.3).

TABLE 7.3. NQR spectra of ^{35}Cl and ^{81}Br for group VB halides at 77°K

Compound	NQR frequencies ν, MHz		Compound	NQR frequencies ν, MHz	
	Hal_A (uncoordinated)	Hal_B (coordinated)		Hal_A (uncoordinated)	Hal_B (coordinated
$NbCl_5$	—	13.28	$TaBr_5$	55.31	91.58
$TaCl_5$	12.30	13.6		55.03	91.43
	doublet	triplet		54.49× 2	91.31
$NbBr_5$	49.43	89.37		54.41	
	49.95	89.83		54.28	
		90.25	$NbOBr_3$	60.59	
				61.05× 2	78.57× 2
				61.57	

The structure of $NbCl_5$ (and of $TaCl_5$ and $NbBr_5$ which are isomorphous to it) was studied by X-ray analysis [102]. There are two crystallographically nonequivalent dimeric molecules in the unit cell, which occupy positions with different multiplicities (2 and 4), the molecules with nonequivalent bridge halogen atoms having the larger multiplicity. Therefore, in the NQR spectrum, the bridge halogen atoms yield three lines of equal intensity.

The change in the distribution pattern of the NQR frequencies of the bridge and terminal halogen atoms is due to the fairly large multiplicity of the Hal—M bond for uncoordinated halogen atoms. This reduces the population of the p_x and p_y orbitals, and also reduces the electric field gradient. The weakening of the bond for uncoordinated atoms means that the populations N_x and N_y and, therefore, also the NQR frequencies all increase.

The NQR spectra of group VA halides also show anomalous behavior which may be due to a coordination of a different type than in the dimers of group III. For these halides two different structures are characteristic: a distorted pyramid ($SbCl_3$, $SbBr_3$, $AsBr_3$, AsI_3 — high-temperature modification) or a layer structure in the form of alternating layers of metal and halogen atoms (AsI_3, SbI_3, BiI_3).

From the first group of compounds the most interesting is $SbCl_3$. In $SbCl_3$ the NQR frequencies of ^{35}Cl have a sufficiently large splitting and different asymmetry parameters (19.304 MHz, 15.3% and 20.912 MHz, 5.7%) [103], and the NQR of the antimony indicates a large asymmetry parameter of the field on the antimony atom. The temperature dependence of the NQR frequencies of ^{35}Cl combined with the data obtained indicate that there is a coordination interaction between the chlorine atom of one molecule of antimony trichloride and the antimony atom of the other molecule (for more detail see Section 3 of Chapter 3).

The structure of the iodides of group VA elements (As, Sb, Bi) is shown on Figure 7.10. Kojima et al. [104] found that the quadrupole coupling constant of ^{127}I in AsI_3 at 83°K is equal to 1328.24 MHz and the asymmetry parameter $\eta = 18.4\%$. This large asymmetry parameter is difficult to explain if we assume that the iodine atom in the compound forms a simple single bond. If we want to ascribe the large value of η to the unequal participation of the p_x and p_y orbitals in the coupling, we will have to assume that there is a large degree of double bonding. Accordingly, Kojima assumed that in the solid state the iodine atom forms two equal bonds with the two nearest equidistant arsenic atoms. The electric field gradient along each of these bonds is equal to half the field gradient eq along the As—I bond in the free molecule. The field gradients along $I—As_1$ and $I—As_2$ were assumed to be axially symmetrical (Figure 7.10,b). Considering the system of xyz coordinates where the x and y axes lie in the plane $As_1—I—As_2$. Then

$$q_{xx} = \frac{q}{4}(1 - 3\cos\theta)$$

$$q_{yy} = \frac{q}{4}(1 + 3\cos\theta) \qquad (7.14)$$

$$q_{zz} = -\frac{q}{2}; \qquad q_{xz} = q_{yz} = q_{xy} = 0,$$

and

$$\eta = \frac{q_{xx} - q_{yy}}{q_{zz}} = 3\cos\theta. \qquad (7.15)$$

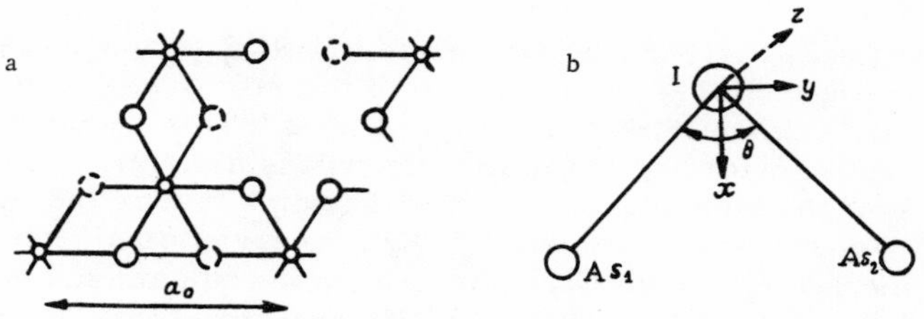

FIGURE 7.10. Arrangement of the atoms in iodide crystals of group V elements [104]: a) diagram of the coordination environment of the iodine atoms; b) orientation of the axes of the field gradient on the nuclei of ^{127}I in AsI_3.

The value of η obtained yields an angle $\theta = 86°30'$. If we assume that part of the field asymmetry is due to double bonding, then we obtain $\theta = 88°$ (this value agrees with that obtained by X-ray analysis) [46]. Since $eq_{zz} = -\frac{eq}{2}$, the observed value of eq_{zz} gives $e^2Qq = 2656$ MHz for the free molecule. However, e^2Qq cannot be larger than $e^2Qq_{at} = 2292$ MHz. The difference observed may be due to the d hybridization of the iodine atom.

For the iodides of group VA elements it was found that the intermolecular interaction is equal in magnitude to the intramolecular interaction. However, weaker intermolecular interactions were found for solid halogens and a number of hetero-organic compounds.

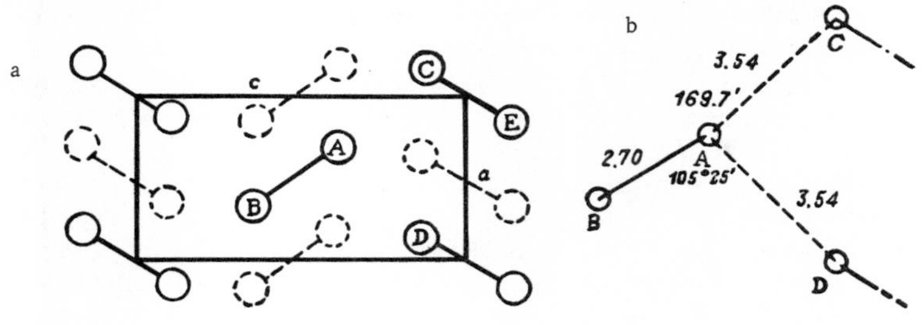

FIGURE 7.11. Structure of I_2 crystals [108]:

a) arrangement of the molecules in the unit cell; b) diagram of the coordination environment of the iodine atom.

Solid halogens crystallize in an orthorhombic cell [105]. The molecules are arranged in planes parallel to the *ac* plane of the unit cell. In the plane *ac* there are two series of molecules making angles 70 ± 5°, 64 ± 12°, and 64 ± 12° for the molecules, Cl_2, Br_2, and I_2, respectively (Figure 7.11). For the I_2 molecule the immediate environment of the iodine atom in position A is shown in Figure 7.11,b.

The distance AB for I_2 in the solid state (2.70 Å) is larger than in the gaseous phase (2.67 Å), i. e. the intramolecular interactions in the solid state are weakened. The distances AC and AD are smaller than twice the Van der Waals radius of the iodine atom ($R = 2.15$ Å), so that we may assume that there is a weak intermolecular interaction between the iodine atom A and the iodine atoms C and D [106]. Bersohn assumes that the energies of the AC and AD bonds are not equal, since the angles BAC and BAD are not equal. It is assumed that there are three field gradients, namely, eq, eKq, and $e\lambda q$ on the nucleus of the iodine atom A along the three directions AB, AC and AD. The value of eq is slightly different from the value of the electric field gradient in the free molecule because of the different distance AB in the solid state. The resulting tensor of the electric field gradient (in the principal axes) will have an axis z turned through a small angle α with respect to the axis of the molecule. The angle between the z axes of the molecules AB and CE (Figure 7.11,a) will differ from the angle between the molecules themselves by 2α. According to Tsukada's measurements [108] it is equal to 63°16' ± 2°. The difference between the X-ray and NQR data lies within the limits of experimental error, and it is difficult to state anything definite about the value of the angle α.

Calculations show that we are able to determine K, λ, and e^2Qq if η and α are known [107].

Robinson et al. [109] and Stevens [110] propose that the coordination bonds are due to the d hybridization of the iodine atom. Stevens showed that the radial distribution of the hybrid orbital sp_zd_{zz} of the iodine atom A has maxima in directions which form angles 0 and 105° with the axis AB of the molecule. The sp_zd_{zz} and p_yd_{yz} orbitals possibly shift under the influence of the Van der Waals forces between the adjacent iodine atoms. This shifts the extreme directions of the electron density closer to the directions of the bonds. If we assume that the *d* hybridization and the Van der Waals forces are responsible for the molecular interaction, this interaction should become weaker on passing from I_2 to Br_2 and Cl_2. Indeed, the asymmetry parameter of ^{127}I in ICl is equal to $\eta = 3.3\%$ [111], and that of ^{35}Cl in Cl_2 is equal to $\eta \leqslant 3\%$ [112].

The formation of the intermolecular coordination and its destruction were also studied from the ^{35}Cl NQR spectra of trichloroacetates [99, 113, 114]. An anomalously large splitting was observed in the NQR frequencies of chemically equivalent chlorine atoms in the group CCl_3 in addition to the overall shift of the NQR frequencies of ^{35}Cl in the trichloromethyl group, which depends on the nature of the metal (Figure 7.12). The spectrum of CCl_3COOAg is characteristic and the largest splitting is observed here due to the coordination interaction between the silver atom and one of the chlorine atoms [114]. This is not contradicted by the dissociation of CCl_3COOAg with the liberation of dichlorocarbene [115]. The high stability of CCl_3COONa is reflected by the smaller splittings in the NQR spectrum of ^{35}Cl.

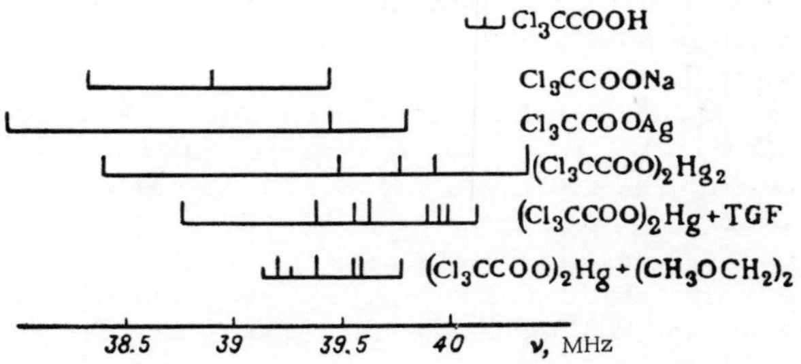

FIGURE 7.12. ^{35}Cl NQR spectra of trichloroacetates and their complexes

The spectrum of $(Cl_3CCOO)_2Hg_2$ (Figure 7.12) has five lines. One of these lines is twice the intensity of the others, which means that there are two independent molecules in the unit cell. From the six chlorine atoms in the unit cell only one participates in the coordination bond with the mercury atom. This is difficult to understand from the point of view of intramolecular association (in which case there would have been two lines shifted to a lower frequency). Comparison of the NQR spectra of $(Cl_3CCOO)_2Hg_2$ and $(Cl_3CCOO)_2Hg + (CH_3OCH_2)_2$ shows that the spectrum of the latter is the spectrum of an ordinary molecular crystal. Thus, we can assume that the formation of the complex with the polar molecule "blocked" the formation of the coordination bond Cl···Hg in $(Cl_3CCOO)_2Hg_2$. The addition of tetrahydrofuran (THF) led to the formation of the complex, but only weakened and did not completely destroy the coordination bond.

Another example of a compound with an intermolecular associative bond Cl···Hg is the "Biginelli complex" (trans-ClCH=CHHgCl) in which the bond Cl···Hg is situated between two of the three molecules in the unit cell.

The ^{35}Cl NQR spectrum of the chlorine atoms near the carbon atom consists of three lines, one of which is strongly shifted to the high-frequency range [99, 114]. The intermolecular character of the coordination C—Cl···Hg has been proved by X-ray analysis of this crystal [116]. When the chlorine atom bonded to the mercury is substituted by a bromine atom, the coordination Hg—Cl(Br)···Hg, which also exists in the crystal, becomes dominant, and the NQR frequency of ^{35}Cl in trans-ClCH=CHHgBr approaches the frequency of the uncoordinated (meaning C—Cl···Hg) chlorine atom. The disappearance of the intermolecular coordination C—Cl···Hg can be clearly followed on the NQR spectra of the complexes of this compound with tetrahydrofuran and dimethylformamide [117].

The compound Cl_3SnCH_2Cl is another example where coordination interaction can be detected in the NQR spectra [118]. The coordination interaction Sn···ClH_2C makes the chlorine atoms of the $SnCl_3$ group sharply asymmetrical, and this is reflected by a sharp change in the ^{35}Cl NQR frequencies for this group.

7. INTRAMOLECULAR COORDINATION INTERACTIONS IN NQR SPECTRA

The question of the possibility of intramolecular interaction between the mercury and the halogen atoms was raised long ago [119, 120], but direct experimental evidence of this phenomenon has not been found.

NQR and X-ray analytical methods were used [114] to obtain the form of the distribution of electron density in α-substituted organomercury compounds.

The characteristic NQR spectra of ^{35}Cl and ^{81}Br of a number of compounds, namely, $(Cl_3C)_2Hg$, Cl_3CHgCl, and Cl_3CHgBr are shown in Figure 7.13. For almost all these compounds the splittings in the NQR spectra are greater than or close to the extreme values possible for molecular crystals [93]. This is due to the coordination bond Hg···Cl—C.

The crystals Cl_3CHgBr and Cl_3CHgCl possess rhombic syngony. The mean distances in the molecule are: Hg—Br = 2.39 ± 0.03 Å, Hg···Cl_1 = = 3.15 ± 0.02 Å, Hg···$Cl_{2,3}$ = 3.21 ± 0.02 Å. The NQR spectrum of ^{35}Cl in the Cl_3CHgBr crystals (for the C—Cl bond) has two lines, ν_1 = 37.580 and $\nu_{1'}$ = 37.898 MHz, which belong to the chlorine atoms of the CCl_2 group, and two lines of half the intensity, ν_2 = 38.794 and $\nu_{2'}$ = 39.032 MHz, which are quite broad, are shifted to the high-frequency range, and belong to the chlorine atoms in the plane of symmetry. The NQR spectrum of ^{81}Br in this compound contains two lines, ν_1 = 122.420 and ν_2 = 130.825 MHz (splitting ~ 6%). The large shift of the NQR lines of the chlorine atoms lying in the plane of symmetry is due to the intramolecular coordination of these chlorine atoms with the mercury atom. X-ray analysis shows that this corresponds to the different distances between the X_3C group halogens and the mercury atom. The considerable splitting of the NQR lines of ^{81}Br corresponds to the intermolecular coordination of the bromine atom of one of the two molecules in the cell with the mercury atom of the other one and, thus, corresponds to the very different environments of the two mercury atoms.

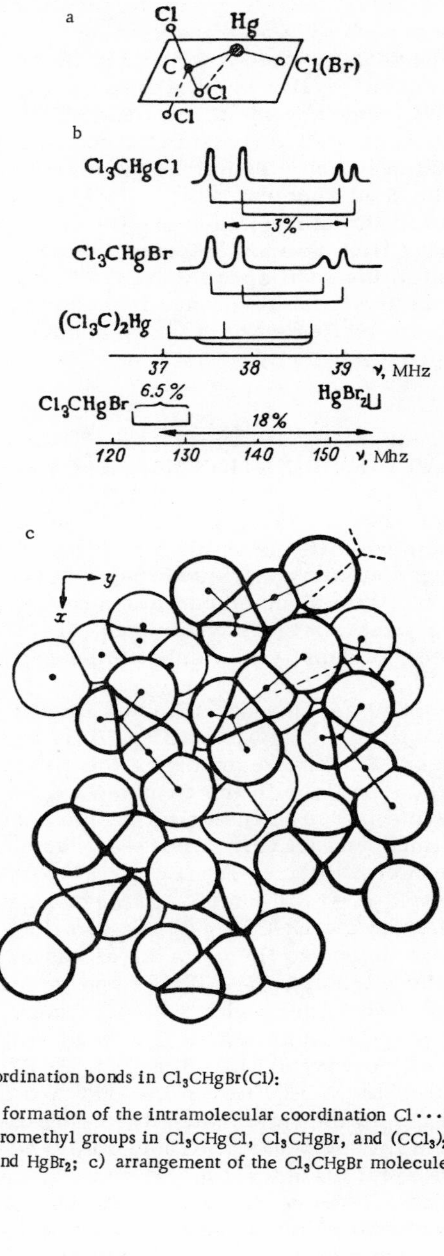

FIGURE 7.13. Coordination bonds in $Cl_3CHgBr(Cl)$:

a) diagram of the formation of the intramolecular coordination Cl $\cdots$ Hg; b) NQR spectra of ^{35}Cl in the chloromethyl groups in Cl_3CHgCl, Cl_3CHgBr, and $(CCl_3)_2Hg$, and the NQR of ^{81}Br in Cl_3CHgBr and $HgBr_2$; c) arrangement of the Cl_3CHgBr molecules in the crystal.

The ^{35}Cl NQR spectrum of the C—Cl bond in Cl_3CHgCl is analogous to the ^{35}Cl spectrum of the Cl_3CHgBr crystal: two lines, corresponding to the four chlorine atoms of the two Cl_2C groups, have frequencies $\nu_1 = 37.604$ and $\nu_{1'} = 37.968$ MHz, while the other two chlorine atoms interact by coordination with the mercury atoms and have much higher frequencies, namely, $\nu_2 = 39.082$ and $\nu_{2'} = 38.172$ MHz.

A similar shift of one of the three lines to the high-frequency range is observed in the spectrum of $(Cl_3C)_2Hg$. This can likewise be explained by the coordination bond Cl···Hg. The crystal belongs to the monoclinic syngony with four molecules per unit cell.

This explanation of the NQR spectra was based on the results obtained by comparing the NQR structure and frequencies for Cl_3CHgBr. It is important to mention that, according to X-ray data, the distances between the bromine atoms in the C—Br bond and the mercury atom in the $ClBr_2CHgBr$ molecule are not equal: Hg···Br_1 = 3.35 Å and Hg···Br_2 = 3.19 Å. The latter distance is much shorter than the sum of the Van der Waals radii. Obviously, this is also associated with the intramolecular coordination of one of the bromine atoms of the $ClBr_2C$ group with the mercury atom.

This specific interaction of the ortho atoms of the halogen with the electron shell of the mercury atom is a reasonable explanation for the NQR spectra of ^{35}Cl [121] and the peculiar chemical properties [122] of pentachlorophenyl mercury derivatives.

The NQR spectra of a number of derivatives of 2,6-dichlorophenol were studied to elucidate the capacity of organometallic compounds containing mercury, tin and lead to undergo intramolecular coordination with halogen atoms [123]. Figure 7.14 shows diagrammatically the position of the NQR signals of ^{35}Cl for some organometallic derivatives of 2,6-dichlorophenol.

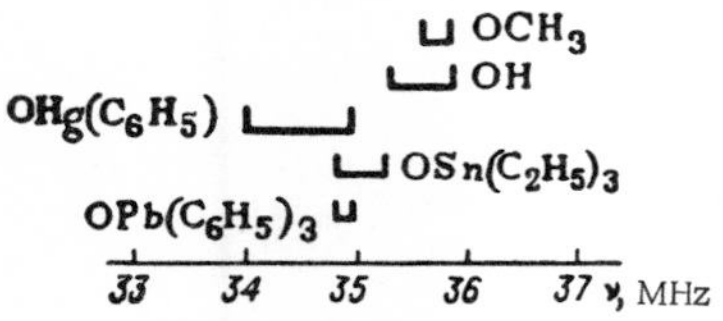

FIGURE 7.14. Diagram of the arrangement of the NQR spectral lines of ^{35}Cl in organometallic 2,6-dichlorophenolates

The splitting between the resonance signals fluctuates between 0.130 and 1.240 MHz, i.e. it remains within the range of the crystal splitting for some compounds, but is greater than this splitting for most compounds. Kravtsov et al. [123] hold that the splitting is caused by intramolecular coordination of the metal with the halogen atom:

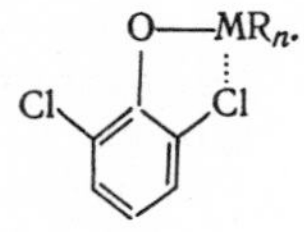

This splitting is not due to the Van der Waals interaction alone, since it quite obviously also depends on the nature of the metal. If we assume that the magnitude of the splitting is proportional to the energy of the coordination interactions, we must draw the conclusion that for the systems studied, organomercury radicals (compared with the organotin and organolead radicals) have the largest coordination capacity. This result is completely unexpected, since, as a rule [124], the tin and lead organo-derivatives have a higher coordination capacity than the analogous organomercury compounds.

It seems most probable [123] that the differences in the coordination capacities of the organometallic radicals in the given systems are due to the specific geometry of the coordinated state of each metal. It is known that in the organotin complexes $R_3SnX \cdot L$, the metal atom has the configuration of a trigonal bipyramid in which the organic substituents lie in the plane of the base, and the tin atom and the two ligands lie on a line perpendicular to the base [125]. Obviously, for intramolecular coordination of the organotin radical in the ortho-substituted phenolate the arrangement of the ligands is not the optimum for the formation of the coordination bond.

Unlike tin and lead, the mercury atom in complexes of organomercury compounds with coordination number 3 has sp^2 hybridization and a planar triangle-shaped configuration with angles close to 120°. This arrangement of the ligands is near to the optimum, i.e. the geometry of the molecule favors intramolecular coordination. The nature of the organic radical bonded to the mercury atom does not seem to affect the coordination capacity of the mercury.

A similar pattern is observed for cis-β-chlorovinylmercury chloride. We have already noted that for the trans derivative an intermolecular interaction of the β chlorine atom with the mercury atom was observed and confirmed by X-ray analysis. It is, therefore, useful to compare the distances between the nonvalence bonded atoms in the cis and trans compounds. In the ideal model of an isolated molecule of cis-chloride, the distance between the β chlorine atom and the mercury is 3.25 Å [117]. The shortest intermolecular distance Cl ··· Hg in trans-chloride, corresponding to the strongest coordination, is equal to 3.40 Å. We can, therefore, assume for cis-β-chlorovinylmercury chloride the existence of an intramolecular coordination:

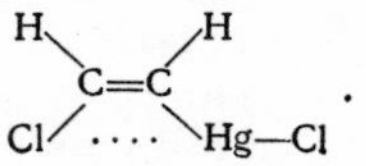

This assumption agrees well with the NQR spectra of ^{35}Cl in the cis-β-chloride.

The intramolecular interaction of the sulfur atom (II) with the ortho substituents in arylsulfenyl chlorides [126] is of interest. It is a known fact that arylsulfenyl chlorides containing a nitro group in an ortho position differ from other derivatives of this series by their capacity to produce solutions in 100% sulfuric acid which conduct electricity [127]. The conductivity of such solutions presupposes the presence of a sulfenium ion which is stabilized by the intramolecular coordination interaction with the oxygen of the nitro group:

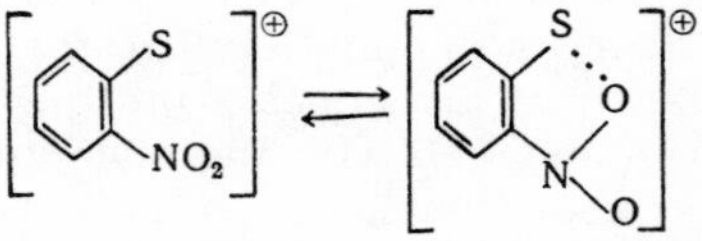

The NQR spectra of ^{35}Cl in the S—Cl bond confirm the existence of such a coordination in the solid state. The NQR frequency of ^{35}Cl in $2\text{-}NO_2C_6H_4SCl$

is lower than in C_6H_5SCl (36.15 and 37.02 MHz, respectively). The NQR frequency in 4-$NO_2C_6H_4SCl$ is equal to 38.74 MHz. In view of the stronger acceptor effect from the ortho position for 2-$NO_2C_6H_4SCl$ we would predict a NQR frequency equal to about 39—39.5 MHz. The sharp drop observed in the NQR frequency when a nitro group is introduced in an ortho position can only be due to intramolecular coordination of the sulfur atom with the oxygen of the nitro group [126].

Intramolecular coordination is also observed when a chlorine atom is introduced ortho to the SCl group. However, the donor effect of the chlorine atom is weaker than that of the oxygen atom of the nitro group. As a result it does not overlap the acceptor effect of the chlorine atom along the valence bonds and, consequently, the NQR frequency of ^{35}Cl in the S—Cl bond in 2-ClC_6H_4SCl is higher than in C_6H_5SCl (38.47 and 37.02 MHz). Nevertheless, the existence of intramolecular interaction is indicated by the very low frequency of the ortho chlorine atom (33.864 MHz, while the NQR frequency of the C—Cl bond in 4-ClC_6H_4SCl is 34.979 MHz), and the population of its p_x orbital is decreased as a result of this interaction.

Thus, it follows from the analysis of the available experimental data that NQR is a convenient and effective method for recognizing and studying different types of coordination bonds.

8. HYDROGEN BONDS IN NQR SPECTRA

The magnitudes of splittings in NQR spectra caused by intra- or intermolecular coordination interactions, are close to the splittings caused by hydrogen bonds.

A close parallel can be drawn between the 2,6-dihalo-4-X-substituted phenols and the 2,6-dichloro-substituted phenolates. In the NQR spectra of 2,6-dihalophenols, a characteristic difference in the frequencies of the ortho halogen atoms is observed. This is apparently due to the nonequivalent interactions of these atoms with the hydroxyl group. The concept of intramolecular hydrogen bonds in such compounds was first advanced by Ludwig [131]. In the crystalline state three variants in principle are possible in substituted 2,6-dihalophenols from the point of view of the nature of the hydrogen bond: 1) only intramolecular hydrogen bonds O—H···Hal; 2) combination of intramolecular and intermolecular bonds; 3) only intermolecular hydrogen bonds O—H···O. For the second variant we need a favorable arrangement of the molecules in the crystal lattice, and this might not be possible according to the close-packing principle. However, the results obtained by Sacurai [143, 144], when he investigated the structure of C_6Cl_5OH and 1,4-$(OH)_2C_6Cl_4$ using X-ray and NQR analytical methods, indicate that this variant may also be possible.

It was established that the proton of the hydroxyl group interacts with the adjacent chlorine and oxygen atoms to form a "forked" hydrogen bond of the type $\mathrm{O{-}H}\langle{}^{\mathrm{O}}_{\mathrm{Hal}}$ [143, 144].

Substitution of the hydrogen of the hydroxyl group by a methyl group or an alkali metal (K, Na) sharply reduces the splitting in the NQR spectra

[99]. In all the NQR spectra of 2,6-dihalo-4-X-phenols a lower intensity of the low-frequency line, a smaller value of the spin-spin relaxation time T_2 and a stronger dependence of T_2 on the temperature are observed. This indicates an additional magnetic broadening of this line (the low-frequency NQR line) as a result of the interaction of the protons with the chlorine atom. Examination of the tensor of the electric field gradient in the example of 2,6-I_2-4-$NO_2C_6H_2OH$ enables us to estimate the change of anisotropy of the C—I bond caused by the hydrogen bond. The values of the quadrupole coupling constants for the two iodine atoms and the asymmetry parameters indicate that the most substantial changes in the electric field gradient occur in the y direction (Table 7.4). (The z axis is directed along the C—I bond, the y axis is perpendicular to z in the plane of the benzene ring, and the x axis is perpendicular to the y and z axes).

TABLE 7.4. Quadrupole coupling constants of ^{127}I in 2,6-I_2-4-$NO_2C_6H_2OH$

e^2Qq_{xx}	e^2Qq_{yy}	e^2Qq_{zz}	η, %
-947.45	-978.77	1957.54	2.3
-877.73	-1123.94	2001.67	12.4

Thus, for 2,6-dihalo-4-X-substituted phenols we can quite accurately establish the spatial structure of the hydrogen bond.

The existence of an intermolecular hydrogen bond explains the splittings in the spectra of chloral hydrate and its semiacetal [128, 129]. The crystalline structure of the chloral hydrate $Cl_3CCH(OH)_2$ was investigated by Kondo and Nitta [130], but their work turned out to be inaccurate. Ogawa carried out a thorough X-ray investigation [145]. In addition to the three different distances C—Cl in the molecule (1.72, 1.87 and 1.86 Å) he also found two short intermolecular distances O···Cl_1 (3.07 and 3.12 Å). Accordingly, the existence of two forked hydrogen bonds, namely, $O_1—H\langle^{Cl''_1}_{O'_2}$ and $O'_2—H\langle^{Cl''_1}_{O''_1}$ was assumed. This agrees with the NQR spectrum of this compound, which consists of a narrow doublet (39.429 and 39.515 MHz) and a line at a much lower frequency (38.190 MHz) which corresponds to the atom participating in the hydrogen bond.

The hydrogen bond in chloral hydrate and its semiacetal was also examined by Grechishkin and Soifer [132].

When intermolecular hydrogen bonds are formed in solutions, quite frequently the phenomenon of migration of protons from one potential well to another is observed. NQR has been used frequently to investigate the possibility of such migrations in the solid state, although their probability is much smaller than in solution.

In [133] the NQR spectra of ^{35}Cl in a series of tautomeric systems possessing a high proton mobility and their organometallic derivatives were studied. We give as an example 4,5-dichloro-2-methylimidazole, for which in the solid state one of the following three structures can be expected:

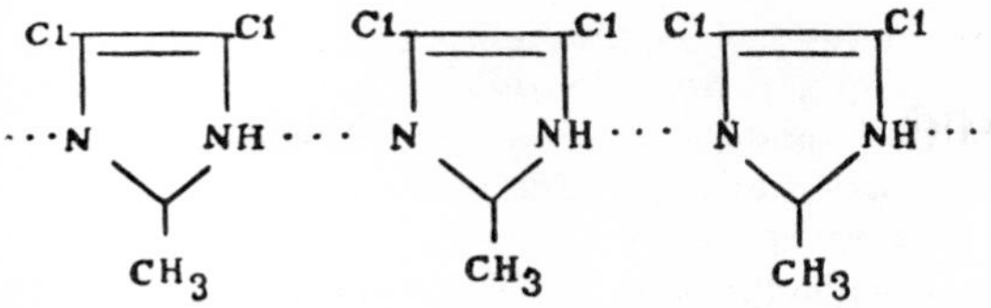

(I)

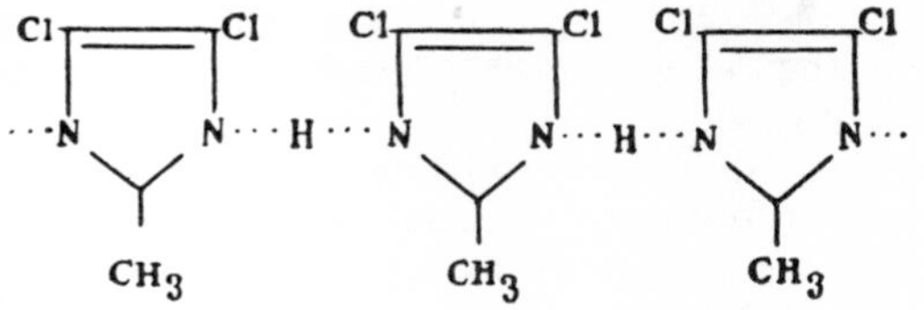

(II)

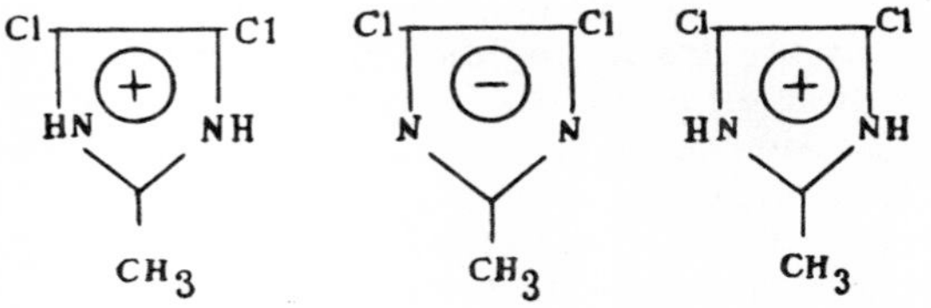

(III)

For the asymmetrical structure (I), two strongly split signals would be expected in the NQR spectrum. In the symmetrical structure (II), one signal would be expected. It was found that the NQR spectrum of this compound has the form of a doublet with splitting 1.32 MHz (36.070 and 37.394 MHz). The NQR frequencies for the anion and cation of structure (III) lie in quite a different frequency range (in the anion a strong shift of the NQR frequency to 34.7 MHz is observed, and in the cation it increases to 39.03 MHz). Therefore, this compound in the crystalline state has structure (I). However, for the phenylmercury derivative of 4,5-dichloro-2-methylimidazole:

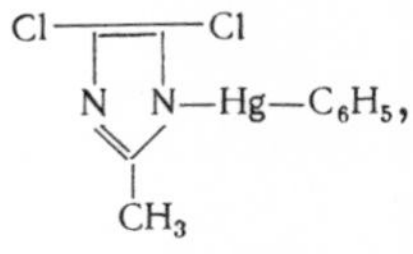

the distance between the signals of the observed doublet is considerably smaller (0.15 MHz) and lies within the crystal splitting. We can, therefore, say that this compound has a structure analogous or close to the symmetrical structure (II).

One of the most characteristic examples of the influence of proton migration on the NQR frequency is the NQR spectrum of ^{75}As in dihydroarsenates MH_2AsO_4 [134–136] (where M = K, Rb, Cs, Li, Ag, NH_4, Na). The NQR frequency of ^{75}As can serve as a measure of the distortion of the tetrahedron $[AsO_4]$. These distortions result from the formation of intermolecular hydrogen bonds O–H···O and the possible coordination interactions with the cation M. The contribution of the cation M to the electric field gradient on the ^{75}As nucleus is noticeable, since the NQR frequencies of ^{75}As depend on M (Cs > Rb > K). An indirect proof of the considerable influence of the hydrogen bonds on AsO_4 is the failure to detect the NQR signal of ^{75}As in Na_3AsO_4, where hydrogen bonds are known to be absent.

The temperature dependence of the NQR frequencies of ^{75}As is anomalous compared with the temperature dependence for the usual molecular crystals and complex ions. With an increase in temperature the NQR frequency drops sharply, and above the Curie point (for M = K, Rb, Cs, i.e. for ferroelectric substances) the NQR frequency decreases by almost one order of magnitude. Thus, in the ferroelectric phase at 77°K the NQR frequency of ^{75}As in KH_2AsO_4 is equal to 35.10 MHz [134—136], and in the para phase at room temperature it is equal to 3.5 MHz [146]. Such a strong dependence of the NQR frequency on temperature can be explained on the basis of the proton migration model in a two-well potential minimum, asymmetrical at a temperature below the Curie point and symmetrical above the phase transition temperature. With an increase in temperature the potential minimum becomes symmetrical and the protons previously situated in the deeper potential well about one of the oxygen atoms start to move toward the other one. This leads to a decrease in the electric field gradient on the ^{75}As nucleus and to a decrease in the NQR frequency.

The considerable isotopic effect on deuterization (the NQR frequency changes by 6—7 MHz) also indicates the substantial influence of the hydrogen bonds and the proton migration. With an increase in temperature, the NQR frequency of ^{75}As in deuterated dihydroarsenates remains practically constant. However, near the Curie point the frequency decreases sharply, just as with the undeuterated analogs.

Thus, NQR can be successfully used to study intra- and intermolecular hydrogen bonds and proton migrations.

The hydrogen bond can also be examined from the point of view of charge transfer [1]. Here, the acid protons play the part of σ acceptors. The transfer of the effect of complex formation on the halogen atoms, usually far from the coordinating center, takes place by an inductive mechanism which leads to an increase in the ionic character of the R—Hal bond. Therefore, the NQR frequencies of halogen-containing compounds must be shifted to the low-frequency range.

The hydrogen bonds in dilute solutions are frequently investigated by using acids capable of self-association. In the solid state, such acids form polymeric structures where the same molecule is simultaneously a donor and an acceptor. When a base is added to such a solution (for experiments with NQR, usually pure components are mixed) a mobile self-associate—complex equilibrium is established in the liquid. On crystallization the two forms of the associates are preserved, and the NQR spectrum as a rule simultaneously shows the frequencies of both the self-associate and the complex changed by steric and crystalline effects. It is difficult to completely understand and assign the frequencies in these spectra. Therefore, it is more convenient to use the mean frequency shift when analyzing the data. The magnitude of the mean shift will increase with increasing basicity of the donor. In spite of this coarse approach, the results are of interest.

A large number of molecular compounds of trichloroacetic acid with oxygen-containing molecules have been studied [137]. The multiplicity of the spectrum of trichloroacetic acid in these compounds increases considerably compared with the NQR spectrum of the pure acid, and the limits of the frequency range also increase. These changes are due to crystalline effects alone [137]. However, the components added to the acid can be referred to two different classes: either acids or bases. In the light of these facts the mean frequencies can also be divided into two groups: "acid" and "base." Figure 7.15 shows the distribution of the mean frequencies N of the two groups on a frequency scale ν. It is seen from the figure that the two groups are shifted with respect to each other and overlap very little. The maximum of the distribution of the "acid" group (a) is near the mean frequency of pure trichloroacetic acid (40.010 MHz), while the maximum of the "base" group (b) is shifted by 0.3 MHz to the low-frequency range. This shift indicates that the ionic character of the C—Cl bond in trichloroacetic acid increases on passing from the self-associate to the complex with base.

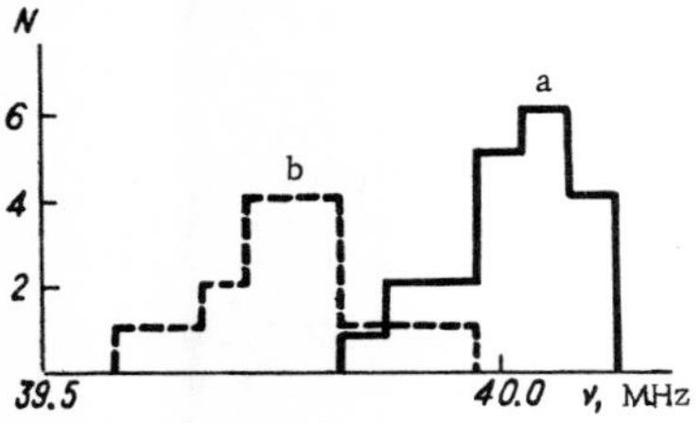

FIGURE 7.15. Statistical scatter of the NQR frequencies of ^{35}Cl in the CCl_3 group in salts of trichloroacetic acid and its complexes with bases (the dashed line represents the salts)

Numerous authors [6, 37, 38, 138] have studied the complexes of chloroform with different n and π donors. In pure chloroform there is no self-association, but the weak acidity contributes to the large degree of dissociation of the complexes. Therefore, in most cases the lines of the original compound, in addition to the frequencies of the bonded chloroform, appear in the NQR spectra of ^{35}Cl in complexes of chloroform. The frequencies of the bonded chloroform in all the complexes studied are shifted to low frequency and usually lie within 1—2.5%. A more detailed analysis is complicated by the different symmetry of the donors, since the geometrical factor plays an important part. We can only note that on passing from aromatic donors to oxygen-containing ones and then to amines the mean value of the low-frequency shift increases.

Interesting results were obtained from the NQR spectra of the hydrochlorides of some nitrogen-containing bases [131, 139—142]. The chlorine atom of hydrogen chloride in these complexes has a very high ionic character and, therefore, the chlorine frequencies should be close to zero, where the possibility of detecting them is limited. In the spectra, the frequencies of the halogens which were part of the donors were observed. In complexes of hydrogen chlorides with aliphatic amines the frequencies are shifted to high frequency (in accordance with the decrease in the ionic character of the C—Hal bond) by about 10%. In the examples of the hydrochlorides of chloroethylamine and chloropropylamine [139] it was shown that the inductive effect of the nitrogen atom on the chlorine in the complexes decreases sharply with the appearance of a third methylene bridge between them.

In complexes of hydrogen chloride with p-haloanilines the transmission effect during complex formation becomes apparent in two ways. The ionic character of the C—Hal bond decreases due to the inductive mechanism, and the frequency is shifted to high frequency. However, during complex formation the conjugation of the π electrons of the ring with the unshared pair of electrons on the nitrogen atom, which participates in the complex formation, decreases. As a result, the conjugation of the halogen with the ring increases, and this must lead to a low-frequency shift. The increase in the degree of double bonding is indicated by the increase in the asymmetry parameter of the iodine atom to 7% in the hydrochloride of p-iodoaniline compared with the original amine (3%) [131].

Many researchers have shown that the influence of the inductive effect is dominant, and that the values of the high-frequency shift increase in the order Cl, Br, I [131, 141, 142].

Thus, we see that all the factors for complexes of σ acceptors are also operative for hydrogen bonds. The phenomena of self-association and dissociation can complicate the interpretation of NQR data, but can also serve as an additional source of information on the influence of the hydrogen bonds in solids.

BIBLIOGRAPHY

1. Mullikem, R. S. — J. Am. Chem. Soc. 72 (1950), 600; 74 (1952), 811.
2. Maksyutin, Yu. K., E. N. Gur'yanova, and G. K. Semin. — Zh. Strukt. Khim. 9 (1968), 701.
3. Semin, G. K., V. I. Robas, V. D. Shteingarts, and G. G. Yakobson. — Zh. Strukt. Khim. 6 (1965), 160.
4. Semin, G. K., V. I. Robas, V. D. Shteingarts, and G. G. Yakobson. — In: "Radiospektroskopiya tverdogo tela," p. 239. Atomizdat, 1967.
5. Maksyutin, Yu. K., E. A. Kravchenko, E. N. Gur'yanova, and G. K. Semin. — Izv. AN SSSR, chem. ser., p. 1271, 1968.
6. Grechishkin, V. S. and I. A. Kyuntsel'. — Zh. Strukt. Khim. 7 (1966), 119.
7. Douglass, D. C. — J. Chem. Phys. 32 (1960), 1882.
8. Kitaigorodskii, A. I. and E. I. Fedin. — Zh. Strukt. Khim. 2 (1961), 216.
9. Maksyutin, Yu. K., T. A. Babushkina, E. N. Gur'yanova, and G. K. Semin. — Zh. Strukt. Khim. 10 (1969), 1025.
10. Maksyutin, Yu. K., T. A. Babushkina, E. N. Gurjanova, and G. K. Semin. — Theoret. chim. Acta 14 (1969), 48.
11. Briegleb, G. Elektronen-Donator-Acceptor Komplexe. — Berlin-Hannover-Hamburg, 1961.
12. Kitaigorodskii, A. I. and A. A. Frolova. — Izv. Sekt. Fiz.-Khim. Analiza Inst. Gen. Inorg. Chem. 19 (1949), 307.
13. Ainbinder, N. E. and A. N. Osipenko. — Trudy ENI Permsk. State Univ. 11, No. 3 (1966), 101.
14. Voronkov, M. G. — Teor. Eksperim. Khim. 1 (1965), 663.
15. Bhattachariya, S. N. and A. V. Anantaraman. — Physica 28 (1962), 633.
16. Watanabe, K. — J. Chem. Phys. 26 (1957), 542.
17. Marsh, R. E. — Acta crystallogr. 15 (1962), 809.
18. Gol'der, G. A., G. S. Zhdanov, and M. M. Umanskii. — Doklady AN SSSR 92 (1953), 113.
19. Arzamanova, I. G. and E. N. Gur'yanova. — Zh. Obshch. Khim. 36 (1965), 1157.
20. Hooper, H. — J. Chem. Phys. 41 (1964), 599.
21. Read, M., R. Cahay, P. Cornil, and J. Duchesne. — Compt. rend. 257 (1963), 1778.
22. Cornil, P., J. Duchesne, M. Read, and R. Cahay. — Bull. belge Phys., p. 89, 1964.
23. Cornil, P., M. Read, J. Duchesne, and R. Cahay. — Acad. Belge Bull. Chem. Sci. 50 (1964), 239.
24. Keefer, R. M. and L. J. Andrews. — J. Am. Chem. Soc. 72 (1950), 4677.
25. Maksyutin, Yu. K. — Author's Summary of Candidate Thesis. Phys.-Chem. Sci. Res. Inst. im. L. Ya. Karpov, 1968. (Russian)
26. Gilson, D. and C. T. O'Konski. — J. Chem. Phys. 48 (1968), 2767.
27. Hassel, O. and K. Stromme. — Acta chem. scand. 8 (1954), 873.
28. Hassel, O. and K. Stromme. — Acta chem. scand. 12 (1958), 1146.

29. Hassel, O. and J. Hvoslef. — Acta chem. scand. 13 (1959), 275.
30. Hassel, O. and C. Romming. — Acta chem. scand. 10 (1956), 696.
31. Jones, L. T. Jr. — Diss. Abstr. 27 (1966), 130.
32. Thieme, H. — Diss. München, 1965.
33. Shirmzan, M. G. — Author's Summary of Candidate Thesis. Phys.-Chem. Sci. Res. Inst. im. L. Ya. Karpov, 1960. (Russian)
34. Hassel, O. and K. Stromme. — Acta chem. scand. 13 (1959), 1781.
35. Person, W. B., R. E. Haphrey, W. A. Diskin, and A. I. Popov. — J. Am. Chem. Soc. 80 (1958), 2049.
36. Arzamanova, I. G. and E. N. Gur'yanova. — Zh. Obshch. Khim. 36 (1966), 1157.
37. Babushkina, T. A. et al. — Izv. AN SSSR, chem. ser., p. 1055, 1969.
38. Grechishkin, V. S. and I. A. Kyuntsel'. — Trudy ENI Permsk. State Univ. 12, No. 1 (1966), 15.
39. Kojima, S., K. Tsukada, S. Ogawa, A. Shimachi, and J. Abe. — J. Phys. Soc. Japan 9 (1954), 805.
40. Barnes, R. and P. J. Gordy. — J. Chem. Phys. 23 (1955), 1177.
41. Ogawa, S. — J. Phys. Soc. Japan 13 (1958), 618.
42. Robinson, H., H. G. Dehmelt, and P. J. Gordy. — J. Chem. Phys. 22 (1954), 511.
43. Bjowatten, T., O. Hassel, and C. Romming. — Nature 189 (1961), 137.
44. Hawes, L. L. — Nature 196 (1962), 766.
45. Bjowatten, T., O. Hassel, and A. Lindheim. — Acta chem. scand. 17 (1963), 689.
46. Wells, A. F. Structural Inorganic Chemistry. 3rd edition. — Oxford, Univ. Press, 1962.
47. Biedenkapp, D. and A. Weiss. — Z. Naturf. 19a (1964), 1518.
48. Maksyutin, Yu. K., E. N. Gur'yanova, and G. K. Semin. — Uspekhi Khimii, p. 301, 1970.
49. Kuple, S. — Z. Chemie, Lpz. 5 (1965), 306.
50. Dehmelt, H. G.— J. Chem. Phys. 21 (1953), 380.
51. Maksyutin, Yu. K., E. V. Bryukhova, E. N. Gur'yanova, and G. K. Semin. — Izv. AN SSSR, chem. ser, p. 2658, 1968.
52. Gol'dshtein, I. P., E. N. Kharlamova, and E. N. Gur'yanova. — Zh. Obshch. Khim. 38 (1968), 1982.
52a. Tong, D. A. — J. Chem. Soc. D (1969), 790.
53. Rogers, M. and J. A. Rian. — J. Phys. Chem. 72 (1968), 134.
54. Bronden, C. I. — Acta chem. scand. 17 (1965), 759.
55. Beatie, I. E. and R. Rule. — J. Chem. Soc., p. 3267, 1964.
56. Beatie, I. R. — Q. Rev. Chem. Soc. 17 (1963), 382.
57. Schneider, R. F. and J. V. Di Lorenzo. — J. Chem. Phys. 47 (1967), 2343.
58. Negita, H., T. Okuda, and M. Kashima. — J. Chem. Phys. 45 (1966), 1076.
59. Negita, H., T. Okuda, and M. Kashima. — J. Chem. Phys. 46 (1967), 2450.
60. Grechishkin, V. S. and I. A. Kyuntsel'. — Zh. Strukt. Khim. 4 (1963), 269.
61. Grechishkin, V. S. and I. A. Kyuntsel'. — Optika Spektrosk. 15 (1963), 832.
62. Grechishkin, V. S. and I. A. Kyuntsel'. — Zh. Strukt. Khim. 5 (1964), 53.
63. Grechishkin, V. S. and I. A. Kyuntsel'. — Trudy ENI Permsk. State Univ. 11, No. 2 (1964), 119.
64. Grechishkin, V. S. and I. A. Kyuntsel'. — Optika Spektrosk. 16 (1964), 161.
65. Grechishkin, V. S. and I. A. Kyuntsel'.— Trudy ENI Permsk. State Univ. 11, No. 3 (1966), 9.
66. Grechishkin, V. S. and I. A. Kyuntsel'. — Trudy ENI Permsk. State Univ. 12, No. 4 (1966), 29.
67. Nakamoto, Kazuo. Infrared Spectra of Inorganic and Coordination Compounds. — New York, Wiley, 1963.
68. Okuda, T., A. Nakao, M. Shiroyama, and H. Negita. — Bull. Chem. Soc. Japan 41 (1968), 61.
69. Srivastava, T. S. — Curr. Sci. 37 (1968), 253.
70. Biedenkapp, D. and A. Weiss. — Z. Naturf. 23b (1968), 174.
71. Nakamura, D. — Bull. Chem. Soc. Japan 36 (1963), 1662.
72. Nakamura, D., Y. Uehara, Y. Kurita, and M. Kubo. — J. Chem. Phys. 31 (1959), 1433.
73. Nakamura, D., K. Ito, and M. Kubo. — J. Am. Chem. Soc. 84 (1962), 163.
74. Nakamura, D., K. Ito, and M. Kubo. — Inorg. Chem. 1 (1962), 592; 2 (1963), 61.
75. Nakamura, D. and M. Kubo. — J. Phys. Chem. 68 (1964), 2986.
76. Nakamura, D., Y. Kurita, K. Ito, and M. Kubo. — J. Am. Chem. Soc. 82 (1960), 5783.
77. Ikeda, R., D. Nakamura, and M. Kubo. — J. Phys. Chem. 69 (1965), 2101; Bull. Chem. Soc. Japan 36 (1963), 1056.

78. Ito, K., D. Nakamura, Y. Kurita, Ka. Ito, and M. Kubo. — J. Am. Chem. Soc. 83 (1961), 4526.
79. Ito, Ko., K. Ito, D. Nakamura, and M. Kubo. — Bull. Chem. Soc. Japan 35 (1962), 518.
80. Ito, Ko., D. Nakamura, K. Ito, and M. Kubo. — Inorg. Chem. 2 (1963), 690.
81. Ikeda, R., A. Sasane, D. Nakamura, and M. Kubo. — J. Phys. Chem. 70 (1966), 2926.
82. Sasane, A., D. Nakamura, and M. Kubo. — J. Phys. Chem., Vol. 71, 1967.
83. Kashiwagi, H., D. Nakamura, and M. Kubo. — J. Phys. Chem. 71 (1967), 4443.
84. Marram, E. P., E. J. McNiff, and J. L. Ragle. — J. Phys. Chem. 67 (1963), 1719.
85. Caglioti, V., C. Furlani, F. Orestano, and F. M. Capece. — Atti Accad. naz. Lincei Rc. 35 (1963), 10.
86. Caglioti, V., G. Sartori, and C. Furlani. — Proc. Intern. Conf. on Coordination Chemistry, p. 81. Vienna, 1964.
87. Kubo, M. and D. Nakamura. — Advances in Inorganic Chemistry and Radiochemistry 8 (1966), 257.
88. Cippollini, E., J. Owen, J. H. M. Thornloy, and C. Windsor. — Proc. Phys. Soc. 79 (1962), 1083.
89. Ikeda, R., D. Nakamura, and M. Kubo. — J. Phys. Chem. 72 (1968), 2982.
90. Ikeda, R., D. Nakamura, and M. Kubo. — J. Phys. Chem. 70 (1966), 3626.
91. Ikeda, R., D. Nakamura, and M. Kubo. — Bull. Chem. Soc. Japan 40 (1967), 701.
92. Furlani, G., F. Capece, and C. Sartori. International Summer School on Theoretical Chemistry. Nuclear Quadrupole Resonance Spectroscopy, Italy, 1967.
93. Babushkina, T. A., V. I. Robas, and G. K. Semin. — In: "Radiospektroskopiya tverdogo tela," p. 221. Atomizdat, 1967.
94. Barnes, R. G. and S. L. Segel. — J. Chem. Phys. 25 (1956), 180.
95. Barnes, R. G. and S. L. Segel. — J. Chem. Phys. 25 (1956), 578.
96. Barnes, R. G., S. L. Segel, P. J. Bray, and P. A. Casabella. — J. Chem. Phys. 26 (1957), 4345.
97. Casabella, P. A., P. J. Bray, and R. G. Barnes. — J. Chem. Phys. 30 (1959), 1393.
98. Renes, P. A. and C. H. McGilvary. — Recl. Trav. chim. Pays-Bas Belg. 63 (1944), 283.
98a. Segel, S. and R. Barnes. Catalog of Nuclear Quadrupole Interactions and Resonance Frequencies in Solids, Part 1. — US Atomic Energy Commission, 1962.
99. Semin, G. K. and V. I. Robas. — In: "Radiospektroskopiya tverdogo tela," p. 229. Atomizdat, 1967.
100. Evans, J. C. and G. Y. S. Lo. — Inorg. Chem. 6 (1967), 836.
101. Buslaev, Yu. A., E. A. Kravchenko, S. M. Sinitsyna, and G. K. Semin. — Izv. AN SSSR, chem. ser., p. 2816, 1969.
102. Zalkin, A. and D. E. Sands. — Acta crystallogr. 11 (1958), 615.
103. Okuda, T., A. Nakao, M. Shiroyama, and H. Negita. — Bull. Chem. Soc. Japan 41 (1968), 61.
104. Kojima, S., K. Tsukada, S. Ogawa, A. Shimachi, and Y. Abe. — J. Phys. Soc. Japan 9 (1954), 805.
105. Harris, P., E. Mack, and F. C. Blake. — J. Am. Chem. Soc. 50 (1928), 1583.
106. Townes, C. H. and B. P. Dailey. — J. Chem. Phys. 20 (1952), 35.
107. Bersohn, R. Cited after T. P. Das and E. L. Hahn, "Nuclear Quadrupole Resonance Spectroscopy." — In: "Solid state physics," Suppl. 1. New-York—London, 1958.
108. Tsukada, K. — J. Phys. Soc. Japan 11 (1956), 956.
109. Robinson, H. G., H. G. Dehmelt, and W. Gordy. — J. Chem. Phys. 22 (1953), 511.
110. Stevens, K. W. — Techn. Report No. 197, Cruft Laboratory, Harvard Univ., 1954.
111. Kojima, S., K. Tsukada, S. Ogawa, and A. Shimachi. — J. Chem. Phys. 23 (1955), 1963.
112. Adrian, F. J. — J. Chem. Phys. 38 (1963), 1258.
113. Velichko, F. K., L. A. Nikonova, and G. K. Semin. — Izv. AN SSSR, chem. ser., p. 84, 1967.
114. Babushkina, T. A., E. V. Bryukhova, F. K. Velichko, V. I. Pakhomov, and G. K. Semin. — Zh. Strukt. Khim. 9 (1968), 207.
115. Joan, V., F. Badea, E. Coranescu, and C. D. Nenitzescu. — Angew. Chem. 72 (1960), 416.
116. Pakhomov, V. I. and A. I. Kitaigorodskii. — Zh. Strukt. Khim. 7 (1966), 862.
117. Bryukhova, E. V., M. A. Kashutina, T. A. Babushkina, O. Yu. Okhlobystin, and G. K. Semin. — Doklady AN SSSR 183 (1968), 827.
118. Semin, G. K., T. A. Babushkina, A. K. Prokof'ev, and P. G. Kostyanovskii. — Izv. AN SSSR, chem. ser., p. 1401, 1968.
119. Ledwith, A. and L. Phillips. — J. Chem. Soc., p. 3796, 1962.
120. Seyferth, D. and J. Y.-P. Mui. — J. Am. Chem. Soc. 86 (1964), 2961.

121. Bregadze, V. I., T. A. Babushkina, O. Yu. Okhlobystin, and G. K. Semin. — Teor. Eksperim. Khim. 3 (1967), 547.
122. Paulik, F. E., S. I. E. Green, and R. E. Dessy. — J. Organometall. Chem. 3 (1965), 229.
123. Kravtsov, D. H., A. P. Zhukov, B. A. Fainyur, E. M. Rokhlina, G. K. Semin, and A. N. Nesmeyanov. — Izv. AN SSSR, chem. ser., p. 1703, 1968.
124. Kravtsov, D. N. — Zh. Obshch. Khim. 12 (1967), 53.
125. Hulme, R. — J. Chem. Soc., p. 1524, 1963.
126. Babushkina, T. A. and M. I. Kalinkin. — Izv. AN SSSR, chem. ser., p. 157, 1969.
127. Kharash, N. Organic Sulfur Compounds 1 (1961), 379.
128. Allen, H. — J. Am. Chem. Soc. 74 (1952), 1074.
129. Bray, P. J. and P. J. Ring. — J. Chem. Phys. 21 (1953), 2226.
130. Kondo, S. and I. Nitta. — X Sen (X-Rays) 6 (1950), 53.
131. Ludwig, G. — J. Chem. Phys. 25 (1956), 159.
132. Greichishkin, V. S. and G. B. Soifer. — Trudy ENI Permsk. State Univ. 11, No. 2 (1964), 125.
133. Nesmeyanov, A. N. et al. — Doklady AN SSSR 179 (1968), 102.
134. Zhukov, A. P., L. S. Golovchenko, and G. K. Semin. — Izv. AN SSSR, chem. ser., p. 1399, 1968.
135. Zhukov, A. P., I. S. Rez, V. I. Pachomov, and G. K. Semin. — Physica Status Solidi 27 (1968), K129.
136. Zhukov, A. P., L. S. Golovchenko, and G. K. Semin. — Chem. Communs., p. 854, 1968.
137. Weiss, A. and D. Biedenkapp. — Ber. Bunsengesellschaftphys. Chem. 70 (1966), 738.
138. Bennet, R. A. and H. Hooper. — J. Chem. Phys. 47 (1968), 4855.
139. Hooper, H. and P. J. Bray. — J. Chem. Phys. 33 (1960), 334.
140. Segel, S. L. and R. G. Barnes. — J. Chem. Phys. 25 (1956), 1286.
141. Dewar, M. J. S. and E. A. C. Lucken. — J. Chem. Soc., p. 426, 1959.
142. Bray, P. J. — J. Chem. Phys. 22 (1954), 950.
143. Sacurai, T. — Acta crystallogr. 15 (1962), 443.
144. Sacurai, T. — Acta crystallogr. 15 (1962), 1164.
145. Ogawa, K. — Bull. Chem. Soc. Japan 36 (1963), 610.
146. Bjorkstam, K. — Bull. Am. Phys. Soc. 12 (1962), 442.

Chapter 8

APPLICATION OF NQR TO HIGH-MOLECULAR-WEIGHT COMPOUNDS

1. INVESTIGATION BY NQR OF MOLECULAR CRYSTALS HAVING A STRUCTURE WITH BASICALLY NONREMOVABLE ELEMENTS OF DISORDER

The use of pulsed methods to observe NQR signals of ^{35}Cl enables us to detect very wide lines. Detection of line widths in the order of 0.02% of the resonance frequency is the most which can be achieved using the stationary method. With the pulsed NQR spectrometer we are able to observe line widths in the order of 2% [1].

Such broad lines are observed in essentially unordered structures, such as molecular crystals with nonremovable elements of disorder. It is very important to study these structures, since crystalline polymers can be considered as molecular crystals with a certain degree of disorder.

We shall begin with a discussion of the elementary types of disorder in elementary molecular crystals, considering the static and dynamic consequences of the disorder of the molecules in the crystal [2—8]. Statistically unordered structures can be formed from molecules when similar but not identical substituents are arranged about a center of symmetry. It was found that the elementary forms of statistical disorientation of the molecules in the crystal can be conveniently studied using hexasubstituted chloro- and bromobenzenes [2, 4—7], where the chlorine and bromine atoms are alternated, or using a group with a similar size, such as the methyl group. We can also note cases of statistical disorientation in benzene molecules containing amino or nitro groups (for example, $Cl_5C_6NH_2$ and $Br_5C_6NH_2$ [2, 8], p-$ClC_6H_4NO_2$ [9], $Cl_5C_6NO_2$ [10]). Similar effects were also observed in tetrahalomethanes $Cl_{4-n}CBr_n$ [2], and in halo-derivatives of naphthalene [11].

In the series of hexahalo-substituted benzenes of the type $Br_nC_6Cl_{6-n}$ the compounds C_6Cl_6 and C_6Br_6 are isomorphous and crystallize in the space group $P2_1/c$ with two molecules per unit cell (the symmetry of the molecule in the crystal is $\bar{1}$).

The other compounds in the series form statistically unordered phases, possess a high packing density, and are isostructural with C_6Cl_6.

The symmetry of the statistically "averaged" molecule for compounds without a center of symmetry is greater than the intrinsic symmetry, and for compounds with a center of symmetry it is identical to the intrinsic symmetry.

The character of the statistical disorder was studied by X-ray and NQR analytical methods. The results obtained by the two methods show good agreement.

All the unordered structures can be split into two basic types according to the character of the statistical disorder.

1a. Averaging takes place over all atoms (groups) with different atomic weights. The intrinsic symmetry of the statistically "averaged" molecule is maximum and all orientations are equally probable. All the molecules participate in the averaging.

1b. The orientation of part of the molecules is fixed, while the remaining ones are disoriented according to type 1a.

2a. Averaging does not take place over all atoms (groups) with different atomic weights. The intrinsic symmetry of the statistically "averaged" molecule is higher than the initial symmetry. All the molecules participate in the averaging.

2b. The orientation of part of the molecules is fixed, while the remaining ones are disoriented according to type 2a.

Let us examine 1,4-$Br_2C_6Cl_4$ [6] (Figure 8.1). This molecule has a center of symmetry. Crystallization in the phase isostructural to hexachlorobenzene, which has three crystallographically nonequivalent positions of the chlorine atom, namely, Hal_1, Hal_2, Hal_3, does not necessitate the appearance of disorder. However, the NQR spectrum of ^{35}Cl contains two broad lines ($\Delta\nu_{1/2} \approx 100$ kHz) of unequal intensity. It was established by X-ray analysis that one of the crystallographically nonequivalent halogen atoms (Hal_3) is 67% bromine and 33% chlorine, while the other two (Hal_1 and Hal_2) are 83.5% chlorine and 16.5% bromine. To explain this it is necessary to assume that 67% of the molecules in the crystal are fixed, and the remaining 33% are rotated through $\pm 60°$ (the two rotations being equally probable). In the NQR spectrum of ^{35}Cl this implies a decrease in the intensity of the lines for which the Hal_1 and Hal_2 atoms are responsible and the appearance of a third line for the Hal_3 atom. The lines determined by Hal_2 and Hal_3 atoms are unresolved since they are very close. The number of rotated molecules, calculated from the intensity ratio of the NQR lines, is equal to 32.3%.

In many cases, such as for $Cl_5C_6NH_2$ [8], we can establish the number of "averaged" molecules more accurately from the NQR spectrum than from X-ray data. In fact, the difference between the frequencies is larger for chemically nonequivalent atoms than for crystallographically nonequivalent ones. Thus, from the shape of the NQR line corresponding to a given chemically nonequivalent atom we can calculate the contribution of the other atoms mixed with the given one by statistical "averaging." X-ray data give no information on the chemical nonequivalence of averaged atoms.

The broadening of the NQR lines in statistically unordered structures is apparently due to a number of factors [2]:

1) broadening caused by the disordering effect of the crystal field of the nearest environment;

2) additional "broadening" caused by the coalescence of lines from crystallographically nonequivalent atoms when the distance between the lines does not exceed the half-width of the resonance lines from the corresponding atoms; for chemically nonequivalent atoms with small splittings the lines can also coalesce.

The dynamics of these unordered molecular crystals is also peculiar, since the NQR signals disappear relatively rapidly with an increase in temperature, and the spin-lattice relaxation time T_1 simultaneously becomes considerably shorter. Apparently, the rotation of the molecules is facilitated in these structures and, since the crystals are defective, the rotation starts near the defects and propagates gradually to the whole crystal. In fact, in many of the crystals investigated a reorientation motion of the molecules about the sixth-order axis occurs at room temperature. (This was shown by the NMR of ^1H in the solid [12, 13]). Obviously, at lower temperatures, when the motion of the molecules is arrested, the distribution among the potential wells is preserved, each well possessing a certain probability of finding the molecules there.

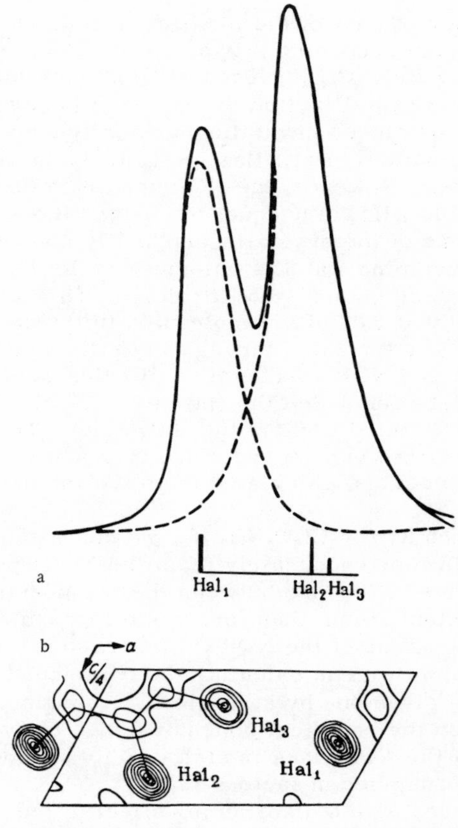

FIGURE 8.1. NQR spectrum of ^{35}Cl in 1,4-$Br_2C_6Cl_4$ and a diagram of the frequency shifts for three nonequivalent positions (a) and projection (b) of the electron density of the structure on face *ac*

A large dipole-dipole interaction (for example in p-$ClC_6H_4NO_2$, $C_6Cl_5NO_2$, $C_6Cl_5NH_2$, $C_6Br_5NH_2$) may also be one of the factors contributing to the formation of statistically unordered structures. In this case, the unordered structure probably corresponds to the minimum dipole-dipole interaction. If the packing of the molecules in the crystal is not too loose, a statistically unordered structure will possibly be formed, which is less efficient from the point of view of close-packing considerations.

To confirm this assumption, ordered crystals of *n*-nitrochlorobenzene were grown on crystal "matrices" of *n*-nitrobromo- and *n*-nitroiodobenzenes [14]. The "matrix" acted as a "substrate" orienting the dipoles of *n*-$ClC_6H_4NO_2$ in an unusual epitaxial process. In the NQR spectra of ^{35}Cl an oriented phase appears together with a narrow line (10—20 kHz) and the broad line (~350 kHz) from the unordered phase which remains. In the course of time (about 1 year) the crystals oriented on the "matrix" of *n*-bromonitrobenzene become fully ordered (the broad line in the NQR spectrum disappears). At the end of this time the crystals oriented on the matrix of *n*-iodonitrobenzene give a spectrum consisting of a broad (~350 kHz) and a narrow (~60 kHz) line, i.e. the ordering is not complete [14].

2. POSSIBILITY OF APPLYING NQR TO POLYMERS

We have already mentioned in Section 1 that crystalline polymers can be treated as molecular crystals with a certain degree of disorder. It would, therefore, be natural to expect broad NQR lines in the polymers as a result of random differences in the crystal field at resonance nuclei belonging to the same chemical type.

The information obtained from the NQR spectra enables us to predict the properties of the polymer molecules, the character of their arrangement, and the mobility of the polymer molecules in the crystal.

So far the NQR spectra of ^{35}Cl in polymers such as poly(vinylidene chloride) [15, 16] and its copolymers with polystyrene and the polyester of hydroxyenanthic acid, the polymerization products of trichloropropene [15—17], polychloral, polytrifluorochloroethylene and some other chlorine-containing polymers have been studied [16].

The shifts of the NQR frequencies indicate a change in the chemical nature of the atoms when macromolecules are formed from the monomers. Comparative data on the frequencies of the polymers and the corresponding monomers are given in Table 8.1.

Poly(vinylidene chloride) and its copolymers. Structural analytical data [18, 19, 20] show that this crystalline polymer has a monoclinic unit cell with $a = 22.54$, $b = 12.53$, $c = 4.68$ Å, $\beta = 84°10'$, with two monomer units along the axis and eight monomer units in the unit cell. As indicated by Bann [19], structural data are not a sufficient basis for assuming that a polymer has a "head-tail, head-tail" type structure. At the same time, the data obtained from proton magnetic resonance [21] indicate that the content of "head-head" units is about 10—14%.

The NQR spectrum of poly(vinylidene chloride) has eight components (Figure 8.2,a). This means that there must be eight types of chemically or crystallographically nonequivalent chlorine atoms. The region in the spectrum from 36.56 to 38.18 MHz excludes only the crystallographically nonequivalent atoms.

TABLE 8.1. NQR frequencies (MHz) of some chlorine-containing polymers

Monomer			Polymer	
formula	ν_{exp}	$\nu_{av.exp}$	ν	$\Delta\nu_{1/2}$
$CH_2{=}CCl_2$	36.253 36.526 36.526 36.872	36.544	38.18 38.08 38.08 37.97	0.40
			37.97 37.27 37.27	0.20
			36.58	0.16
$CF_2{=}CFCl$	38.017 38.171	38.094	39.94	0.65
Cl_3CCOH	38.794 39.172 39.291	39.026	39.77	0.40
$Cl_3CCH{=}CH_2$	38.150 38.272 38.476	38.299	37.01	0.5

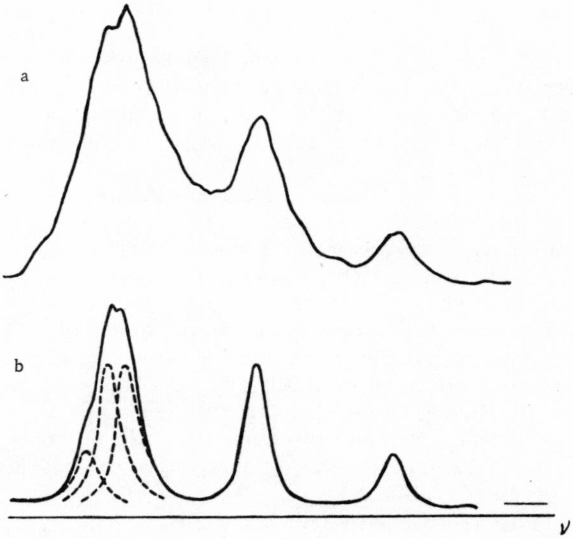

FIGURE 8.2. NQR spectra of ^{35}Cl:

a) in pure poly(vinylidene chloride); b) in the copolymer of poly(vinylidene chloride) with the polyester of hydroxyenanthic acid.

The selection of the structure of the polymer chain was based on the empirical method of calculation for polyhalogen-substituted aliphatic compounds [22, 23]. The frequency of a halogen inside the chain (secondary, tertiary, etc.) is equal to:

$$\nu = \nu_{0II} + \Delta_1\nu \sum_i \xi^{n_i - 1},$$

where ν_{0II} is the origin of the coordinates, $\Delta_1\nu$ the primary shift for a halogen near the secondary carbon, ξ the coefficient of the conduction effect of the i-th substituent separated from the given halogen atom by n carbon atoms. By comparing the calculated frequencies with the measured spectra we were able to determine the structure of the polymeric chain fragment.

The following structure is the most probable for poly(vinylidene chloride): to every seven monomeric units joined by a "head-tail" type structure, there is one "erroneous" one ("head-head"). This structure of the fragment, determined from the NQR spectra, agrees with the data obtained from PMR spectra [21].

The determination of the number of components in the spectrum of poly(vinylidene chloride) is complicated by the large width of the overlapping lines. Only after investigating its copolymers were we able to accurately establish the number of components. By comparing the NQR spectrum of ^{35}Cl in pure poly(vinylidene chloride) with the spectra of its block copolymers (polystyrene and the polyester of hydroxyenanthic acid, Figure 8.2,b) we were able to establish that the number and arrangement of the lines is preserved, although the lines of the pure polymer are much broader. This means that the structure of the fragment and the character of the basic pattern of arrangement of the molecules remained the same, but the degree of order in the block copolymers was higher. Since each unit cell contains four polymer chains, we can assume that the statistical disorder has a translational character, i.e. the arrangement of the "erroneous" monomeric unit in each chain with respect to the adjacent chains is statistical. The formation of block copolymers involves the ordering of this translational disorder.

The course of formation of ordered copolymers of poly(vinylidene chloride) tentatively takes the following form. Polymerization apparently starts in the solvate shell formed by vinylidene chloride about the copolymer macromolecules dissolved in it. The very ordered fragments of poly(vinylidene chloride) which are formed are used as substrates for the growth of the following macromolecules. When the molecular weight reaches a certain value, poly(vinylidene chloride) is precipitated from the solution of its monomer.

Polymerization products of trichloropropene. The amorphous (soluble) phase PTKhP-1 and the amorphous-crystalline (insoluble) phase PTKhP-II were used to study the polymerization of trichloropropene. The NQR data of ^{35}Cl for PTKhP-II show that the mean frequency of the broad line basically corresponds to the calculated frequency of the fragment with the vinylidene rearrangement $[-CH_2-CHCl-CCl_2-]_n$. A poorly resolved fine structure, which may have arisen due to differences between crystallographically nonequivalent atoms, was also observed [15–17].

The amorphous phase PTKhP-1 has a broad NQR line of ^{35}Cl ($\Delta\nu_{1/2} \approx \approx 0.80$ MHz or along the base ~ 2 MHz). This large line width indicates the presence of a large set of nonequivalent chlorine atom positions (including positions with large steric hindrances). This is due to the random rolled shape of the chain which consists mainly of fragments containing the group CCl_3. However, the long spin-lattice relaxation time ($T_1 = 1.5$ sec) reveals the very compact nature of the globular formations in which the macromolecules containing fragments with CCl_3 groups are packed. The combination of data from NQR and electron microscopy [17] leads us to assume that the polymerization occurs at the same time as the vinylidene rearrangement. This leads to the formation of a continuous range of products: from macromolecules consisting entirely of fragments formed after the rearrangement, to macromolecules whose chains are built entirely of nonrearranged fragments with the CCl_3 groups.

It is natural to expect well formed crystals only from linear macromolecules containing the $[-CH_2-CHCl-CCl_2-]_n$ grouping. Even if a small number of fragments containing the group CCl_3 are present, this must lead to the loss of linearity on certain portions of the chain. If a large number of these fragments are present, the polymer may become completely unstable and globular formations will appear [17].

NQR spectra of ^{35}Cl in polytrifluorochloroethylene and polychloral. The increase in the NQR frequency of ^{35}Cl in polytrifluorochloroethylene compared with the monomer (see Table 8.1) is regular [22, 23], since the number of atoms of the neighboring halogens increases as the polymer chain is built. The large width of the NQR line of ^{35}Cl indicates the low order in the arrangement of the macromolecules of polytrifluorochloroethylene. Apparently, an azimuthal statistical disorientation of the molecules is possible in this case. The short spin-lattice relaxation time ($T_1 = 2$ msec) indicates that the polymer molecules are very mobile with respect to each other. The results obtained by NQR for polytrifluorochloroethylene do not contradict the data obtained from X-ray analysis [24] and IR spectra [25] on the spiral structure of the polymer chain.

The relatively small increase in the NQR frequency of ^{35}Cl compared with the monomer is due to the fact that the inductive effect is strongly weakened by the intermediate atoms between the chlorine atoms of the CCl_3 groups. The line widths and the spin-lattice relaxation time indicate that slightly ordered and fairly mobile macromolecules are formed in this polymer. Thus, on the basis of the NQR data we can state that the CCl_3 groups suffer very small changes on polymerization. The difference in the mobility of polychloral and amorphous trichloropropene containing the CCl_3 group is significant. The much higher mobility in polychloral is due to the existence of "hinged" bonds around the oxygen atom.

3. NQR SPECTRA OF INORGANIC POLYMERS

A large number of metal halides form layered ionic structures with infinite one-, two- and three-dimensional grids, in which the halogen atoms are bonded to two or more metal atoms. Such structures are formed by $MHal_2$ (where M = Cd, Zn, Ti, Cr, Fe) [26–28] and $MHal_3$ (where M = Ti, Cr, V, Fe, rare-earth elements*) [26, 29–31]. The NQR spectra of these compounds have narrow lines at very low frequencies. This corresponds to the large

* The trichlorides of rare-earth metals are discussed in detail in Section 3 of Chapter 9.

ionic character of the M—Hal bonds and also to the low degree of disorder in the arrangement of the atoms in these polymers (apparently, in layered structures the degree of freedom for translational and orientational motions is low).

The coordination number of the halogen atoms in most of these compounds is equal to two or more and, therefore, it is difficult to estimate the bond multiplicity; we note, however, that the frequencies of $TiCl_3$ are higher than the frequencies of $TiCl_4$. For $TiCl_4$, according to the Townes-Dailey theory, the bond Ti—Cl has a 30% triple character (see Chapter 5). The higher frequency in $TiCl_3$ means that the halogen atom participates in additional bonds with the other metal atoms through the p_x and p_y orbitals, just as in the group V dimers (see Section 6 of Chapter 7).

These kinds of infinite polymeric chains and networks are formed by compounds of the type $A_2^V B_3^{VI}$ [32–37] and $A^V B^{VI}Hal$ [38—40] (where A = As, Sb, Bi; B = O, S, Se), but the bonds are essentially covalent. The crystals formed are not ionic but coordination crystals in which the intra- and intermolecular interactions are of the same magnitude. These compounds exist in both the crystalline and glass state. Therefore, the NQR spectra may have both narrow and broad lines and, just as in the case of organic polymers, the width of the NQR lines can serve as a criterion of the crystallinity and order. For these compounds there are NQR signals of both the halogens and the group V elements (see Section 4 of Chapter 9).

A typical example of inorganic polymers which do not have layered structures are the cyclic polymers of phosphonitrile chlorides $(PNCl_2)_n$ (n = 3, 4, 5, 6). Craig and Paddock et al. [41] showed by molecular orbital calculations that the ring aromaticity must increase with an increase in the dimensions of the ring. However, the NQR spectra of ^{35}Cl [42—49] do not confirm these calculations, apparently because the chlorine atoms are not conjugated with the ring, but are located "above" or "below" the plane of the ring. This is indicated by the small value of the asymmetry parameter of the tensor of the electric field gradient for $(PNCl_2)_3$, which is equal to $0 \leqslant \eta \leqslant 0.02$ [50]. For even n (n = 4, 6) two forms exist, which are characterized by different conformations of the molecules namely, the "chair" and "boat" forms [41].

BIBLIOGRAPHY

1. Pavlov, B. N., I. A. Safin, G. K. Semin, E. I. Fedin, and D. Ya. Shtern. — Vest. Akad. Nauk SSSR, No. 11 (1964), 40.
2. Khotsyanova, T. L., V. I. Robas, and G. K. Semin. — In: "Radiospektroskopiya tverdogo tela," p. 223. Atomizdat, 1967.
3. Khotsyanova, T. L., G. K. Semin, and V. I. Robas. — Zh. Strukt. Khim. 5 (1964), 644.
4. Khotsyanova, T. L. — Zh. Strukt. Khim. 7 (1966), 472.
5. Khotsyanova, T. L., T. A. Babushkina, and G. K. Semin. — Zh. Strukt. Khim. 7 (1966), 634.
6. Khotsyanova, T. L., T. A. Babushkina, and G. K. Semin. — Zh. Strukt. Khim. 9 (1968), 148.
7. Babushkina, T. A., T. L. Khotsyanova, S. I. Kuznetsov, and G. K. Semin. — Kristallografiya 17 (1972), 221.
8. Khotsyanova, T. L., T. A. Babushkina, S. I. Kuznetsov, and G. K. Semin. — Zh. Strukt. Khim. 10 (1969), 525.
9. Mak, T. C. W. and J. Trotter. — Acta crystallogr. 15 (1962), 1078.
10. White, A., B. Biggs, and S. Morgan. — J. Am. Chem. Soc. 62 (1940), 16.
11. Myasnikova, R. M., V. I. Robas, and G. K. Semin. — Zh. Strukt. Khim. 6 (1965), 474.

12. Brot, C. and J. Darmon. — J. Chim. phys. 63 (1966), 100.
13. Eveno, M. and J. Meinnel. — J. Chim. phys. 63 (1966), 108.
14. Semin, G. K., T. A. Babushkina, and V. I. Robas. — Doklady AN SSSR 169 (1966), 816.
15. Babushkina, T. A., V. I. Robas, and G. K. Semin. — Doklady AN SSSR 159 (1964), 164.
16. Semin, G. K., V. I. Robas, and T. A. Babushkina. — In: "Radiospektroskopiya tverdogo tela," p. 218. Atomizdat, 1967.
17. Suprun, A. P., A. S. Shashkov, T. A. Soboleva, G. K. Semin, T. T. Vasil'eva, G. P. Lopatina, T. A. Babushkina, and R. Kh. Freidlina. — Doklady AN SSSR 173 (1967), 1356.
18. Fuller, C. S. — Chem. Rev. 26 (1940), 143.
19. Bann, C. W. — J. Chem. Soc., p. 297, 1947.
20. Narita, S. and K. Okuda. — J. Polym. Sci. 58 (1959), 270.
21. Chen, H. Y. — J. Polym. Sci. 16 (1962), 12.
22. Semin, G. K. — Doklady AN SSSR 158 (1964), 1169.
23. Semin, G. K. — In: "Radiospektroskopiya tverdogo tela," p. 205. Atomizdat, 1967.
24. Kaufmann, H. S. — J. Am. Chem. Soc. 75 (1953), 1477.
25. Liang, S. and S. Krimm. — J. Chem. Phys. 25 (1953), 563.
26. Barnes, R. G., S. L. Segel, and W. H. Jones. — J. Appl. Phys., Suppl. 1, 33 (1962), 296.
27. Barnes, R. G. and S. L. Segel. — J. Chem. Phys. 37 (1962), 1895.
28. Kojima, S., K. Tsukada, S. Ogawa, and A. Shimauchi. — J. Chem. Phys. 23 (1955), 1963.
29. Narath, A. — J. Chem. Phys. 40 (1964), 1169.
30. Hughes, W. E., C. G. Montgomery, W. G. Moulton, and E. H. Carlson. — J. Phys. 41 (1964), 3470.
31. Carlson, E. H. and H. S. Adams. — J. Chem. Phys. 51 (1969), 388.
32. Pen'kov, I. N. and I. A. Safin. — Fiz. Tverd. Tela 7 (1965), 190.
33. Safin, I. A. — Zh. Strukt. Khim. 4 (1963), 267.
34. Pen'kev, I. N. and I. A. Safin. — Doklady AN SSSR 156 (1964), 139.
35. Safin, I. A. — Doklady AN SSSR 147 (1962), 410.
36. Safin, I. A. — Kristallografiya 13 (1968), 330.
37. Kravchenko, E. A. and G. K. Semin. — Izv. AN SSSR, ser. inorg. mater. 5 (1969), 1161.
38. Kravchenko, E. A., S. A. Dembovskii, A. P. Chernov, and G. K. Semin. — Physica Status Solidi 31 (1968), K19.
39. Kravchenko, E. A., S. A. Dembovskii, A. P. Chernov, and G. K. Semin. — Izv. AN SSSR, phys. ser. 33 (1969), 279.
40. Krainik, N. N., S. N. Popov, and N. E. Mel'nikova. — Izv. AN SSSR, phys. ser. 33 (1969), 271.
41. Craig, D. P. and N. L. Paddock. — J. Chem. Soc., p. 4118, 1962.
42. Whitehead, M. A. — Can. J. Chem. 42 (1964), 1212.
43. Negita, H. and S. Satou. — J. Chem. Phys. 24 (1956), 621.
44. Torizuka, K. — J. Phys. Soc. Japan 11 (1956), 84.
45. Dixon, M., H. D. B. Jenkins, J. A. S. Skuth, and D. A. Tong. — Trans. Faraday Soc. 63 (1967), 2852.
46. Kaplansky, M. and M. A. Whitehead. — Can. J. Chem. 45 (1967), 1669.
47. Negita, H. — J. Sci. Hiroshima Univ. 21A (1958), 261.
48. Negita, H. and S. Satou. — Bull. Chem. Soc. Japan 29 (1956), 426.
49. Negita, H. and Z. Hirano. — Bull. Chem. Soc. Japan 31 (1958), 660.
50. Lucken, E. A. C. — Proc. Colloque AMPERE 11 (1962), 678.

Chapter 9

NUCLEAR QUADRUPOLE RESONANCE OF ELEMENTS OTHER THAN HALOGENS

Compared with the halogens, the data on the quadrupole interactions obtained from NQR and NMR spectra of other elements are scarce and uncoordinated. An exception is nitrogen ^{14}N, and compounds containing ^{14}N have recently been studied intensively. Accordingly, in this chapter we shall describe in detail examples of quadrupole interaction data employed in solving some chemical problems. These examples are mainly crystallographical investigations of ionic compounds, since NQR has only just recently been used to study organometallic compounds.

1. GROUP I ELEMENTS

Deuterium ^2D. The quadrupole coupling constants of deuterium are small (due to the small quadrupole moment of the deuterium and the spherical symmetry of its valence electron). Thus, all the data on e^2Qq were obtained either from the hyperfine structure of the purely rotational microwave spectra in the gaseous phase or from the quadrupole splittings in the spectra of deuteric magnetic resonance in the solid state or in nematic liquids. Evaluations of e^2Qq based on the spin-lattice relaxation times T_1 of the NMR signals in liquids also exist. All the above methods for measuring e^2Qq of deuterium can also be applied to measure e^2Qq of the other elements of groups I and II and also, to a considerable extent, to e^2Qq of ^{14}N.

It must be kept in mind that the electric field gradient on the deuterium is determined almost entirely by the charge distribution on the atom to which it is bonded [1]. Deuterium has only one $1s$ valence electron. The absence of internal closed electron shells means that the Sternheimer factor is absent. Thus, accurate calculations of the electric field gradient on deuterium are possible. However, the overlapping term $\langle\psi_a|q|\psi_b\rangle$ here is more significant than in the other cases dealt with up till now.

The dependence of e^2Qq of deuterium on the nature of the atom to which it is bonded can be shown by the example in Figure 9.1, where the Y axis represents e^2Qq of deuterium in compounds with the M—D bond, and the X axis the force constants K_e of this bond [2]. The dependence of e^2Qq of deuterium on the hybridization of the carbon atom to which it is bonded has been determined in [3]. The calculations show that $e^2Qq_D = 153.6$, 189, and 208.7 kHz for the bond D—C, where carbon has sp^3, sp^2, and sp hybridization respectively. The experiment also confirms that e^2Qq_D increase in unsubstituted hydrocarbons with a decrease in the p character of the hybrid orbital of carbon.

The available data on the values of the quadrupole coupling constants indicate that the quadrupole coupling constants are weakly dependent on the field of the external charges. Thus, the values of e^2Qq and η obtained at low temperatures for $Ba(ClO_3)_2 \cdot D_2O$, $Li_2SO_4 \cdot D_2O$, $CuSO_4 \cdot 5D_2O$, etc., are similar. Chiba [1], who measured the temperature dependence of e^2Qq_D for $Ba(ClO_3)_2 \cdot D_2O$, showed that the approximation to 0°K yields $e^2Qq_D = 270$ kHz, which is only 13% lower than the value for the free D_2O molecule. This small difference can be caused by the increase in ionic character of the O—D bond when the substance passes from the gaseous phase to the solid state, and also by the influence of the adjacent Ba^{2+} ions.

The quadrupole coupling constant of a single crystal of D_2O is smaller by almost 30% than that of the free molecule. Such a sharp difference is attributed to changes in the electronic structure due to the formation of hydrogen bonds. MO LCAO calculations showed the validity of this assumption [4, 5]. When calculating the field gradient on the deuteron in the free D_2O molecule, a tetrahedral hybridization of the $2s$ and $2p$ orbitals of oxygen was assumed. The following charge distributions were considered: a) a point charge of +6 on the oxygen nucleus, which includes the internal electron shell; b) a pair of electrons on the molecular orbital of the considered OD bond; c) one electron on the tetrahedral hybrid orbital along the other OD bond; d) two pairs of electrons on the remaining tetrahedral hybrid orbital. For the experimental distance OD, equal to $d = 0.96$ Å, the quadrupole coupling constant was calculated to be $e^2Qq_D = 385$ kHz ($\eta = 0.08$), which agrees well with the experimental values ($e^2Qq_D = 312$ kHz and $\eta = 0.12$).

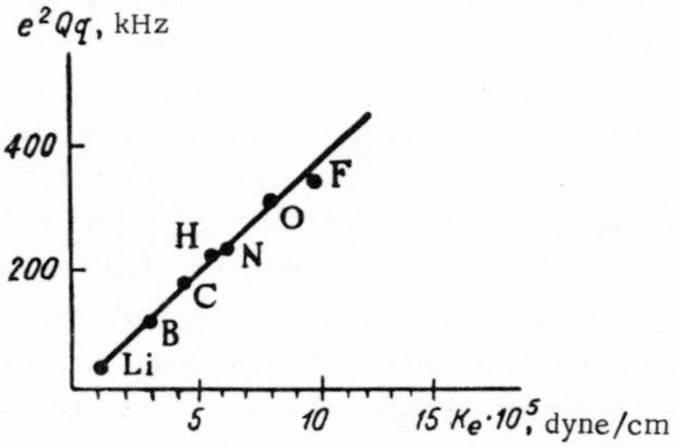

FIGURE 9.1. e^2Qq of deuterium in the M—D bond as a function of the force constant K_e of this bond

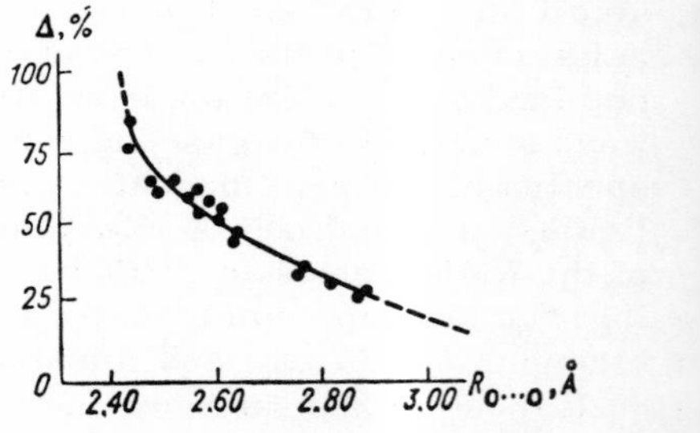

FIGURE 9.2. Chemical shift of the quadrupole coupling constant of deuterium

$$\Delta = \left(1 - \frac{e^2Qq_X}{e^2Qq_{HDO}}\right) \cdot 100\%$$

as a function of the length of the hydrogen bond O—H···O [6]

In the presence of two D_2O molecules with a bond of the type $-O_1-D \cdots O_2-$ the charge distribution changes: two positive point charges on the oxygen nuclei, nine electrons on the tetrahedral hybrid orbitals, and four electrons participating in the hydrogen bonds on the two molecular orbitals. When $d = 0.96$, the value obtained for e^2Qq_D is only 10% smaller than for the free molecule. An even smaller value of e^2Qq is obtained when the electron delocalization and the influence of the adjacent molecules are allowed for. The contributions to the quadrupole coupling constant can be divided into two: a positive one, determined by the influence of the nuclei of O_1, proportional to d^{-3}, and a negative one, due to the influence of the electrons and nuclei of O_2, linearly dependent on d. The experimental value of $e^2Qq = 215$ kHz is obtained when $d = 1.015$ Å.

The complete agreement with experimental data shows that the formation of the hydrogen bond and the lengthening of the O_1D bond by a distance equal to d is more important than the change in the electronic structure, although the two effects are additive.

It is well known that the stronger hydrogen bonds have longer distances in O_1D. The calculations predict a decrease in the quadrupole coupling constants for the stronger couplings.

Hydrogen bonds in crystals were studied by Blinc and Chiba [6, 7]. Figure 9.2 shows the chemical shifts of the quadrupole coupling constants as a function of the distance $O_1 \cdots O_2$, where the chemical shift is determined with respect to the free molecule HDO:

$$\Delta = \left(1 - \frac{e^2Qq_X}{e^2Qq_{HDO}}\right)\cdot 100\%. \qquad (9.1)$$

The extrapolation shows that $\Delta \rightarrow 0$ for $R_{O\cdots O} = 3.3$ Å. This value is usually taken as the lower limit for the existence of hydrogen bonds. A chemical shift equal to 100%, i.e. $e^2Qq_X = 0$, is obtained for $R_{O\cdots O} = 2.42$ Å, where both contributions to the field gradient on the deuterium become equal. Blinc assumed that e^2Qq of deuterium is very sensitive to a change in the bridge length, and for strong hydrogen bonds (for $R_{O\cdots O} \approx 2.6$ Å) it is even more sensitive than the valence OH vibrations. Even for symmetrical hydrogen bonds e^2Qq may be a positive number, but with a very small magnitude (e.g., for KD-maleate). If $e^2Qq \equiv 0$, then the investigated nucleus coincides with a center of sufficiently high symmetry.

Blinc assumed that knowing the values of the quadrupole coupling constants of deuterium would enable us to solve one of the most important problems of the hydrogen bond, namely, to determine the form of the potential surface. The quadrupole coupling constant seems fairly sensitive to the length of the O—H bond. It is important to note that $R_{O-H} = {}^1/_2 R_{O\cdots O}$ in a symmetrical potential well and $R_{O-H} < {}^1/_2 R_{O\cdots O}$ in a potential well with two minima. Therefore, e^2Qq must be smaller for a potential well than for a surface with two minima.

This principle is illustrated by the values of e^2Qq_D in triglycinesulfate and KD-maleate, in both of which the distance $R_{O\cdots O} = 2.44$ Å. However, the KD-maleate has a symmetrical hydrogen bond and $e^2Qq = 56$ kHz, while for triglycinesulfate $e^2Qq = 78.8$ kHz. This indicates that the O—H distance is smaller, and that the deuterium in the bond is not in the center. Obviously, it is impossible to determine beyond doubt the shape of the potential well using this example alone. The experiments in which the spin-lattice relaxation time T_1 and the quadrupole coupling constant are determined simultaneously are the most promising.

Deuteron quadrupole interactions are also an important source of information on the types of motion in the crystal and on the orientation of the X—D bonds in the crystal.

The slowed down internal motion of the molecules or of their different parts causes an averaging of the electric field gradient at the site of the investigated nucleus and this in turn causes a change in the quadrupole interaction constants [9, 10]. A good example is the decrease in e^2Qq_D for D_2O molecules when the temperature is increased. Reorientation motions of the D_2O molecules about the bisectrix of the angle O⟨D D at a sufficiently high temperature occur in almost all the crystal hydrates studied.

For deuterated benzene the quadrupole coupling constant decreases by a factor of almost two during the reorientation motion about the sixth order axis. Rotation of the NH_4^+ ion about the tetrahedral axis (e.g., for $(ND_4)_2SO_4$) causes the quadrupole coupling constant to decrease from 175 kHz to the negligible amount of ~4 kHz.

Lithium ^7Li. Most lithium-containing compounds (for which the value of the quadrupole coupling constant of ^7Li was determined) are ionic compounds and the quadrupole coupling constants in them are small ($e^2Qq_{Li} \leqslant 120$ kHz in the solid state [11]). The reason for this phenomenon is the small quadrupole moment [11] and the small Sternheimer factor [12], which is much larger than unity for ions having p and d electrons. The data on the quadrupole coupling constants for most compounds of group 1 elements are used to ascertain the structure of the compounds in the crystal more accurately and to elucidate the type of chemical bonds in them by comparing the calculated and measured values of the electric field gradient.

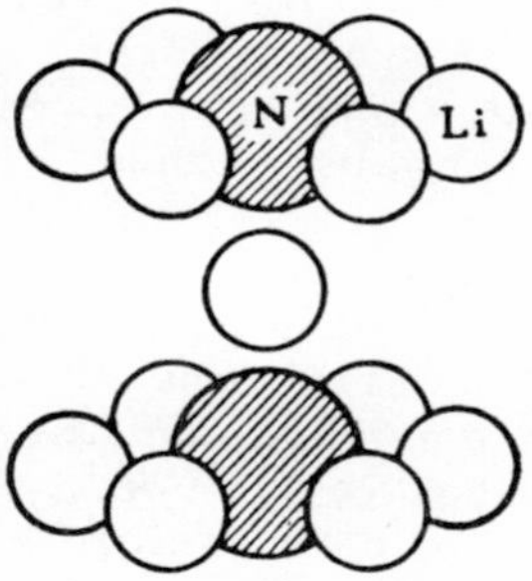

FIGURE 9.3. Structure of the Li_3N crystal [12]

The largest quadrupole coupling constant of ^7Li was observed in Li_3N [14, 15]. The crystal structure of Li_3N (Figure 9.3) indicates the existence of two nonequivalent lithium atom positions [16]. The quadrupole coupling constants are equal to 288 kHz for the lithium atoms between the planes and 584 kHz for the atoms at the apices of the hexagonal units.

A considerable part of the lithium compounds studied are ferroelectric crystals (e.g., $LiNbO_3$, $LiTaO_3$, LiTa-tartrate, $LiNH_4$-tartrate, $LiN_2H_5SO_4$, etc.). Burns [17] studied the temperature dependence of e^2Qq in these compounds to find a relationship between the changes in the ferroelectric properties at the Curie point and the quadrupole coupling constants. However, in all the compounds studied, there was a negligible change in e^2Qq with the change in temperature. The only exception was $LiN_2H_5SO_4$ in which the asymmetry parameter of the electric field gradient increases from 10 to 40% on heating from −196 to +23°C. The ferroelectric phase transition in these compounds does not seem to be associated with considerable changes in the field gradient at the site of the lithium nucleus. The calculation of the electric field gradients in ionic crystals and the comparison with the experimental values of e^2Qq yield very important information on the crystal structure. The electric field gradient on the considered nucleus in ionic crystals is mainly determined by the charges of the surrounding ions and the induced dipoles. Accordingly, the gradients are usually calculated using the electrostatic model of point charges and dipoles [25].

Thus, the calculations of the electric field gradients on the ^7Li and ^{27}Al nuclei in the alluminate $LiAl_5O_8$ [18], isomorphous to $LiFe_5O_8$ [19], having the structure of a spinel with four formula units per unit cell, enabled us to evaluate the distortions of the $LiAl_5O_8$ crystal compared with the ideal spinel lattice.

AlkLi organometallic lithium compounds have been studied (where Alk = CH_3, C_2H_5, C_4H_9) [21, 22].

The CH_3Li molecules form tetrameric structures in the crystal with lithium atoms at the apices of regular tetrahedrons and carbon atoms above the centers of the tetrahedral faces [20]. The structure of C_2H_5Li is distorted compared to the structure of CH_3Li. One of the Li—C distances is much longer than the others, and one of the Li—Li distances is much shorter. In the investigations of quadrupole splittings in alkyllithium compounds [21, 22] much smaller quadrupole coupling constants were detected than expected from the calculations of e^2Qq of ^7Li for C_2H_5Li in toluene solutions [23] (98 KHz in the solid state and 570 kHz in the liquid). The comparison of e^2Qq for CH_3Li (16 kHz) with C_2H_5Li (98 kHz) reveals that the small quadrupole coupling constant for CH_3Li seems to be due to the presence of large field gradients with opposite signs.

S o d i u m ^{23}Na. The quadrupole coupling constants of ^{23}Na are much larger than the constants of ^7Li in corresponding compounds. This difference is apparently caused by the large differences in the antiscreening factors of the electron shells and the quadrupole moments of these nuclei.

The quadrupole coupling constant of ^{23}Na is strongly dependent on the symmetry of the charge environment. If the coordination environment consists of six oxygen atoms arranged in the form of a slightly distorted octahedron, then the value of e^2Qq of ^{23}Na will lie in the range 330—860 kHz. One of the largest quadrupole interaction constants of ^{23}Na was found in $Na_2S_2O_3 \cdot 5H_2O$ [24] (2258 kHz). This large value of e^2Qq is due to the asymmetry of the nearest environment (five oxygen atoms and one sulfur atom).

Bersohn [25] calculated the quadrupole coupling constants of ^{23}Na in a number of crystals, namely, $NaNO_3$, $NaClO_3$, and $NaBrO_3$ by direct summation of the charge contribution of the lattice sums:

$$e^2Q\,(1-\gamma_\infty)\,(\Delta E)_{\text{ext}} = e^2Q\,(1-\gamma_\infty)\sum_j e_j \sum_i \frac{3z_{ij}^2 - r_{ij}^2}{r_{ij}^5}. \qquad (9.2)$$

The summation with respect to i is carried out over all the atoms of the crystal of type j with charge e_j.

The theoretical analysis for the complex ions XO_3^- (where X = N, Cl, Br) is somewhat tentative due to the absence of detailed data on the electronic charge distribution in these ions.

For $NaNO_3$ $e^2Qq_{Na} > 0$. The salts $NaBrO_3$ and $NaClO_3$ are isomorphous and, therefore, one should not assume that the signs of the field gradients on the ^{23}Na nuclei in these compounds will be different. However, starting from the requirement that the charges on the chlorine and bromine atoms must assume reasonable values, z_{Cl}, z_{Br}, and $(1 - \gamma_\infty)_{Na^+}$, Bersohn proposed [25] that $e^2Qq_{Na} < 0$ in $NaClO_3$ and $e^2Qq_{Na} > 0$ in $NaBrO_3$. Thus, $z_{Br} = +2$, which corresponds to Pauling's structure [26]:

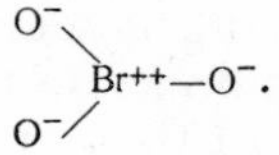

Bernheim and Gutowsky [27] studied the dependence of the quadrupole interaction constants of ^{23}Na in these compounds on the pressure and temperature (and, therefore, also on the volume). They found that $e^2Qq \cdot V^{-n} = \text{const.}$ (where $n = 3.8$, 1.9, 2.0 for $NaNO_3$, $NaClO_3$, and $NaBrO_3$, respectively). For the first two compounds, the authors of [27] explained that the results are due to the uniform increase in the distances between adjacent ions. For $NaBrO_3$ no reasonable assumptions concerning the charge distribution in the BrO_3^- ion can explain the value of n which ought to be smaller than for $NaClO_3$.

The method developed by Bersohn was also used successfully to calculate the charge distribution in $NaBF_4$ [28], for which the quadrupole interaction constants of both ^{23}Na and ^{11}B were determined.

Unlike the lithium-containing ferroelectrics, the quadrupole interaction constants of ^{23}Na change considerably on passing through the Curie point, in both deuterated [29] and nondeuterated [30] rochelle salts. The similarity between the values of $e\,Qq$ and η in deuterated and nondeuterated samples for the paraelectric state and their noticeable difference for the ferroelectric phase indicate the different spontaneous polarizations of these two crystals (see Table 9.1). The behavior of one of the quadrupole satellites during a gradual change in temperature from −40 to +40°C was also studied for the nondeuterated sample. The absence of discontinuous jumps of the satellite frequencies during the phase transition possibly indicates that this is a phase transition of the second order. For $NaNO_3$ [31] and $NaNO_2$ [32], a sharp change in the quadrupole interaction constant of ^{23}Na is observed during the phase transition.

TABLE 9.1. Quadrupole coupling constants of ^{23}Na in the rochelle salt and its deuterated analog in the para- and ferroelectric phases

T, °K	Nondeuterated		Deuterated	
	e^2Qq, MHz	η, %	e^2Qq, MHz	η, %
303	1.313	80.9	1.313	80.0
273	1.205	87.2	1.323	80.3
	1.407	67.5	1.419	63.9

A comparative examination of the quadrupole coupling constants of the halides of group 1 elements is possible.

Alkali metal halides were investigated in the gaseous phase by the microwave and molecular beam methods. Since diatomic molecules MHal (where M is an alkali metal) are purely ionic compounds, the electric field gradient on the nuclei of the ions can be calculated from the electrostatic model [34]:

$$eq_x = \pm \frac{2}{R^3}(1-\gamma_\infty)\left(1+\frac{3\alpha_y}{R^3}\right), \qquad (9.3)$$

where the sign depends on which nucleus (of the positive or negative ion) is considered, α_y is the dipole polarizability of the other ion, and R is the distance M—Hal.

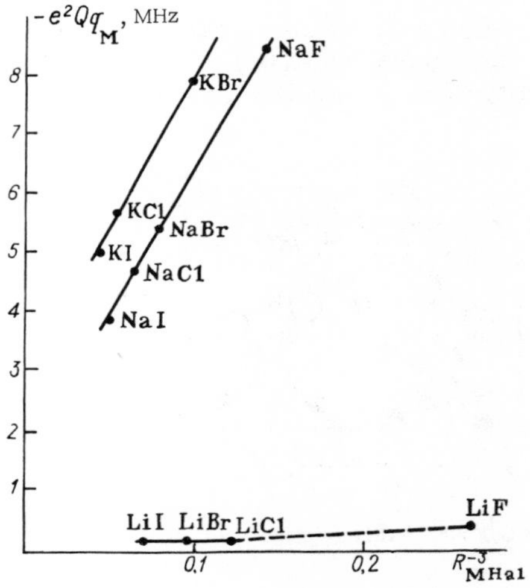

FIGURE 9.4. e^2Qq of alkali metals for MHal compounds as a function of R^{-3} (where R is the distance M—Hal)

For most alkali metal halides, the contribution of the dipole term is small and a linear dependence of e^2Qq_M on R^{-3} is observed (Figure 9.4). e^2Qq_{Li} in LiF does not lie on the line. In this compound, the bond is strongly covalent (see the quantum-chemical calculation of e^2Qq_{Li} in LiF [33]). For M^+Hal^- compounds we must distinguish between two types of covalence. The first is due to overlapping between the orbitals of the ions, which leads to a distortion of the spherical symmetry of the ions. The second is due to partial charge transfer from Hal^- to M^+, as a result of which the ionic charges are no longer equal to ±1. This latter type of covalence can be treated by the Townes-Dailey theory (see Chapter 5). The calculations show that the first type of interaction is the most important in alkali metal halides [34]. Thus, agreement between the theoretical and experimental results for e^2Qq_{Cl} in KCl is obtained only after we allow for the overlapping of the orbitals of the ions [39]. For the field gradient on the K^+ nucleus these effects are much less important.

Copper $^{63,65}Cu$, Gold ^{197}Au. Data on the quadrupole interactions of group IB elements are known only for a few compounds of copper and for one compound of gold.* The NQR in Cu_2O, including the temperature dependence of the NQR frequencies [36–38], the spin-lattice relaxation time T_1 [39, 40], and the Zeeman splitting of the NQR spectra at room temperature [41] have been studied in some detail. The data were interpreted on the basis of the ionic model of Cu_2O. The cubic unit cell of Cu_2O contains two molecules [42]. Each oxygen atom has a tetrahedral environment of four copper atoms, and each copper atom has two oxygen atoms at equal distances (see Figure 9.5). The calculation of the electric field gradient on the copper atoms from the point charges of the surrounding Cu^+ and O^- ions [25, 43] yields the value $\frac{e^2Qq}{h} = -\frac{e^2}{ha^3} \cdot 42.49\,(1-\gamma_\infty)Q$. On comparison with the empirical value of e^2Qq we obtain $\gamma_\infty = -16.45$. The theoretical value of γ_∞ is equal to -16.0. The difference between the calculated and measured values is usually due to the covalence.

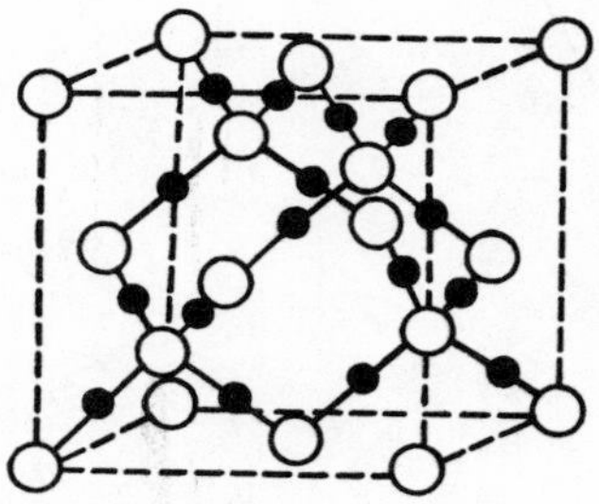

FIGURE 9.5. Structure of the Cu_2O crystal [42]

The temperature dependence of the NQR frequencies of Cu_2O can also be explained on the basis of the ionic model. The motion of the copper ions about the nearest oxygen atoms, i. e. the deformation vibrations of the Cu—O bonds, are taken into account. Satisfactory agreement with the experiment was obtained when the frequency of the transverse vibrations was equal to 97 cm^{-1}. This value is close to that obtained when using data from Raman spectroscopy.

The determination of the NQR frequencies of ^{63}Cu in $KCu(CN)_2$ [44] and the asymmetry parameter of the electric field gradient on the copper atoms [45], and the study of the crystal structure [46] enabled us to determine the type of bond in this compound [47]. The structure consists of spiral chains of copper atoms and CN groups. Each copper atom is bonded to the carbon

* Data for evaluating the electric field gradients on gold nuclei can also be obtained from Mössbauer data (cf. "Khimicheskie primeneniya messbauerovskoi spektroskopii," [Chemical Applications of Mössbauer Spectroscopy], Izd. "Mir," 1970).

atom of one CN group and to the nitrogen atom of another. The third CN group is bonded to the copper atom by the carbon atom, and its nitrogen atom does not participate in the coordination (Figure 9.6).

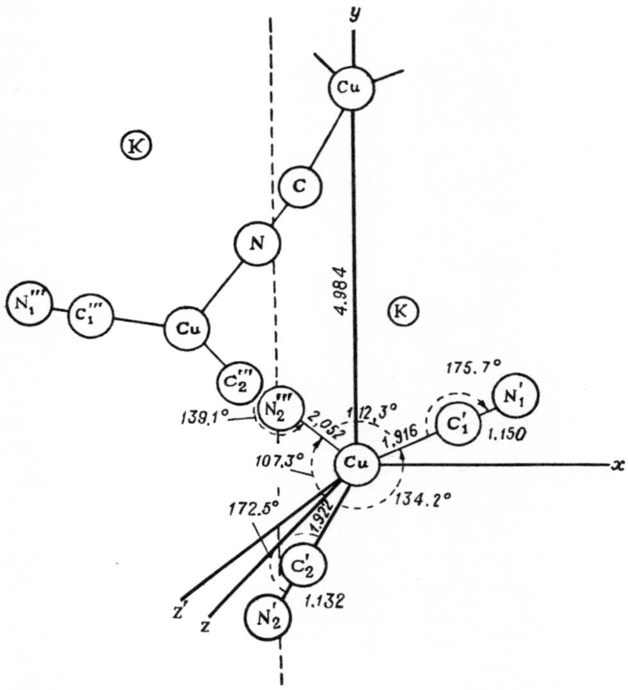

FIGURE 9.6. Arrangement of atoms in the molecule $KCu(CN)_2$ [47]

The asymmetry parameter of the field gradient is large and equal to 78%. The components of the tensor of the electric field gradient were calculated on the basis of the following model [47]: 1) the copper atom forms three sp^2 hybridized σ bonds with the adjacent CN groups; 2) the Cu—C bonds have a considerable degree of double bonding of the type d—π^*, in which the electrons of the full copper d orbitals are transferred to the empty antibonding π^* orbital of the CN group; 3) the C—N bond has a considerable ionic character.

Comparison of the measured and calculated values of e^2Qq yields a Sternheimer factor $\gamma_\infty = -12.2$. Since $KCu(CN)_2$ is not an ionic compound, this value can be considered quite satisfactory (cf. Cu_2O).

The data on the quadrupole interaction constants of ^{63}Cu also confirm the existence of δ bonds in the dimeric copper acetate $Cu_2(C_2H_3O_2)_4 \cdot 2H_2O$ [48]. The existence of a strong Cu—Cu interaction is indicated by the short Cu—Cu distance, equal to 2.64 Å [49], and by the singlet ground state [50].

In copper acetate the d orbitals of the two copper atoms form three molecular orbitals: $a_{2u}(\sigma^*)$, formed by the atomic d_{z^2} orbitals and directed along the Cu—Cu axis; $b_{1g}(\delta)$ and $b_{1u}(\delta^*)$, formed by the $d_{x^2-y^2}$ orbitals and perpendicular to the Cu—Cu axis [51, 52].

Vacant δ^* orbitals correspond to δ bonds between the two copper atoms, and vacant σ^* orbitals correspond to σ bonds. The data on the values of the quadrupole interaction constants for the ground state and the lowest triplet state $e^2Qq_{gr} > e^2Qq_{tr}$ indicate the following sequence of the orbitals: $\delta < \sigma^* < \delta^*$.

The only compound in which a NQR signal of ^{197}Au was observed is AuCl: $e^2Qq_{Au} \approx 514$ MHz [53]. The fairly large electric field gradient on the gold nucleus ($13 \cdot 10^{15}$ e.s.u $\cdot$ cm^{-1}) is due to the *sp* hybridization of the orbitals. Taking into account that the difference between the electronegativities of gold and chlorine is not large, we can assume that the ionic character of the Au—Cl bond is also small.

2. GROUP II ELEMENTS

The number of compounds in which the quadrupole interaction of group II elements has been studied is very small. All of them, except mercury chloride in which the NQR of both ^{201}Hg and ^{35}Cl were studied, are ionic compounds. Nuclear quadrupole interaction was also studied in metallic beryllium, magnesium, and zinc. These metals have hexagonal crystal structures. The electric field gradient on the nuclei of the atoms of these metals can be considered as consisting of two parts [54]:

$$eq = eq_1 (1 - \gamma_\infty) + eq_2 (1 - R_Q), \tag{9.4}$$

where the first term is the electric field gradient of the surrounding ions, and the second term is the contribution of the conduction electrons to the field gradient. R_Q is the atomic screening factor. The calculation shows that the contribution of the conduction electrons to the field gradient is considerable [54] (see Table 9.2). The effect of the screening of the conduction electrons appears to lead to a decrease in the total ionic valence of the metal.

The quadrupole interaction constants in the beryllium oxides beryl and chrysoberyl were interpreted on the basis of the ionic model [55—58]. However, allowing for only the contribution of the point charges in the electric field gradient in BeO [55, 56] leads to $e^2Qq_{Be} = 7$ kHz (the experimental value is $e^2Qq_{Be} = 41$ kHz). Since the symmetry of the arrangement of the Be^{2+} and O^{2-} ions differs from ideal tetrahedral symmetry, it is necessary to take into account the contribution of the dipole and quadrupole moments of the oxygen ion in the electric field gradient [55]:

$$\vec{E}(\vec{\mu}) = \frac{\vec{\mu}}{\alpha},$$

where E is the electric field created by the induced dipole μ. It was found that in BeO, $\alpha_{dip} = 2.19$ Å^3 and $\alpha_{quad} = 7.0$ Å^5 for the dipole and quadrupole polarizability of the O^{2-} ion. The dipole contribution is greater than the quadrupole contribution which is much greater than the contribution from the point charges.

TABLE 9.2. Theoretical and experimental values of e^2Qq for the nuclei of group II metals

Metal	$1-\gamma_\infty$	e^2Qq_{exp}	$e^2Qq_1\ (1-\gamma_\infty)$
^{9}Be	0.81	0.062	0.068
^{25}Mg	0.22	0.23	0.105
^{67}Zn	0.16	79	5.7

The values of the quadrupole interactions for ^{9}Be and ^{27}Al were determined in beryl [$Be_3Al_2(SiO_3)_6$] and chrysoberyl $BeAl_2O_4$. The results of the calculation of e^2Qq and η in chrysoberyl [57] are shown in Table 9.3 (α_{dip}= 1.00 Å^3 for O^{2-}).

TABLE 9.3. Experimental and calculated values of e^2Qq and η for aluminum and beryllium nuclei in chrysoberyl

Ion	η exp	η calc	e^2Qq exp	e^2Qq calc
Al (1)	0.76	0.80	2.846	2.60
Al (2)	0.94	0.75	2.850	2.26
Be	0.90	0.98	0.318	0.35

The quadrupole and higher orders of polarization were neglected in the calculations. However, the agreement with the experimental data was quite good.

The $HgCl_2$ crystal has an orthorhombic symmetry and consists of linear molecules arranged in parallel planes [42]. However, the two atoms are nonequivalent and, therefore, their NQR frequencies are different. Assuming that the hybridization of the chlorine atom in the Hg—Cl bond is equal to 15%, the ionic character of the Hg—Cl bond is estimated to be ~50% [59]. If the Hg—Cl bond is formed with the 6s6p hybrid orbital of the mercury atom, then the *p* character of this bond is equal to 50%. The evaluation of the quadrupole moment of ^{201}Hg based on these data gives $eQ = 0.6$ barn. The value $eQ = 0.559$ barn was obtained from the hyperfine splitting of the atomic absorption spectra.

The NQR frequencies of barium were studied either by double resonance NMR—NQR [60, 61] or by pulsed NQR. The investigation of the NQR of ^{135}Ba and ^{137}Ba is complicated by the small natural abundance of these isotopes.

3. GROUP III ELEMENTS

The elements of group III have a stronger trend to form covalent bonds than those of groups I and II. The ionic character of their compounds increases down the periodical table, so that boron compounds are the least ionic and indium compounds are the most ionic.

Boron ^{11}B. The trihalogen derivatives of boron in both the solid and gaseous phase have a planar D_{3h} structure. Lucken [64] suggests the following explanation of this fact: the formation of a strong $p_\pi - p_\pi$ bond and, therefore, the filling of the vacant $2p_z$ orbital of the boron atom through the formation of a structure of the type:

$$\begin{array}{c} \text{Hal} \\ \diagdown \\ \quad B^- = Hal^+ \\ \diagup \\ \text{Hal} \end{array}$$

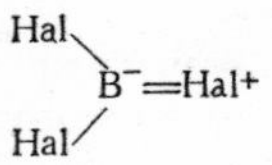

is the most stable situation for the boron atom.

The assumption that the B—Hal bond possesses a considerable degree of double bonding is confirmed by the large value of the asymmetry parameter of the electric field gradient on the halogen nuclei. Thus, in BHal$_3$ $\eta = 45\%$ for Br [65], $\eta = 45\%$ for I [66], which corresponds to a degree of double bonding of 13.3 and 15.6%, respectively. The other elements of group III (Al, In, Ga) fill their vacant orbitals by forming bonds through the bridge halogen atoms.

Love [67] measured the quadrupole interaction constants of ^{10}B and ^{11}B in the series R$_3$B (where R = CH$_3$, C$_2$H$_5$, C$_3$H$_7$, C$_4$H$_9$). It is assumed that these compounds are planar, although there are no data on their crystal structure. The asymmetry parameter of the electric field gradient on the boron atom is close to zero, i.e. these compounds have a third order axis along the z axis. Therefore, the mean electron densities N_x and N_y will be equal and the density of the uncompensated p electrons will be a function of two variables only, $U_p = f(N_x, N_z)$. Love further assumes that the inductive effects of the alkyl groups are very similar and, therefore, the changes in N_x, reflecting the change in σ^* of the different alkyls, will be much smaller than the changes in N_z, i.e. $U_p = f(N_z)$. When the boron configuration changes from planar trigonal sp^2 to tetrahedral sp^3, the electron density of N_z changes from 0 to 1. Then U_p changes from 1 to 0 (for $N_x = N_y = 1$). The electron density on the boron atom is determined by the degree of delocalization of the B—C bond. It also depends on the degree of overlap of the C—H bond with the p_z orbital of the boron (Figure 9.7). Love held that the change in the quadrupole coupling constants of ^{11}B is due to the change in the degree of delocalization of the C—H bond.

Figure 9.7, a shows that for different conformations of the methyl group in the trimethylboron molecule the p_z orbital of boron overlaps with the 1.7—2.0 C—H bonds on each methyl group (the methyl group can rotate freely about the B—C bond). Figure 9.7, b, c shows that for the ethyl, propyl, and butyl groups the p_z orbital of boron overlaps with the 1.0—1.7 C—H bonds in the group. However, in this case the potential function will have unequal minima when the CH$_2$R group rotates about the BC bond.

Figure 9.7, d, e shows the conformations of (iso-C$_3$H$_7$)$_3$B. Taking into account the steric hindrances the most probable variant is e where the p_z orbital of the boron does not overlap with the C—H bond.

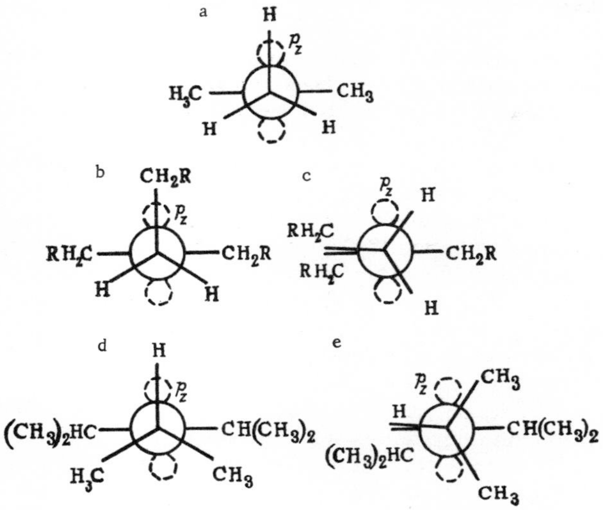

FIGURE 9.7. Possible conformations of trialkylboron molecules [60]: R = H, CH_3, C_2H_5.

Therefore, $(CH_3)_3B$ has 5.7—6.0 overlaps with the C—H bonds; $(C_2H_5)_3B$, $(C_3H_7)_3B$, and $(C_4H_9)_3B$ have 3.0—5.7 overlaps; and (iso-$C_3H_7)_3B$ apparently has no overlap of the p_z orbital of the boron with the C—H bond. If we take into account that e^2Qq of ^{11}B for $(C_2H_5)_3B$, $(C_3H_7)_3B$, and $(C_4H_9)_3B$ are practically equal, and the difference Δe^2Qq of ^{11}B, equal to 0.125 MHz for $(CH_3)_3B$ and $(C_2H_5)_3B$, increases to 0.249 MHz for $(CH_3)_3B$ and (iso-$C_3H_7)_3B$, it is obvious that the phenomena observed can be explained by the classical theory of delocalization of the carbon-hydrogen bond.

The number of inorganic compounds of boron in which the nuclear quadrupole interaction of ^{11}B has been studied is fairly large. Most of them have similar structures. In most cases, NQR was used to determine the variation in the coordination number of boron [68, 69].

If the boron atom is in a highly symmetrical environment of $BO_4(BPO_4)$ or $B(OH)_4$ [$Na_2B(OH)_4Cl$] groups, the quadrupole interaction constant is close to zero. Large values of e^2Qq equal to about 0.3—0.7 MHz are observed in hydrated polyborates for the tetrahedral groups $BO_2(OH)_2$ and BO_3OH. In these cases the asymmetry parameter can be quite large.

The quadrupole interaction constants of ^{11}B in trigonal borates lie in the range 2.5—2.8 MHz. These values of e^2Qq can be explained on the basis of the sp^2 hybridization of the boron atom which causes the B—O bond to have a considerable degree of ionic character. For the different types of trigonal borates only small changes in e^2Qq are observed and the groups BO_3, BO_2OH, and $B(OH)_3$ cannot be distinguished by this method [69]. In general, the coordination of the boron atoms in most of the compounds studied is determined by the arrangement of the boron atom and the atoms surrounding it, which are bonded covalently to the boron [69].

The quadrupole interaction constants of boron decrease when the oxygen is replaced by sulfur or selenium ($e^2Qq = 2.76$, 2.7, 2.5 MHz for B_2O_3, B_2S_3, and B_2Se_3, respectively). This may be due to two phenomena: a decrease in the inverse donor effect X—B (increasing the population of the p_z orbital of boron), and a decrease in the ionic character of the B—X bond [87].

The determination of the quadrupole interaction constants of ^{11}B is aided by studying boric glasses obtained from B_2O_3. Both two-component [70] and three-component [71] glasses were studied. The addition of an alkali metal to the boron oxide leads to the formation of a BO_4 configuration in the glasses. Figure 9.8 shows the number of boron atoms in the tetrahedral BO_4 configuration as a function of the content of alkali metal oxide. The maximum is observed at 40 mole % alkali metal oxide.

Three component glasses of composition $xBaO \cdot B_2O_3 \cdot ySiO_2$ were studied in [71]. It was shown that the number of boron atoms in the tetrahedral BO_4 configuration is maximum for a 35% molar fraction of BaO and the quadrupole interaction constant is equal to 0.430 MHz. If the content of BaO is less than 33%, then the number of BO_4 tetrahedrons increases by 2.12—2.20 on each additional oxygen atom. The quadrupole interaction constant in the trigonal configuration is equal to 2.76 MHz. In the presence of SiO_2 several broad lines are observed, i. e. there is a series of different types of boron in a trigonal environment.

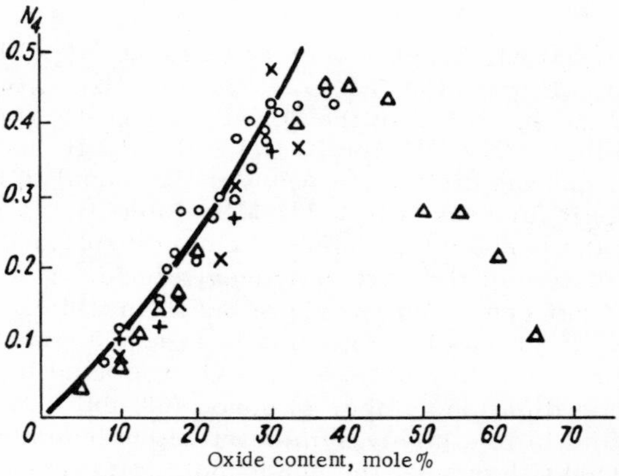

FIGURE 9.8. Number of boron atoms (N_4) in the tetrahedral BO_4 configuration as a function of the content of alkali metal oxide [62]:

● — Na_2O; ○ — K_2O; △ — Li_2O; + Rb_2O; × — Cs_2O.

The data on the quadrupole interaction constants of ^{11}B in KBF_4 [70] and $NaBF_4$ [28], where the boron atom is in the tetrahedral environment of the B—F bonds, are also of interest. The electric field gradient on the boron nucleus is due to the point charges at the sites of the sodium and fluorine nuclei. The calculation established that the effective charge for $NaBF_4$ on the fluorine atom in the tetrahedron is equal to −0.5 and, therefore the charge formula can be written in the form $Na^+B^+F_4^{-1/2}$ [28]. The point charge model, and the data on the quadrupole interaction and the crystal structure agree with a 55% ionic character for the B—F bonds. The high-temperature phases for KBF_4 and $NaBF_4$ are nonisomorphous. However, KBF_4 has a simple crystal structure above 300°C, and $NaBF_4$ above 250°C does not have a cubic point symmetry at the sites of the B atoms.

Aluminum ^{27}Al, gallium $^{69,71}Ga$, indium ^{115}In. The characteristics of the structure of the trihalogen derivatives of group III elements (Al, Ga, In) and their NQR spectra were examined in Chapter 7. $AlCl_3$ is an exception (and possibly also $InCl_3$ whose NQR spectrum for both ^{115}In and ^{35}Cl is unknown). The crystal structure of $AlCl_3$ is very different from the structure of the other $AlHal_3$ (which are dimeric) and consists of parallel layers of Cl^- and Al^{3+} [62] (see Figure 9.9). The quadrupole interaction constant of ^{27}Al is very small (0.4 MHz), and the asymmetry parameter is equal to $\eta = 0$ [63]. The calculation for a fully ionic layered structure of $AlCl_3$ with uniformly charged planes gives the value $e^2Qq = -2.78$ MHz and $\eta = 0$ [64]. The best agreement with experimental data is obtained for $(1 - I) = 1/3$ (where I is the degree of ionic character of the Al—Cl bond).

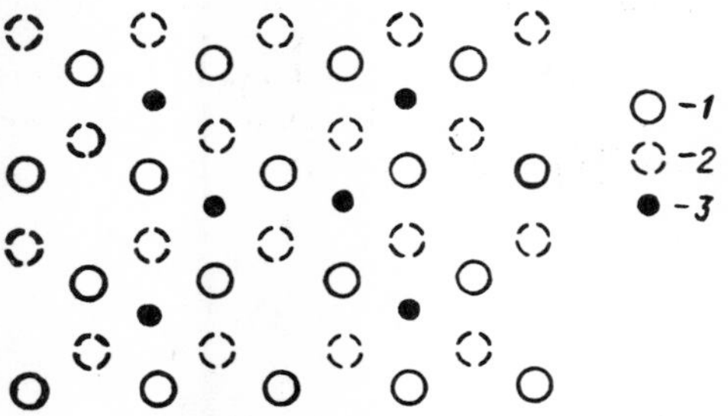

FIGURE 9.9. Structure of the $AlCl_3$ crystal [42]: 1) upper layer of Cl^-; 2) lower layer of Cl^-; 3) intermediate layer of Al^{3+}.

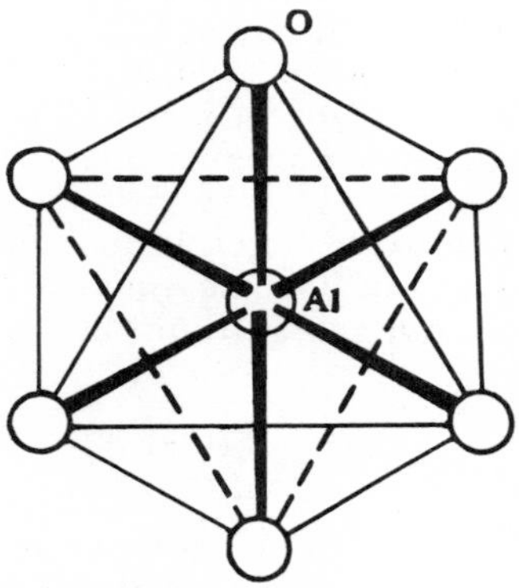

FIGURE 9.10. Coordination environment of the aluminum atom in Al_2O_3 [42]

The Al_2O_3 crystal has a rhombohedral unit cell with two Al_2O_3 molecules. The six oxygen atoms nearest to the aluminum atom (Figure 9.10) are arranged in the form of an irregular octahedron and form two parallel equilateral triangles on each side of the aluminum atom. For the aluminum atom there is a third order axis perpendicular to the two oxygen triangles. Bersohn calculated the field gradient on the Al^{3+} nucleus and assumed that the structure is 100% ionic [25]. Using the Sternheimer factor $\gamma_\infty = -2.59$ for the Al^{3+} ion, he obtained a quadrupole interaction constant 35% smaller than the observed value. This points to a certain degree of covalence in the Al—O bond.

Nuclear quadrupole interactions for ^{27}Al were observed in a number of zeolites, such as natrolite [75], alum [80] and other minerals [76—79].

Here, as for ^{11}B, it is important to determine the coordination number of the aluminum atom, which can be equal to 4 or 6. However, for ^{27}Al it is much more difficult than for ^{11}B to uniquely determine the position of the aluminum ion and its coordination number from the value of the quadrupole interaction constant. Nevertheless, it was shown that spinel $MgAl_2O_4$ [77], and yttriumaluminum garnet $Y_3Al_5O_{12}$ [81] have aluminum atoms with both octahedral and tetrahedral environments.

The Al—O bond is more ionic than the B—O bond and, therefore, the influence of the surrounding ions on the AlO_4 tetrahedron or the AlO_6 octahedron is stronger. The configuration of the oxygen atoms surrounding the aluminum atom is easily distorted and this causes large changes in the electric field gradient on the ^{27}Al nucleus even when the coordination number of the aluminum atom is preserved. As a result, the study of the aluminum-containing glasses and the determination of the coordination number of the aluminum atom in them on the basis of measuring the quadrupole interaction constants of ^{27}Al present considerable difficulties [82].

The study of the quadrupole interaction of ^{27}Al in corundum at different concentrations of Cr^{3+} showed that the content of Cr^{3+} does not affect the electric field gradient [83]. The combined use of NQR and EPR was found to be particularly useful for studying the physical properties of corundum and a number of alums [84, 85].

The NQR of ^{69}Ga and ^{71}Ga was studied in a much smaller number of inorganic compounds than the NQR of boron- or aluminum-containing compounds.

The oxide Ga_2O_3 is isomorphous with Al_2O_3, and the quadrupole interaction constant of ^{71}Ga is equal to 8.20 MHz. By comparing the values of e^2Qq for aluminum and gallium the following expression was obtained:

$$(1-\gamma_\infty)_{Ga^{3+}} = (1-\gamma_\infty)_{Al^{3+}} \frac{e^2Qq_{^{71}Ga}\,(a_i^3)_{Ga_2O_3}\,Q_{^{27}Al}}{e^2Qq_{^{27}Al}\,(a_i^3)_{Al_2O_3}\,Q_{^{71}Ga}}, \qquad (9.5)$$

where a_i is the parameter of the unit cell [73].

Using the known data for $\gamma_\infty^{Al^{3+}}$, a_i, and Q, Veigele [73] obtained a Sternheimer factor $\gamma_\infty = -14.4$ for Ga^{3+}, which is different from the value calculated by Sternheimer himself in [74]: $\gamma_\infty = -9.5$. Possibly, as for Al_2O_3, the difference between the calculated and measured values of γ_∞ is due to covalence.

The structure of In_2O_3 is different from the structure of Al_2O_3 and Ga_2O_3 [86]. The observed NQR spectrum completely agrees with the structure: there are two values of the quadrupole interaction constant e^2Qq of ^{115}In with different values of the asymmetry parameter. The NQR lines for $e^2Qq = 128$ MHz ($\eta = 85\%$) are more intense than for $e^2Qq = 182$ MHz ($\eta \approx 1\%$). This fits in with the fact that there are two In_2O_3 molecules with different multiplicities in the unit cell [87].

The double chalcogenides of indium, namely In_2S_3, In_2Se_3, In_2Te_3 are ionic crystals. Compounds of indium with sulfur, selenium, and tellurium exist in several crystalline modifications [88]. However, direct research on their structure is difficult due to the existence of vacancies at the nodes of the cationic sublattice, narrow regions of homogeneity, deviations from stoichiometry, and a tendency to twinning. NQR can serve as a criterion for the selection of a structural model. It follows from the NQR data that in crystals of β-In_2S_3, α- and β-In_2Se_3, and α-In_2Te_3 the indium atoms occupy two crystallographically nonequivalent positions. The higher asymmetry parameters of the electric field gradient in β-In_2Se_3 ($\eta_1 = 42.5\%$, $\eta_2 = 74.4\%$) compared with α-In_2Se_3 ($\eta_1 = 1.9\%$, $\eta_2 = 3.9\%$) reflect the higher symmetry of the environment of the indium ions in the second case.

The oxyhalides of indium InOHal (where Hal = F, Cl, Br) have layered structures [89], in which the covalently bonded layers of indium and oxygen alternate with layers of halogen ions. The calculations of e^2Qq of ^{115}In and η using the electrostatic model of charges and ion dipoles [90] indicate that for the following charge distribution: $In^{+1.5}O^{-0.5}Hal^{-1}$, the electric field gradient and its asymmetry are determined within 10–30% by the ionic contribution. The large asymmetry parameters of the electric field gradient on the metal atom depend on the symmetry of the oxygen environment of the indium atom. Thus, for InOBr the angles In—O—In are equal to 136 and 114°, for InOCl the angles In—O—In are equal to 139 and 109°, and η = 70.2% and 85.0%, respectively.

Organometallic indium and gallium compounds have been studied recently for the first time. X-ray crystal analysis shows that the $In(CH_3)_3$ molecule forms a tetrameric structure with the indium atoms at the center of a strongly distorted bipyramid [91]. At the same time, IR data indicate that the crystal has a planar monomeric structure [92]. The values of e^2Qq for ^{115}In and ^{69}Ga in $In(CH_3)_3$, $In(C_2H_5)_3$, $Ga(C_2H_5)_3$ and $Ga(C_6H_5)_3$ are very large (larger than e^2Qq_{at}). To explain this we assume that the groups at the apices of the bipyramid contribute little to the electric field gradient, i.e. the central atom is in an sp^2 hybrid state [93].

D i a t o m i c h a l i d e s o f g r o u p III e l e m e n t s. The presence of an inert pair of electrons ns^2 in group III elements enables them to form aluminum, gallium, indium and thallium monohalides in the gaseous phase.

The quadrupole interactions in them were studied by microwave methods (Table 9.4).

TABLE 9.4. Characteristics of the bonds in MHal compounds (M = Al, Ga, In, Tl) according to data on the quadrupole interaction constants

Compound	e^2Qq_{exp}, MHz		e^2Qq/e^2Qq_0 [64]		I, % [94]	M, % [94]	$a_\pi + a_\sigma$ [64]	sa_σ [64]
	Hal	M	Hal	M				
AlF	. . .	—37.6						
AlCl	—8.8	—29.2	0.081	0.778	91	27	1.919	0.697
GaCl	—20	—84.7	0.183	0.677	79	21	1.817	0.494
GaBr	—134	—74	0.174	0.595	80	18	1.826	0.421
GaI	—549	—66	0.246	0.527	72	13	1.754	0.281
InF		—724						
InCl	—13.7	—657	0.165	0.729	81	24	1.835	0.564
InBr	138	—642	0.179	0.714	79	23	0.821	0.555
TlCl	—15.8							
TlBr	13.0							
TlI	—537							

Barrett and Mandel [94], in their discussion on e^2Qq in these compounds, assumed a 15% *sp* hybridization and the absence of a π bond in the halogen atoms. The interaction constant e^2Qq_0 is equal to −37.5, −125, and −899 MHz for ^{27}Al, ^{71}Ga, and ^{115}In, respectively (Table 9.4). However, Lucken [64] holds that the Townes-Dailey rule of 15% *s* hybridization of the negative ion of the diatomic molecule presupposes a zero hybridization of the positive ion of this molecule, which contradicts the results of [94]. Accordingly, Lucken makes other assumptions regarding the bond in these compounds. He assumes that the hybridization of the halogen is zero, and that the $p_\pi - p_\pi$ M−Hal bond has a definite constant value [64]. Then, if a_σ is the population of the σ orbital of the halogen and a_π is the population of the π orbital of the metal, we obtain:

$$\left(\frac{e^2Qq}{e^2Qq_0}\right)_{\text{Hal}} = 2 - (a_\sigma + a_\pi)$$

$$\left(\frac{e^2Qq}{e^2Qq_0}\right)_{\text{M}} = 2 - (a_\sigma + a_\pi) + sa_\sigma,$$

where *s* is the *s* character of the σ orbital of the metal.

The results of the calculation, given in Table 9.4, indicate a considerable hybridization of the metal orbitals. For the possible value of $a_\sigma \approx 1.5$, $s \approx 0.2-0.3$, which is close to the value obtained by Barrett and Mandel [94]

With regard to the data on the quadrupole interaction in InCl and TiCl, Delvigne and Wijn [95] point out the considerable contribution of the ionic configuration M^+Cl^- and hold that the electric field gradient on the two ions is created as a result of the dipole and quadrupole polarizabilities induced by the charges of the adjacent ion. The evaluation of the contribution of the ionic configuration yields the value of 75% for TlCl and 77% for InCl [95]

Trihalides of rare-earth elements. The trichlorides of rare-earth elements, in which the NQR of ^{35}Cl has been studied have three crystal forms: the chlorides of light rare-earth elements from $LaCl_3$ to $GdCl_3$ are isomorphous with hexagonal UCl_3; the chlorides of heavy rare-earth elements from $DyCl_3$ to $YbCl_3$ have the layered structure of monoclinic $AlCl_3$; $TbCl_3$, $NdBr_3$, and also the low-temperature phase of $DyCl_3$ are isomorphous with orthorhombic $PuBr_3$ [86]. The NQR frequencies of the halogens and the asymmetry parameters of the electric field gradients of these compounds are shown in Table 9.5 which also shows the NQR data for the halides of chromium, ruthenium, and uranium.

All the chlorides isomorphous with UCl_3 have one NQR line (all the chlorine atoms are equivalent) and a large asymmetry parameter of the electric field gradient [96]. This is what we would have expected in view of the structure.

In the $AlCl_3$ structure, the Cl^- ions are in positions with the following point symmetries: 4Cl(1): (*i*)*m*, 8Cl (2): (*j*)1. As a result, each of the trichlorides of this group have two NQR lines, one twice as intense as the other. Note that there are chlorine atoms in position *j* which have a lower frequency in all compounds except $CrCl_3$ and $RuCl_3$ [96, 97].

$PuBr_3$ also has a structure in which there are two positions of the halogen atoms with different multiplicities: position *c* (multiplicity 1) and position *f* (multiplicity 2). However, the symmetry of the environment of the halogen atoms in position c is much higher ($\eta \approx 7-15\%$) than in f ($\eta \approx 50-70\%$) [97, 98].

TABLE 9.5. NQR frequencies of the halogens and asymmetry parameters of the electric field gradients of the trihalides of rare-earth elements

Compound	T, °K	ν, MHz	η, %
		UCl_3 structure	
$LaCl_3$	77	4.167	50
$CeCl_3$	77	4.387	. . .
$PrCl_3$	77	4.5667	58 (49)
$NdCl_3$	77	4.7290	. . .
$SmCl_3$	77	5.0330	. . .
$CdCl_3$	77	5.315	42.5
UCl_3	77	4.51	. . .
		$AlCl_3$ structure	
$RuCl_3$	. . .	16.815	. . .
		16.880×2	. . .
$CrCl_3$	300	12.9080	. . .
		12.9472×2	. . .
YCl_3	4.2	4.2122×2	. . .
		4.2731	. . .
$DyCl_3$	300	4.203	. . .
		4.206	. . .
$HoCl_3$	77	4.3027×2	. . .
		4.3647	. . .
$ErCl_3$	77	4.4495×2	52
		4.506	53
$YbCl_3$	77	4.7817×2	. . .
		4.8366	. . .
$TlCl_3$		14.756	. . .
		14.812	. . .
		$PuBr_3$ structure	
$TbCl_3$	77	7.45	. . .
		4.30×2	. . .
$NdBr_3$	Room	$e^2Qq = 89.076$ MHz	6.8
		$e^2Qq = 45.348$ »	48.7
UI_3	6	$e^2Qq = 365.35$ »	15.4
		$e^2Qq = 183.49$ »	70.6

All the trichlorides have an ionic structure. Assuming that the principal contribution to the electric field gradient on the Cl^- ion is due to the two nearest M^{3+} ions, we can use the theory of [96], which was developed for the iodides of group VA elements [99] (see equations (7.12) and (7.13)). According to this theory, each M^{3+} ion creates an axially symmetrical field gradient on the Cl^- ion, which has a zz component equal to eq_{at}. Thus,

$\eta = -3\cos\theta$, where θ is the angle 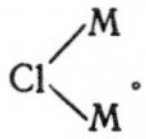 . For $ErCl_3$, $\theta^{(1)}_{NQR} = 100 \pm 0.6°$, $\theta^{(2)}_{NQR} = 100.2 \pm 0.4°$, while X-ray data yield $\theta^{(1)} = 97.5°$, $\theta^{(2)} = 98.1°$.

The calculation using the ionic model, which allows only for the charge term in compounds with a UCl_3 structure, is not very satisfactory [96]. The observed NQR frequencies increase smoothly from $LaCl_3$ to $GdCl_3$ by 27%, while the calculated frequencies increase by only 11%. However, according to the experiments, the point charge models show that the z axis of the field gradient tensor must be directed along the C_3 axis [99].

The calculations of the electric field gradient in $ErCl_3$ [96] are also inadequate. Obviously, allowing for the covalence (when the electrons are not localized on one ion) can lead to agreement with the experimental data.

4. GROUP VA ELEMENTS

Arsenic ^{75}As, antimony $^{121, 123}Sb$, bismuth ^{209}Bi. The inorganic compounds of group VA elements, especially their oxides, sulfides and selenides, and also the ternary compounds based on them, have been studied in greatest detail by NQR.

It follows from the NQR spectra of M in all the compounds of type $M_2^V X_3^{VI}$ (where M = As, Sb, Bi) that there are two nonequivalent positions of the metal atoms whose electric field gradients have sharply different asymmetry parameters (Table 9.6). An atom in position M_1 has a large value of e^2Qq and a small (or near to zero) asymmetry parameter. An atom in position M_2 has a considerably smaller quadrupole interaction constant e^2Qq and a large asymmetry parameter η. This indicates the low-symmetry environment of the atom M_2. In the case of As_2O_3 and Sb_2O_3 the two different positions of the metal correspond to different phases of the substance. In the other compounds, the two positions of the metal exist simultaneously in the same crystal [42].

TABLE 9.6. Quadrupole interaction constants of ^{75}As, ^{121}Sb and ^{209}Bi and asymmetry parameters of their electric field gradients in compounds of type M_2X_3 (where X = O, S, Se) at 77°K

Compound	^{75}As	^{121}Sb		^{209}Bi	
	ν, MHz	e^2Qq, MHz	η, %	e^2Qq, MHz	η, %
M_2O_3	110.91 (claudetite)	541.43 (valentinite)	36	482.71	41.6
	116.78 (arsenolite)	554.83 (senarmontite)	0 [100]	556.99 (bismite)	9.7 [100, 101]
M_2S_3	70.34	249.76	38.2	152.75	88.9
	72.80 [100, 102]	317.14 [103, 105]	1.4	339.80 [105]	1.7
M_2Se_3	55.96	198.87	41.7		
	60.12 [104, 105]	272.75 [105]	0.8		

The structures of arsenolite and senarmontite are isomorphous (Table 9.6) and similar to the structure of hexamethylenetetramine $N_4(CH_2)_6$ (depicted in Figure 9.11) with As(Sb) atoms instead of the nitrogen atoms and oxygen atoms instead of the CH_2 groups. In this structure the metal atoms are located on a third-order axis and, therefore, the zero asymmetry parameter for senarmontite is due to the highly symmetrical environment of the metal atom.

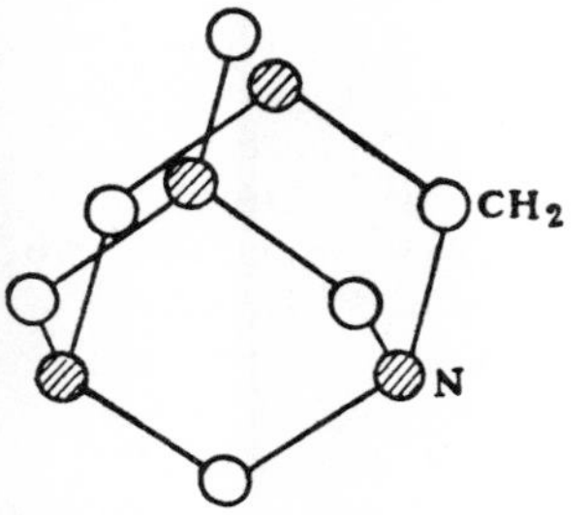

FIGURE 9.11. Arrangement of the atoms in the $N_4(CH_2)_6$ molecule [42]

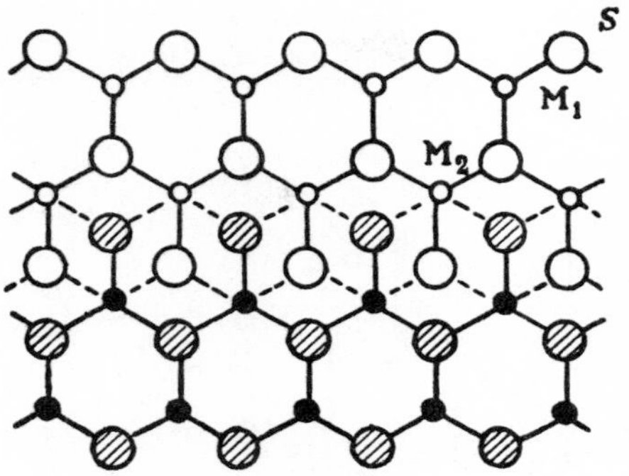

FIGURE 9.12. Arrangement of the atoms in the crystal lattice of Sb_2S_3 and As_2S_3 [42]

Claudetite and valentinite are the high-temperature phases of As_2O_3 and Sb_2O_3, but they can be easily supercooled to 77°K. Valentinite has a structure consisting of infinite zigzag chains of alternating antimony and oxygen atoms. Each pair of adjacent chains is connected by a chain of oxygen atoms lying in another plane. The three angles $\mathrm{Sb}\langle^{\mathrm{O}}_{\mathrm{O}}$ for valentinite are not equal (81, 93, and 99°). This agrees with the large asymmetry parameter $\eta = 36\%$. The angles $\mathrm{As}\langle^{\mathrm{O}}_{\mathrm{O}}$ in claudetite are equal and close to 102°, but they differ from the corresponding angles in arsenolite (100°). We can, therefore, assume that $\eta \neq 0$ here as well, but will be smaller than for the antimony atoms of valentinite.

The bismuth atoms have four phases: monoclinic, tetragonal, and α and β cubic. NQR data are known for only two of them: bismite and sillenite. It is noteworthy that for the first of them there are two positions of the metal atoms with a different symmetry of the environment ($\eta = 9.7$ and 41.6%).

The sulfides of arsenic, antimony and bismuth are isostructural and have double chains (Figure 9.12). Each metal atom is covalently bonded to three sulfur atoms which lie in the plane parallel to the axis of the chain. There are two types of metal atoms, namely, M_1 and M_2. Three sulfur atoms are equally located around the metal atom M_1 at a distance of 2.50 Å; the other atoms are located at distances greater than 3.14 Å. The M_2 atom has one sulfur atom at a distance of 2.37 Å, two at a distance of 2.67Å, and two belonging to the adjacent chain at a distance of 2.83 Å. Therefore, the main difference between M_1 and M_2 lies in the coordination numbers, which are equal to 3 and 5, respectively. This leads to two different values of the quadrupole interaction constant e^2Qq, the smaller of which corresponds to the atom with the larger coordination number. The asymmetry parameter is larger for the atom with coordination number 5. This indicates that this atom is in a position with low symmetry.

From the NQR data we can predict that the selenium derivatives of group VA, namely As_2Se_3 and Sb_2Se_3, are isomorphous with Sb_2S_3.

A fairly large number of ternary compounds based on the oxides and sulfides of group VA elements have been studied by NQR. Among them are a number of natural minerals [104–109] and chalcogenides [110]. The complex bismuth oxides [111] are worth noting. In the spectrum of $Bi_4Ti_3O_{12}$ there are two sets of NQR frequencies of ^{209}Bi. These correspond to the bismuth atoms in the layers of $(Bi_2O_2)^{2+}$ and to the bismuth atoms in the perovskite-like layers of $(Bi_2Ti_3O_{12})^{2+}$ (Figure 9.13). For atoms of the first type we must expect a larger quadrupole interaction constant and a layer asymmetry parameter, since the second position of the bismuth atom has the higher coordination number (12) and a higher symmetry. However, the asymmetry parameter $\eta = 20.3\%$ obtained for the second type of bismuth atom indicates the absence of axial symmetry for this position. This can probably be associated with the continuous change of the lattice symmetry from rhombic to tetrahedral, as established in [111].

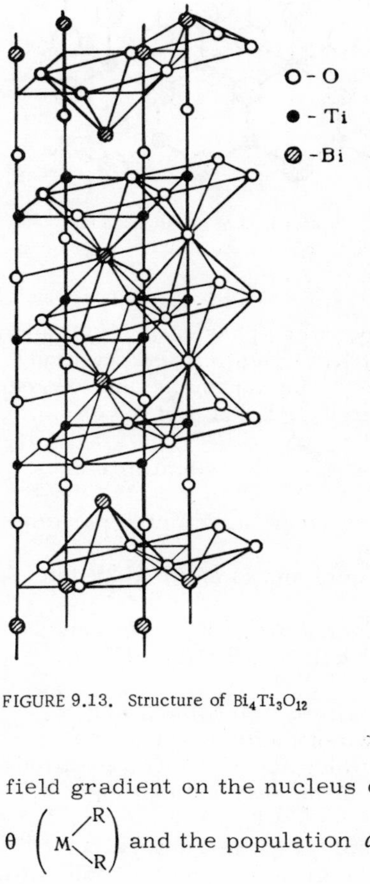

FIGURE 9.13. Structure of $Bi_4Ti_3O_{12}$

The trivalent MR_3 compounds of group V elements have the configuration of a trigonal pyramid. The central atom M is sp^3 hybridized with three hybrid orbitals participating in the formation of the σ M–R bonds, while the fourth is occupied by an unshared pair of electrons. According to the Townes-Dailey theory, the electric field gradient on the nucleus of atom M is determined by the angle $\theta \left(\mathrm{M}\!<^{R}_{R}\right)$ and the population a of the orbital R –M:

$$e^2Qq = e^2Qq_0\left(-\frac{3\cos\theta}{1-\cos\theta}\right)(2-a). \tag{9.6}$$

For equivalent substitution, if the magnitude of the $\mathrm{M}\!<^{R}_{R}$ angle in MR_3 is preserved, the changes in the electric field gradient on the central atom are relatively small. This can be seen from the e^2Qq of ^{209}Bi for the series of triphenyl-substituted derivatives of bismuth $(p\text{-}XC_6H_4)_3Bi$ [112].

The nonequivalent substitution $R'R''_2M$, if the configuration of the molecule is preserved, has a stronger effect on the electric field gradient [113].

In such molecules a sharp increase in the NQR frequencies is observed (for example, 98.50 and 98.90; 161.90; 170.807 MHz for $(C_6H_5)_3As$, $(C_6H_5)_2ClAs$ and $(CH_3)Cl_2As$, respectively). This change in the electric field gradient of the central atom can be caused by at least two factors. The first is the change in the degree of *sp* hybridization of the lone pair (which distorts the angles of the pyramid) and the deflection of this hybrid orbital from its initial direction. The second is the distortion of the tensor of the electric field gradient, in which the *z* axis (the axis of the maximum gradient) is no longer directed along the hybrid orbital occupied by the lone pair of electrons, and the tensor itself with the previous selection of the coordinate axes ceases to be diagonal. It is noteworthy that for an equal degree of distortion of the MR_3 pyramid, for example, in the series $AlkAsCl_2$, the electric field gradients on the central atom (arsenic) and on the ligands (chlorine) change in the usual manner. This means that a decrease in the ionic character of the As—Cl bond causes an increase in the NQR frequency of ^{35}Cl and a decrease in the frequency of ^{75}As [113].

The quadrupole interaction constants change most sharply when the coordination number of the central atom changes. Thus, the NQR frequency of ^{75}As in AsI_3, when AsI_3 has a layered structure, is equal to 29.34 MHz. If AsI_3 has the configuration of a pyramid, the frequency of ^{75}As is equal to 116.86 MHz [99].

From the pentavalent compounds of group VA elements the pentahalides $MHal_5$ (where M = Sb and P) and the series of compounds of type R_3SbHal_2, already mentioned in Chapter 6, have been studied. At a temperature above 228°K, $SbCl_5$ forms molecular crystals in which the molecules have a trigonal bipyramid configuration [117]. The pentachloride of phosphorus forms an ionic lattice with cations $[PCl_4]^+$ and anions $[PCl_6]^-$ [114—116], the pentabromide of phosphorus has a cation $[PBr_4]^+$ and an anion Br^-. The mean NQR frequency of ^{35}Cl in the cations $[PCl_4]^+$ is little affected by the nature of the anion, just as the NQR frequency of ^{35}Cl in the anions $[PCl_6]^-$ and $[SbCl_6]^-$ is little affected by the nature of the cation [114]. Di Lorenzo and Schneider discussed in detail the question of the probability of $d_\pi - p_\pi$ conjugation of the M—Cl bond in the cations $[PCl_4]^+$ and $[AsCl_4]^+$, and also in the anions $[PCl_6]^-$ and $[SbCl_6]^-$ [114].

Nitrogen ^{14}N. NQR investigations of ^{14}N have always attracted considerable interest. However, the quadrupole moment of the nitrogen nucleus is not large ($Q = +0.016 \cdot 10^{-24}$ cm^2) and the quadrupole interaction constants of ^{14}N are also small (smaller than 6 MHz).

The low intensity of the NQR lines (low frequency) and the large spin-lattice relaxation time (integer spin $I = 1$) complicate the search for NQR signals. The data obtained by microwave spectroscopy in the gaseous state usually give quadrupole interaction constants for the principal axes of inertia of the molecule. The nondiagonal elements of the tensor of the electric field gradient are difficult to measure due to the small value of e^2Qq. As a result, these data usually enable us to obtain only tentative values of e^2Qq for the principal axes of the electric field gradient.

All the experimentally obtained values of e^2Qq can be classified according to the type of nitrogen bond.

Assuming that the direction of the valence hybrid orbitals of the nitrogen atom are known, they can very often be represented by the wave functions of the atomic orbitals. By specifying the population of the molecular orbitals in a general form, we can express, on the basis of the Townes-Dailey theory, e^2Qq_{zz} and η by quantities which are equal to the populations of the molecular orbitals [136]:

$$e^2Qq_{zz} = -e^2Qq_{\text{at}}\left(\frac{N_x + N_y}{2} - N_z\right). \tag{9.7}$$

The minus sign before the right-hand side of expression (9.7) shows that the electric field gradient on the nitrogen is created by the p electrons on the external electron shell. This is different from the electric field gradient on halogen atoms which lack one electron for a closed shell.

The electric field gradient along the z axis is created by the valence electrons:

$$q_{zz} = \sum_i^m (q_z)_i, \tag{9.8}$$

where $(q_z)_i$ is the field gradient created by the i-th electron.

If the molecular wave function Ψ_i is expressed in terms of the atomic wave functions φ_j, then:

$$\Psi_i = \sum_j a_j \varphi_j. \tag{9.9}$$

The electric field gradient created by the electron on the Ψ_i orbital is equal to:

$$\begin{aligned}(q_z)_i &= \int \Psi_i \frac{3\cos^2\theta_{iz} - 1}{r^3} \Psi_i^* \, d\tau = \\ &= \int \sum_j a_j \varphi_j \frac{3\cos^2\theta_{iz} - 1}{r^3} \sum_k a_k^* \varphi_k^* \, d\tau = \\ &= \sum_j a_j^2 \int \varphi_j \frac{3\cos^2\theta_{iz} - 1}{r^3} \varphi_j^* \, d\tau + \\ &+ \sum\sum_{k \pm j} a_k^* a_j \int \varphi_k^* \frac{3\cos^2\theta_{iz} - 1}{r^3} \varphi_j d\tau,\end{aligned} \tag{9.10}$$

where θ_{iz} is the angle between the z axis and the vector joining the i-th electron and the nucleus, and r is the length of this vector. The second term is close to zero, since the $3d$ orbitals of nitrogen have too large an energy and do not play much part in the formation of the bond. The summation $a_{2s}a_{2p}\int \varphi_{2s}\frac{3\cos^2\theta - 1}{r^3}\varphi_{2p}d\tau \equiv 0$ (for more detail see Ch. 5). In addition, since the s electrons are spherically symmetrical and do not create a field gradient:

$$\begin{aligned}(q_z)_i = a_x^2 \int \varphi_{p_x} \frac{3\cos^2\theta_{iz} - 1}{r^3} \varphi_{p_x}^* \, d\tau + a_y^2 \int \varphi_{p_y} \frac{3\cos^2\theta_{iz} - 1}{r^3} \varphi_{p_y}^* \, d\tau + \\ + a_z^2 \int \varphi_{p_z} \frac{3\cos^2\theta_{iz} - 1}{r^3} \varphi_{p_z}^* \, d\tau = a_x^2 q_z^x + a_y^2 q_z^y + a_z^2 q_z^z.\end{aligned} \tag{9.11}$$

In equation (9.11), q_z^i is the electric field gradient along the z axis created by one p_i electron. It was shown in Ch. 5 that:

$$q_x^x = q_y^y = q_z^z = q_p, \tag{9.12}$$

and

$$q_x^y = q_y^x = q_x^z = q_z^x = q_y^z = -\frac{1}{2} q_p.$$

Thus, the electric field gradient in the z direction on the nucleus can be calculated if the number of electrons in the i-th molecular orbitals and the coefficients a_j are known.

The wave functions of the valence orbitals of nitrogen in the sp^2 hybrid state (for example, the nitrogen atom in pyridine) are:

$$\begin{aligned} \Psi_\pi &= \varphi_{p_y} \\ \Psi_{\sigma_1} &= \sqrt{\cot^2\gamma}\,\varphi_s + (1-\cot^2\gamma)\,\varphi_{p_z} \\ \Psi_{\sigma_2} &= \frac{1}{\sqrt{2}}\left(\sqrt{1-\cot^2\gamma}\,\varphi_s - \cot\gamma\,\varphi_{p_z} + \varphi_{p_x}\right) \\ \Psi_{\sigma_3} &= \frac{1}{\sqrt{2}}\left(\sqrt{1-\cot^2\gamma}\,\varphi_s - \cot\gamma\,\varphi_{p_z} - \varphi_{p_x}\right). \end{aligned} \tag{9.13}$$

Here, the y axis is perpendicular to the plane of the molecule, the x and z axes lie in the plane of the molecule, and the z axis is directed along the bissectrix of the angle $\angle$CNC. The molecular orbital Ψ_{σ_1} is occupied by the lone pair of electrons, and the angle 2γ is the angle $\angle$CNC. If a is the population of the π orbital, and b_1 and b_2 are the populations of the σ orbitals of the N—C bond, then:

$$\begin{aligned} \frac{1}{2}(b_1+b_2) - a &= \frac{2}{3}\eta\frac{e^2Qq}{e^2Qq_p} \\ \left(2-\frac{b_1+b_2}{2}\right) &= \frac{1}{1-\cot^2\gamma}\left(1-\frac{\eta}{3}\right)\frac{e^2Qq}{e^2Qq_p}. \end{aligned} \tag{9.14}$$

Equations (9.14) show that the electric field gradient depends very strongly on the angular term $\frac{1}{1-\cot^2\gamma}$, i.e. on the degree of sp hybridization of the lone pair ($s = \cot^2\gamma$) [136—138]. The angle $\angle$CNC is known for many six-membered nitrogen-containing heterocyclic rings and as a rule is smaller than 120°. For symmetrical azabenzenes (pyridine, pyrazine, sym-triazine) $b_1 = b_2$ and we can calculate the populations of the π and σ orbitals from the experimental values of e^2Qq and η (Table 9.7) [137].

TABLE 9.7. Populations of the π and σ orbitals calculated from the NQR data for $\gamma = 60°$ and by the self-consistent field method (SCF)

Substance	$\gamma = 60°$		SCF	
	a	*b*	*a*	*b*
Pyridine	1.100	1.250	1.100	1.268
Pyrazine	1.033	1.251	1.080	1.280
sym-Triazine	1.116	1.290	1.116	1.293
Pyridazine	1.045	0.935	1.055	
		1.270		

The values of a and b obtained by this method agree quite well with the results of the theoretical calculations carried out by Peacock [139] using the self-consistent field method. It is noteworthy that the σ population changes much less than the π population.

Equations (9.14) also hold for nonsymmetrical azabenzenes (pyrimidine and pyridazine), but all q_{ii} correspond to the system of coordinates in which the z axis is not directed along the bisectrix of the angle 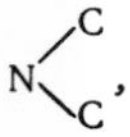, but is rotated through an angle θ with respect to it. If the value of θ is known, then we are able to calculate b_1 and b_2 separately [137]:

$$\frac{b_1 - b_2}{4 - b_1 + b_2} = \frac{\tan 2\theta}{\tan 2\gamma}. \qquad (9.15)$$

Only for pyridazine do we have experimental data on the direction of the axes of the tensor of the electric field gradient ($\theta = 9 \pm 1°$) [140]. The values of a, b_1 and b_2 are given in Table 9.7 [137]. The changes in b_1 and b_2 are logical, since we expect the σ population of the N—N bond to be smaller than that of the N—C bond. The z axis of the electric field gradient is rotated in the direction of the second nitrogen atom.

Substituted pyridines have undergone much investigation [118, 137, 138, 141]. For symmetrical derivatives of pyridine the σ and π populations of the bonds can be investigated using equations (9.14), where $b_1 = b_2 = b$ [137]. Most of the data show that the substituents alter the electric field gradient on the nitrogen by only 5—10%. However, the σ and π electron densities vary uniformly as a function of the nature and position of the substituent. The electron densities of the σ bonds remain practically constant, which indicates a weak damping of the inductive effect along the ring. This proves that the σ bonds are not as strongly localized as usually assumed [118].

Moreover, the π electron density for pyridine with a substituent in position 3 is close to the π electron density of unsubstituted pyridine [141], i.e. the nature of the substituent in position 3 has hardly any effect on the distribution of the π electrons in the ring. In accordance with this, the NQR frequencies of ^{35}Cl in 3-chloropyridine are larger than in 2- and 4-chloropyridine, which indicates a lower degree of double bonding of the C—Cl bond in 3-chloropyridine [119].

The data for aminopyridines [138] indicate that the amino group gives the ring a roughly constant charge independent of its position in the ring (e^2Qq and η of ^{14}N in the amino group remain practically unchanged). However, the nitrogen atom in the ring accepts this charge only when the amino group is in the α or γ position.

The first attempts to correlate the populations a and b with the Hammett σ constants of the substituents were only tentative [118].

Different investigators used different values for e^2Qq_p (from 8 to 12 MHz), and it is possible that the values of e^2Qq_p for the p_σ and p_π orbitals do differ [137].

Equations (9.14) were derived using a number of simplifying assumptions which could lead to considerable errors in the determination of the absolute values of a and b [120]. Thus, we must take into account that: the orbitals $2p_x$, $2p_y$, and $2p_z$ actually contain contributions from the excited states; the population density of the nonbonding nitrogen atom is not constant and not equal to two; and the overlap between adjacent orbitals cannot be neglected [120].

NQR has also been widely used to study five-membered heterocyclic rings [121—123]:

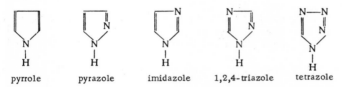

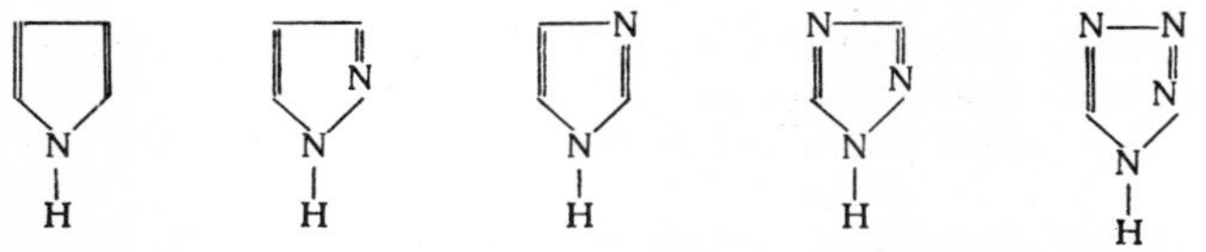

pyrrole pyrazole imidazole 1,2,4-triazole tetrazole

These heterocyclic rings (except pyrrole) contain two types of nitrogen atom, namely, with two- and three-fold coordination. For pyrrole we have data on e^2Qq and η for nitrogen with three-fold coordination both in the solid [122] and gaseous states [124].

It follows from the symmetry of the pyrrole molecule that the z axis is perpendicular to the plane of the ring along the p_π orbital, and the x axis along the N—H bond, since the electron density of the N—C σ orbital is lower than the density of the N—H σ orbital. If the population of the Ψ_π, Ψ_{σ_1}, Ψ_{σ_2}, Ψ_{σ_3} orbitals (the Ψ_σ orbitals are sp hybridized) in equations (9.13)* are denoted by a, b, c, and c, respectively, then [122]:

$$(a-c) = \left(1+\frac{\eta}{3}\right)\frac{e^2Qq}{e^2Qq_p}$$
$$\frac{3}{2}(b-c)(1-\cot^2\gamma) = \frac{e^2Qq}{e^2Qq_p}. \qquad (9.16)$$

Even if we assume that the angular term is known, we still have two equations with three unknowns. Accordingly in [122] a population of the N—H bond close to the population of N—H in amides (b = 1.45) was selected and the values a = 1.62 and c = 1.37 were obtained, which agree with those calculated by Hückel theory.

* Equations (9.13) must be rewritten with φ_{p_z}, φ_{p_x}, and φ_{p_y} instead of φ_{p_y}, φ_{p_z}, and φ_{p_x}, respectively.

For the nitrogen with two-fold coordination in these compounds, the values of e^2Qq and η are described by equations (9.14), taking into account that the axes of the tensor of the electric field gradient deviate from those specified for the symmetrical case. (Pyrazole can be compared with pyridazine in this respect.) However, we must keep in mind that imidazole and pyrazole can form hydrogen bonds in the solid state.

The difference between the asymmetry parameters of the electric field gradients on the nitrogen atoms with two-fold coordination in pyrazole ($\eta = 65.7\%$) and imidazole ($\eta = 12.9\%$), and also for the two nitrogen atoms with two-fold coordination in 1,2,4-triazole ($\eta_2 = 66.5\%$ and $\eta_4 = 16.9\%$) are due to differences in environment. These differences arise because the populations of the N—C orbitals of N_2 in imidazole and N_4 in triazole are similar, but the populations of the N—C orbitals of N_2 in pyrazole and N_2 in triazole are sharply different [122].

We must also note that if $1 - \cot^2 \gamma \approx 0.5$ (where 2γ is the angle of a regular pentahedron, equal to 108° which is close to a tetrahedral angle), equations (9.14) will yield $a = 1.27$ and $b = 1.30$ or 1.24, depending on whether the x or y axis will be perpendicular to the plane of the ring. In any case, the π electron density of a is greater than for the six-membered heterocyclic ring (compare Table 9.7).

Equations analogous to (9.13), with $\varphi_{p_y} \to \varphi_{p_z}$, $\varphi_{p_z} \to \varphi_{p_y}$, $\varphi_{p_x} \to \varphi_{p_x}$ as for the nitrogen atoms with three-fold coordination in the five-membered rings, describe the wave functions of the hybrid orbitals in the amides $RCONH_2$. The angle 2γ is the angle N<H H. The molecular orbital Ψ_π is occupied by a lone pair of electrons, but its population $a \neq 2$. The orbital Ψ_{σ_1} is directed along the N—C bond b, and Ψ_{σ_2} and Ψ_{σ_3} along the N—H bonds c. Then,

$$a - c - \left(1 - \frac{\eta}{3}\right) \cdot \frac{e^2Qq}{e^2Qq_p}, \qquad (9.17)$$

$$\frac{3}{2}(c - b)(1 - \cot^2 \gamma) = \eta \frac{e^2Qq}{e^2Qq_p}. \qquad (9.18)$$

Formamide, urea, and thiourea have also been studied [125, 126]. It was shown for thiourea [125] that the z axis is indeed perpendicular to the plane of the molecule. Equations (9.17) can be analyzed only if we assume that the population of one of the bonds is known (for example N—H) [127]. For a given value c, the population of the Ψ_π orbital for urea is larger than the population of the Ψ_π orbital for thiourea. This agrees with the large barrier of rotation about the C—N axis for the latter [127].

In aromatic amines the main problem is to determine the configuration of the amino group. The conjugation between the amino group and the benzene ring requires them to be coplanar. However, it was shown [129] that in aniline the angle between the π orbital of ^{14}N and the z axis of the ring is equal to 25°18'. This deviation of the π orbital of nitrogen is noticeable and considerably alters the conjugation. Besides, the angle N<H H differs for different substituted anilines, which means that the hybridization of the

itrogen atom also varies. However, calculations of the asymmetry para-
neter by the method of equalization of the orbital electronegativities
128], varying the magnitude of the resonance integral β_{CN}, yield results
which agree well with the experimental values of $\eta_{^{14}N}$. For p-chloroaniline
he calculation of the NQR frequency of ^{35}Cl establishes the low sensitivity
f the electric field gradient on ^{35}Cl to β_{CN}, i. e. to the degree of coplanarity
f the molecule [128].

The hybridization of the nitrogen atom in the NH_2 group in aminopyri-
ines is also unknown. However, analysis of the relations linking e^2Qq and η
vith the parameters of the *sp* hybridization of the orbital occupied by the
one pair of electrons and the population of the hybrid orbitals of the nitro-
en atom shows [138] that the small difference in $e^2Qq_{^{14}N}$ for the NH_2 group
n 3-aminopyridine and 4-aminopyridine is associated with a difference in the
opulation of the orbital of the lone pair of electrons. The decrease in the
opulation of this orbital for 4-aminopyridine is due to the large transfer
f charge from the amine to the ring and the resulting increase in the
tructure of type:

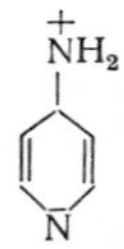

The degree of *sp* hybridization of the lone pair also determines the
lectric field gradient on the nitrogen atom in compounds of type NR_3 with
pyramid structure [127].

In this case the hybridization of the nitrogen atom is nearly sp^3 and
epends on the angle 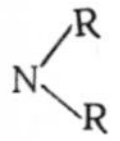 (α). The orbital occupied by the lone pair of
lectrons is directed along the third-order *z* axis. The expression for the
uadrupole interaction constant of the central atom [127] was given when
ve discussed the pyramidal compounds of arsenic (equation (9.6)). The
uadrupole data in amines are analogous to e^2Qq_M in MR_3 compounds of
rsenic and antimony. The electric field gradient on nitrogen depends on
ne angle 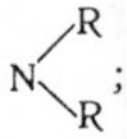; when α changes by 4° (if α is close to tetrahedral) the
ngular term changes by 15% [127]. Comparison of the data on the popula-
ion of the σ bond in NH_3 and $N(CH_3)_3$ ($a_{(CH_3)_3N} < a_{NH_3}$) indicates the consider-
ble part played by the conjugation through which the methyl groups supply
ne electrons:

	e^2Qq, MHz	a
NH_3	-4.084 (gas)	1.33
$N(CH_3)_3$	-5.194 (gas)	1.21
NF_3	-7.07 (gas)	0.53
$N_2(C_2H_4)_3$	3.695	1.45
$N_4(CH_2)_6$	4.543	1.33

The magnitude of this conjugation depends strongly on the orientation o the protons of the methyl group and the lone pair of electrons on the nitrogen.

The large difference in e^2Qq of NH_3 for the solid and gaseous states is due to the formation of hydrogen bonds. The recent calculations of O'Kon [130] have shown that the proton of the hydrogen bond $H_3N \cdots H$ affects the electric field gradient on the nitrogen atom mainly by polarization of the electron shell of the atom, and not through charge transfer.

The influence of nonequivalent substitution on the value of e^2Qq of nitrogen in molecules of the type $R'R_2''N$ is analogous to the influence of that effect on the ^{75}As frequency in $R'R_2''As$ compounds. Pierce [131] and Lide [132] gave a detailed analysis of the change in e^2Qq of nitrogen in CH_3NH_2 and HNF_2. They allowed for deviations of the axes of the tensor of the electric field gradient from the molecular axes.

The NQR of ^{14}N in nitriles of the organic acids RCN [133, 134] and the cyano-derivatives of pyridine [118, 123, 133] have been studied extensively.

The nitrogen atom in these compounds is sp hybridized. If the z axis is along the C—N bond, then the triple bond will consist of the σ bond form by the sp_z orbital of the nitrogen and the sp_z orbital of carbon, and the π bo formed by overlapping of the $2p_x$ and $2p_y$ orbitals of nitrogen with the $2p_x$ and $2p_y$ orbitals of carbon. The lone pair of electrons on the nitrogen occupy the sp_z antibonding orbital.

In the nitriles of organic acids, the CN group is bonded to the hydrocarbon radical R by a single bond and, therefore, the population of the p_x and p_y orbitals of nitrogen is the same ($a_1 = a_2$) and the asymmetry parameter of the electric field gradient vanishes. Assuming that the population of the antibonding orbital of the nitrogen is equal to 2, and the population of the σ bond is equal to b, we have:

$$2s + b(1 - s) - a = \frac{e^2Qq}{e^2Qq_p}, \tag{9.18}$$

where s is the degree of sp hybridization of the σ orbital [127]. Even assuming that there is pure sp hybridization ($s \approx 0.5$) there are still two unknow in the equation. Since carbon is less electronegative than nitrogen, $a > 1$ and $b > 1$ and $e^2Qq/e^2Qq_p < 0.5$, which is in fact observed in most cases. T effect of the nature of the substituent R on the quadrupole interaction constant of ^{14}N in RCN is difficult to analyze, since both a and b and the s character of the σ orbital can change.

It is important to note that in nitriles and dinitriles the quadrupole interaction constant of nitrogen tends to a constant (3.80 MHz) when the length of the hydrocarbon chain is increased [134]. It follows that the asymmetry parameter of the electric field gradient ($\eta \approx 5-7\%$) in these compounds is determined by both the crystal field and the electronic effects [134].

In those cases when the nitrile group is bonded to a conjugated system (benzene ring, pyridine), the π system of the substituent interacts with the π electrons of the CN group. This leads to an unequal population of the p_x and p_y orbitals of nitrogen ($a_1 \neq a_2$) and to the appearance of an asymmetr parameter of the electric field gradient. The fact that η in benzonitrile ($\eta = 10.4\%$) is larger than in cyanopyridines ($\eta = 7.2$, 10.0, 1.4% for

2-, 3-, and 4-cyanopyridines, respectively) indicates that in the cyanopyridines electrons are transferred from the ring to the CN group. Actually one would have expected a large asymmetry parameter for cyanopyridines, since the nitrogen atom of the ring acts as an additional acceptor of electrons [118].

Assuming that $a_1 = 1$, i.e. that the π orbital of nitrogen in the plane of the aromatic ring is occupied by one electron, we have [118]:

$$\frac{3}{2}(a_2 - 1) = \eta \frac{e^2Qq}{e^2Qq_p}$$
$$b(1 - s) + 2s - \frac{1}{2}(a_2 - 1) = \frac{e^2Qq}{e^2Qq_p}. \tag{9.19}$$

Evaluations of the parameters in (9.19) give the following characteristics of the N—C bond [118]: $s \approx 30\%$, ionic character of the σ bond ~15%, ionic character of the π bond ~1.5%, total ionic character of the bond ~17%. The C—N bond itself in pyridines is less ionic than in benzonitrile. This corresponds to the acceptor properties of the nitrogen atom in the heterocyclic ring. The same situation is characteristic of 4-methoxybenzonitrile.

Numerous calculations by the MO method [135] show that positions 2 and 4 in pyridine have a lower π electron density than position 3. This agrees with the results of the analysis of the values of e^2Qq and η for cyanopyridines [118] (Table 9.8). (Compare the sequence of changes of Δa and the calculated data in this table.)

TABLE 9.8. Characteristics of the C≡N bond calculated from the NQR data for cyanopyridines

Compound	b	a_2	$\Delta a = a_2 - a_1$	Calculated π electron charge
C_6H_5CN	1.305	1.033	0.033	—
2-CNC_5H_4N	1.309	1.022	0.022	0.963
3-CNC_5H_4N	1.305	1.032	0.032	0.997
4-CNC_5H_4N	1.288	1.004	0.004	0.984
4-$CH_3OC_6H_4CN$	1.305	1.000	0.000	—

Note the existence of large asymmetry parameters in the series of aliphatic nitriles: CH_3OCH_2CN $\eta = 18.25\%$; $ClCH_2CN$ $\eta = 14.1\%$; CH_3OCOCH_2CN $\eta = 8.59\%$, in which superconjugations and other additional interactions are possible.

Thus, the NQR data of ^{14}N can yield much valuable information on the nature of the chemical bond, the distribution of the electron density, and the inductive effects of the substituents.

5. GROUP VIIB AND VIII ELEMENTS. MANGANESE ^{55}Mn, RHENIUM ^{185,187}Re, COBALT ^{59}Co

Very few transition metal compounds have been studied by NQR. The NQR data for cyclopentadienyl-substituted tricarbonylmanganese (rhenium were examined in detail in Ch. 6 [142, 143]. For most substituents in the cyclopentadienyl ring the asymmetry parameter of the electric field gradi ent on the nucleus of the manganese (rhenium) atom is not large. However when there is a substituent of the type COR in the ring, the asymmetry parameter increases noticeably (see Table 9.9).

TABLE 9.9. NQR data for cyclopentadienyl COR-substituted tricarbonylmanganese (rhenium), where R = CH_3, C_6H_5, CF_3, Cl, OH

X in $C_5H_4XMn(Re)(CO)_3$	^{55}Mn		^{185}Re	
	e^2Qq_{zz}, MHz	η, %	e^2Qq_{zz}, MHz	η, %
H	65.24	1.9	619.72	8.7
$COCH_3$	61.63	6.8	579.38	30.5
COC_6H_5	62.41	11.8	490.91	41.9
$COCF_3$	59.35	13.9	—	—
COCl	—	—	505.24	39.1
COOH	—	—	558.23	28.4

This increase in η cannot be due to the perturbation of the symmetry of the molecule following the introduction of the substituent, since such effect are not observed for all other substituents (except COR). Most probably there is a specific interaction between the substituent and the metal atom in this case.

Khotsyanova conducted a full X-ray investigation into one of these compounds ($(CO)_3ReC_5H_4COCH_3$) and arrived at the following conclusions.

Firstly, she observed no distortion of the molecule, i.e. there is no sh or inclination of the ring and no substantial deviation of the substituent fro the plane of the ring. The latter means that the conjugation of the substituent with the ring is not perturbed. Secondly, since the distance Re—O is too large, there is probably no direct interaction between the lone pair of the oxygen atom and the metal. Finally, all the distances C—C in the ring are equal, which means that a Fulven rearrangement is impossible.

On the other hand, it is known that the carbon atom of the COR group has a positive charge [144]. Therefore, if the distance Re—C (where C is the carbon atom of the substituent COR), were smaller than the sum of the Van der Waals radii of rhenium and carbon, we could assume that the π acceptor carbon orbital interacts with the filled rhenium orbital. Unfortunately, the Van der Waals radius of rhenium has not been determined. However, it is known that the Van der Waals radius of the atom is always larger than its covalent radius. The covalent radius of rhenium can be estimated as follows. The distance Re—C (where C is the carbon atom in the cyclopentadienyl ring) in the compound studied is equal to 2.31 Å.

The covalent radius of carbon equals 0.77 Å. Therefore, the covalent radius of rhenium equals 1.54 Å. Furthermore, the sum of the Van der Waals radius of carbon (1.72—1.8 Å) and the covalent radius of rhenium is 3.26—3.34 Å, which is fairly close to the distance Re—C (where C is the carbon atom of the substituent COR) found for $(CO)_3ReC_5H_4COCH_3$. Since the distance Re—C (3.35 Å) in this compound is noticeably smaller than the sum of the Van der Waals radii of rhenium and carbon, the probability of their direct interaction must be taken into account.

Such interactions are much stronger in cyclopentadienylrheniumtricarbonyls than in the manganese analogs, since the asymmetry parameters for the rhenium analogs are much larger. Apparently, due to the large size of its atom, rhenium has a higher coordination capacity than the manganese atom in compounds of this type.

The quadrupole interaction constants of rhenium and manganese were also studied in the tetrahedral compounds ReO_3F, ReO_3Cl, MnO_3F, and $KReO_4$ [145, 146]. For the first three compounds, e^2Qq_{zz} of ^{185}Re and ^{55}Mn were obtained by the microwave method in the gaseous state. It is interesting to note that the sign of $e^2Qq_{^{185}Re}$ in ReO_3F is negative, whereas $e^2Qq_{^{185}Re}$ in ReO_3Cl and $e^2Qq_{^{55}Mn}$ in MnO_3F are positive [145]. Lotspeich [145] holds that the configuration of the rhenium and manganese atoms in these compounds is tetrahedral: the angle $\mathrm{M}\langle^{\mathrm{O}}_{\mathrm{Hal}}$ is equal to 109°31', 109°22', and 108°27' for ReO_3F, ReO_3Cl, and MnO_3F, respectively (the ideal tetrahedral angle is 109°28'3"). He further assumes that the hybrid orbitals of rhenium are spd^5 orbitals, each of which has a $^9/_{32}$ p character and a $^{15}/_{32}$ d character. For manganese in MnO_3F, the combination of spd^5 and sp tetrahedral orbitals is most probable. This combination includes a 65% p character on one bond. Charges external to this atom contribute considerably to the electric field gradient on the central atom (see Table 9.10).

It was assumed in the calculation that the ionic character of the Re—O bonds is ~50%, Re—F ~95%, Re—Cl ~65%, Mn—O ~45—50%, Mn—F ~95% [145].

TABLE 9.10. Calculated values of the electric field gradients on rhenium and manganese atoms due to point charges $q(e)$ and the p component of the valence hybrid orbitals $q(p)$

Compound	$q_{obs}\cdot10^{-14}$, e.s.u	Point charges	$q(e)\cdot10^{-14}$, e.s.u	% of q obs	$q(p)\cdot10^{-14}$, e.s.u
ReO_3F	-2.6	F	-1.49	+57	+1.03
		O_3	+1.95	-75	
ReO_3Cl	+14	C	-0.87	-6	+0.624
		O_3	+1.95	+14	
MnO_3F	+5.8	F	-1.88	-32	+0.70
		O_3	+2.52	+43	
$KReO_4$	±11.2	O_4	-0.456		-3.54

In addition to the direct contribution of the external electric charges to the field gradient, they also have an indirect influence because they polarize the atomic core of the rhenium (or manganese). The calculation shows that $\gamma_\infty(\text{Re}) = -28$ and $\gamma_\infty(\text{Mn}) = -3$. Then:

$$q_{\text{obs}} = q(p) + q_e(1-\gamma_\infty) = \frac{9}{32}[(1-I_{\text{Hal}}) - (1-I_{\text{O}_3})]\, q_p + [q(e)_{\text{Hal}}\, I_{\text{Hal}} + q(e)_{\text{O}_3}\, I_{\text{O}_3}](1-\gamma_\infty), \tag{9.20}$$

where $q(p)$ is the contribution of the p component of the valence hybrid orbitals to the gradient, q_e is the contribution of the partially ionic atoms of oxygen and halogen to the gradient, I_A is the ionic character of the Re—. bond, q_p is the gradient from one p electron: $q_p(\text{Re}) = -8.13 \cdot 10^{15}$ e.s.u, $q_p(\text{Mn}) = -2.4 \cdot 10^{15}$ e.s.u [147], and $q(e)$ is the field gradient on the rhenium (or manganese) atoms created by a unit negative charge at the position of the halogen or oxygen atoms. The ratios of the last two terms of equation (9.20) for the total gradient q on the central atom determine the sign of e^2Qq in the compounds ReO_3Cl, ReO_3F, and MnO_3F.

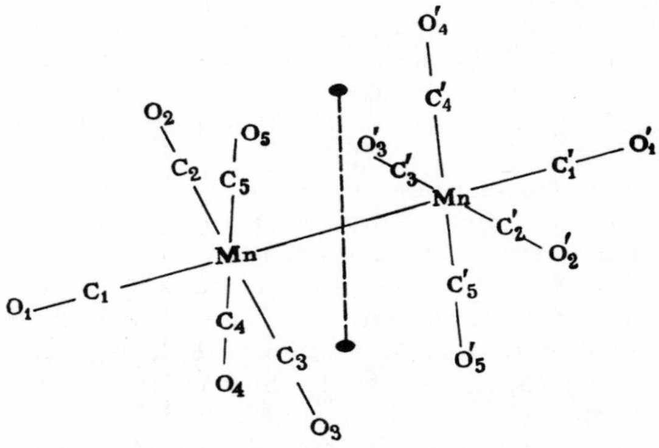

FIGURE 9.14. Structure of $Mn_2(CO)_{10}$

In $KReO_4$ the tetrahedral environment of the rhenium atom is more distorted: the angles $\text{Re}\langle^{\text{O}}_{\text{O}}$ are equal to 105 and 112° [148]. It is assumed that the hybrid orbitals of rhenium here are of the same type as for ReO_3H. The formal charges on Re^+, K^+, and O^- are equal to +4, +1 and −1.25, respectively [148]. The uncertainty in the evaluation of γ_∞ $(-20 < \gamma_\infty < -1$ for $KReO_4$ depends on the degree of tetrahedral distortion of ReO_4^-.

Metal carbonyls are one of the most important groups of organometallic compounds. It is, therefore, natural that the values of the electric field gradient on the metal atom and the symmetry of the metal atom in these components have attracted the attention of many research workers. These compounds $Mn_2(CO)_{10}$ [149, 150], $Re_2(CO)_{10}$ [142, 151—154], and $Co_2(CO)_8$ [155, 156] have been studied by NQR.

The decacarbonyls of manganese and rhenium are at present the subject)f much research. X-ray analysis [157] indicates that these molecules have ı high symmetry (close to D_{2h}) in the crystal (Figure 9.14). This must lead o a zero asymmetry parameter of the field gradient. However, the experi- nent indicates that $\eta \neq 0$ (Table 9.11). Moreover, for $Re_2(CO)_{10}$ η increases vith a decrease in temperature, whereas for $Mn_2(CO)_{10}$ it decreases. Harris 153] proposed that the increase in η with a decrease in temperature is due o the considerable difference in the amplitudes of the thermal motion of :is carbonyl groups. He noted that the thermal motion increased with a lecrease in temperature. However, since $Re_2(CO)_{10}$ and $Mn_2(CO)_{10}$ are somorphous, the thermal vibrations of the carbonyl groups should lead o the same result, and this is not observed.

TABLE 9.11. NQR parameters of ^{55}Mn, ^{185}Re and ^{59}Co in carbonyls of manganese, rhenium and cobalt

Compound	77°K		Room temperature	
	e^2Qq, MHz	η, %	e^2Qq, MHz	η, %
$^{55}Mn_2(CO)_{10}$	4.38	21.1	3.30	36.2
$^{185}Re_2(CO)_{10}$	147.11	87.35	140.17	66.1
$^{59}Co_2(CO)_8$	90.23	31.5	89.61	32.9
	89.23	47.7	89.16	44.6

Obviously, the appearance and behavior of η for these two carbonyls is lue to the electron distribution on the molecular orbitals.

The structure of $Co_2(CO)_8$ is much less symmetrical [158] (see Fig- ıre 9.15). The cobalt atoms in this compound are in an octahedral environ- nent with two distortions along the third- and fourth-order axes C_3 and C_4. Therefore, the total electric field gradient on the nucleus of the cobalt ıtom can be described by the sum of two axially symmetrical tensors vhose z axes are directed along C_3 and C_4 [156]. The z axis of the result- ng tensor of the electric field gradient is deflected through an angle φ rom the Co—Co axis. The calculation shows that $\varphi \approx 62$ and 65° for two :rystallographically independent molecules in the unit cell [156].

The main contribution to the field gradient q_a (z_a is directed along the hird-order axis Co—Co) is caused by the lack of $3d$ electrons in the anti- onding $3d_z^2$ orbital of the cobalt atom. The gradient q_b (z_b is directed along he fourth-order axis) results from the difference in population between the ridge and the terminal carbonyl groups. (The bridge CO groups are similar o the carbonyl groups in the ketone $R_2C{=}O$, while the terminal CO groups ıre similar to free carbon monoxide [159].)

The NQR of ^{59}Co was also studied in bimetallic carbonyl complexes of :obalt $X_3MCo(CO)_4$ (where X = Cl, Br, I, C_6H_5; M = Si, Ge, Sn, Pb [155, 60]) and $X_iM\,[Co(CO)_4]_{4-i}$ [155].

In [155] the character of the transmission effect along the metal-metal ond in $R'R''_2SnCo(CO)_4$ compounds was studied by the correlation method:

$$e^2Qq^{59}\text{Co} = (74.59 + 39.65\Sigma\sigma_I - 71.35\Sigma\sigma_c) \pm 1.97 \text{ MHz}, \qquad (9.21)$$

where $\Sigma\sigma_I$ and $\Sigma\sigma_c$ are the sums of the induction and conjugation constants of the substituents on the tin atom. The high value of the coefficient of $\Sigma\sigma_c$ indicates the multiplicity of the Sn—Co bond (obviously, a δ type conjugation).

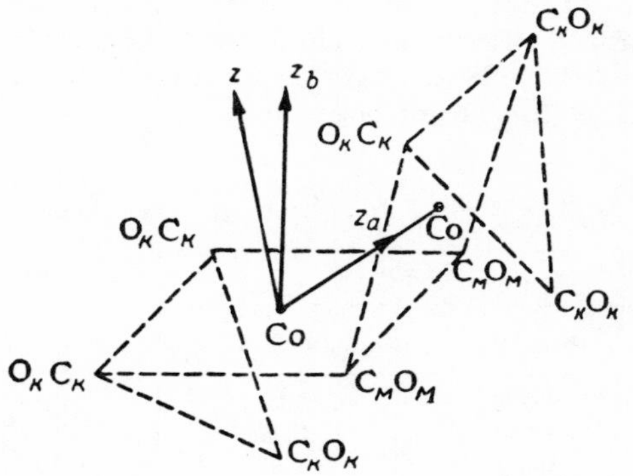

FIGURE 9.15. Environment of the cobalt atom in $Co_2(CO)_8$.

C_KO_K and C_MO_M are the terminal and bridge carbonyl groups, and z_a and z_b are the third- and fourth-order axes of the octahedron.

Brown et al. [160] hold that, since the asymmetry parameter of the elec tric field gradient on the cobalt nucleus in $X_3MCo(CO)_4$ compounds is close to zero, all the molecules preserve their trigonal bipyramid configuration, in which the CO and MX_3 groups occupy axial positions.

The relative populations of the $3d$ orbitals of cobalt in these compounds are assumed to be equal to the populations in the isostructural compound $Fe(CO)_5$ [160]. The electric field gradient on the nucleus of the cobalt ator depends on the population of: 1) the $3d_{z^2}$ orbital (on varying the σ donor properties of the MX_3 group); 2) the $3d_{xz}$, $3d_{yz}$ orbitals (on varying the π bond with the MX_3 group); 3) the σ M—Co bond (on varying the π M—Co bond (π induced σ interaction)). Brown et al. [160] also hold that the M—Co bond is a multiple bond and that there is a $d_\pi - p_\pi$ Si—Cl interaction in the silicon derivative.

The quadrupole interaction constants of ^{59}Co are much smaller in octahedral complexes ($e^2Qq \approx 15-60$ MHz, depending on the degree of distortic of the octahedron [161—163]) than in tetrahedral ones ($e^2Qq \approx 100-160$ MH In the electrostatic approximation (which takes into account the influence c the charges of the adjacent ions), the quadrupole interaction constant for octahedral complexes must be equal to zero. To explain the observed valu of e^2Qq we must assume that the bond in these complexes has some covale character. In this case, the electric field gradient on the nucleus of the cobalt atom will depend on the difference in the populations of the metal-ligand bonds.

BIBLIOGRAPHY

1. Chiba, T. — J. Chem. Phys. 39 (1963), 947.
2. Merchant, S. Z. and B. M. Fung. — J. Chem. Phys. 50 (1969), 2265.
3. Olimpia, P. L., I. Y. Fui, and B. M. Fung. — J. Chem. Phys. 51 (1969), 1610.
4. Weissmann, M. — J. Chem. Phys. 44 (1966), 422.
5. Weissmann, M. and N. V. Cohan. — J. Chem. Phys. 43 (1965), 119.
6. Blinc, R. and D. Hadži. — Nature 212 (1967), 1307.
7. Chiba, T. — J. Chem. Phys. 41 (1964), 1352.
8. Tokuhiro, T. — J. Chem. Phys. 41 (1964), 439.
9. Lotfullin, R. Sh. and G. K. Semin. — Zh. Strukt. Khim. 9 (1968), 813.
10. Chiba, T. — Bull. Chem. Soc. Japan 38 (1965), 259.
11. Hon, J. F. and P. J. Bray. — Phys. Rev. 110 (1958), 824.
12. Kahabas, S. L. and R. K. Nesbet. — Phys. Rev. Lett. 6 (1961), 549.
13. Das, T. P. and R. Bersohn. — Phys. Rev. 102 (1956), 733.
14. Bishop, S. G., P. J. Ring, and P. J. Bray. — J. Chem. Phys. 45 (1966), 1525.
15. Haigh, P., F. Forman, and R. Frisch. — J. Chem. Phys. 45 (1966), 812.
16. Zintl, E. and G. Brauer. — Z. Elektrochem. 41 (1935), 102.
17. Burns, G. — Phys. Rev. 127 (1962), 1193.
18. Stauss, G. H. — J. Chem. Phys. 40 (1964), 1988.
19. Braun, P. B. — Nature 170 (1952), 1123.
20. Dietrich, H. — Acta crystallogr. 16 (1963), 681.
21. Bernheim, R., I. Adler, B. Lavery, D. Lini, B. Scott, and J. Dixon. — J. Chem. Phys. 45 (1966), 3370.
22. Lucken, E. A. C. — J. Organometall. Chem. 4 (1965), 252.
23. Brown, T. L. and J. A. Ladd. — J. Organometall. Chem. 2 (1964), 373.
24. Itoh, J., R. Kusaka, and Y. Yamagata. — J. Phys. Soc. Japan 9 (1954), 289.
25. Bersohn, R. — J. Chem. Phys. 29 (1958), 326.
26. Poling, L. The Nature of the Chemical Bond. — Goskhimizdat, 1947. (Russian)
27. Bernheim, R. A. and H. S. Gutowsky. — J. Chem. Phys. 32 (1960), 1072.
28. Weiss, A. and K. Zohner. — Physica Status Solidi 21 (1967), 257.
29. Blinc, R., J. Petrovsec, and I. Zupančič. — Phys. Rev. 136 (1964), 1684.
30. Miller, H. C. and P. A. Casabella. — Phys. Rev. 152 (1966), 228.
31. Andrew, E. R. — Ber. Bunsengesellsch. phys. Chem. 67 (1963), 295.
32. Weiss, A. and D. Biedenkapp. — Z. Naturf. 17A (1962), 794.
33. Chong, D. P. and W. B. Brown. — J. Chem. Phys. 45 (1966), 392.
34. Das, T. P. and M. Carplus. — J. Chem. Phys. 42 (1965), 2885.
35. Wijn, H. W. de. — J. Chem. Phys. 44 (1966), 810.
36. Wijn, H. W. de and J. L. de Wildt. — Phys. Rev. 150 (1966), 200.
37. Baisa, D. F. and A. F. Prikhot'ko. — Ukr. Fiz. Zh. 9 (1964), 573.
38. Jeffrey, K. and R. Armstrong. — Can. J. Phys. 44 (1966), 2315.
39. Safin, I. A. — Fiz. Tverd. Tela 3 (1961), 2374.
40. Woessner, D. and H. Gutowsky. — J. Chem. Phys. 39 (1963), 440.
41. Cox, H. L. and D. Williams.— J. Chem. Phys. 32 (1960), 633.
42. Wells, A. F. Structural Inorganic Chemistry, 3rd edition. — Oxford, Univ. Press, 1962.
43. Wikner, E. G. and T. P. Das. — Phys. Rev. 109 (1958), 360.
44. Krüger, H. and U. Meyer-Berhout. — Z. Phys. 132 (1952), 171.
45. McKown, G. L. and J. D. Graybeal. — J. Chem. Phys. 44 (1966), 610.
46. Cromer, D. T. — J. Phys. Chem. 61 (1957), 1388.
47. Graybeal, J. D. and G. L. McKown. — Inorg. Chem. 5 (1966), 1909.
48. Royer, D. J. — Inorg. Chem. 4 (1965), 1830.
49. Niekerk, J. N. van, and F. R. L. Schaening. — Acta crystallogr. 6 (1953), 227.
50. Martin, R. L. and H. Watermann. — J. Chem. Soc., p. 2545, 1957.
51. Tonnett, M. L., S. Yamada, and I. G. Ross. — Trans. Faraday Soc. 60 (1964), 84.
52. Forster, L. S. and C. J. Ballhausen. — Acta chem. scand. 16 (1962), 1385.
53. Machmer, P., M. Read, and P. Cornil. — Compt. rend. 262AB (1966), 650.
54. Barnes, R. G. International Summer School on Theoretical Chemistry. Nuclear Quadrupole Resonance Spectroscopy, Italy, 1967.

55. Taylor, T. T. and T. P. Das. — Phys. Rev. 133A (1964), 1327.
56. Hon, J. F. — Phys. Rev. 124 (1961), 1368.
57. Rao, L. V. D. and D. V. L. N. Rao. — Phys. Rev. 160 (1967), 274.
58. Belford, G. G., R. A. Bernheim, and H. S. Gutowsky. — J. Chem. Phys. 35 (1961), 1032.
59. Dehmelt, H. G., H. G. Robinson, and W. Gordy. — Phys. Rev. 93 (1954), 480.
60. McGrath, J. W. and A. A. Silvidi. — J. Chem. Phys. 36 (1962), 1496.
61. Torizuka, K. — J. Phys. Soc. Japan 17 (1962), 239.
62. Ketelaar, J. A. A., C. H. Gilvary, and P. A. Renes. — Recl. Trav. chim. Pays-Bas Belg. 66 (1947), 512.
63. Casabella, P. A. and N. C. Miller. — J. Chem. Phys. 40 (1964), 1363.
64. Lucken, E. A. C. International Summer School on Theoretical Chemistry. Nuclear Quadrupole Resonance Spectroscopy. Italy, 1967.
65. Chiba, T. — J. Phys. Soc. Japan 13 (1958), 860.
66. Laurita, W. G. and W. S. Koski. — J. Am. Chem. Soc. 81 (1959), 3179.
67. Love, P. — J. Chem. Phys. 39 (1963), 3044.
68. Das, T. P. — J. Chem. Phys. 26 (1957), 753.
69. Bray, P. J., J. Edwards, J. O'Keefe, V. Ross, and I. Tatsuzaki. — J. Chem. Phys. 35 (1961), 435.
70. Bray, P. J., M. Leventhal, and H. Hooper. — Physics Chem. Glasses 4 (1963), 47.
71. Greenblatt, S. and P. J. Bray. — Physics Chem. Glasses 8 (1967), 190.
72. Bray, P. J. and A. H. Silver. — Bull. Am. Phys. Soc. 1 (1956), 323.
73. Veigele, W. J. — J. Chem. Phys. 39 (1963), 2389.
74. Sternheimer, R. — Phys. Rev. 130 (1963), 1423.
75. Petch, H. E. and K. S. Pennington. — J. Chem. Phys. 36 (1962), 1216.
76. Brun, E., S. Hafner, and P. Hartmann. — Helv. phys. Acta 33 (1960), 495.
77. Brun, E., S. Hafner, P. Hartmann, and F. Laves. — Naturwissenschaften 47 (1960), 277.
78. Ghose, S. — Solid State Communs. 2 (1964), 361.
79. Mkrtchyan, A. R. and A. S. Marfutin. — Doklady AN SSSR 163 (1965), 609.
80. Burns, G. — J. Chem. Phys. 32 (1960), 1585.
81. Brog, K., W. Jones, and C. Verber. — Phys. Rev. Lett. 20 (1966), 258.
82. Schultz, G. W., J. Sheerer, W. Muller-Warmuth, and W. Poch. — In: "Magnetic resonance and relaxation," p. 408. Amsterdam, 1967.
83. Veigele, W. J., W. H. Tantilla, and S. Verber. — Bull. Am. Phys. Soc. 5 (1960), 344.
84. Walsh, W. — Phys. Rev. 114 (1959), 1485.
85. Burns, G. — Bull. Am. Phys. Soc. 5 (1960), 404.
86. Bokii, G. G. A Practical Course of X-ray Diffraction Analysis. — Moscow State Univ., 1966. (Russian)
87. Svergun, V. I., V. M. Kuznets, Ya. G. Grinberg, and T. A. Babushkina. — Izv. AN SSSR, chem. ser., p. 1448, 1970.
88. Medveda, Z. S. Chalcogenides of Group IIIB Elements, 1968. (Russian)
89. Forsberg, H. E. — Acta chem. scand. 10 (1956), 1287.
90. Svergun, V. I., A. F. Volkov, P. Khagenmyuller, I. Port'e, and G. K. Semin. — Izv. AN SSSR, chem. ser., p. 1676, 1970.
91. Amma, E. L. and R. E. Rundle. — J. Am. Chem. Soc. 80 (1958), 4141.
92. Hall, J. R., L. A. Woodward, and E. A. Ebsworth. — Spectrochim. Acta 20 (1964), 1249.
93. Svergun, V. I., L. M. Bednova, O. Yu. Okhlobystin, and G. K. Semin. — Izv. AN SSSR, chem. ser., p. 1449, 1970.
94. Barrett, A. H. and M. Mandel. — Phys. Rev. 109 (1958), 1572.
95. Delvigne, G. A. and H. W. de Wijn. — J. Chem. Phys. 45 (1966), 3318.
96. Carlson, E. H. and H. S. Adams. — J. Chem. Phys. 51 (1969), 388.
97. Carlson, E. H. — Phys. Rev. Lett. 29A (1969), 696;
98. Parks, S. I. and W. G. Moulton. — Phys. Rev. Lett. 26A (1967), 63.
99. Kojima, S., K. Tsukada, S. Ogawa, A. Shimauchi, and Y. Abe. — J. Phys. Soc. Japan 9 (1954), 805.
100. Safin, I. A. — Zh. Strukt. Khim. 4 (1963), 267.
101. Pen'kov, I. N. and I. A. Safin. — Fiz. Tverd. Tela 7 (1965), 190.
102. Pen'kov, I. N. and I. A. Safin. — Doklady AN SSSR 156 (1964), 139.
103. Safin, I. A. — Doklady AN SSSR 147 (1962), 410.
104. Safin, I. A. — Kristallografiya 13 (1968), 330.

105. Kravchenko, E. A. and G. K. Semin. — Izv. AN SSSR, ser. inorg. mater. 5 (1969), 1161.
106. Pen'kov, I. N. and I. A. Safin. — Doklady AN SSSR 170 (1966), 626.
107. Pen'kov, I. N. and I. A. Safin. — Doklady AN SSSR 168 (1966), 1148.
108. Pen'kov, I. N. and I. A. Safin. — Doklady AN SSSR 161 (1965), 1404.
109. Pen'kov, I. N. and I. A. Safin. — Fiz. Tverd. Tela 6 (1964), 2467.
110. Kravchenko, E. A., S. A. Dembovskii, A. P. Chernov, and G. K. Semin. — Physica Status Solidi 31 (1968), K19.
111. Buslaev, Yu. A., E. A. Kravchenko, V. I. Pachomov, V. M. Skorikov, and G. K. Semin. — Chem. Phys. Letters 5 (1969), 455.
112. Petrov, L. N., I. A. Kyuntsel', and V. S. Grechishkin. — Vest. Leningr. State Univ., No. 1 (1969), 167.
113. Svergun, V. I., T. A. Babushkina, G. N. Shvedova, L. V. Kudryavtseva, and G. K. Semin. — Izv. AN SSSR, chem. ser., p. 482, 1970.
114. Di Lorenzo, J. V. and R. F. Schneider. — Inorg. Chem. 6 (1967), 766.
115. Chihara, H., N. Nakamura, and S. Scki. — Bull. Chem. Soc. Japan 40 (1967), 50.
116. Semin, G. K. and T. A. Babushkina. — TEKh 4 (1968), 835.
117. Schneider, R. and J. Di Lorenzo. — J. Chem. Phys. 47 (1967), 2343.
118. Schempp, E. and P. J. Bray. — J. Chem. Phys. 49 (1968), 3450.
119. Dewar, M. J. S. and E. A. C. Lucken. — J. Chem. Soc., p. 2653, 1958.
120. Guibe, L. and E. A. C. Lucken. —Molec. Phys. 10 (1966), 273.
121. Schempp, E. and P. J. Bray. — Phys. Rev. Lett. 25A (1967), 414.
122. Guibe, L. and E. A. C. Lucken. — Molec. Phys. 14 (1968), 73.
123. Lucken, E. A. C. and C. Mazeline. — Proc. Colloque AMPERE, Amsterdam 13 (1965), 241.
124. Davies, D. W. and W. C. Mackrodt. — Chem. Communs., p. 1226, 1967.
125. Smith, D. and R. M. Cotts. — J. Chem. Phys. 41 (1964), 2403.
126. Minematsu, M. — J. Phys. Soc. Japan 14 (1959), 1030.
127. Lucken, E. A. C. International Summer School on Theoretical Chemistry. Nuclear Quadrupole Resonance Spectroscopy. Lecture 2. Italy, 1967.
127a. Osokin, D. Yu., I. A. Safin, and I. A. Nuretdinov. — Doklady AN SSSR 186 (1969), 1128.
128. Yim, C. T., M. A. Whitehead, and D. H. Lo. — Can. J. Chem. 46 (1968), 3595.
129. Clark, D. T. and J. W. Emsley. — Molec. Phys. 12 (1967), 365.
130. Ha Tae Kyu and C. T. O'Konski. — J. Chem. Phys. 51 (1969), 460.
131. Pierce, L., R. G. Hayes, and J. F. Beecher. — J. Chem. Phys. 46 (1967), 4352.
132. Lide, D. R. Jr. — J. Chem. Phys. 38 (1963), 456.
133. Colligiani, H., L. Guibe, P. J. Haigh, and E. A. C. Lucken. — Molec. Phys. 14 (1968), 89.
134. Onda, S., R. Ikeda, D. Nakamura, and M. Kubo. — Bull. Chem. Soc. Japan 42 (1969), 1771.
135. Brown, D. J. Pyrimidines, p. 27. — New York, 1962.
136. Lucken, E. A. C. — Trans. Faraday Soc. 57 (1961), 729.
137. Guibe, L. and E. A. C. Lucken. — Molec. Phys. 14 (1968), 79.
138. Marino, R., L. Guibe, and P. J. Bray. — J. Chem. Phys. 49 (1968), 5104.
139. Peacock, T. E. — J. Chem. Soc., p. 1946, 1960.
140. Werner, W., H. Dreizler, and H. D. Rudolph. — Z. Naturf. 22A (1967), 531.
141. Ikeda, R., S. Onda, D. Nakamura, and M. Kubo. — J. Phys. Chem. 72 (1968), 2501.
142. Nesmeyanov, A. N., G. K. Semin, E. V. Bryukhova, T. A. Babushkina, K. N. Anisimov, N. E. Kolobova, and Yu. V. Makarov. — Izv. AN SSSR, chem. ser., p. 1953, 1968.
143. Nesmeyanov, A. N., G. K. Semin, E. V. Bryukhova, T. A. Babushkina, K. N. Anisimov, N. E. Kolobova, and Yu. V. Makarov. — Tetrahedron Lett., p. 3987, 1968.
144. Temnikova, T. I. A Course of Theoretical Principles in Organic Chemistry, p. 467. — "Khimiya," 1968. (Russian)
145. Lotspeich, J. F. — J. Chem. Phys. 31 (1959), 643.
146. Rogers, M. T. and K. V. S. Rama Rao. — J. Chem. Phys. 49 (1968), 1229.
147. Jovan, A. and A. Engebrecht. — Phys. Rev. 96 (1954), 649.
148. Morrow, J. C. — Acta crystallogr. 13 (1960), 443.
149. Lucken, E. A. C. — In: "Magnetic resonance and relaxation," p. 1121. Amsterdam, 1967.
150. Segel, S. L. — J. Chem. Phys. 51 (1969), 849.
151. Segel, S. and R. Barnes. — Phys. Rev. 107 (1957), 638.
152. Semin, G. K. and E. V. Bryukhova. Nuclear Physics, p. 1346, 1968. (Russian)

153. Harris, C. — Inorg. Chem., p. 1517, 1968.
154. Segel, S. and L. Anderson. — J. Chem. Phys. 49 (1968), 1407.
155. Nesmeyanov, A. N., G. K. Semin, E. V. Bryukhova, K. N. Anisimov, N. E. Kolobova, and V. N. Khandozhko. — Izv. AN SSSR, chem. ser., p. 1936, 1969.
156. Mooberry, E. S., H. W. Spiess, B. B. Garrett, and R. K. Sheline. — J. Chem. Phys. 51 (1969), 1970.
157. Dahl, L. F. and R. E. Rundle. — Acta crystallogr., 16 (1963), 419.
158. Gardner-Summer, G., H. P. Klug, and L. E. Alexander. — Acta crystallogr. 17 (1964), 732.
159. Orchin, M. and H. J. Jaffé. Antibonding Orbital. — New York, Wiley, 1967.
160. Brown, T. L., P. A. Edwards, C. B. Harris, and J. L. Kirsch. — Inorg. Chem. 8 (1969), 763.
161. Hartmann, H. and H. Silescu. — Theoret. chim. Acta 2 (1964), 371.
162. Spiess, H. W., H. Haas, and H. Hartmann. — J. Chem. Phys. 50 (1969), 3057.
163. Watanabe, I. and Y. Yamagata. — J. Chem. Phys. 46 (1967), 407.

Appendix I

NUCLEAR CONSTANTS OF THE ELEMENTS

The table contains the required characteristics of nuclei and atoms which possess a quadrupole moment. The isotope, the nuclear spin, and the natural abundance are listed in the first three columns. The quadrupole moments in barns (1 barn = 10^{-24} cm^2) and the gyromagnetic ratios $\gamma_I/2\pi$ in Hz·cm^{-1} are listed in the fourth and last columns, respectively.

The Sternheimer factors for the ions of the corresponding elements and the ionic charges are listed in column five. QM^n/QM^{n^o} has been determined for some elements: 0.019 (Li$^{6/7}$), 2.084 (B$^{10/11}$), 1.2688 (Cl$^{35/37}$), 1.220 (K$^{39/41}$), 1.0806 (Cu$^{63/65}$), 1.5867 (Ga$^{69/71}$), 1.19707 (Br$^{79/81}$), 2.07 (Rb$^{87/85}$), 1.0146 (In$^{115/113}$), 1.27475 (Sb$^{123/121}$), 3.0 (La$^{138/139}$), 1.0565 (Re$^{185/187}$).

All the data were taken from periodicals. The following sources were also used: A. Leshe, Nuclear Induction, IL, 1963; A. Weiss, International Summer School on Theoretical Chemistry, Frascati (Rome), 11—21 October 1967.

Nucleus	Spin I	Natural abundance, %	Electric quadrupole moment Q, barn	Sternheimer factor γ_∞	$\frac{\gamma_I}{2\pi}$, Hz·cm^{-1}
^{2}H	1	$1.56\cdot10^{-2}$	$2.82\cdot10^{-3}$	1.149(⁻)	653.5
^{6}Li	1	7.43	$4.6\cdot10^{-4}$	0.257(+)	626.7
^{7}Li	$^3/_2$	92.57	$-4.2\cdot10^{-2}$	0.257(+)	1655
^{9}Be	$^3/_2$	100	$2\cdot10^{-2}$	0.186(2+)	598.7
^{10}B	3	18.83	0.111	0.145(3+)	457.6
^{11}B	$^3/_2$	81.17	$3.55\cdot10^{-2}$	0.145(3+)	1367
^{14}N	1	99.635	$1.6\cdot10^{-2}$	0.101(5+)	307.7
^{17}O	$^5/_2$	$3.7\cdot10^{-2}$	$-2.6\cdot10^{-2}$	—429(2⁻)	577.2
^{21}Ne	$^3/_2$	0.257	$9.3\cdot10^{-2}$	...	...
^{23}Na	$^3/_2$	100	$9.7\cdot10^{-2}$	—4.5(+)	1126.7
^{25}Mg	$^5/_2$	10.05	0.14	—3.04(2+)	260.6
^{27}Al	$^5/_2$	100	0.149	—2.24(3+)	1110
^{33}S	$^3/_2$	0.74	$-6.4\cdot10^{-2}$	...	326.7
^{35}Cl	$^3/_2$	75.4	$-7.97\cdot10^{-2}$	—53.9(⁻)	417.3
^{37}Cl	$^3/_2$	24.6	$6.35\cdot10^{-2}$	—53.9(⁻)	347.4
^{39}K	$^3/_2$	93.08	0.09	—12.2(+)	198.7
^{41}K	$^3/_2$	6.91	0.11	...	109.2
^{45}Sc	$^7/_2$	100	—0.22	—9.46(3+)	1034
^{47}Ti	$^5/_2$	7.75	...	...	240.0
^{49}Ti	$^7/_2$	5.51	...	...	240.1
^{50}V	6	0.24	0.4	...	424.4
^{51}V	$^7/_2$	~100	0.3	—6.50(5+)	1121
^{53}Cr	$^3/_2$	9.54	$-3\cdot10^{-2}$	...	300

(contd.)

Nucleus	Spin I	Natural abundance %	Electric quadrupole moment Q, barn	Sternheimer factor γ_∞	$\frac{\nu_I}{2\pi}$, Hz · cm^{-1}
^{55}Mn	$^5/_2$	100	0.355	—6.81(2^-)	1056
^{59}Co	$^7/_2$	100	0.404	...	1011
^{63}Cu	$^3/_2$	69.09	—0.15	—13.77$(^+)$	1131
^{65}Cu	$^3/_2$	30.91	—0.14	—13.77$(^+)$	1213
^{67}Zn	$^5/_2$	4.12	0.18	...	267
^{69}Ga	$^3/_2$	60.2	0.2318	—8.75(3^+)	1023
^{71}Ga	$^3/_2$	39.8	0.1461	—8.75(3^+)	1299
^{73}Ge	$^9/_2$	7.61	—0.2	...	148.7
^{75}As	$^3/_2$	100	0.27	...	729.3
^{79}Br	$^3/_2$	50.57	0.33	—99(—40)$(^-)$	1070
^{81}Br	$^3/_2$	49.43	0.28	—99(—40)$(^-)$	1153
^{83}Kr	$^9/_2$	11.55	0.26	...	164
^{85}Rb	$^5/_2$	72.8	0.286	—49.3$(^+)$	411.3
^{87}Rb	$^3/_2$	27.2	0.1431	...	1397
^{87}Sr	$^9/_2$	7.02	0.36	...	190
^{93}Nb	$^9/_2$	100	—0.1···—0.7	—15(5^+)	1045
^{113}In	$^9/_2$	4.16	1.114	—15.3(3^+)	931.1
^{115}In	$^9/_2$	95.84	1.125	—15.3(3^+)	946
^{121}Sb	$^5/_2$	57.25	—0.53	...	1019
^{123}Sb	$^7/_2$	42.75	—0.4	...	551.9
^{127}I	$^5/_2$	100	—0.62	−179(−90)$(^-)$	855.5
^{131}Xe	$^3/_2$	21.24	—0.12	...	410
^{133}Cs	$^7/_2$	100	—0.0024	—110$(^+)$	561.7
^{135}Ba	$^3/_2$	6.59	0.18	...	425
^{137}Ba	$^3/_2$	11.32	0.28	...	476
^{139}La	$^7/_2$	99.911	0.23	—68(3^+)	605
^{141}Pr	$^5/_2$	100	—5.4·10^{-2}	—78.5(3^+)	1160
^{143}Nd	$^7/_2$	12.2	2.06	...	220
^{145}Nd	$^7/_2$	8.30	—0.255	...	140
^{147}Sm	$^7/_2$	15.07	—0.208	...	147
^{149}Sm	$^7/_2$	13.84	0.060	...	119
^{151}Eu	$^5/_2$	47.77	1.16	...	1000
^{153}Eu	$^5/_2$	52.23	2.92		460
^{155}Gd	$^3/_2$	14.68	1.1		
^{157}Gd	$^3/_2$	15.64	1.0		
^{159}Tb	$^3/_2$	100	1.32		
^{161}Dy	$^5/_2$	18.73	—1.13		
^{163}Dy	$^5/_2$	24.97	1.62		
^{165}Ho	$^7/_2$	100	2.83		
^{167}Hr	$^7/_2$	22.82	2.82		
^{173}Yb	$^5/_2$	16.08	3.9	—79(3^+)	198
^{175}Lu	$^7/_2$	97.40	5.68	...	570
^{177}Hf	$^7/_2$	18.39	3		
^{179}Hf	$^9/_2$	13.78	3		
^{181}Ta	$^7/_2$	100	3···6	...	460
^{185}Re	$^5/_2$	37.07	2.8	...	958.4
^{187}Re	$^5/_2$	62.93	2.6	...	968.1
^{189}Os	$^3/_2$	16.1	0.6	...	360
^{191}Ir	$^3/_2$	38.5	1.2	...	81
^{193}Ir	$^3/_2$	61.5	1.0	...	86
^{197}Au	$^3/_2$	100	0.56	...	69.1
^{201}Hg	$^3/_2$	13.24	0.50	—25.3(2^+)	308
^{209}Bi	$^9/_2$	100	0.4	—42.4(3^+)	684.3

Appendix II

EIGENVALUES OF THE ENERGY OF QUADRUPOLE INTERACTIONS AS A FUNCTION OF THE ASYMMETRY PARAMETER, FOR SPINS $I = {}^{5}/_{2}, {}^{7}/_{2}, {}^{9}/_{2}$

We give the results of the calculation of the energy levels (in relative units) for spins $I = {}^{5}/_{2}, {}^{7}/_{2}, {}^{9}/_{2}$ as a function of the asymmetry parameters. The calculation was based on the secular equations (1.5).

The tables are used as follows. The ratios of the experimentally determined frequencies ν_{ij} are compared with the listed values of ΔE_{ij}. From the asymmetry parameter found in this manner the quadrupole interaction constants are calculated from the relations:

$$I = {}^{5}/_{2} \quad e^2Qq = 20\,\frac{\Sigma\nu_{ij}}{\Sigma\Delta E_{ij}}$$

$$I = {}^{7}/_{2} \quad e^2Qq = 28\,\frac{\Sigma\nu_{ij}}{\Sigma\Delta E_{ij}}$$

$$I = {}^{9}/_{2} \quad e^2Qq = 24\,\frac{\Sigma\nu_{ij}}{\Sigma\Delta E_{ij}}.$$

Suppose that for some substance at least one NQR frequency of the different isotopes with different spins is known. Then the quadrupole interaction constants can be found from these tables. Thus, from the frequencies of the i — j transitions for ^{121}Sb and ^{123}Sb we can calculate the asymmetry parameter using the equation:

$$\frac{\Delta E_{ij}^{7/2}}{\Delta E_{i'j'}^{5/2}} = 1.09825\,\frac{\nu_{ij}^{7/2}}{\nu_{i'j'}^{5/2}},$$

where the ratios of quadrupole moments of the antimony isotopes $Q^{123}/Q^{121} = 1.27475$ and the values of the spins $I = {}^{7}/_{2}$ for ^{121}Sb and $I = {}^{5}/_{2}$ for ^{123}Sb are taken into account.

$I = {}^{5}/_{2}$

η	$E_{5/2}$	$E_{3/2}$	$E_{1/2}$	$\Delta E_{3/2-5/2} = E_{3/2}-E_{5/2}$	$\Delta E_{1/2-3/2} = E_{1/2}-E_{3/2}$	$\left\lvert\frac{\Delta E_{3/2-5/2}}{\Delta E_{1/2-3/2}}\right\rvert$
0.00	5.0000	—1.0000	—4.0000	—6.0000	—3.0000	2.0000
0.01	5.0000	—0.9999	—4.0002	—5.9999	—3.0003	1.9997
0.02	5.0001	—0.9994	—4.0007	—5.9995	—3.0013	1.9990
0.03	5.0003	—0.9987	—4.0016	—5.9990	—3.0029	1.9977

$I = {}^5/_2$ (contd.)

η	$E_{5/2}$	$E_{3/2}$	$E_{1/2}$	$\Delta E_{3/2-5/2} = E_{3/2} - E_{5/2}$	$\Delta E_{1/2-3/2} = E_{1/2} - E_{3/2}$	$\left\lvert \frac{\Delta E_{3/2-5/2}}{\Delta E_{1/2-3/2}} \right\rvert$
0,04	5,0004	—0,9976	—4,0028	—5,9980	—3,0052	1.9959
0,05	5.0007	—0.9963	—4,0044	—5,9970	—3.0081	1,9935
0.06	5,0010	—0.9946	—4,0064	—5,9956	—3,0118	1 9907
0.07	5,0014	—0.9927	—4.0087	—5,9940	—3,0160	1,9874
0.08	5.0018	—0.9904	—4.0113	—5.9922	—3,0209	1.9836
0,09	5.0023	—0.9879	—4.0143	—5.9902	—3.0264	1,9793
0.10	5.0028	—0.9851	—4.0177	—5,9879	—3.0326	1.9745
0,11	5,0034	—0.9820	—4.0214	—5,9854	—3.0394	1,9693
0.12	5,0040	—0.9786	—4,0254	—5.9826	—3,0468	1.9636
0,13	5.0047	—0,9749	—4.0298	—5,9796	—3.0549	1,9574
0.14	5,0055	—0.9710	—4,0345	—5.9764	—3.0635	1.9508
0.15	5.0063	—0,9667	—4.0395	—5.9730	—3.0728	1.9438
0.16	5,0071	—0.9622	—4.0449	—5,9693	—3.0827	1.9364
0.17	5.0080	—0,9574	—4.0506	—5.9655	—3.0932	1,9286
0,18	5,0090	—0,9524	—4.0566	—5,9614	—3.1043	1.9204
0.19	5.0100	—0.9471	—4,0630	—5.9571	—3.1159	1.9118
0.20	5,0111	—0,9415	—4,0697	—5.9526	—3,1282	1.9029
0.21	5.0123	—0.9356	—4,0766	—5,9479	—3.1410	1.8936
0.22	5.0135	—0,9295	—4.0839	—5,9430	—3,1544	1.8840
0.23	5.0147	—0.9232	—4,0915	—5.9379	—3,1684	1,8741
0.24	5.0160	—0.9166	—4.0995	—5,9326	—3.1829	1,8639
0,25	5.0174	—0.9098	—4.1077	—5.9272	—3.1979	1,8535
0.26	5.0188	—0,9027	—4,1162	—5,9215	—3.2135	1.8427
0.27	5.0203	—0.8954	—4.1250	—5.9157	—3,2296	1.8317
0.28	5.0219	—0,8878	—4,1340	—5.9097	—3.2462	1.8205
0.29	5,0235	—0.8801	—4.1434	—5,9035	—3.2633	1.8091
0,30	5.0251	—0.8721	—4.1530	—5.8972	—3.2809	1,7974
0,31	5.0268	—0.8639	—4.1629	—5,8907	—3.2991	1,7856
0,32	5.0286	—0.8555	—4.1731	—5.8841	—3,3177	1,7736
0,33	5,0304	—0.8468	—4.1836	—5.8773	—3.3367	1,7614
0.34	5,0323	—0.8380	—4,1943	—5.8703	—3.3563	1,7491
0.35	5.0342	—0.8290	—4,2053	—5,8632	—3.3763	1,7366
0,36	5.0362	—0.8198	—4.2165	—5.8560	—3,3967	1,7240
0,37	5,0383	—0.8104	—4.2279	—5.8487	—3.4176	1,7113
0,38	5.0404	—0,8008	—4.2396	—5.8412	—3,4389	1.6986
0.39	5,0426	—0.7910	—4,2516	—5.8336	—3,4606	1,6857
0.40	5,0448	—0.7810	—4.2638	—5.8258	—3,4827	1,6728
0.41	5.0471	—0,7709	—4.2762	—5.8180	—3.5053	1,6598
0,42	5.0494	—0.7606	—4,2888	—5.8101	—3.5282	1,6467
0,43	5.0519	—0,7502	—4,3017	—5.8020	—3.5515	1,6337
0.44	5.0543	—0,7395	—4.3148	—5.7939	—3.5752	1,6205
0,45	5.0568	—0.7288	—4.3281	—5.7856	—3.5993	1,6074
0.46	5.0594	—0.7178	—4.3416	—5.7773	—3.6237	1.5943
0,47	5.0621	—0.7068	—4,3553	—5.7688	—3.6485	1,5811
0,48	5,0648	—0.6956	—4,3692	—5,7603	—3,6736	1,5680
0,49	5,0675	—0.6842	—4.3833	—5.7517	—3.6991	1,5549
0.50	5,0703	—0.6727	—4.3976	—5.7431	—3,7249	1.5418
0,51	5,0732	—0.6611	—4.4121	—5,7343	—3,7510	1.5287
0.52	5.0762	—0.6494	—4,4268	—5.7255	—3.7774	1,5157

$I = {}^5/_2$ (contd.)

η	$E_{5/2}$	$E_{3/2}$	$E_{1/2}$	$\Delta E_{3/2-5/2} = E_{3/2}-E_{5/2}$	$\Delta E_{1/2-3/2} = E_{1/2}-E_{3/2}$	$\left\| \frac{\Delta E_{3/2-5/2}}{\Delta E_{1/2-3/2}} \right\|$
0.53	5.0792	—0.6375	—4.4417	—5.7167	—3,8042	1.5027
0.54	5,0822	—0.6255	—4.4567	—5.7077	—3.8312	1.4898
0.55	5,0853	—0.6134	—4.4719	—5,6988	—3.8585	1.4769
0.56	5.0885	—0.6012	—4.4873	—5.6897	—3,8861	1.4641
0.57	5,0918	—0,5889	—4.5022	—5.6807	—3.9140	1.4514
0.58	5,0951	—0,5765	—4,5186	—5,6715	—3.9421	1,4387
0.59	5.0984	—0,5640	—4.5345	—5,6624	—3.9705	1,4261
0.60	5,1019	—0.5513	—4.5505	—5.6532	—3,9992	1.4136
0,61	5,1053	—0,5386	—4,5667	—5.6440	—4.0281	1,4012
0.62	5.1089	—0,5258	—4.5831	—5.6347	—4,0572	1,3888
0,63	5.1125	—0,5129	—4.5996	—5.6254	—4.0866	1.3766
0.64	5.1162	—0.4999	—4.6162	—5,6161	—4,1162	1,3644
0.65	5,1199	—0.4869	—4.6330	—5.6068	—4,1461	1.3523
0.66	5,1237	—0.4738	—4.6499	—5,5975	—4.1761	1,3404
0,67	5.1276	—0,4606	—4.6670	—5.5881	—4.2064	1.3285
0.68	5,1315	—0,4473	—4.6842	—5.5788	—4.2368	1,3167
0,69	5.1355	—0,4340	—4,7015	—5,5694	—4.2675	1.3051
0,70	5.1395	—0.4206	—4.7190	—5.5601	—4.2984	1.2935
0,71	5,1436	—0.4071	—4.7365	—5,5507	—4,3294	1,2821
0,72	5.1478	—0,3936	—4,7542	—5,5414	—4,3607	1,2708
0,73	5,1520	—0,3800	—4.7721	—5,5320	—4.3921	1,2595
0.74	5,1563	—0,3663	—4.7900	—5.5227	—4,4237	1,2484
0.75	5,1607	—0,3526	—4.8081	—5.5134	—4,4554	1,2374
0,76	5.1652	—0,3389	—4,8263	—5,5040	—4.4874	1,2266
0,77	5.1696	—0,3251	—4,8445	—5.4947	—4.5194	1.2158
0,78	5,1742	—0,3113	—4,8629	—5,4855	—4.5517	1,2052
0,79	5,1788	—0.2974	—4.8815	—5,4762	—4.5841	1,1946
0,80	5,1835	—0,2835	—4.9001	—5,4670	—4.6166	1.1842
0.81	5,1883	—0.2695	—4.9188	—5,4578	—4,6493	1,1739
0.82	5,1931	—0,2555	—4,9376	—5.4486	—4,6821	1,1637
0.83	5,1980	—0,2415	—4,9565	—5,4395	—4,7150	1,1537
0,84	5,2030	—0,2274	—4,9755	—5,4304	—4,7481	1,1437
0.85	5,2080	—0.2134	—4.9946	—5.4214	—4,7813	1.1339
0,86	5,2131	—0,1993	—5,0138	—5,4123	—4.8146	1,1242
0.87	5,2182	—0,1851	—5,0331	—5,4034	—4,8480	1,1146
0.88	5.2235	—0,1710	—5.0525	—5,3944	—4.8815	1,1051
0.89	5,2288	—0,1568	—5,0720	—5,3855	—4,9152	1,0957
0.90	5,2341	—0,1426	—5,0915	—5,3767	—4,9489	1,0864
0,91	5,2395	—0,1284	—5,1112	—5,3679	—4,9828	1,0773
0,92	5,2450	—0,1142	—5,1309	—5,3592	—5,0167	1,0683
0,93	5,2506	—0,0999	—5.1507	—5,3505	—5,0508	1,0593
0,94	5,2562	—0,0857	—5.1706	—5,3419	—5.0849	1,0505
0,95	5,2619	—0,0714	—5,1906	—5,3333	—5,1192	1,0418
0,96	5.2677	—0,0571	—5,2106	—5,3248	—5,1535	1.0333
0.97	5,2736	—0,0429	—5,2307	—5,3164	—5,1879	1,0248
0.98	5,2795	—0,0286	—5,2509	—5,3080	—5,2223	1,0164
0.99	5.2855	—0,0143	—5,2712	—5,2997	—5,2569	1,0082
1,00	5,2915	—0.0000	—5,2915	—5,2915	—5.2915	1,0000

$I = {}^{7}/_{2}$

η	$E_{7/2}$	$E_{5/2}$	$E_{3/2}$	$E_{1/2}$	$\Delta E_{7/2-5/2} = E_{5/2} - E_{7/2}$	$\Delta E_{5/2-3/2} = E_{3/2} - E_{5/2}$	$\Delta E_{3/2-1/2} = E_{1/2} - E_{3/2}$	$\left\lvert \frac{\Delta E_{7/2-5/2}}{\Delta E_{5/2-3/2}} \right\rvert$	$\left\lvert \frac{\Delta E_{5/2-3/2}}{\Delta E_{3/2-1/2}} \right\rvert$
0.00	7.0000	1.0000	—3.0000	—5.0000	—6.0000	—4.0000	—2.0000	1.5000	2.0000
0.01	7.0000	1.0001	—2.9997	—5.0004	—5.9999	—3.9998	—2.0007	1.5001	1.9992
0.02	7.0001	1.0003	—2.9988	—5.0017	—5.9998	—3.9991	—2.0029	1.5003	1.9966
0.03	7.0002	1.0008	—2.9972	—5.0037	—5.9995	—3.9980	—2.0065	1.5006	1.9925
0.04	7.0004	1.0013	—2.9951	—5.0066	—5.9990	—3.9964	—2.0116	1.5011	1.9867
0.05	7.0006	1.0021	—2.9923	—5.0104	—5.9985	—3.9944	—2.0181	1.5017	1.9793
0.06	7.0008	1.0030	—2.9889	—5.0149	—5.9978	—3.9919	—2.0260	1.5025	1.9704
0.07	7.0011	1.0041	—2.9850	—5.0202	—5.9971	—3.9891	—2.0353	1.5034	1.9600
0.08	7.0015	1.0053	—2.9805	—5.0264	—5.9962	—3.9858	—2.0459	1.5044	1.9482
0.09	7.0019	1.0068	—2.9754	—5.0333	—5.9951	—3.9821	—2.0579	1.5055	1.9350
0.10	7.0023	1.0083	—2.9697	—5.0410	—5.9940	—3.9781	—2.0713	1.5068	1.9206
0.11	7.0028	1.0101	—2.9635	—5.0494	—5.9927	—3.9736	—2.0859	1.5081	1.9050
0.12	7.0034	1.0120	—2.9568	—5.0586	—5.9913	—3.9688	—2.1018	1.5096	1.8883
0.13	7.0039	1.0141	—2.9496	—5.0685	—5.9898	—3.9637	—2.1189	1.5112	1.8706
0.14	7.0046	1.0164	—2.9419	—5.0791	—5.9882	—3.9582	—2.1372	1.5128	1.8521
0.15	7.0053	1.0188	—2.9337	—5.0904	—5.9864	—3.9525	—2.1567	1.5146	1.8327
0.16	7.0060	1.0214	—2.9250	—5.1023	—5.9846	—3.9464	—2.1773	1.5164	1.8125
0.17	7.0067	1.0242	—2.9160	—5.1150	—5.9826	—3.9401	—2.1990	1.5184	1.7918
0.18	7.0076	1.0271	—2.9065	—5.1282	—5.9804	—3.9336	—2.2218	1.5204	1.7705
0.19	7.0084	1.0302	—2.8966	—5.1421	—5.9782	—3.9268	—2.2455	1.5224	1.7487
0.20	7.0093	1.0335	—2.8863	—5.1566	—5.9758	—3.9198	—2.2703	1.5245	1.7266
0.21	7.0103	1.0370	—2.8756	—5.1716	—5.9733	—3.9126	—2.2960	1.5267	1.7041
0.22	7.0113	1.0406	—2.8646	—5.1873	—5.9707	—3.9052	—2.3226	1.5289	1.6814
0.23	7.0124	1.0444	—2.8533	—5.2034	—5.9680	—3.8977	—2.3501	1.5311	1.6585
0.24	7.0135	1.0484	—2.8417	—5.2201	—5.9651	—3.8901	—2.3784	1.5334	1.6356
0.25	7.0146	1.0525	—2.8298	—5.2373	—5.9621	—3.8823	—2.4075	1.5357	1.6126
0.26	7.0158	1.0568	—2.8176	—5.2550	—5.9590	—3.8745	—2.4374	1.5380	1.5896
0.27	7.0171	1.0613	—2.8052	—5.2732	—5.9557	—3.8665	—2.4680	1.5403	1.5666
0.28	7.0183	1.0660	—2.7925	—5.2918	—5.9523	—3.8585	—2.4993	1.5426	1.5438
0.29	7.0197	1.0709	—2.7796	—5.3109	—5.9488	—3.8505	—2.5313	1.5450	1.5211
0.30	7.0211	1.0759	—2.7665	—5.3304	—5.9452	—3.8424	—2.5639	1.5473	1.4986

$I = {}^7/_2$ (contd.)

η	$E_{7/2}$	$E_{5/2}$	$E_{3/2}$	$E_{1/2}$	$\Delta E_{7/2-5/2} = E_{5/2} - E_{7/2}$	$\Delta E_{5/2-3/2} = E_{3/2} - E_{5/2}$	$\Delta E_{3/2-1/2} = E_{1/2} - E_{3/2}$	$\left\lvert \frac{\Delta E_{7/2-5/2}}{\Delta E_{5/2-3/2}} \right\rvert$	$\left\lvert \frac{\Delta E_{5/2-3/2}}{\Delta E_{3/2-1/2}} \right\rvert$
0.31	7.0225	1.0811	—2.7532	—5.3504	—5.9414	—3.8343	—2.5972	1.5495	1.4763
0.32	7.0240	1.0865	—2.7398	—5.3707	—5.9375	—3.8262	—2.6309	1.5518	1.4543
0.33	7.0255	1.0920	—2.7261	—5.3914	—5.9335	—3.8182	—2.6653	1.5540	1.4326
0.34	7.0271	1.0978	—2.7124	—5.4125	—5.9293	—3.8101	—2.7001	1.5562	1.4111
0.35	7.0287	1.1037	—2.6985	—5.4339	—5.9250	—3.8021	—2.7355	1.5583	1.3899
0.36	7.0304	1.1098	—2.6844	—5.4557	—5.9206	—3.7942	—2.7713	1.5604	1.3691
0.37	7.0321	1.1161	—2.6703	—5.4778	—5.9160	—3.7864	—2.8075	1.5624	1.3487
0.38	7.0339	1.1225	—2.6561	—5.5003	—5.9113	—3.7786	—2.8442	1.5644	1.3285
0.39	7.0357	1.1292	—2.6418	—5.5230	—5.9065	—3.7710	—2.8812	1.5663	1.3088
0.40	7.0375	1.1360	—2.6274	—5.5461	—5.9015	—3.7634	—2.9187	1.5681	1.2894
0.41	7.0394	1.1430	—2.6130	—5.5695	—5.8964	—3.7560	—2.9564	1.5699	1.2705
0.42	7.0414	1.1502	—2.5985	—5.5931	—5.8912	—3.7487	—2.9945	1.5715	1.2519
0.43	7.0434	1.1576	—2.5840	—5.6170	—5.8858	—3.7416	—3.0330	1.5731	1.2337
0.44	7.0455	1.1652	—2.5695	—5.6411	—5.8803	—3.7346	—3.0717	1.5745	1.2158
0.45	7.0476	1.1729	—2.5549	—5.6656	—5.8747	—3.7278	—3.1106	1.5759	1.1984
0.46	7.0497	1.1809	—2.5404	—5.6902	—5.8689	—3.7212	—3.1498	1.5771	1.1814
0.47	7.0519	1.1890	—2.5258	—5.7151	—5.8629	—3.7148	—3.1893	1.5783	1.1648
0.48	7.0542	1.1973	—2.5112	—5.7402	—5.8569	—3.7085	—3.2290	1.5793	1.1435
0.49	7.0565	1.2058	—2.4967	—5.7656	—5.8507	—3.7025	—3.2689	1.5802	1.1326
0.50	7.0588	1.2145	—2.4822	—5.7911	—5.8443	—3.6966	—3.3089	1.5810	1.1172
0.51	7.0612	1.2233	—2.4677	—5.8169	—5.8379	—3.6910	—3.3492	1.5816	1.1021
0.52	7.0637	1.2324	—2.4532	—5.8428	—5.8312	—3.6856	—3.3896	1.5821	1.0873
0.53	7.0662	1.2417	—2.4388	—5.8690	—5.8245	—3.6805	—3.4302	1.5825	1.0730
0.54	7.0687	1.2511	—2.4245	—5.8953	—5.8176	—3.6756	—3.4709	1.5828	1.0590
0.55	7.0713	1.2607	—2.4102	—5.9219	—5.8105	3.6709	—3.5117	1.5829	1.0453
0.56	7.0739	1.2706	—2.3959	—5.9486	—5.8034	—3.6665	—3.5526	1.5828	1.0321
0.57	7.0766	1.2806	—2.3818	—5.9754	—5.7960	—3.6624	—3.5937	1.5826	1.0191
0.58	7.0794	1.2908	—2.3677	—6.0025	—5.7886	—3.6585	—3.6348	1.5822	1.0065
0.59	7.0822	1.3012	—2.3537	—6.0297	—5.7810	—3.6549	—3.6760	1.5817	0.9942
0.60	7.0850	1.3118	—2.3398	—6.0570	—5.7732	—3.6515	—3.7172	1.5811	0.9823
0.61	7.0879	1.3225	—2.3259	—6.0845	—5.7654	—3.6485	—3.7586	1.5802	0.9707
0.62	7.0908	1.3335	—2.3122	—6.1122	—5.7573	—3.6457	—3.8000	1.5792	0.9594

0.63	7.0938	1.3447	−2.2985	−6.1400	−5.7492	−3.6432	−3.8414	1.5781	0.9484
0.64	7.0969	1.3560	−2.2850	−6.1679	−5.7409	−3.6410	−3.8829	1.5767	0.9377
0.65	7.1000	1.3676	−2.2716	−6.1960	−5.7324	−3.6391	−3.9244	1.5752	0.9273
0.66	7.1031	1.3793	−2.2582	−6.2242	−5.7238	−3.6375	−3.9659	1.5736	0.9172
0.67	7.1063	1.3912	−2.2450	−6.2525	−5.7151	−3.6362	−4.0075	1.5717	0.9074
0.68	7.1096	1.4033	−2.2319	−6.2809	−5.7063	−3.6352	−4.0490	1.5697	0.8978
0.69	7.1129	1.4156	−2.2189	−6.3095	−5.6972	−3.6345	−4.0906	1.5675	0.8885
0.70	7.1162	1.4281	−2.2061	−6.3382	−5.6881	−3.6342	−4.1321	1.5652	0.8795
0.71	7.1196	1.4408	−2.1934	−6.3670	−5.6788	−3.6341	−4.1737	1.5626	0.8707
0.72	7.1231	1.4536	−2.1808	−6.3959	−5.6694	−3.6344	−4.2152	1.5599	0.8622
0.73	7.1266	1.4667	−2.1683	−6.4250	−5.6599	−3.6350	−4.2567	1.5571	0.8539
0.74	7.1301	1.4799	−2.1560	−6.4541	−5.6502	−3.6359	−4.2981	1.5540	0.8459
0.75	7.1337	1.4933	−2.1438	−6.4833	−5.6404	−3.6371	−4.3396	1.5508	0.8381
0.76	7.1374	1.5070	−2.1317	−6.5127	−5.6304	−3.6387	−4.3810	1.5474	0.8306
0.77	7.1411	1.5208	−2.1198	−6.5421	−5.6204	−3.6405	−4.4223	1.5438	0.8232
0.78	7.1449	1.5347	−2.1080	−6.5716	−5.6102	−3.6427	−4.4636	1.5401	0.8161
0.79	7.1487	1.5489	−2.0964	−6.6013	−5.5998	−3.6453	−4.5049	1.5362	0.8092
0.80	7.1526	1.5632	−2.0849	−6.6310	−5.5894	−3.6481	−4.5461	1.5321	0.8025
0.81	7.1565	1.5778	−2.0735	−6.6608	−5.5787	−3.6513	−4.5872	1.5279	0.7960
0.82	7.1605	1.5925	−2.0623	−6.6907	−5.5680	−3.6548	−4.6283	1.5235	0.7897
0.83	7.1645	1.6074	−2.0513	−6.7206	−5.5572	−3.6586	−4.6694	1.5189	0.7835
0.84	7.1686	1.6224	−2.0404	−6.7507	−5.5462	−3.6628	−4.7103	1.5142	0.7776
0.85	7.1728	1.6377	−2.0296	−6.7808	−5.5351	−3.6673	−4.7512	1.5093	0.7719
0.86	7.1770	1.6531	−2.0190	−6.8111	−5.5239	−3.6721	−4.7920	1.5043	0.7663
0.87	7.1812	1.6687	−2.0086	−6.8414	−5.5125	−3.6773	−4.8328	1.4991	0.7609
0.88	7.1855	1.6845	−1.9983	−6.8717	−5.5011	−3.6827	−4.8735	1.4937	0.7557
0.89	7.1899	1.7004	−1.9881	−6.9022	−5.4895	−3.6885	−4.9140	1.4882	0.7506
0.90	7.1943	1.7165	−1.9781	−6.9327	−5.4778	−3.6946	−4.9546	1.4826	0.7457
0.91	7.1988	1.7328	−1.9683	−6.9633	−5.4660	−3.7011	−4.9950	1.4768	0.7410
0.92	7.2033	1.7492	−1.9586	−6.9939	−5.4541	−3.7079	−5.0353	1.4709	0.7364
0.93	7.2079	1.7659	−1.9491	−7.0247	−5.4420	−3.7149	−5.0756	1.4649	0.7319
0.94	7.2125	1.7826	−1.9397	−7.0555	−5.4299	−3.7223	−5.1158	1.4587	0.7276
0.95	7.2172	1.7996	−1.9305	−7.0863	−5.4176	−3.7301	−5.1558	1.4524	0.7235
0.96	7.2220	1.8167	−1.9214	−7.1173	−5.4053	−3.7381	−5.1958	1.4460	0.7194
0.97	7.2268	1.8339	−1.9125	−7.1482	−5.3928	−3.7464	−5.2357	1.4394	0.7155
0.98	7.2316	1.8514	−1.9037	−7.1793	−5.3803	−3.7551	−5.2755	1.4328	0.7118
0.99	7.2366	1.8689	−1.8951	−7.2104	−5.3676	−3.7641	−5.3152	1.4260	0.7082
1.00	7.2416	1.8867	−1.8867	−7.2416	−5.3549	−3.7734	−5.3549	1.4191	0.7047

$I = {}^{9}/_{2}$

η	$E_{9/2}$	$E_{7/2}$	$E_{5/2}$	$E_{3/2}$	$E_{1/2}$	$\Delta E_{9/2-7/2} = E_{7/2}-E_{9/2}$	$\Delta E_{7/2-5/2} = E_{5/2}-E_{7/2}$	$\Delta E_{5/2-3/2} = E_{3/2}-E_{5/2}$	$\Delta E_{3/2-1/2} = E_{1/2}-E_{3/2}$	$\left\lvert \frac{\Delta E_{9/2-7/2}}{\Delta E_{7/2-5/2}} \right\rvert$	$\left\lvert \frac{\Delta E_{7/2-5/2}}{\Delta E_{5/2-3/2}} \right\rvert$	$\left\lvert \frac{\Delta E_{5/2-3/2}}{E_{3/2-1/2}} \right\rvert$
0.00	6.0000	2.0000	—1.0000	—3.0000	—4.0000	—4.0000	—3.0000	—2.0000	—1.0000	1.3333	1.5000	2.0000
0.01	6.0000	2.0000	—0.9999	—2.9996	—4.0005	—4.0000	—2.9999	—1.9997	—1.0009	1.3333	1.5002	1.9979
0.02	6.0000	2.0002	—0.9996	—2.9985	—4.0021	—3.9999	—2.9998	—1.9989	—1.0036	1.3334	1.5007	1.9918
0.03	6.0001	2.0004	—0.9991	—2.9967	—4.0048	—3.9997	—2.9995	—1.9976	—1.0081	1.3335	1.5015	1.9816
0.04	6.0002	2.0007	—0.9984	—2.9941	—4.0085	—3.9995	—2.9991	—1.9958	—1.0143	1.3336	1.5027	1.9676
0.05	6.0004	2.0012	—0.9974	—2.9909	—4.0132	—3.9992	—2.9986	—1.9935	—1.0223	1.3337	1.5042	1.9500
0.06	6.0005	2.0017	—0.9963	—2.9870	—4.0189	—3.9988	—2.9980	—1.9907	—1.0319	1.3338	1.5060	1.9290
0.07	6.0007	2.0023	—0.9950	—2.9824	—4.0256	—3.9984	—2.9973	—1.9874	—1.0432	1.3340	1.5081	1.9051
0.08	6.0009	2.0030	—0.9934	—2.9772	—4.0333	—3.9979	—2.9964	—1.9838	—1.0560	1.3342	1.5105	1.8785
0.09	6.0012	2.0038	—0.9917	—2.9714	—4.0418	—3.9974	—2.9955	—1.9798	—1.0704	1.3345	1.5130	1.8496
0.10	6.0014	2.0047	—0.9897	—2.9651	—4.0513	—3.9968	—2.9944	—1.9754	—1.0861	1.3347	1.5158	1.8187
0.11	6.0017	2.0056	—0.9876	—2.9583	—4.0615	—3.9961	—2.9932	—1.9707	—1.1033	1.3351	1.5188	1.7863
0.12	6.0021	2.0067	—0.9852	—2.9510	—4.0726	—3.9953	—2.9919	—1.9658	—1.1317	1.3354	1.5220	1.7526
0.13	6.0024	2.0079	—0.9826	—2.9432	—4.0845	—3.9945	—2.9905	—1.9606	—1.1413	1.3357	1.5253	1.7179
0.14	6.0028	2.0091	—0.9798	—2.9351	—4.0971	—3.9936	—2.9889	—1.9553	—1.1621	1.3361	1.5287	1.6826
0.15	6.0032	2.0105	—0.9768	—2.9265	—4.1104	—3.9927	—2.9873	—1.9498	—1.1839	1.3366	1.5321	1.6469
0.16	6.0037	2.0120	—0.9735	—2.9177	—4.1244	—3.9917	—2.9855	—1.9441	—1.2067	1.3370	1.5357	1.6111
0.17	6.0041	2.0135	—0.9701	—2.9085	—4.1390	—3.9906	—2.9836	—1.9384	—1.2305	1.3375	1.5392	1.5754
0.18	6.0046	2.0151	—0.9664	—2.8991	—4.1542	—3.9895	—2.9816	—1.9327	—1.2551	1.3380	1.5427	1.5399
0.19	6.0052	2.0169	—0.9626	—2.8895	—4.1700	—3.9883	—2.9794	—1.9269	—1.2805	1.3386	1.5462	1.5048
0.20	6.0057	2.0187	—0.9585	—2.8797	—4.1863	—3.9870	—2.9772	—1.9212	—1.3066	1.3392	1.5497	1.4703
0.21	6.0063	2.0206	—0.9542	—2.8697	—4.2031	—3.9857	—2.9748	—1.9155	—1.3335	1.3398	1.5530	1.4365
0.22	6.0069	2.0226	—0.9496	—2.8595	—4.2204	—3.9843	—2.9723	—1.9099	—1.3609	1.3405	1.5562	1.4034
0.23	6.0076	2.0247	—0.9449	—2.8493	—4.2382	—3.9828	—2.9696	—1.9044	—1.3889	1.3412	1.5593	1.3712
0.24	6.0082	2.0270	—0.9399	—2.8390	—4.2564	—3.9813	—2.9668	—1.8991	—1.4174	1.3419	1.5622	1.3398
0.25	6.0089	2.0293	—0.9347	—2.8286	—4.2750	—3.9797	—2.9639	—1.8940	—1.4464	1.3427	1.5650	1.3094
0.26	6.0097	2.0316	—0.9292	—2.8181	—4.2940	—3.9780	—2.9609	—1.8889	—1.4758	1.3435	1.5675	1.2799
0.27	6.0104	2.0341	—0.9236	—2.8077	—4.3133	—3.9763	—2.9577	—1.8841	—1.5056	1.3444	1.5698	1.2514
0.28	6.0112	2.0367	—0.9177	—2.7973	—4.3330	—3.9745	—2.9544	—1.8796	—1.5357	1.3453	1.5718	1.2299
0.29	6.0120	2.0394	—0.9116	—2.7868	—4.3530	—3.9726	—2.9510	—1.8753	—1.5662	1.3462	1.5736	1.1973
0.30	6.0129	2.0422	—0.9052	—2.7765	—4.3734	—3.9707	—2.9474	—1.8712	—1.5969	1.3472	1.5751	1.1718
0.31	6.0138	2.0450	—0.8986	—2.7661	—4.3940	—3.9687	—2.9437	—1.8675	—1.6279	1.3482	1.5763	1.1472

0.32	6.0147	2.0480	-0.8918	-2.7559	-4.4149	-3.9667	-2.9399	-1.8641	-1.6590	1.3493	1.5771	1.1236
0.33	6.0156	2.0511	-0.8848	-2.7457	-4.4361	-3.9645	-2.9359	-1.8609	-1.6904	1.3504	1.5776	1.1009
0.34	6.0166	2.0542	-0.8775	-2.7357	-4.4576	-3.9623	-2.9318	-1.8581	-1.7219	1.3515	1.5778	1.0791
0.35	6.0176	2.0575	-0.8700	-2.7257	-4.4793	-3.9601	-2.9275	-1.8557	-1.7536	1.3527	1.5776	1.0582
0.36	6.0186	2.0609	-0.8623	-2.7159	-4.5012	-3.9577	-2.9232	-1.8537	-1.7853	1.3539	1.5770	1.0382
0.37	6.0196	2.0643	-0.8543	-2.7062	-4.5234	-3.9553	-2.9186	-1.8519	-1.8172	1.3552	1.5761	1.0191
0.38	6.0207	2.0678	-0.8462	-2.6967	-4.5457	-3.9529	-2.9140	-1.8505	-1.8491	1.3565	1.5747	1.0008
0.39	6.0218	2.0715	-0.8377	-2.6873	-4.5683	-3.9503	-2.9092	-1.8495	-1.8810	1.3579	1.5729	0.9833
0.40	6.0230	2.0752	-0.8291	-2.6780	-4.5911	-3.9477	-2.9043	-1.8490	-1.9130	1.3593	1.5708	0.9665
0.41	6.0241	2.0791	-0.8202	-2.6690	-4.6140	-3.9451	-2.8993	-1.8488	-1.9450	1.3607	1.5682	0.9506
0.42	6.0253	2.0830	-0.8111	-2.6601	-4.6372	-3.9423	-2.8941	-1.8490	-1.9770	1.3622	1.5652	0.9352
0.43	6.0266	2.0871	-0.8018	-2.6514	-4.6605	-3.9395	-2.8888	-1.8497	-2.0090	1.3637	1.5618	0.9207
0.44	6.0278	2.0912	-0.7922	-2.6429	-4.6839	-3.9366	-2.8834	-1.8507	-2.0410	1.3653	1.5580	0.9068
0.45	6.0291	2.0954	-0.7824	-2.6346	-4.7075	-3.9337	-2.8778	-1.8522	-2.0730	1.3669	1.5538	0.8935
0.46	6.0304	2.0998	-0.7724	-2.6265	-4.7313	-3.9307	-2.8721	-1.8541	-2.1048	1.3685	1.5491	0.8809
0.47	6.0318	2.1042	-0.7622	-2.6186	-4.7552	-3.9276	-2.8664	-1.8564	-2.1367	1.3702	1.5440	0.8688
0.48	6.0331	2.1087	-0.7517	-2.6109	-4.7793	-3.9244	-2.8604	-1.8591	-2.1684	1.3720	1.5386	0.8574
0.49	6.0345	2.1134	-0.7411	-2.6034	-4.8035	-3.9212	-2.8544	-1.8623	-2.2001	1.3737	1.5327	0.8465
0.50	6.0360	2.1181	-0.7302	-2.5961	-4.8278	-3.9179	-2.8483	-1.8659	-2.2317	1.3755	1.5265	0.8361
0.51	6.0374	2.1229	-0.7191	-2.5890	-4.8523	-3.9145	-2.8420	-1.8699	-2.2632	1.3774	1.5198	0.8262
0.52	6.0389	2.1279	-0.7078	-2.5822	-4.8768	-3.9111	-2.8356	-1.8744	-2.2947	1.3793	1.5128	0.8168
0.53	6.0405	2.1329	-0.6963	-2.5756	-4.9015	-3.9076	-2.8292	-1.8793	-2.3260	1.3812	1.5055	0.8079
0.54	6.0420	2.1380	-0.6846	-2.5691	-4.9263	-3.9040	-2.8226	-1.8846	-2.3572	1.3831	1.4977	0.7995
0.55	6.0436	2.1433	-0.6727	-2.5630	-4.9513	-3.9003	-2.8150	-1.8904	-2.3883	1.3851	1.4897	0.7915
0.56	6.0452	2.1486	-0.6605	-2.5570	-4.9763	-3.8966	-2.8092	-1.8965	-2.4193	1.3871	1.4813	0.7839
0.57	6.0469	2.1541	-0.6483	-2.5513	-5.0014	-3.8928	-2.8023	-1.9030	-2.4501	1.3891	1.4726	0.7767
0.58	6.0485	2.1596	-0.6358	-2.5458	-5.0266	-3.8889	-2.7954	-1.9100	-2.4809	1.3912	1.4636	0.7699
0.59	6.0503	2.1653	-0.6231	-2.5405	-5.0519	-3.8850	-2.7884	-1.9174	-2.5115	1.3933	1.4543	0.7634
0.60	6.0520	2.1710	-0.6103	-2.5354	-5.0773	-3.8810	-2.7813	-1.9251	-2.5419	1.3954	1.4447	0.7574
0.61	6.0538	2.1769	-0.5972	-2.5306	-5.1028	-3.8769	-2.7741	-1.9333	-2.5723	1.3975	1.4349	0.7516
0.62	6.0556	2.1828	-0.5840	-2.5259	-5.1284	-3.8727	-2.7669	-1.9419	-2.6025	1.3997	1.4248	0.7462
0.63	6.0574	2.1889	-0.5707	-2.5215	-5.1541	-3.8685	-2.7596	-1.9509	-2.6326	1.4018	1.4146	0.7411
0.64	6.0592	2.1951	-0.5571	-2.5173	-5.1799	-3.8641	-2.7522	-1.9602	-2.6625	1.4040	1.4040	0.7362
0.65	6.0611	2.2014	-0.5434	-2.5134	-5.2057	-3.8598	-2.7448	-1.9699	-2.6923	1.4062	1.3934	0.7317
0.66	6.0631	2.2078	-0.5296	-2.5096	-5.2316	-3.8553	-2.7374	-1.9800	-2.7220	1.4084	1.3825	0.7274
0.67	6.0650	2.2143	-0.5156	-2.5061	-5.2576	-3.8507	-2.7299	-1.9905	-2.7515	1.4104	1.3715	0.7234
0.68	6.0670	2.2209	-0.5015	-2.5027	-5.2837	-3.8461	-2.7223	-2.0013	-2.7809	1.4128	1.3603	0.7196
0.69	6.0690	2.2276	-0.4872	-2.4996	-5.3098	-3.8414	-2.7147	-2.0125	-2.8102	1.4150	1.3490	0.7161
0.70	6.0710	2.2344	-0.4727	-2.4967	-5.3360	-3.8367	-2.7071	-2.0240	-2.8393	1.4172	1.3376	0.7128

$I = {}^{9}/_{2}$ (contd.)

η	$E_{9/2}$	$E_{7/2}$	$E_{5/2}$	$E_{3/2}$	$E_{1/2}$	$\Delta E_{9/2-7/2} = E_{7/2} - E_{9/2}$	$\Delta E_{7/2-5/2} = E_{5/2} - E_{7/2}$	$\Delta E_{5/2-3/2} = E_{3/2} - E_{5/2}$	$\Delta E_{3/2-1/2} = E_{1/2} - E_{3/2}$	$\left\lvert \frac{\Delta E_{9/2-7/2}}{\Delta E_{7/2-5/2}} \right\rvert$	$\left\lvert \frac{\Delta E_{7/2-5/2}}{\Delta E_{5/2-3/2}} \right\rvert$	$\left\lvert \frac{\Delta E_{5/2-3/2}}{\Delta E_{3/2-1/2}} \right\rvert$
0.71	6.0731	2.2413	−0.4582	−2.4940	−5.3623	−3.8318	−2.6995	−2.0358	−2.8683	1.4194	1.3260	0.7098
0.72	6.0752	2.2484	−0.4435	−2.4915	−5.3886	−3.8269	−2.6919	−2.0480	−2.8971	1.4216	1.3144	0.7069
0.73	6.0774	2.2555	−0.4287	−2.4892	−5.4150	−3.8219	−2.6842	−2.0605	−2.9258	1.4238	1.3027	0.7042
0.74	6.0795	2.2628	−0.4138	−2.4870	−5.4415	−3.8168	−2.6766	−2.0733	−2.9544	1.4260	1.2910	0.7017
0.75	6.0817	2.2701	−0.3988	−2.4851	−5.4680	−3.8116	−2.6689	−2.0864	−2.9829	1.4282	1.2792	0.6994
0.76	6.0840	2.2776	−0.3836	−2.4834	−5.4946	−3.8064	−2.6612	−2.0998	−3.0112	1.4303	1.2674	0.6973
0.77	6.0862	2.2852	−0.3684	−2.4819	−5.5212	−3.8010	−2.6536	−2.1135	−3.0394	1.4324	1.2556	0.6954
0.78	6.0885	2.2930	−0.3530	−2.4805	−5.5479	−3.7956	−2.6460	−2.1275	−3.0674	1.4345	1.2437	0.6936
0.79	6.0909	2.3007	−0.3376	−2.4793	−5.5747	−3.7901	−2.6383	−2.1417	−3.0954	1.4366	1.2319	0.6919
0.80	6.0932	2.3087	−0.3221	−2.4783	−5.6015	−3.7846	−2.6308	−2.1562	−3.1232	1.4386	1.2201	0.6904
0.81	6.0956	2.3167	−0.3065	−2.4775	−5.6284	−3.7789	−2.6232	−2.1710	−3.1508	1.4406	1.2083	0.6890
0.82	6.0981	2.3249	−0.2908	−2.4769	−5.6553	−3.7732	−2.6157	−2.1861	−3.1784	1.4425	1.1966	0.6878
0.83	6.1005	2.3331	−0.2750	−2.4764	−5.6822	−3.7674	−2.6082	−2.2014	−3.2058	1.4444	1.1849	0.6877
0.84	6.1030	2.3415	−0.2592	−2.4761	−5.7092	−3.7615	−2.6007	−2.2169	−3.2332	1.4463	1.1732	0.6857
0.85	6.1055	2.3500	−0.2433	−2.4760	−5.7363	−3.7555	−2.5934	−2.2326	−3.2604	1.4481	1.1616	0.6848
0.86	6.1081	2.3587	−0.2274	−2.4760	−5.7634	−3.7494	−2.5860	−2.2486	−3.2874	1.4499	1.1500	0.6840
0.87	6.1107	2.3674	−0.2113	−2.4762	−5.7906	−3.7433	−2.5788	−2.2648	−3.3144	1.4516	1.1386	0.6833
0.88	6.1133	2.3763	−0.1953	−2.4765	−5.8178	−3.7370	−2.5715	−2.2812	−3.3413	1.4532	1.1273	0.6827
0.89	6.1160	2.3852	−0.1792	−2.4770	−5.8450	−3.7307	−2.5644	−2.2978	−3.3680	1.4548	1.1160	0.6823
0.90	6.1186	2.3943	−0.1630	−2.4777	−5.8723	−3.7243	−2.5574	−2.3146	−3.3947	1.4563	1.1049	0.6818
0.91	6.1214	2.4036	−0.1468	−2.4785	−5.8996	−3.7178	−2.5504	−2.3316	−3.4212	1.4578	1.0938	0.6815
0.92	6.1241	2.4129	−0.1306	−2.4794	−5.9270	−3.7112	−2.5435	−2.3488	−3.4476	1.4591	1.0829	0.6813
0.93	6.1269	2.4223	−0.1143	−2.4805	−5.9544	−3.7046	−2.5367	−2.3662	−3.4739	1.4604	1.0721	0.6811
0.94	6.1297	2.4319	−0.0981	−2.4817	−5.9819	−3.6968	−2.5300	−2.3837	−3.5002	1.4616	1.0614	0.6810
0.95	6.1326	2.4416	−0.0817	−2.4831	−6.0094	−3.6910	−2.5234	−2.4013	−3.5263	1.4627	1.0508	0.6810
0.96	6.1355	2.4514	−0.0654	−2.4846	−6.0369	−3.6840	−2.5168	−2.4192	−3.5523	1.4638	1.0404	0.6810
0.97	6.1384	2.4614	−0.0491	−2.4862	−6.0645	−3.6770	−2.5104	−2.4372	−3.5782	1.4647	1.0301	0.6811
0.98	6.1414	2.4714	−0.0327	−2.4880	−6.0920	−3.6699	−2.5041	−2.4553	−3.6040	1.4655	1.0199	0.6813
0.99	6.1444	2.4816	−0.0164	−2.4899	−6.1197	−3.6627	−2.4980	−2.4735	−3.6298	1.4663	1.0099	0.6814
1.00	6.1474	2.4919	−0.0149	−2.4919	−6.1474	−3.6555	−2.4919	−2.4919	−3.6555	1.4669	1.0000	0.6817

Appendix III

MEAN PROBABILITIES OF TRANSITIONS AS A FUNCTION OF THE ASYMMETRY PARAMETER FOR SPINS $I = {}^5/_2, {}^7/_2, {}^9/_2$

The mean probabilities of transition ($\Delta m = 1$ and $\Delta m = 2$) for spins $I = {}^5/_2$, ${}^7/_2$ and ${}^9/_2$ for different asymmetry parameters are listed in the table.

The relative intensity of the observed NQR lines in powder substances containing nuclei with spin ${}^5/_2$, ${}^7/_2$, and ${}^9/_2$ is calculated as follows:

$$\overline{W}_{ij} \cdot \nu_{ij}^2 : \overline{W}_{i'j'} \cdot \nu_{i'j'}^2 = A_{ij} : A_{i'j'} \ldots$$

Here, $\overline{W}_{ij}$ is the mean probability of the transition i — j, ν_{ij} is the experimentally observed NQR frequency of the corresponding transition, and A_{ij} is the observed signal to noise ratio.

In the calculation, it is necessary to allow for the different content of atoms in the various chemical and (or) crystal positions (see Figure 1.3).

The data are taken from M. H. Cohen's paper, Phys. Rev., 96, 1278 (1954).

$I = {}^5/_2$

η	$\overline{W}_{1/2-3/2}$	$\overline{W}_{3/2-5/2}$	$\overline{W}_{1/2-5/2}$
0.1	2.64	1.66	. . .
0.2	2.55	1.66	0.02
0.3	2.44	1.65	0.03
0.4	2.31	1.65	0.05
0.5	2.18	1.65	0.07
0.6	2.06	1.66	0.09
0.7	1.99	1.67	0.1
0.8	1.88	1.70	0.11
0.9	1.82	1.73	0.11
1.0	1.77	1.77	0.12

$I = {}^{7}/_{2}$

η	$\overline{W}_{1/2-3/2}$	$\overline{W}_{3/2-5/2}$	$\overline{W}_{5/2-7/2}$	$\overline{W}_{1/2-5/2}$	$\overline{W}_{3/2-7/2}$
0.1	4.77	3.98	2.33	0.03	. . .
0.2	4.26	3.93	2.34	0.10	. . .
0.3	3.71	3.90	2.34	0.17	0.01
0.4	3.27	3.89	2.35	0.22	0.02
0.5	2.96	3.92	2.35	0.24	0.03
0.6	2.74	3.96	2.36	0.24	0.04
0.7	2.60	4.01	2.37	0.23	0.07
0.8	2.52	4.05	2.38	0.21	0.09
0.9	2.46	4.08	2.40	0.18	0.12
1.0	2.42	4.08	2.42	0.15	0.15

$I = {}^{9}/_{2}$

η	$\overline{W}_{1/2-3/2}$	$\overline{W}_{3/2-5/2}$	$\overline{W}_{5/2-7/2}$	$\overline{W}_{7/2-9/2}$	$\overline{W}_{1/2-5/2}$	$\overline{W}_{3/2-7/2}$
0.1	7.11	6.90	5.33	3.00	0.12	. . .
0.2	5.61	6.77	5.34	3.00	0.30	0.02
0.3	4.54	6.71	5.33	3.02	0.40	0.05
0.4	3.93	6.68	5.33	3.03	0.40	0.10
0.5	3.60	6.61	5.32	3.04	0.35	0.18
0.6	3.42	6.46	5.31	3.06	0.28	0.27
0.7	3.32	6.25	5.34	3.08	0.22	0.36
0.8	3.25	6.02	5.39	3.11	0.16	0.42
0.9	3.20	5.81	5.49	3.13	0.12	0.46
1.0	3.17	5.63	5.63	3.17	0.08	0.47

Appendix IV

CATALOG OF NUCLEAR QUADRUPOLE INTERACTION CONSTANTS AND NUCLEAR QUADRUPOLE RESONANCE FREQUENCIES

The tables contain data on nuclear quadrupole interaction constants and NQR frequencies published in the literature or measured in the NQR group of INEOS AN SSSR.

The data in the tables are arranged according to the types of resonance atoms in the order that they appear in the periodic table. Each group of elements is in a different table. However, the elements of group VII are an exception. They occupy four tables since the amount of NQR data on halogens is so great. The first table contains data on the NQR frequencies of the chlorine atom bonded to various elements except carbon. The spectra of compounds containing the M—Hal bond are arranged according to the groups in the periodic table. The data on the NQR frequencies of the chlorine atoms in the C—Cl bond are given in a separate table. Thus, for C_6Cl_5HgCl the NQR frequencies of ^{35}Cl in the Hg—Cl bond are given in the first of the four tables mentioned, and the data for the pentachlorophenyl group are given in the second table. The third and fourth tables contain data on the NQR frequencies of ^{79}Br and ^{127}I. The NQR frequencies of the metals of groups VII and VIII are also given in the last table.

Within each table the compounds are arranged in increasing order of the atomic number of the resonance atom. The elements of the main group are given first, and then those of the subgroup. Thus, in the first table the elements are arranged in the following order: D, Li, Na, K, Rb, Cs, Cu, Au. The empirical formulas for each element are arranged alphabetically. For all the organic and organometallic compounds (including the salts of organic acids) we start with C, H, and then continue alphabetically. Deuterated compounds are listed after their hydrogen-containing analogs (in those cases where the deuterium is not a resonance atom), for example HC $\equiv$ CCl, DC $\equiv$ CCl. For deuterated compounds which do not have hydrogen-containing analogs, the adopted sequence of the atoms: C, H, D, etc., in alphabetical order is maintained. Combinations of molecules, solvated crystals, etc., are placed after the basic compound on which the measurements are carried out: CCl_4, $CCl_4 + C_6H_6$, $CCl_4 + C_8H_{10}$, etc., or D_2O, $D_2O + \frac{1}{5}CuSO_4$ ($CuSO_4 \cdot 5D_2O$), etc.

The first column in all the tables gives the empirical formula of the substance, and the second column gives its structural formula. The third column lists the temperatures at which the measurements were carried out. For halogens (except iodine) only the results obtained at the temperature of liquid nitrogen (77°K) are given. This is a standard procedure in NQR. Data at other temperatures are given when the measurements at 77°K are missing, and this is indicated in the notes. The next columns list the necessary parameters of the NQR spectra: the frequency ν, e^2Qq, and the asymmetry parameter η. For halogens, the signal to noise ratio (s/n) is also given. Only data for ^{79}Br are given, and in the absence of such data the frequencies of ^{81}Br are converted to ^{79}Br ($\nu_{^{79}Br} = 1.19707\ \nu_{^{81}Br}$). This is noted in the last column by the word "conversion." In the last column the reader will find references and notes indicating the method used for studying the quadrupole interactions. The following symbols are used here: "NMR cr." — e^2Qq and η were obtained from the NMR spectra of the substances in the solid state; "NMR l." — e^2Qq was estimated from the spin-lattice relaxation times from NMR spectra of liquid samples (or solutions); "g" — e^2Qq and η were obtained from microwave spectra in the gaseous state; "beam" — the data were obtained from atomic or molecular beam measurements; "nematic l." — the data were obtained from splittings in the NMR spectra of substances dissolved in nematic liquids. No measurements by γ resonance spectroscopy were carried out. Direct observations of the NQR spectra are not mentioned specifically. For the NQR of ^{35}Cl and ^{79}Br, the values of η obtained by Zeeman analysis of the NQR spectra are given in the order of the frequencies. The existence of a phase transition studied by NQR and $\Delta\nu$ (the line width at half-height) for very broad NQR lines are also recorded in the last column.

If the spectrum of the compound has been studied in many papers, the most reliable data are given, but the other papers are also mentioned as references. The spectra obtained in the NQR group of INEOS AN SSSR are denoted by an asterisk. The compounds whose spectra were checked in this group are similarly denoted by an asterisk.

The tables give spectral data which have appeared in print up to the end of 1969. The supplement also includes 1970 papers which were available to the authors. It was impossible to introduce corrections in the course of printing and, as a result, no references to papers which appeared while the book was in print are given.

The references are given for each table separately.

The following abbreviations are used in the tables:

at.	— atomic
met.	— metallic
mol.	— molecular
p.t.	— phase transition
acac	— acetylacetonate
bipy	— dipyridyl
DG	— dimethyl ether of diethylene glycol
diox	— dioxane
DME	— dimethoxyethane
DMF	— dimethylformamide
DMSO	— dimethylsulfoxide
en	— ethyleneimine
Py	— pyridine (C_5H_5N)
THF	— tetrahydrofuran.

Group I elements

Empirical formula	Structural formula	T, °K	e^2Qq, MHz	η, %	References, notes
		^2D			
BD_4Na	$NaBD_4$	76.8	0.094	...	[1]. NMR cr.
BrD_4N	ND_4Br	Room	0.180	...	[2]. NMR cr.
CDFO	DCOF	—	0.230	...	[2a]. g
$CDKO_3$	$KDCO_3$	Room	0.154	19.4	[3]. NMR cr.
CDN	DCN	—	0.150	...	[4, 5]. g
CD_2O	DCDO	—	0.171	2	[6]. g
CD_3I	CD_3I	Room	0 098	—	[7, 124]. NMR l.
CD_4N_2O	$(ND_2)_2CO$	»	0.2108 0.2107	13.9 14.7	[8]. NMR cr.
CD_4O	CD_3OD	»	0.112	—	[7]. NMR l.
CHDO	HCDO	—	0.170	<15	[9]. g
CH_3D	CH_3D	—	0.100 0.100	... —	[125]. g [10]. NMR l.
C_2DCl	DC≡CCl	—	0.225	...	[11, 12]. g
C_2DF	DC≡CF	—	0.212	...	[12]. g
C_2D_2	DC≡CD	—	0.200	...	[13, 14]. g
$C_2D_2O_4 + 2D_2O$	$(COOD)_2 \cdot 2D_2O$	Room	0.1392	9.8	[3, 15, 16]. NMR cr. cf. D_2O
C_2D_6O	C_2D_5OD	»	0.158	—	[7]. NMR l.
C_2D_6OS	$(CD_3)_2SO$	»	0.172	—	[7]. NMR l. .

Empirical formula	Structural formula	T, °K	e^2Qq, MHz	η, %	References, notes
$C_2H_2D_3NO_2$	ND_2CH_2COOD	150	0.168 0.188	0 0	} COOD
		Room	0.05110 0.05146	26.3 28.2	} ND_2 {$[ND_3]^+$} [18, 19]. NMR cr.
$C_2H_3DO_2$	CH_3COOD	»	0.148	—	[7]. NMR l.
C_3H_3D	$CH_3C\equiv CD$	—	0.208	...	[12]. g
C_3D_6O	CD_3COCD_3	295	0.135	—	[7, 20]. NMR l.
$C_3H_4D_3NO_2$	L-$CH_3CHCOOD$ (with ND_2 on the CH)	Room	0.04927 0.04923 0.04906 0.04933	17.8 17.4 17.8 17.6	[21]. NMR cr.
$C_4H_2DKO_4$	cis-DOOCCH=CHCOOK	»	0.0560	53	[3]. NMR cr.
C_5D_5N	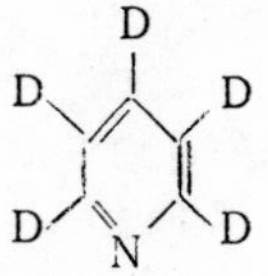	»	0.154	—	[7]. NMR l.
C_6D_6	C_6D_6	108 263	0.193 0.0933	... ...	[7, 20, 22—25, 124]. NMR cr.
C_6D_{12}	cyclo-C_6D_{12}	157—168	0.174	...	[7, 23]. NMR cr.
$C_6H_6D_{11}N_3O_{10}S$	$[ND_3CH_2COOD]_2\cdot[\overset{+}{N}D_3CH_2CO\bar{O}]SO_4$	293	0.1310 0.0788	16.0 19.2	} O—D
			0.0470 0.0548 0.0528	24.7 13.3 14.7	} ND_3

		346	0.1305 0.850	18.0 22.3	} O—D
			0.0373 0.0519	2.1 14.8	} ND_3 [17, 26]. NMR cr.
$C_7H_5D_3$	$C_6H_5CD_3$	111—120	0.165	...	[23]. NMR cr.
$C_{10}H_{11}DO_2$	4-$C_3H_7C_6H_4COOD$	418—427	0.184	...	[23]. Dimer in nematic phase
$C_{14}D_{10}$	Deuteroanthracene	Room	0.181	6.4	[27]. NMR cr.
ClD_4N	ND_4Cl	77 113	0.1757 0.1801	20 0	[2, 28, 136]. NMR cr.
DF	DF	—	0.340	...	[29]. Beam
DHO	HDO	—	0.3186	6	[9, 30—32]. g
DHS	HDS	— —	0.149 0.1547	... 12	[33]. Beam [9]. g
DLiO	LiOD	77	0.327	...	[34]. NMR cr. cf. Li
DLiO + D_2O	$LiOD \cdot D_2O$	77 273	0.2715 0.2677	4.3 4.4	[35]. NMR cr.
D_2KO_4P	KD_2PO_4	173 303	0.1195 0.1275	11.6 4.9	[17, 42, 43]. NMR cr.
D_2O	D_2O	— Room 263	0.312 0.208 0.2132 0.2164	<12 — 10 ...	[20, 32]. g [24]. NMR l. [36, 126]. NMR cr.
$D_2O + BaCl_2O_6$	$Ba(ClO_3)_2 \cdot D_2O$	77 143 293	0.24717 0.2435 0.1215	10.3 7.4 97.6	[15, 37, 38]. NMR cr.
$D_2O + C_2D_2O_4$	$(COOD)_2 \cdot D_2O$	203 340	0.2303 0.219 0.121	6.2 16 93	[3, 15, 16]. NMR cr. cf. $(COOD)_2$

Empirical formula	Structural formula	T, °K	e^2Qq, MHz	η, %	References, notes
$D_2O + C_2K_2O_4$	$(COOK)_2 \cdot D_2O$	223	0.2097	9.0	[39]. NMR cr.
		298	0.1142	85.1	
$D_2O + {}^1/_5CuO_4S$	$CuSO_4 \cdot 5D_2O$	133.2	0.2256	11.7	[40, 127]. NMR cr.
			0.2116	13.4	
			0.2324	8.3	
			0.2031	10.6	
			0.2083	11.0	
			0.2083	11.3	
			0.2002	11.7	
			0.2122	14.2	
			0.2198	11.4	
			0.2415	9.6	
		294.8	0.1183	85.6	
			0.1147	87.5	
			0.1118	95.7	
			0.1108	97.5	
			0.1242	81.8	
$D_2O + Li_2O_4S$	$Li_2SO_4 \cdot D_2O$	148	0.237	14	[41, 42]. NMR cr.
		Room	0.123	80	
$D_3LiO_6Se_2$	$LiD_3(SeO_3)_2$	130	0.1078	17.2	[44, 128]. NMR cr.
			0.1279	15.4	
			0.1811	10.1	
D_3N	ND_3	Room	0.200	...	[42, 45, 46]. NMR cr.
$D_3NaO_6Se_2$	$NaD_3(SeO_3)_2$	134	0.1795	13	[47, 129, 130]. NMR cr. cf. Na
			0.1414	17.2	
			0.1671	16.5	
			0.1212	21.6	
			0.1426	14.8	
			0.1113	24.8	
		173	0.167	18.4	
			0.138	13.4	
			0.148	30.2	

			0.111	20.9	
			0.147	37.7	
			0.138	18.8	
		293	0.127	9.8	
			0.147	17.2	
		298	0.1455	15.6	
			0.1297	13.5	
D_4INO_3	ND_4IO_3	Room	0.039	8—10	[48]. NMR cr.
$D_4N_2O_3$	ND_4NO_3	76,8	0.159	...	[17, 49]. NMR cr.
		135	0.0036	90	
D_4Si	SiD_4	Room	0.095	—	[50]. NMR l.
$D_5LiN_2O_4S$	$Li[ND_2ND_3]SO_4$	»	0.047	...	[51]. NMR cr.
D_6NO_4P	$[ND_4]D_2PO_4$	233	0.006	89	[3, 52]. NMR cr.
		Room	0.0034	0	ND_4
			0.1196	5.3	D_2PO_4
$D_8N_2O_4S$	$[ND_4]_2SO_4$	77	0.174	...	[28, 48, 53, 54]. NMR cr.
		133	0.0079	53.1	
			...	...	
		173	0.0071	63.3	
			0.0048	54.1	
		213	0.0064	87.5	
			0.0043	95.3	
		233	0.0055	96.4	
			0.0029	79.3	
		263	0.0044	100	
			...	...	
		298	0.0040	95	
			0.0021	90.5	
		^{7}Li			
Al + Li	LiAl	Room	0.180	...	[55]. NMR cr. cf. Al
$AlLiO_4P$	$LiAlPO_4(OH, F)$	»	0.04424	62	[56]. NMR cr.
			0.03244	35.9	

Empirical formula	Structural formula	T, °K	e^2Qq, MHz	η, %	References, notes
$AlLiO_6Si_2$	$LiAl(SiO_3)_2$	293	0.0757	79	[57—60]. NMR cr. cf. Al
Al_5LiO_8	$LiAl_5O_8$	Room	0,041	...	[61]. NMR cr. cf. Al
BH_4Li	$LiBH_4$	293	0,0	0	[62]. NMR cr.
BrLi	LiBr	—	0.184	...	[63—65]. g. cf. Br
CH_3Li	CH_3Li	298	0.016	0	[66, 67]. NMR cr.
CD_3Li	CD_3Li	Room	0.048	0	The same
CLi_2O_3	Li_2CO_3	293	<0.092	...	[68]. NMR cr.
C_2H_5Li	C_2H_5Li	298	0.098	62	[66, 67]. NMR cr.
C_2D_5Li	C_2D_5Li	77	0.083	79	The same
$C_2Li_2O_4$	COOLi | COOLi	293	0.0	0	[62]. NMR cr.
$C_4H_4LiO_6Tl + H_2O$	CH(OH)COOLi | CH(OH)COOTl · H_2O	77	0,093	0	[69]. NMR cr.
$C_4H_8LiNO_6 + H_2O$	CH(OH)COOLi | CH(OH)COO[NH_4] · H_2O	77	0.093	0	The same
ClLi	LiCl	—	0,192	...	[65]. g
FLi	LiF	Room —	0,0073 0.408	... ...	[70] [63—65]. g
HLi	LiH	—	0,346	...	[65, 71]. g
HLiO	LiOH	293	0.110	...	[62]. NMR cr.
DLiO	LiOD	Room	0,0945	...	[34]. NMR cr. cf. D
$HLiO + H_2O$	$LiOH \cdot H_2O$	293	0,084	...	[62]. NMR cr.
H_2LiN	$LiNH_2$	Room	0,119	...	[62, 72]. NMR cr.

H_4LiN	$LiNH_4$ $(LiH \cdot NH_3)$	77	0.093	0	[73]. NMR cr.
$H_5LiN_2O_4S$	$Li(NH_2NH_3)SO_4$	77 297	0.054 ...	10 40	[69]. NMR cr.
ILi	LiI	—	0,172	...	[63—65]. g cf. I
$ILiO_3$	$LiIO_3$	293	0,0452	...	[62]. NMR cr. cf. I
$LiNO_3$	$LiNO_3$	293	0,0392	0	[62, 74]. NMR cr.
$LiNO_3 + 3H_2O$	$LiNO_3 \cdot 3H_2O$	293	0.0394	...	[62]. NMR cr.
$LiNbO_3$	$LiNbO_3$	77—423	0.0460	0	[69, 73, 75, 76]. NMR cr.
LiO_3Ta	$LiTaO_3$	Room 598	0.081 0.094	0 ...	[69, 75, 76, 131, 132]. NMR cr.
Li_2	Li_2	—	0.06	...	[65, 71]. Beam
Li_2HN	Li_2NH	Room	0	0	[72]. NMR cr.
Li_2O_3P	Li_2PO_3	293	0.0916	...	[62]. NMR cr.
Li_2O_3Si	Li_2SiO_3	293	0.0	0	The same
Li_2O_3Ti	Li_2TiO_3	293	0.0704	...	»
Li_2O_3Zr	Li_2ZrO_3	293	0.108 0,0656	... ...	»
$Li_2O_4S + H_2O$	$Li_2SO_4 \cdot H_2O$	—	<0.09	...	[68, 78—80]. g
$Li_2O_4S + D_2O$	$Li_2SO_4 \cdot D_2O$	Room	0,03805 0.05695	87 22	[77]. NMR cr. cf. D
Li_3N	Li_3N	293	0.296 0.200	... ...	[72]. NMR cr.
		293	0,288 0,584	... ...	[72a]. NMR cr.

Empirical formula	Structural formula	T, °K	e^2Qq, MHz	η, %	References, notes
		^{23}Na			
$AlNaO_8Si_3$	$NaAlSi_3O_8$	Room	2.62	25	[81]. NMR cr. cf. Al
$Al_2Na_2O_{10}Si_3 + 2H_2O$	$Na_2Al_2Si_3O_{10} \cdot 2H_2O$	»	1.7593	64.27	[82]. NMR cr. cf. Al
BF_4Na	$NaBF_4$	296	1.008	9.5	[83]. NMR cr. cf. B
$B_4Na_2O_7 + 4H_2O$	$Na_2B_4O_7 \cdot 4H_2O$	Room	ν=1.560 MHz		[84]. cf. B
$B_4Na_2O_7 + 5H_2O$	$Na_2B_4O_7 \cdot 5H_2O$	»	0.539 0.785 1.299	74.1 0 0	[140]. NMR cr.
$B_4Na_2O_7 + 10H_2O$	$Na_2B_4O_7 \cdot 10H_2O$	»	0.541 0.849	44.9 14.3	[140]. NMR cr.
BrNa	NaBr	—	4.68	...	[65, 85]. g cf. Br
$BrNaO_3$	$NaBrO_3$	4.1 Room	0.886 0.864	... 0	[86, 133]. NMR cr. cf. Br
$CHNaO_3$	$NaHCO_3$	»	2.46	99	[134]. NMR cr.
$CDNaO_3$	$NaDCO_3$	»	2.46	99	The same
$C_4H_4KNaO_6 + 4H_2O$	CH(OH)COOK | · $4H_2O$ CH(OH)COONa	276 301	1.285 1.313 1.407	87.2 80.9 67.5	[78, 87]. NMR cr.
ClNa	NaCl	—	5.40	—	[65, 85, 88, 89]. g. cf. Cl
$ClNaO_3$	$NaClO_3$	299	0.813 ν=0.396 MHz	0	[90, 135]. cf. Cl
FNa	NaF	—	8.4	...	[65, 88, 89]. g
HNaS	NaSH	Room	<0.02	...	[137]
$H_2NaO_4P + 2H_2O$	$NaH_2PO_4 \cdot 2H_2O$	»	1.179	46.7	[91]. NMR cr.
HgNa	NaHg	»	0.57 1.30	... ...	[92]. NMR cr.

Hg_2Na_3	Na_3Hg_2	»	0,02 1,27	... ...	The same
Hg_2Na_5	Na_5Hg_2	»	0,011	...	»
Hg_4Na	$NaHg_4$	»	0.002	...	»
INa	NaI	—	3.88	...	[65, 85, 93]. g cf. I
$NNaO_2$	$NaNO_2$	Room	1.1003	10.9	[138]. NMR cr. cf. N
$NNaO_3$	$NaNO_3$	» 598	0.334 0.3297	... ...	[68, 94—96]. NMR cr. cf. N
N_3Na	NaN_3	Room	0.0068	0	[139]. NMR cr.
Na_2	Na_2 (met.)	—	0.423	...	[65, 97]. g
$Na_2O_3S_2 + 5H_2O$	$Na_2S_2O_3 \cdot 5H_2O$	Room	2.260 0.830	33.4 40.9	[98]. $\nu = 1.137$ MHz
$Na_2O_6S_2 + 2H_2O$	$Na_2S_2O_6 \cdot 2H_2O$	293	2.3697	35.09	[99]. NMR cr.
$Na_2O_6S_2 + 2D_2O$	$Na_2S_2O_6 \cdot 2D_2O$	Room	2.362	35.1	[141]
$Na_3O_4V + 14H_2O$	$Na_3VO_4 \cdot 14H_2O$	»	0.320	...	[142]. cf. V
$Na_4O_7V_2 + 16H_2O$	$Na_4V_2O_7 \cdot 16H_2O$	»	0.750	...	[142]. cf. V
		^{39}K			
BrK	KBr	—	5.002	...	[100—102]. g cf. Br
ClK	KCl	—	5.656 6.899[1]	... ...	[65, 101, 102]. g cf. Cl
$ClKO_3$	$KClO_3$	299	1.058 1.292[1]	ν=0.526 MHz ν=0.639MHz*	[90, 103]. cf. Cl
FK	KF	Room	1.138 0.916	0 61	[104]. NMR cr.
		—	7.933	...	[105]. g
H_2KO_4P	KH_2PO_4	77 Room	1.479 1.68	69.6 ...	[107, 107a, 107b] NMR cr.
K_2	K_2 (met.)	—	0,158	...	[65, 97]. g

* Data for ^{41}K.

Empirical formula	Structural formula	T, °K	e^2Qq, MHz	η, %	References, notes
			^{85}Rb		
ClRb	RbCl	—	52.675	...	[108]. g cf. Cl
		—	25.485[1]	...	
ClO_3Rb	$RbClO_3$	299	13.3	0	[90]. cf. Cl
			$\nu_{1/2-3/2} = 2.00$ MHz, $\nu_{3/2-5/2} = 4.00$ MHz		
		—	6.44[1]	0	$\nu = 3.22$ MHz*
FRb	RbF	—	70.31	...	[109—111]. g
		—	33.96[1]	...	
Rb_2	Rb_2	—	0.580[1]	...	[65]. g
$MnO_8Rb_2S_2 + 6H_2O$	$Rb_2Mn(SO_4)_2 \cdot 6H_2O$	Room	3.141	47	[112]. NMR cr.
			^{133}Cs		
ClCs	CsCl	—	<4	...	[113, 114]. g cf. Cl
$ClCsO_3$	$CsClO_3$	299	0.2475	9.2	[90]. cf. Cl
			$\nu_{1/2-3/2} = 0.0182$MHz; $\nu_{3/2-5/2} = 0.0352$ MHz		
			$\nu_{5/2-7/2} = 0.0530$MHz; $\nu_{\pm 1/2-\mp 3/2} = 0.009$ MHz		
			$\nu_{1/2-5/2} = 0.0273$ MHz; $\nu_{3/2-7/2} = 0.0440$ MHz		
CsF	CsF	—	1.240	...	[106]. g
Cs_2	Cs_2	—	0.23	...	[65, 97]. g
$Cs_2MgO_8S_2$	$Cs_2Mg(SO_4)_2$	293	1.3	...	[112]. NMR cr.

* Data for ^{87}Rb.

Empirical formula	Structural formula	T, °K	e^2Qq, MHz	ν, MHz	References, notes
		^{63}Cu			
C_2CuKN_2	$K[Cu(CN)_2]$	77	...	32.661	[115—117]
			...	30,221*	
		84	...	32.647	
			...	30,212*	
		289	60.9	33.479	η = 78%
			...	30.294[1]	
$C_4CuK_3N_4$	$K_3[Cu(CN)_4]$	293	1.113	—	[78]. NMR cr.
$C_8H_{12}CuO_8$	$Cu_2(CH_3COO)_4$	77	2.19	—	[118]. NMR cr.
			1,46	—	
Cu_2O	Cu_2O	87	...	26.697	[115, 119—121]
			...	24.705*	
		289	52.06	26.020	η = 0.12%
			...	24.079*	
		^{197}Au			
AuCl	AuCl	300	514 av.	257.075	[122, 123]
				257.173	
				257.347	

* Data for ^{65}Cu.

BIBLIOGRAPHY

1. Pyykkö, P. — Annls. Univ. fenn. Åbo, Sar A1 (1967), 46.
2. Hovi, V. and P. Pyykkö. — Phys. kondens. matter. 5 (1966), 1.
2a. Maclean, C., E. Mackon, and C. Hilbers. — J. Chem. Phys. 46 (1967), 3393.
3. Chiba, T. — J. Chem. Phys. 41 (1964), 1352.
4. White, R. — Phys. Rev. 94 (1954), 789A.
5. White, R. — J. Chem. Phys. 23 (1955), 2531.
6. Flygar, W. H. — J. Chem. Phys. 41 (1964), 206.
7. Zeiler, M. — Ber. Bunsengesellsch. phys. Chem. 69 (1965), 659.
8. Chiba, T. — Bull. Chem. Soc. Japan 38 (1965), 490.
9. Thaddeus, P., L. Krisher, and J. Loubser. — J. Chem. Phys. 45 (1966), 1670.
10. Coves, T. and M. Karplus. — J. Chem. Phys. 45 (1966), 1670.
11. White, R. — J. Chem. Phys. 23 (1954), 253.
12. Weiss, V. and W. Flugare. — J. Chem. Phys. 45 (1966), 810.
13. Ramsey, N. — Am. Scient. 49 (1961), 509.
14. Kern, C. and M. Karplus. — J. Chem. Phys. 42 (1965), 1062.
15. Chiba, T. — International Symposium on Molecular Structure and Spectroscopy. Tokyo, 1962.
16. Chiba, T. — Techn. Rept. ISSP, A, No. 100 (1964), 8.
17. Bjorkstam, J. — Phys. Rev. 153 (1967), 599.
18. Derbyshire, W., T. Gorvin, and D. Warner. — Molec. Phys. 12 (1967), 299.
19. Vijayaraghavan, R. and V. Sarasvati. — J. Phys. Soc. Japan 20 (1965), 2290.
20. Bonera, G. and A. Rigamonti. — J. Chem. Phys. 42 (1965), 175.
21. Warner, D., T. Gorvin, and W. Derbyshire. — Molec. Phys. 14 (1968), 281.
22. Phillips, W., J. Rowell, and L. Melby. — J. Chem. Phys. 41 (1964), 2551.
23. Rowell, J., W. Phillips, L. Melby, and M. Panar. — J. Chem. Phys. 43 (1965), 3442.
24. Woessner, D. — J. Chem. Phys. 40 (1964), 2341.
25. Pyykkö, P. and U. Lähteenmäki. — Annls. Univ. fenn. Åbo A1 (1961), 7.
26. Blinc, R., M. Pintar, and I. Zupancic. — J. Phys. Chem. Sol. 28 (1967), 405.
27. Ellis, D. — Bull. Am. Phys. Soc. 12 (1967), 468; Ellis, D. and J. Bjorkstam. — J. Chem. Phys. 46 (1967), 4460.
28. Rabideau, S. and P. Waldstein. — J. Chem. Phys. 42 (1965), 3822.
29. Mark, N. and J. Leavitt. — Phys. Rev. 122 (1961), 856.
30. Weisbaum, S., G. Beers, and G. Hermann. — J. Chem. Phys. 23 (1955), 1601.
31. Posener, D. — Aust. J. Phys. 10 (1957), 276.
32. Treacy, E. and G. Beers. — J. Chem. Phys. 36 (1962), 1473.
33. Thaddeus, P., J. Loubser, and L. Krisher. — Bull. Am. Phys. Soc. 5 (1960), 74.
34. Chiba, T. — J. Chem. Phys. 47 (1967), 1592.
35. Clifford, J., M. Dixon, and J. Smith. — Chem. Phys. Letters 1 (1967), 185.
36. Waldstein, P., S. Rabideau, and J. Jackson. — J. Chem. Phys. 41 (1964), 3407.
37. Chiba, T. — J. Chem. Phys. 39 (1963), 947.
38. Gabriel, C. — Cited after E. Hahn, XIII Colloque AMPERE, p. 42, 1964.
39. McGroth, J. and G. Ossman. — J. Chem. Phys. 46 (1967), 1824.
40. Clifford, J. and J. Smith. — Molec. Phys. 13 (1967), 297.
41. Ketudat, S. and R. Pound. — J. Chem. Phys. 26 (1957), 708.
42. Bersohn, R. — J. Chem. Phys. 32 (1960), 85.
43. Bjorkstam, J. and E. Uehling. — Phys. Rev. 114 (1959), 961.
44. Anderson, D. — XIV Colloque AMPERE, Ljubljana, 1966.
45. Hermann, G. — J. Chem. Phys. 29 (1958), 875.
46. Kato, Y. — J. Phys. Soc. Japan 16 (1961), 121.
47. Cvikl, B. — cited after R. Blinc and D. Hadzi. Nature 212 (1966), 922.
48. Chiba, T. — J. Chem. Phys. 36 (1962), 1122.
49. Hovi, V., U. Järvinen, and P. Pyykkö. — J. Phys. Soc. Japan 21 (1966), 2742.
50. Lähteenmäki, U., L. Niemelä, and P. Pyykkö. — Phys. Rev. Lett. 25A (1967), 460.
51. Howell, F., R. Knispel, and V. Schmidt. — Bull. Am. Phys. Soc. 12 (1967), 924.
52. Chiba, T. — Bull. Chem. Soc. Japan 38 (1965), 490.
53. O'Reilly, D. and Tung Tsang. — J. Chem. Phys. 46 (1967), 1291.

54. Kydon, D., M. Pintar, and H. Petch. — J. Chem. Phys. **47** (1967), 1185.
55. Schone, H. and W. Knight. — Bull. Am. Phys. Soc. **6** (1961), 104.
56. Mkrtchyan, A. R. and A. S. Marfunin. — Doklady AN SSSR 163 (1965), 609.
57. Belford, G. — J. Chem. Phys. **35** (1961), 1032.
58. Granna, N. — Can. J. Phys. **31** (1953), 1185.
59. Petch, H. — Can. J. Phys. **31** (1953), 837.
60. Petch, H. — Phys. Rev. **88** (1952), 1201.
61. Stauss, G. — J. Chem. Phys. **40** (1964), 1988.
62. Hon, J. and P. Bray. — Phys. Rev. **110** (1958), 624.
63. Kusch, P. — Phys. Rev. **75** (1949), 887.
64. Kusch, P. — Phys. Rev. **76** (1949), 138.
65. Logan, R., R. Coté, and P. Kusch. — Phys. Rev. **86** (1952), 280.
66. Bernheim, R., I. Adler, B. Lavery, D. Lini, B. Scott, and J. Dixon. — J. Chem. Phys. **45** (1966), 3442.
67. Lucken, E. — J. Organometall. Chem. **4** (1965), 252.
68. Pound, R. — Phys. Rev. **79** (1950), 685.
69. Burns, G. — Phys. Rev. **127** (1962), 1193.
70. Whorton, L., L. Gold, and W. Klemperer. — Phys. Rev. **133** (1964), 270.
71. Kahala, S. — Phys. Rev. Lett. **6** (1961), 549.
72. Haigh, P., R. Forman, and R. Frisch. — J. Chem. Phys. **45** (1966), 812.
72a. Bishop, S., P. Ring, and P. Bray. — J. Chem. Phys. **45** (1966), 1525.
73. Robinson, W., W. O'Sullivan, and W. Simmons. — Bull. Am. Phys. Soc., ser. 2, **6** (1961), 105.
74. Anderson, D. — J. Chem. Phys. **35** (1961), 1353.
75. Burns, G. — Bull. Am. Phys. Soc., ser. 2, **6** (1961), 105.
76. Peterson, G. and P. Bridenbaugh. — J. Chem. Phys. **46** (1967), 4009.
77. Brun, E. and B. Derighetti. — Helv. phys. Acta **34** (1961), 382.
78. Becker, G. — Z. Phys. **130** (1951), 415.
79. Pound, R. — Bull. Am. Phys. Soc., ser. 1, **23** (1948), 161.
80. Murakami, M. — J. Phys. Soc. Japan **11** (1956), 607.
81. Brun, E., S. Hafner, and P. Hartmann. — Helv. phys. Acta **33** (1960), 495.
82. Petch, H. and K. Pennington. — J. Chem. Phys. **36** (1962), 1216.
83. Weiss, A. and K. Zohner. — Physica Status Solid 21 (1967), 257.
84. Haering, R. and G. Volkoff. — Can. J. Phys. **34** (1956), 577.
85. Nierenberg, W. and N. Ramsey. — Phys. Rev. **72** (1947), 1075.
86. Gutowsky, H. and G. Williams. — Phys. Rev. **105** (1957), 464.
87. Miller, N. and P. Casabella. — Phys. Rev. **152** (1966), 228.
88. Zeiger, H. and D. Boleff. — Phys. Rev. **85** (1952), 788.
89. Sazonov, A. M. and S. B. Grigor'ev. — Fiz. Tverd. Tela **7** (1965), 1389.
90. Emshwiller, M., E. Hahn, and D. Kaplan. — Phys. Rev. **118** (1960), 414.
91. Holuj, F. and H. Petch. — Can. J. Phys. **34** (1956), 1169.
92. Radhakrishna, S. — J. Phys. Chem. Sol. **27** (1966), 1567.
93. Anderson, D. — Bull. Am. Phys. Soc. ser. 2, **6** (1961), 363.
94. Eades, R. — Proc. Phys. Soc. **79** (1962), 954.
95. Andrew, E. — Proc. Phys. Soc. **79** (1962), 954.
96. Gourdji, M. and L. Guibe. — Compt. rend. **260** (1965), 1131.
97. Kusch, P. and S. Millman. — Phys. Rev. **55** (1939), 1176.
98. Robinson, L. — Can. J. Phys. **35** (1957), 1344.
99. Berthold, G. and A. Weiss. — Ber. Bunsengesellsch. phys. Chem. **68** (1964), 640.
100. Fabrikand, B., R. Carlson, C. Lee, and I. Rabi. — Phys. Rev. **91** (1953), 1403.
101. Lee, C., B. Fabrikand, R. Carlson, and I. Rabi. — Phys. Rev. **86** (1952), 607.
102. Lee, C., B. Rabrikand, R. Carlson, and I. Rabi. — Phys. Rev. **91** (1953), 1395.
103. Kaplan, D. and E. Hahn. — Bull. Am. Phys. Soc., ser. 2, **2** (1957), 10.
104. Hartland, A. — Phys. Rev. Lett. **20** (1966), 567.
105. Grabner, L. and V. Hughes. — Phys. Rev. **79** (1950), 819.
106. Trischka, J. — Phys. Rev. **74** (1948), 718.
107. Kunitomo, M., T. Terao, Y. Tsutsumi, and T. Haschi. — J. Phys. Soc. Japan **22** (1967), 945.
107a. Tsutsumi, Y., M. Kunitomo, T. Terao. — J. Phys. Soc. Japan **24** (1968), 429.
107b. Tsutsumi, Y., M. Kunitomo, T. Terao, and T. Haschi. — J. Phys. Soc. Japan **26** (1969), 16.

108. Trischka, J. and R. Braunstein. — Phys. Rev. 96 (1954), 968.
109. Hughes, V. and L. Grabner. — Phys. Rev. 79 (1950), 814.
110. Boleff, D. and H. Zeiger. — Phys. Rev. 85 (1952), 799.
111. Bonczyk, P. and V. Hughes. — Phys. Rev. 161 (1967), 15.
112. Kiddle, R. — Phys. Rev. 104 (1951), 932.
113. Luce, R. and J. Trischka. — Phys. Rev. 82 (1951), 323.
114. Luce, R. and J. Trischka. — Phys. Rev. 83 (1951), 851.
115. Kruger, H. and U. Meyer-Berkhout. — Z. Phys. 132 (1952), 171.
116. Graybeal, D. and G. McKown. — Inorg. Chem. 5 (1966), 1909.
117. McKown, G. and J. Graybeal. — J. Chem. Phys. 44 (1966), 610.
118. Royer, D. — Inorg. Chem. 4 (1965), 1830.
119. Cox, H. and D. Williams. — J. Chem. Phys. 32 (1960), 633.
120. Jeffrey, K. and R. Armstrong. — Can. J. Phys. 44 (1966), 2315.
121. Wijn, H. and J. Wildt. — Phys. Rev. 150 (1966), 200.
122. Machmer, P. — Compt. rend. 262 (1966), 650.
123. Machmer, P., M. Read, and P. Cornil. — Inorg. Nucl. Chem. Letters 3 (1967), 215.
124. Caspary, W., F. Millet, M. Reichbach, and B. Dailey. — J. Chem. Phys. 51 (1969), 623.
125. Anderson, C. and N. Ramsey. — Phys. Rev. 149 (1966), 14.
126. Rabideau, S., E. Finch, and A. Denison. — J. Chem. Phys. 49 (1968), 4660.
127. Soda, G. and T. Chiba. — J. Chem. Phys. 50 (1969), 439.
128. Soda, G. and T. Chiba. — J. Phys. Soc. Japan 26 (1969), 717.
129. Soda, G. and T. Chiba. — J. Phys. Soc. Japan 26 (1969), 723.
130. Blinc, R., J. Stepišnik, and I. Zupančič. — Phys. Rev. 176 (1968), 732.
131. Peterson, G. and P. Bridenbaugh. — J. Chem. Phys. 48 (1968), 3402.
132. Vladimirtsev, Yu. V., V. A. Golenishchev-Kutuzov, U. Kh. Kopvillem, N. A. Shamukov and V. V. Upalova. — Fiz. Tverd. Tela 10 (1968), 3500.
133. Whidden, C., C. Williams, and R. Tipsword. — J. Chem. Phys. 50 (1969), 507.
134. McGrath, J. — J. Chem. Phys. 47 (1967), 4276.
135. Cvikl, B. and J. McGrath. — J. Chem. Phys. 49 (1968), 3323.
136. Linzer, M. and R. Forman. — J. Chem. Phys. 46 (1967), 4690.
137. Coogan, C., G. Belford, and H. Gutowsky. — J. Chem. Phys. 39 (1963), 3061.
138. Weiss, A. — Z. Naturf. 15A (1960), 536.
139. Kampbell, I. and C. Coogan. — J. Chem. Phys. 44 (1966), 2075.
140. Cuthbert, J. and H. Petch. — J. Chem. Phys. 39 (1963), 1247.
141. Ketudat, S., J. Berthold, and A. Weiss. — Z. Naturf. 22a (1967), 135.
142. Saraswati, V. — J. Phys. Soc. Japan 23 (1967), 761.

Group II elements

Empirical formula	Structural formula	T, °K	e^2qQ, MHz	η, %	References, notes
			^{9}Be		
Al_2BeO_4	$BeAl_2O_4$	Room	0.318	90	[1, 2]. NMR cr. cf. Al
$Al_2Be_3O_{18}Si_6$	$Be_3Al_2(SiO_3)_6$	»	0.504	9	[3, 4]. NMR cr. cf. Al
Be	Be (met.)	4.2	0.057	...	[5—8, 19, 20]. NMR cr.
		293	0.048	...	
BeO	BeO	293	0.0377	0	[9—11]. NMR cr.
			^{25}Mg		
Mg	Mg (met.)	4.2	0.324	...	[7, 21]. NMR cr.
		296	0.23	...	
			^{137}Ba		
$BaBr_2+2H_2O$	$BaBr_2\cdot 2H_2O$	77	ν=25.835 MHz		*[12]
		Room	44.78	73	
			ν=24.186 MHz		
$BaClO_3+6H_2O$	$BaClO_3\cdot 6H_2O$	293	25.4	48	[13]
				$\nu_{^{137}Ba}$ =13.18 MHz;	$\nu_{^{135}Ba}$=8.55 MHz
			Zn		
Zn	Zn (met.)	4	79	...	[7, 14, 15]. NMR cr.
		293	74	...	
			Cd		
Cd	Cd (met.)	293	5.4—9.3	...	[7, 16]. NMR cr.
CdIn	Cd—In (alloy)	293	14	...	[7, 17]. NMR cr.
			^{201}Hg		
Cl_2Hg	$HgCl_2$	83	ν=361.966 MHz		[18]. cf. Cl
		299.1	ν=354.207 MHz		

BIBLIOGRAPHY

1. Reaves, H. and T. Gilmer. — J. Chem. Phys. 42 (1965), 4138.
2. Farrell, E. — Am. Miner. 48 (1963), 804.
3. Brown, L. and D. Williams. — J. Chem. Phys. 24 (1956), 751.
4. Brown, L. and D. Williams. — Phys. Rev. 95 (1954), 1110.
5. Knight, W. — Phys. Rev. 92 (1953), 539.
6. Pomerantz, M. — Phys. Rev. 119 (1960), 70.
7. Das, T. — Phys. Rev. 123 (1961), 2070.
8. Barnes, R., D. Barnaal, B. McCart, and D. Torgeson. — Bull. Am. Phys. Soc. 12 (1967), 314; Phys. Rev. 157 (1967), 510.
9. Hon, J. — Phys. Rev. 124 (1961) 1368.
10. Hon, J. — Bull. Am. Phys. Soc. 4 (1959), 354.
11. Troup, G. and J. Walter. — J. Nucl. Mater. 14 (1964), 272.
12. McGrath, J. and A. Silvidi. — J. Chem. Phys. 36 (1962), 1496.
13. Nakamura, S. and H. Enokiva. — J. Phys. Soc. Japan 18 (1963), 183.
14. Seidel, G. and P. Keesom. — Phys. Rev. Lett. 2 (1959), 261.
15. Lien, W. — Phys. Rev. 118 (1960), 958.
16. Pound, R. — Phys. Rev. 92 (1953), 522.
17. Heer, G. — Physica Status Solidi 9 (1955), 217.
18. Dehmelt, H., H. Robinson, and W. Gordy. — Phys. Rev. 93 (1954), 480.
19. McCart, B. and R. Barnes. — J. Chem. Phys. 48 (1968), 127.
20. Alloul, H. and C. Fradevaux. — J. Phys. Chem. Sol. 29 (1968), 1623.
21. Dougan, P., S. Sharma, and D. Williams. — Can. J. Phys. 47 (1969), 1047.

Group III elements

Empirical formula	Structural formula	T, °K	Transition	ν, MHz	e^2Qq, MHz	η, %	References, notes
			^{11}B				
$AsBO_4$	$BAsO_4$	Room	—	—	0.4	...	[1]. NMR cr.
B	B (at.)	»	$^1/_2$—$^3/_2$	2.695	5.39	0	[2]
BC_3	BC_3	293	—	—	10.31	...	[3]. NMR cr.
$BCaHO_5Si$	$HCaBSiO_5$	Room	—	—	0.172	63.3	[4]. NMR cr.
$BClH_4Na_2O_4$	$Na_2B(OH)_4Cl$	»	—	—	0.09	...	[5]. NMR cr.
BF_2H	HBF_2	—	—	—	3.30	0.6	[6]. g
		—	—	—	6.871	0.6*	
BF_3	BF_3	77	—	—	2.64	...	[7, 86]. NMR cr.
$BNaF_4$	$NaBF_4$	296	—	—	0.0599	65	[89]. NMR cr.
BHO_2	HBO_2	Room	—	—	0.85	...	Phase 1, rhombic [1]. NMR data on the two other phases [8]
BN	BN	293	—	—	2.96	...	[9]. NMR cr.
$BNaO_3+H_2O$	$NaBO_3\cdot 4H_2O$	Room	—	—	<0.06	...	[1]. NMR cr.
BO_4P	BPO_4	»	—	—	0.0504	...	[1]. NMR cr.
B_2CaO_4	CaB_2O_4	»	—	—	2.58	55	[8, 87]. NMR cr.
B_2Cr	CrB_2	»	—	—	0.9	...	[10]. NMR cr.
B_2O_3	B_2O_3 (glass)	293	—	—	2.76	0	[1, 11]. NMR cr.
B_2S_3	B_2S_3	293	—	—	2.70	0	*[93]. NMR cr.

* Data for ^{10}B.

Empirical formula	Structural formula	T, °K	Transition	ν, MHz	e^2Qq, MHz	η, %	References, notes
B_2Se_3	B_2Se_3	293	—	—	2.50	25	*[93]. NMR cr.
B_2Ta	TaB_2	293	—	—	0.114	...	[10]. NMR cr.
B_2Ti	TiB_2	293	—	—	0.360	...	[9]. NMR cr.
			—	—	0,127	...	[10]. NMR cr.
B_2Zr	ZrB_2	293	—	—	0,118	...	[9]. NMR cr.
			—	—	0.084	...	[10]. NMR cr.
$B_3CaH_3O_7 + H_2O$	$CaB_3O_4(OH)_3 \cdot H_2O$ (colemanite)	Room	—	—	0,309	82.5	[12]. NMR cr.
			—	—	0.436	48,7	
			—	—	2.541	5,9	
$B_3K_3O_6$	$K_3B_3O_6$	»	—	—	<0.95	...	[9]. NMR cr.
B_3La	LaB_3	»	—	—	2.74	0	[1]. NMR cr.
B_4C	B_4C	293	—	—	0,2	...	[11]. NMR cr.
			—	—	1.3	...	
			—	—	5,58	...	
$B_4Na_2O_7 + 4H_2O$	$Na_2B_4O_7 \cdot 4H_2O$ (kernite)	Room	$1/2$—$3/2$	1.287	2.567	11.7	[13, 14]. NMR cr. cf. Na
			$1/2$—$3/2$	1,286	2.563	16.3	
			...	...	0,645	54	
			...	...	0.588	60	
$B_4Na_2O_7 + 5H_2O$	$Na_2B_4O_7 \cdot 5H_2O$	»	—	—	0,478	47,3	[15]. NMR cr.
			—	—	2,528	3,9	
			—	—	<0,72	...	[1]. NMR cr.
			—	—	2.75	...	
$B_4Na_2O_7 + 10H_2O$	$Na_2B_4O_7 \cdot 10H_2O$	»	—	—	0,487	71.4	[15]. NMR cr. cf. Na
			—	—	2,528	8,9	
$B_5CaH_6NaO_{12} + 5H_2O$	$NaCaB_5O_6(OH)_6 \cdot 5H_2O$	»	—	—	$0,050 < e^2Qq < 0,600$	...	[16]. NMR cr.
			—	—	2.530	...	

$B_5CaNaO_9 + 8H_2O$	$NaCaB_5O_0 \cdot 8H_2O$	»	—	—	<0.06	...	[1]. NMR cr.
			—	—	2.62	...	
$B_5H_4KO_{10} + 2H_2O$	$KB_5O_6(OH)_4 \cdot 2H_2O$	»	—	—	0.234	9.3	[16]. NMR cr.
			—	—	2.536	21.2	
			—	—	2.521	19.5	
$B_5KO_8 + 4H_2O$	$KB_5O_8 \cdot 4H_2O$	»	—	—	<0.6	...	[1]. NMR cr.
					2.51		
B_5Mo_2	Mo_2B_5	»	—	—	0.3	...	[10]. NMR cr.
$B_6Ca_2O_{11} + 5H_2O$	$Ca_2B_6O_{11} \cdot 5H_2O$	»	—	—	0.309	83	[17]. NMR cr.
			—	—	0.436	48	
			—	—	2.540	5.8	
$B_6Ca_2O_{11} + 7H_2O$	$Ca_2B_6O_{11} \cdot 7H_2O$	»	—	—	<0.5	...	[1]. NMR cr.
			—	—	2.62	...	
$B_6Mg_2O_{11} + 15H_2O$	$Mg_2B_6O_{11} \cdot 15H_2O$ lesserite	»	—	—	0.207	92.7	[18]. NMR cr.
			—	—	0.406	39.7	
			—	—	2.529	7.5	
	inderite	»	—	—	0.355	51	[19]. NMR cr.
			—	—	0.517	76	
			—	—	2.546	6.6	
$B_6O_{10}Sr + 4H_2O$	$SrO \cdot 3B_2O_3 \cdot 4H_2O$	»	—	—	0.612	15.7	[20]. NMR cr.
			—	—	0.673	38.6	
			—	—	0.705	9.0	
			—	—	2.503	15.6	
			—	—	2.505	12.3	
			—	—	2.509	9.3	
$B_7Mg_3O_{13}$	$Mg_3B_7O_{13}$	»	—	—	<0.77	...	[1]. NMR cr.
					2.70	...	
$B_{10}H_{12}N_2$	$B_{10}H_8(NH_2)_2$	»	—	—	1.9	...	[21]. NMR cr.
$B_{10}H_{18}N_2 + xH_2O$	$[(NH_4)]_2B_{10}H_{10} \cdot xH_2O$	»	—	—	2.0	...	The same
$B_{12}Cs_2H_{12}$	$Cs_2B_{12}H_{12}$	»	—	—	0	0	»
$B_{12}H_{12}Na_2 + 2H_2O$	$Na_2B_{12}H_{12} \cdot 2H_2O$	»	—	—	0.070	...	»

Empirical formula	Structural formula	T, °K	Transition	ν, MHz	e^2Qq, MHz	η, %	References, notes
CH_3BO	CH_3BO	Room	$^1/_2$—$^3/_2$	0.775	1.55	...	[2,22—24]
		—	—	—	1.55 (g)	...	
		—	—	—	3.4 (g)*	...	
C_2H_5BO	C_2H_5BO	—	—	—	3.08	...	[25]. g
$C_2H_{15}B_{10}CsS$	$Cs[1\text{-}B_{10}H_9S(CH_3)_2]$	Room	—	—	2.3	...	[21]. NMR cr.
	$Cs[2\text{-}B_{10}H_9S(CH_3)_2]$	»	—	—	2.0	...	[21]. NMR cr.
C_3H_9B	$B(CH_3)_3$	77	$^1/_2$—$^3/_2$	2.438	...	...	[26]
			2—3	2.538*	...	...	
	Phase I	83	$^1/_2$—$^3/_2$	2.4367	4.8734	0	[3]. p. t. at 112°K
			1—2	1.5225*	10.151*	0	
			2—3	2.5378*			
	Phase II	83	$^1/_2$—$^3/_2$	2.4845	4.969	0	
			1—2	1.5536*	10.356*	0	
			2—3	2.5890*			
C_3D_9B	$B(CD_3)_3$	77	$^1/_2$—$^3/_2$	2.452	...	...	[26]
			2—3	2.554*	...	...	
$C_3H_{12}B_3N_3$	$(HBNCH_3)_3$	Room	—	—	1.71	0	[27]. NMR cr.
$C_6H_{15}B$	$(C_2H_5)_3B$	77	$^1/_2$—$^3/_2$	2.502			[3, 26]
		83	$^1/_2$—$^3/_2$	2.5016	5.003	0.9	
		»	0—1	1.5554*			
			1—2	1.5740*	10.427*	0.9[1]	
			2—3	2.6068*			
$C_6H_{32}B_{12}N_2$	$[(CH_3)_3NH]_2B_{12}H_{12}$	Room	—	—	2.2	...	[21]. NMR cr.
$C_8H_{24}B_{12}Cl_{12}N_2$	$[(CH_3)_4N]_2B_{12}Cl_{12}$	»	—	—	0.022	...	»
$C_8H_{30}B_{10}I_4N_2$	$[(CH_3)_4N]_2B_{10}H_6I_4$	»	—	—	1.8	...	»
$C_8H_{35}B_{12}IN_2$	$[(CH_3)_4N]_2B_{12}H_{11}I$	»	—	—	0.160	...	»
$C_8H_{24}B_{12}D_{11}IN_2$	$[(CH_3)_4N]_2B_{12}D_{11}I$	»	—	—	0.160	...	»

$C_8H_{24}D_{12}B_{12}N_2$	$[(CH_3)_4N]_2B_{12}D_{12}$	»	—	—	<0.003	...	The same
$C_9H_9B_3O_{12}$	$[OB(OCH)_3]_3$	»	—	—	2.14	0	[27]. NMR cr.
$C_9H_{21}B$	$(C_3H_7)_3B$	77	$1/2$—$3/2$	2,500	...	...	[26]
	$(iso\text{-}C_3H_7)_3B$	77	$1/2$—$3/2$	2.625	...	...	[26]
$C_9H_3D_{18}B$	$[(CD_3)_2CH]_3B$	77	$1/2$—$3/2$	2,623	...	...	[26]
$C_{12}H_{27}B$	$(C_4H_9)_3B$	77	$1/2$—$3/2$	2,500	...	...	[26]
			^{27}**Al**				
Al	Al (met.)	—	—	$eq=8.3\cdot10^{14}$ V/cm^2			[28, 29]
	Al (in corundum)	Room	—	—	0,180	...	[30]. NMR cr.
$AlBeHO_5Si$	$HBeAlSiO_5$	»	—	—	5.173	69.8	[36]. NMR cr.
$AlBr_3$	$AlBr_3$	77	$1/2$—$3/2$ $3/2$—$5/2$	3.0319 3.8326	13,858	72.85	[37]. cf. Br
		296	$1/2$—$3/2$ $3/2$—$5/2$	3.017 3,742	13,57	74,1	
AlCl	AlCl	—	—	—	29.2	...	[39]. g cf. Cl
$AlCl_3$	$AlCl_3$	Room	—	—	0.472	0	[38]. NMR cr.
AlF	AlF	—	—	—	37.6	...	[39]. g
$AlH_4NO_8S_2 + 12H_2O$	$Al[NH_4](SO_4)_2\cdot12H_2O$	Room	—	—	0,447	0	[40]. NMR cr.
$AlKO_8S_2 + 12H_2O$	$AlK(SO_4)_2\cdot12H_2O$	»	—	—	0,400	0	[40]. NMR cr.
$AlLiO_6Si$	$LiAl(SiO_3)_2$	»	$1/2$—$3/2$ $3/2$—$5/2$	0.7515 0,7935	2,950	93	[41, 42, 42a]. cf. Li
$AlNaO_8Si_3$	$NaAlSi_3O_8$	»	—	—	3.37	63,4	[43]. NMR cr. cf. Na
AlO_4P	$AlPO_4$	»	—	—	4.088	36.7	[44]. NMR cr.
Al_2BeO_4	$BeAl_2O_4$	»	— —	— —	2,846 2.850	... ...	[88]. NMR cr. cf. Be
$Al_2Be_3O_{18}Si_6$	$Be_3Al_2(SiO_3)_6$	»	—	—	3.093	0	[45]. NMR cr. cf. Be

* Data for ^{10}B.

Empirical formula	Structural formula	T, °K	Transition	ν, MHz	e^2Qq, MHz	η, %	References, notes
Al_2Ce	$CeAl_2$	Room	—	—	4.72	...	[31, 32]. NMR cr.
Al_2Er	$ErAl_2$	»	—	—	3.56	...	[31]. NMR cr.
$Al_2FeMgO_{12}Si_3$	$(Fe^{II}, Mg)_3Al_2(SiO_4)_3$	»	—	—	1.510	0	[47]. NMR cr.
Al_2Gd	$GdAl_2$	»	—	—	4.27	...	]46]. NMR cr.
Al_2Lu	$LuAl_2$	»	—	—	2.85	...	[31]. NMR cr.
Al_2MgO_4	$MgAl_2O_4$	»	—	—	3.72	...	[48, 49]. NMR cr.
$Al_2Na_2O_{10}Si_3 + 2H_2O$	$Na_2Al_2Si_3O_{10} \cdot 2H_2O$	»	—	—	1.663	50.29	[50]. NMR cr. cf. Na
Al_2Nd	$NdAl_2$	»	—	—	4.56	...	[31, 32]. NMR cr.
Al_2O_3	Al_2O_3	293	—	—	2.393	0	[51—53]. NMR cr.
		297.1	$^1/_2$—$^3/_2$	0.17842	...	...	
Al_2Pr	$PrAl_2$	Room	—	—	4.60	...	[32, 33]. NMR cr.
Al_2Tm	$TmAl_2$	»	—	—	3.45	...	[31]. NMR cr.
Al_2U	UAl_2	»	—	—	4.31	...	[35]. NMR cr.
Al_2Y	YAl_2	»	—	—	3.97	...	[31]. NMR cr.
Al_3La	$LaAl_3$	»	—	—	0.8	...	[32]. NMR cr.
Al_3Sm	$SmAl_3$	»	—	—	0.8	...	[34]. NMR cr.
Al_3Tb	$TbAl_3$	»	—	—	0.8	...	[32]. NMR cr.
Al_5LiO_8	$LiAl_5O_8$	»	—	—	0.284	...	[54]. NMR cr. cf. Li
			—	—	0.683	...	
$Al_5O_{12}Y_3$	$Y_3Al_5O_{12}$	»	—	—	0.632	...	[55]. NMR cr.
			—	—	6.107	...	
$CH_6AlN_3O_8S_2 + 6H_2O$	$C(NH_2)_3Al(SO_4)_2 \cdot 6H_2O$	296	—	—	±0.023	...	[56]. NMR cr. cf. ^{14}N (supplement)
			—	—	∓0.030	...	

$CD_6AlN_3O_8S_2 + 6D_2O$	$C(ND_2)_3Al(SO_4)_2 \cdot 6D_2O$	296	—	—	±0.023	...	[56]. NMR cr.
			—	—	∓0.030	...	
$CH_6AlN_3O_8Se + 6H_2O$	$C(NH_2)_3Al(SeO_4)_2 \cdot 6H_2O$	296	—	—	±0.300	...	[56]. NMR cr. cf. ^{14}N (supplement)
			—	—	±0.200	...	
$C_6H_{15}Al + C_5H_5N$	$Al(C_2H_5)_3 \cdot Py$	Room	—	—	17	—	[57]. NMR cr.
$C_9H_{21}Al + C_5H_5N$	$(C_3H_7)_3Al \cdot Py$	»	—	—	18.3	—	[57]. NMR cr.
$C_{18}H_{39}Al + C_5H_5N$	$(C_6H_{13})_3Al \cdot Py$	»	—	—	20.7	—	[57]. NMR cr.
			^{69}Ga				
BrGa	GaBr	—	—	—	74	...	[57 a]. g cf. Br
Br_3Ga	$GaBr_3$	77	$^1/_2$—$^3/_2$	26.494	...	...	[58]. cf. Br
$CH_6GaN_3O_8S + 6H_2O$	$C(NH_2)_3Ga(SO_4)_2 \cdot 6H_2O$	Room	—	—	∓0.485	...	[56]. NMR cr. cf. ^{14}N (supplement)
					±0.635	...	
$CH_6GaN_3O_8Se + 6H_2O$	$C(NH_2)_3Ga(SeO_4)_2 \cdot 6H_2O$	»	—	—	±0.078	...	[56]. NMR cr.
			—	—	∓0.197	...	
ClGa	GaCl	—	—	—	84.7	...	[57 a]. g cf. Cl
Cl_3Ga	$GaCl_3$	77	$^1/_2$—$^3/_2$	29.365	...	...	[59, 90]. cf. Cl
			$^1/_2$—$^3/_2$	18.510*	...	...	
		300	$^1/_2$—$^3/_2$	29.080	51.93	86.7	
			$^1/_2$—$^3/_2$	18.153*	32.46	86.7	
Ga	Ga (met.)	4.2	$^1/_2$—$^3/_2$	11.3122	22.5	17.9	[61, 91]
		78	$^1/_2$—$^3/_2$	11.2532	22.4	17.9	
		273	$^1/_2$—$^3/_2$	10.908	...	...	
		285.3	$^1/_2$—$^3/_2$	10.8772	21.7	17.1	
		273	$^1/_2$—$^3/_2$	6.866*	...	...	
GaI	GaI	—	—	—	66	...	[57, a]. g cf. I
GaI_3	GaI_3	77.3	$^1/_2$—$^3/_2$	21.586	...	...	[58, 60]. cf. I
		298	$^1/_2$—$^3/_2$	21.441	...	...	
Ga_2O_3	α-Ga_2O_3	Room	—	—	8.20*	0	[62]. NMR cr.

* Data for ^{71}Ga.

Empirical formula	Structural formula	T, °C	Transition	ν, MHz	e^2Qq, MHz	η, %	References, notes
			^{115}In				
BiIn	InBi	Room	—	—	6.6	...	[71]. NMR cr.
ClIn	InCl	—	—	—	657.48	...	[63, 64]. g cf. Cl
		—	—	—	646.9†	...	
ClInO	InOCl	77	1/2—3/2	17.45	128.4	85.1	*[92]
			3/2—5/2	11.95			
			5/2—7/2	13.88			
			7/2—9/2	20.10			
$Cl_5H_{10}InN_2O$	$[NH_4]_2[InCl_5 \cdot H_2O]$	110	1/2—3/2	...	108.16	10.3	[65]. cf. Cl
			3/2—5/2	8.903			
			5/2—7/2	13.501			
			7/2—9/2	17.998			
		300	1/2—3/2	10.205	90.58	65.4	
			3/2—5/2	7.453			
			5/2—7/2	10,346			
			7/2—9/2	14.552			
FInO	InOF	77	1/2—3/2	33.78	252.87	83.1	*[92]
			3/2—5/2	23.20			
			5/2—7/2	27.47			
			7/2—9/2	39.68			
I_3In	InI_3	4.2	3/2—5/2	26.860	328.94	63.96	[58, 72]. cf. I. p.t. at 278°K and 1034 bar, 303 °K and 758 bar, 343 °K and 1172 bar.
			5/2—7/2	37.725			
		54	3/2—5/2	26.861	...	...	
		77.3	1/2—3/2	36.718	327.20	65.04	
			3/2—5/2	26.864			
			5/2—7/2	37.420			
		195	1/2—3/2	37.154	323.97	67.04	
			3/2—5/2	26.876			
			5/2—7/2	36.842			
		274	1/2—3/2	37.422	321.77	68.35	
			3/2—5/2	26.885			
			5/2—7/2	36.465			

		298	$^1/_2$—$^3/_2$	37.506	320.41	69.0	
			$^3/_2$—$^5/_2$	26.985			
			$^5/_2$—$^7/_2$	36.265			
			$^7/_2$—$^9/_2$	51.238			
			$^3/_2$—$^7/_2$	63.2			
		300	$^1/_2$—$^3/_2$	37.492	321.08	68.77	
			$^3/_2$—$^5/_2$	26.888			
			$^5/_2$—$^7/_2$	36.340			
		328	$^3/_2$—$^5/_2$	26.893	...	...	
		363	$^3/_2$—$^5/_2$	26.901	...	...	
In	In (met.)	4.2	$^1/_2$—$^3/_2$	1.886	...	...	[66—70]
			$^3/_2$—$^5/_2$	3.770			
			$^5/_2$—$^7/_2$	5.655			
			$^7/_2$—$^9/_2$	7.533			
		77	$^1/_2$—$^3/_2$	1.786	...	...	
			$^3/_2$—$^5/_2$	3.575			
			$^5/_2$—$^7/_2$	5.364			
			$^7/_2$—$^9/_2$	7.148			
		4.2	$^5/_2$—$^7/_2$	5.57†	...	...	
In_2S_3	β-In_2S_3	77	$^1/_2$—$^3/_2$	11.28	269.0	⩽1	*[93]
				...	108.0	⩽1	
			$^3/_2$—$^5/_2$	22.47			
				...			
			$^5/_2$—$^7/_2$	33.71			
				13.5			
			$^7/_2$—$^9/_2$	45.00			
				18.05			
			^{203}**Tl**				
ITl	TlI	—	—	—	—523	...	[73]. g cf. I

† Data for ^{113}In

Empirical formula	Structural formula	T, °K	Transition	ν, MHz	e^2Qq, MHz	η, %	References, notes
			^{45}Sc				
Sc	Sc (met.)	Room	$^1/_2$—$^3/_2$	0.144 (three pairs of lines)	2.02	...	[74]. NMR cr.
			^{139}La				
$C_6H_{15}LaO_{12}S_3$	$La(C_2H_5OSO_3)_3$	77.3	—	—	16.595	...	[75, 76]. NMR cr.
		194.6	—	—	15.126	...	
		273.2	—	—	14.154	...	
		293.2	—	—	13.952	...	
Cl_3La	$LaCl_3$	Room	—	—	15.327	...	[76]. NMR cr.
F_3La	LaF_3	88	$^1/_2$—$^3/_2$	2.73	17.08	78.45	[77, 78]
			$^3/_2$—$^5/_2$	2.23			
			$^5/_2$—$^7/_2$	3.52			
		112			16.99	78.84	
		137			16.91	78.97	
		260			16.33	80.29	
		284			16.22	80.36	
		294			16.14	80.54	
		304			16.09	80.53	
		382			15.61	80.33	
		447			15.17	79.93	
La	La (met.)	1—210	—	—	7.8		
		293	—	—	<1.540	...	[79—81, 94]. NMR cr.
		300	$^1/_2$—$^3/_2$	0.101	1.41		
$LaMgN_5O_{15}$	$LaMg(NO_3)_5$	77.3	—	—	15.638	...	[75]. NMR cr.
		194.6	—	—	14.932	...	
		273.2	—	—	14.045	...	
		293.2	—	—	13.801	...	
$LaMgO_{16}S_4$	$La_2Mg(SO_4)_4$	Room	—	—	13.079	...	[76]. NMR cr.

			^{141}Pr				
$O_{12}S_3Pr + 8H_2O$	$Pr(SO_4)_3 \cdot 8H_2O$	Room	—	—	21.80	47.7	[82]. NMR cr.
			Nd				
Gd + Nd	NdGd	Room	—	—	64.4 (^{143}Nd)	...	[83]. NMR cr.
			—	—	56.4 (^{145}Nd)	...	
			Sm				
Gd + Sm	SmGd	Room	—	—	134.4 (^{147}Sm)	...	[83]. NMR cr.
			—	—	70 (^{149}Sm)	...	
			^{159}Tb				
Gd + Tb	TbGd	Room	—	—	1404	...	[83]. NMR cr.
Tb	Tb (met.)	»	—	—	2674	...	[84], NMR cr.
			^{161}Dy				
Dy	Dy (met.)	Room	—	—	2318	33.2	[85]. NMR cr.
Dy + Gd	DyGd	»	—	—	2666	...	[83]. NMR cr.
		»	—	—	2813*	...	
			^{167}Er				
Er + Gd	ErGd	Room	—	—	1596	...	[83]. NMR cr.
			^{175}Lu				
$C_6H_{15}LuO_{12}S_3$	$Lu(C_2H_5OSO_3)_3$	20	$1/2—3/2$	65.115	911.61	...	[75]
		77.3	$1/2—3/2$	65.013	910.185	...	
		194.6	$1/2—3/2$	63.082	883.154	...	
		273.2	$1/2—3/2$	61.152	856.125	...	
		296	$1/2—3/2$	60.438	846.135	...	

* Data for ^{163}Dy.

BIBLIOGRAPHY

1. Bray, P., J. Edwards, J. O'Keefe, V. Ross, and J. Tatsuzaki. — J. Chem. Phys. **35** (1961), 435.
2. Das, T. — J. Chem. Phys. **27** (1957), 1.
3. Dehmelt, H. — Z. Phys. **134** (1953), 642.
4. Lal, K. and H. Petch. — J. Chem. Phys. **43** (1965), 1834.
5. Ross, V. and J. Edwards. — Am. Miner. **44** (1959), 775.
6. Kasuya, T., W. Lafferty, and D. Lide. — J. Chem. Phys. **48** (1968), 1.
7. Casabella, P. — J. Chem. Phys. **41** (1964), 3793.
8. Silver, A. — J. Chem. Phys. **32** (1960), 959.
9. Silver, A. and P. Bray. — J. Chem. Phys. **32** (1960), 288.
10. Malyuchkov, O. T. and V. A. Povitskii. — Fizika Metall. Metalloved. **13** (1962), 676.
11. Silver, A. and P. Bray. — J. Chem. Phys. **31** (1959), 247.
12. Holej, F. and H. Petch. — Can. J. Phys. **38** (1960), 515.
13. Haering, R. and G. Volkoff. — Can. J. Phys. **34** (1956), 577.
14. Water, H. and G. Volkoff. — Can. J. Phys. **33** (1955), 156.
15. Cuthebert, J. and H. Petch. — J. Chem. Phys. **38** (1963), 1912.
16. Lal, K. and H. Petch. — J. Chem. Phys. **43** (1965), 178.
17. Holej, F. and H. Petch. — Can. J. Phys. **36** (1958), 145.
18. Pennington, K. and H. Petch. — J. Chem. Phys. **36** (1962), 2151.
19. Pennington, K. and H. Petch. — J. Chem. Phys. **38** (1960), 329.
20. Cuthebert, J., W. McFarlane, and H. Petch. — J. Chem. Phys. **43** (1965), 173.
21. Eaton, D. — Inorg. Chem. **4** (1965), 1520.
22. Gordy, W., H. Ring, and A. Burg. — Phys. Rev. **75** (1949), 1325.
23. Gordy, W., H. Ring, and A. Burg. — Phys. Rev. **78** (1950), 512.
24. Weiss, T., M. Strandberg, R. Lawrence, and C. Loomis. — Phys. Rev. **78** (1950), 202.
25. Hand, J. and R. Schwendeman. — J. Chem. Phys. **45** (1966), 3349.
26. Love, P. — J. Chem. Phys. **39** (1963), 3044.
27. Ring, L. and W. Koski. — J. Chem. Phys. **35** (1961), 381.
28. Rowland, T. — Acta metall. **3** (1955), 74.
29. Dowley, M. — Solid State Communs. **3** (1965), 351.
30. Lawrence, N. and J. Lambe. — Phys. Rev. **132** (1963), 1029.
31. Jones, W., T. Graham, and R. Barnes. — Phys. Rev. **132** (1963), 1891.
32. Van Diepen, A., H. de Wijn, amd K. Buschow. — J. Chem. Phys. **46** (1967), 3489.
33. Barnes, R. and R. Lecandr. — J. Phys. Soc. Japan **22** (1967), 930.
34. Wijn, H. de, A. Van Diepen, and K. Buschow. — Phys. Rev. **161** (1967), 253.
35. Jaccarino, V., B. Mattias, M. Peter, H. Suhl, and J. Wernick. — Phys. Rev. Lett. **5** (1960), 25
36. Eades, R. — Can. J. Phys. **33** (1955), 286.
37. Casabella, P. and P. Bray. — J. Chem. Phys. **30** (1959), 1393.
38. Casabella, P. and N. Miller. — J. Chem. Phys. **40** (1964), 1363.
39. Lide, D. — J. Chem. Phys. **42** (1965), 1013.
40. Segleken, W. and H. Torrey. — Phys. Rev. **98** (1955), 1537.
41. Robinson, L. — Can. J. Phys. **36** (1958), 1295.
42. Robinson, L. — Can. J. Phys. **35** (1957), 1344.

42a. Belford, G., R. Bernheim, and H. Gutowsky. — J. Chem. Phys. **35** (1961), 1032.

43. Brun, E., S. Hafner, and P. Hartmann. — Helv. phys. Acta **33** (1960), 495.
44. Brun, E., P. Hartmann, F. Laves, and D. Schwarzenbach. — Helv. phys. Acta **34** (1961), 388.
45. Brown, C. and D. Williams. — J. Chem. Phys. **24** (1956), 751; Phys. Rev. **95** (1954), 1110.
46. Jones, E. — J. Appl. Phys. **37** (1966), 1250.
47. Subrata, C. — Solid State Communs. **2** (1964), 361.
48. Brun, E., S. Hafner, and P. Hartmann. — Helv. phys. Acta **33** (1960), 496.
49. Brun, E., S. Hafner, P. Hartmann, and F. Laves. — Naturwissenschaften **47** (1960), 277.
50. Petch, H. and K. Pennington. — J. Chem. Phys. **36** (1962), 1216.
51. Veigele, W. — Bull. Am. Phys. Soc., ser. 2, **5** (1960), 344.
52. Kushida, T. and A. Silver. — Phys. Rev. **130** (1963), 1692.
53. Veigele, W., A. Bevan, and W. Tanttila. — J. Chem. Phys. **38** (1963), 1596.
54. Stauss, G. — J. Chem. Phys. **40** (1964), 1988.
55. Brog, K., W. Jones, and C. Verber. — Phys. Rev. Lett. **20** (1966), 258.

56. Burns, G. — Phys. Rev. 123 (1961), 1634.
57. Petrakis, L. and H. Swift. — J. Phys. Chem. 72 (1968), 546.
57a. Barrett, A. and M. Mandel. — Phys. Rev. 109 (1958), 1572.
58. Barnes, R., S. Segel, P. Bray, and P. Casabella. — J. Chem. Phys. 26 (1957), 1345.
59. Dehmelt, H. — Phys. Rev. 92 (1953), 1240.
60. Brooker, H. — J. Chem. Phys. 47 (1967), 1864.
61. Knight, W. D., R. R. Hewitt, and M. Pomerantz. — Phys. Rev. 104 (1956), 271.
62. Veigele, W. — J. Chem. Phys. 39 (1963), 2389.
63. Hofft, J. — Z. Phys. 163 (1961), 262.
64. Delvigne, G. and H. de Wijn. — J. Chem. Phys. 45 (1966), 3318.
65. Carr, S., B. Garrett, and W. Moulton. — J. Chem. Phys. 47 (1967), 1170.
66. Hewitt, R. and W. Knight. — Phys. Rev. Lett. 3 (1959), 18.
67. Herlach, F. — Helv. phys. Acta 34 (1961), 305.
68. Hewitt, R. and T. Taylor. — Phys. Rev. 125 (1962), 524.
69. Hewitt, R. and D. Williams. — Phys. Rev. 129 (1963), 1188.
70. O'Sullivan, W., W. Robinson, and W. Simmons. — Bull. Am. Phys. Soc., ser. 2, 5 (1960), 413.
71. Radhakrishna, S. and B. Mangurwadi. — Proc. Phys. Soc. 90 (1967), 495.
72. Brooker, H. and T. Scott. — J. Chem. Phys. 41 (1964), 475.
73. Happ, H. — Z. Phys. 147 (1957), 567.
74. Barnes, R., F. Borsa, and S. Segel. — Phys. Rev. 137 (1965), 1828.
75. Edmonds, D. and A. Lindop. — Proc. Phys. Soc. 87 (1966), 721.
76. Edmonds, D. — Phys. Rev. Lett. 10 (1963), 129.
77. Andersson, L. and W. Proctor. — Helv. phys. Acta 38 (1965), 360.
78. Lee, K., A. Sher, L. Andersson, and W. Proctor. — Phys. Rev. 150 (1966), 168.
79. Blumberg, W. and J. Gizinger. — Phys. Rev. Lett. 5 (1960), 52.
80. Torgelson, D. and R. Barnes. — Phys. Rev. 136 (1964), 738.
81. Das, T. — Phys. Rev. 123 (1961), 2070.
82. Teplov, M. A. — Zh. Eksper. Teoret. Fiz. 53 (1967), 1510.
83. Kobayashi, S., N. Sano, and J. Itoh. — J. Phys. Soc. Japan 23 (1967), 474.
84. Kobayashi, S., N. Sano, and J. Itoh. — J. Phys. Soc. Japan 22 (1967), 676.
85. Kobayashi, S., N. Sano, and J. Itoh. — J. Phys. Soc. Japan 21 (1966), 1456.
86. Casabella, P. and T. Oja. — J. Chem. Phys. 50 (1969), 4814.
87. Kriz, H., S. Bishop, and P. Bray. — J. Chem. Phys. 49 (1969), 557.
88. Hockenber, J., C. Brown, and D. Williams. — J. Chem. Phys. 28 (1958), 367.
89. Weiss, A. and K. Sohner. — Physica Status Solidi 21 (1967), 257.
90. Peterson, G. and P. Bridenbaugh. — J. Chem. Phys. 51 (1969), 238.
91. Valič, M. and D. Williams. — J. Phys. Chem. Sol. 30 (1969), 2337.
92. Svergun, V. I., P. Khagenmyuller, I. Port'e, A. F. Volkov, and G. K. Semin. — Izv. AN SSSR, chem. ser., p. 1676, 1970.
93. Svergun, V. I., Ya. Kh. Grinberg, V. M. Kuznets, and T. A. Babushkina. — Izv. AN SSSR, chem. ser., p. 1448, 1970.
94. Narath, A. — Phys. Rev. 179 (1969), 359.

Group IV elements

Empirical formula	Structural formula	T, °K	e^2Qq, MHz	References, notes
		^{73}Ge		
CH_6Ge	CH_3GeH_3	—	3.00	[1, 2]. g
		—	<0.500*	
C_2H_4Ge	$H_3GeC{\equiv}CH$	—	+32.5	[3]. g
$ClGeH_3$	H_3GeCl	—	—95	[4, 5]. g
$FGeH_3$	H_3GeF	—	—92	[6]. g
Ge	Ge (met.)		—54.566	^3p_z state
		—		[7]. Beam
GeSe	GeSe	—	+175	[8]. 1.
		Pb		
Pb + Tl	Pb Tl	293	12	[9, 10]. NMR cr.
		^{47}Ti, ^{49}Ti		
Ti	Ti (met.)	4.2	7.7	[11]. NMR cr.

*Data for ^{72}Ge.

BIBLIOGRAPHY

1. Barchukov, A. I. and A. M. Prokhorov. — Optika Spektrosk. 5 (1958), 530.
2. Laurie, V. — J. Chem. Phys. 30 (1959), 1210.
3. Thomas, E. and V. Laurie. — J. Chem. Phys. 44 (1966), 2602.
4. Townes, C., J. Mays, and B. Dailey. — Phys. Rev. 76 (1949), 137, 700.
6. Griffiths, J. and K. McAfee. — Proc. Chem. Soc., p. 456, 1961.
7. Childs, W. and L. Goodman. — Phys. Rev. 141 (1966), 15.
8. Hoeft, J. — Z. Naturf. 21a (1966), 1 240.
9. Das, T. and M. Pomerantz. — Phys. Rev. 123 (1961), 2070.
10. Wertheim, G. and R. Pound. — Phys. Rev. 102 (1956), 185.
11. [illegible]

Group V elements

Empirical formula	Structural formula	T, °K	ν_+, MHz	ν_-, MHz	e^2Qq, MHz	η, %	References, notes
			^{14}N				
BrCN	BrCN	77	2.5203	2.5109	3.3541	0.18	[1, 2]. cf. Br
		199	2.4963	2.4910	3.3249	0.106	
		273	2.4650	2.4650	3.2867	0	
		297.2	2.4650	2.4627	3.2851	0.046	
					3.83 (g)	...	
CClN	ClCN	77	2.428	2.400	3.219	1.57	[1, 3, 4]. cf. Cl
		—	—	—	3.63 (g)	...	
$CCoH_{12}N_5O_6$ + + $^1/_2H_2O$	$[Co(NH_3)_4CO_3]NO_3\cdot{}^1/_2H_2O$	Room	—	—	2.98	—	[5]. NMR 1. cf. Co
CHN	HCN	77	3.0223	3.0052	4.0183	0.85	[6—10, 33]
		195	2.9178	2.9178	3.8904	0	
		—	—	—	—4.714 (g)	...	
CHNO	HCNO	—	—	—	2.00	...	[11]. g
CHNS	HCNS	—	—	—	+1.20	...	[11, 12]. g
CH_2FNO	NH_2COF	—	—	—	—4.05	16.5	[13]. g
CH_2N_4	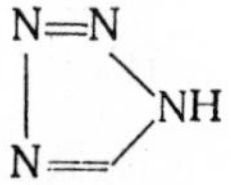	77	3.3023	...	...	...	[14]
			2.9782	...	...	...	
CH_3NO	$HCONH_2$	77	1.920	1.490	2.274	37.8	[15, 16]
		—	—	—	—2.7 (g)	...	
	$H_2C{=}NOH$	—	—	—	3.0	70	[17]. g
CH_3NO_2	CH_3NO_2	298	—	—	1.35	—	[18]. NMR 1.
CH_3NO_3	CH_3ONO_2	Room	—	—	0.7	—	[168]. NMR 1.
		—	—	—	<0.3	...	[19]. g

Empirical formula	Structural formula	T, °K	ν_+, MHz	ν_-, MHz	e^2Qq, MHz	η, %	References, notes
CH_3NSi	H_3SiCN	—	—	—	—4.7	...	[20]. g
CH_4N_2O	$CO(NH_2)_2$	77	2.9137	2.347	3.507	32.3	[21—26, 169, 170]
		199	2.8741	2.3260	3.4667	31.5	
		294	2.8264	2.2921	3.4123	31.3	
CD_4N_2O	$CO(ND_2)_2$	77	2.9410	...	...	...	[26]
CH_4N_2S	$CS(NH_2)_2$	77	2.6497	2.0327	3.1216	39.54	[22, 27—29]
			2.6291	2.0201	3.0996	39.30	
CH_5N	CH_3NH_2	77	3.3622	2.6117	3.9862	37.4	[30, 171, 172]
		—	—	—	—4.9 (g)	14 (g)	
$CH_{15}ClCoN_6O_8$ + + $^1/_2H_2O$	$[Co(NH_3)_5CN]\cdot[ClO_4]_2\cdot{}^1/_2H_2O$	298	—	—	2.67	—	[5]. NMR 1. cf. Co
ClN	ICN	199	2.5512	...	3.4016	0	[1, 2]. cf. I
		273	2.5451	...	3.3935	0	
		299.8	2.5424	...	3.3899	0	
		—	—	—	—3.40 (g)	0	
CKNS	KSCN	77	1.8406	1.8065	2.4314	2.81	[31]
CKNSe	KSeCN	77	2.1690	2.0980	2.8449	4.97	[31]
C_2Cl_3N	CCl_3CN	77	3.0444	3.0337	4.0521	0.53	[32, 33]. cf. Cl
C_2F_3N	CF_3CN	—	—	—	—4.70	...	[34]. g
C_2HCl_2N	Cl_2CHCN	77	2.9986	2.9895	3.947	3.4	[33]. cf. Cl
			2.9343	2.9173			
C_2H_2ClN	$ClCH_2CN$	77	3.0582	2.7832	3.8943	1.4	[33]. cf. Cl
C_2H_3N	CH_3CN						[6, 33, 35—40]
	Phase I	77	2.8078	2.7992	3.7380	0.46	
	Phase II	77	2.8108	2.7954	3.7375	0.82	
		—	—	—	4.40 (g)	...	
	CH_3NC	Room	—	—	0.27	—	[173]. NMR 1.

		--	—	—	0.483	...	[36, 38—40]. g
C_2H_3NO	CH_3NCO	—	—	—	2,3	...	[41, 42]. g
$C_2H_3NO_2$	$H_2C{=}CHNO_2$	Room —	— —	— —	1.25 0.883	— ...	[168]. NMR 1. [43]. g
C_2H_3NS	CH_3SCN	77 198 298 —	3.0524 3,0117 — —	2.2207 2.2065 — —	3.5154 3.4786 3,75 —4.04	47.32 46.29 — 53.4	[18, 42, 44, 174] NMR 1. g
	CH_3NCS	298	—	—	2.4	—	[18, 42]. NMR 1.
$C_2H_3N_3$	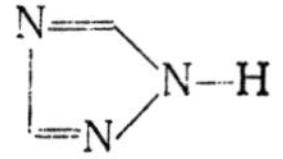	77	3.8759 2.5510 2.1648	2.4690 2.2790 ...	4.2301 3.220 ...	66.5 16.9 ...	[14, 45]
$C_2H_5NO_2$	C_2H_5ONO	298	—	—	3.8	—	[18]. NMR 1.
	$C_2H_5NO_2$	298	—	—	1.5	—	[18]. NMR 1..
$C_2H_5NO_3$	$C_2H_5ONO_2$	298	—	—	0.9	—	[18, 168]. NMR 1.
$C_2H_8N_2$	$H_2NCH_2CH_2NH_2$	77 90 205 252.2 275.7	3.3092 3,3884 3.3079 3.2928 3,2841	2.6846 2.6844 2.6597 2.6488 2.6433	3.996 4.048 3,978 3,961 3.952	31.25 34.78 33.13 32.52 32.43	[46, 47]. p.t.
C_2N_2	$N{\equiv}C{-}C{\equiv}N$	4.2 77	3.2381 3.225	3.1886 3.178	4.2845 4.269	2.3 2.2	[33, 48]
C_2N_2S	$S(CN)_2$	—	—	—	—3,48	30.1	[49]. g
$C_3Cl_3N_3$	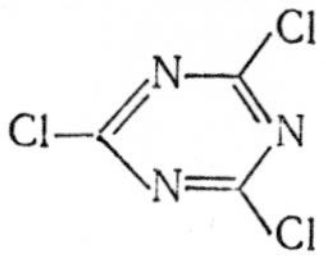	77	3.0799	3.0445	4.083	1,7	[50]. cf. Cl
C_3HN	$HC{\equiv}CCN$	—	—	—	—4.2	...	[51]. g

Empirical formula	Structural formula	T, °K	ν_+, MHz	ν_-, MHz	e^2Qq, MHz	η, %	References, notes
$C_3H_2N_2$	$CH_2(CN)_2$	77	3.0154	2.8670	3.9216	7.57	[33, 35, 52, 175]. p. t. at 140°, 260°, 295 °K
C_3H_3N	$H_2C{=}CHCN$	77	2.9063 2.8852	2.8085 2.8015	3.800	4.77	[33, 53]
		—	—	—	—4.21	...	g
$C_3H_3N_3$	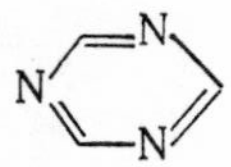	77			4.572	44.3	[54, 55]
$C_3H_3N_3O_3$	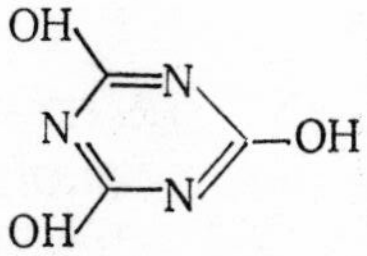	77	2.7915 2.7829	... ...	... ...	>14	[26]
$C_3H_4N_2$	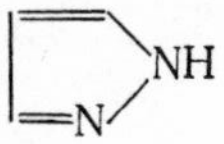	77	3.6649 3.6397	2.3653 2.3148	3.9949	65.7	[14, 45]
			2.1182 2.0737	... ...	... ...	... ...	
	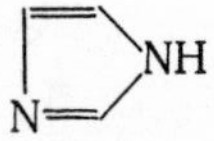	77	2.559 1.417	2.348 ...	3.271 ...	12.9 ...	[14, 45]
C_3H_5N	C_2H_5CN	77	2.8513	2.8121	3.7756	2.08	[33, 35, 56, 176]
		—	—	—	—4.14 (g)	...	
C_3H_5NO	CH_3OCH_2CN	77	3.1122	2.7552	3.9116	18.3	[33]
	C_2H_5NCO	298	—	—	3.3	—	[18]. NMR 1.
C_3H_5NS	C_2H_5SCN	77	3.1178	2.2677	3.5903	47.35	[44]

$C_3H_6N_6$	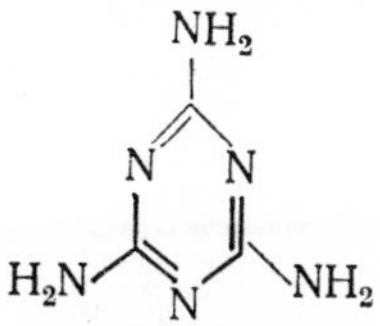	77	2.5865	...	...	...	[26]
C_3H_7NO	$(CH_3)_2NCHO$	298	—	—	4.25	—	[18]. NMR 1.
$C_3H_7NO_2$	$C_3H_7NO_2$	298	—	—	1.75	—	The same
	iso-$C_3H_7NO_2$	298	—	—	1,55	—	»
C_3H_9N	$N(CH_3)_3$	77 —	3.8954 —	 —	5.1939 5.47 (g)	0 ...	[57, 58, 172]
$C_4H_2Cl_2N_2$	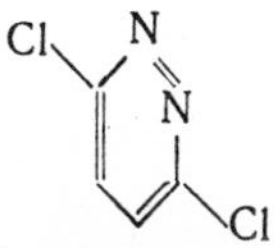	77	4.102	3.347	4.966	30.0	[59]
$C_4H_4N_2$	$NCCH_2CH_2CN$	77	2,9952 2.9625	2.8441 2.8169	3,8728	7.66	[33, 177]
	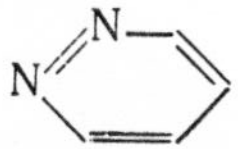	77	4.0118 3.9930	3.7843 3.7777	5.1889	8.53	[60]
	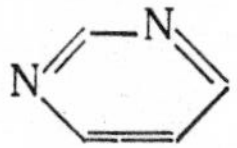	77	3.7579 3.7524	2.9004 2.8979	4.4362	38.6	[60]
		77	4.2945	2.9923	4.8578	53.6	[54, 55, 60]
$C_4H_4N_2S_2$	$NCSCH_2CH_2SCN$	77 198 299	3,0726 3.0576 3.0387	2.2446 2.2284 2.2120	3.5448 3.5240 3.5005	46.71 47,06 47.24	[44]
C_4H_5N	$H_2C{=}C(CH_3)CN$	77	2.9157	2.8303	3.8307	4.5	[33]

Empirical formula	Structural formula	T, °K	ν_+, MHz	ν_-, MHz	e^2Qq, MHz	η, %	References, notes
C_4H_5N	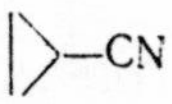	77	2.9297	2.6211	3.7005	16.7	[33]
	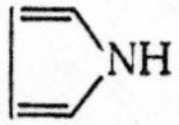	77	1.6851	1.4079	2.062	26.88	[14, 61, 178]
$C_4H_5NO_2$	CH_3OCOCH_2CN	77	2.9731	2.8075	3.8537	8.6	[33]
$C_4H_5N_3$	$NCCH_2NHCH_2CN$	77	2.9328	2.745	3.7852	9.9	[33]
$C_4H_6N_2$	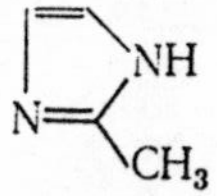	77	2.5416	...	...	...	[45]
C_4H_7N	C_3H_7CN	77	2.9355	2.8322	3.8451	5.4	[33, 177]
	iso-C_3H_7CN	77	2.8804	2.8296	3.8066	2.7	[27, 33, 62, 177]
		—	—	—	—3.905 (g)	6.41 (g)	
$C_4H_{10}Cl_3CoN_2$	trans-$[Co(en)_2Cl_2]Cl$	Room	—	—	2.53	—	[5]. NMR 1. cf. Co
$C_4H_{10}N_2$	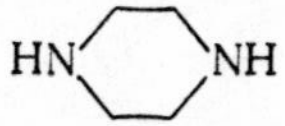	270	3.6036	3.0161	4.4131	26.6	[63, 179]
$C_5H_3Br_2N$	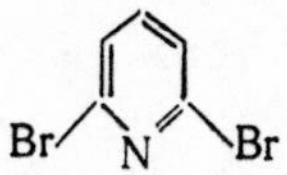	77	3.3661	3.1285	4.3297	11.9	[59]
$C_5H_3Cl_2N$	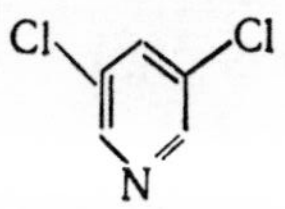	77	3.8974	3.0474	4.6299	36.7	[59]. cf. Cl
	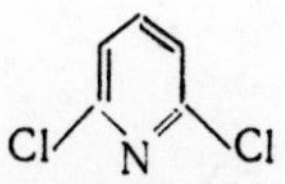	77	3.3052	3.0793	4.2563	10.6	[59, 180]. cf. Cl

C_5H_4ClN	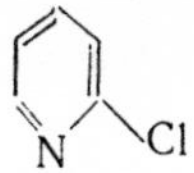	77	3.6268	3,0508	4.4517	25.9	[59, 180, 181]. cf. Cl
	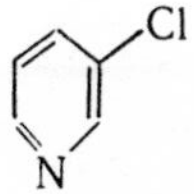	77	3,9126 3,9054	3,0365 3,0286 }	4.628	37.9	The same
	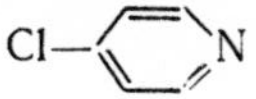	77	3.8409 3.8297	3.0105 3.0043 }	4.562	36.3	[59, 181]. cf. Cl
C_5H_5N	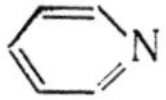	77 —	3.9143 3.9048 3.8938 3.8560 —	3.0016 2.9920 2.9872 2.9566 —	4.610 4.598 4.587 4.541 —4.88 (g)	41.5 39.7 39.5 37.4 40.5 (g)	[22, 27, 59, 64, 65]
$C_5H_6N_2$	$NC(CH_2)_3CN$	77	2.9231	2.7995	3.8150	6.5	[33, 177]
	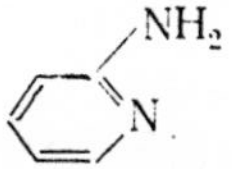	77	2.842 2.970	2.776 2.335	3.745 3.550	3.52 34.65	[27, 180, 182]
$C_5H_6N_2O$	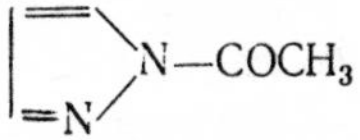	77	4.0954 2.5836	2.5828 1.5122	4.4524 2.7305	67.9 78.5	[45]
	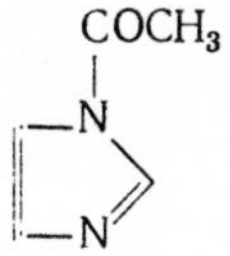	77	2.9840 1.6977	2.7964 1,5868	3.8535 2.1897	9.74 10.12	[45]
C_5H_7N	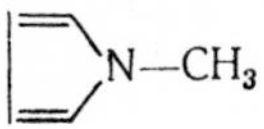	—	—	—	—2.16	...	[66]. g

Empirical formula	Structural formula	T, °K	ν_+, MHz	ν_-, MHz	e^2Qq, MHz	η, %	References, notes
$C_5H_8N_2$	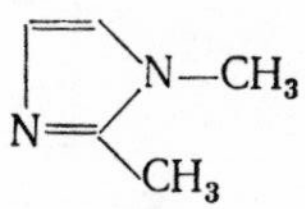	77	2.7348	2,5552	3,5267	10.2	[45]
C_5H_9N	C_4H_9CN	77	2.9118	2.8090	3.8138	5.4	[33, 177]
	$(CH_3)_2CHCH_2CN$	77	2,9075	2.8080	3.8103	5,2	[33]
	$(CH_3)_3CCN$	77	2,8956	2.8845	3.8534	0.6	[33]
$C_5H_9NO_2$	iso- C_5H_9ONO	298	—	—	3.7	—	[18]. NMR 1.
$C_5H_{11}N$	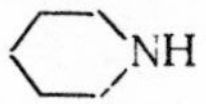	77 141 195	3.6668 3,6385 3.6085	2.9802 2.9803 2.9793	4.4313 4.4125 4.3919	31.0 29.8 28.7	[22, 178, 183]
$C_6CoK_3N_6$	$K_3[Co(CN)_6]$	77	2.7945 2.7892 2.7878	2.7099 2.7363 2.6888 }	3.684	3.0	[184]. cf. Co
		298	—	—	3.62	—	[5]. NMR 1.
$C_6H_4N_2$	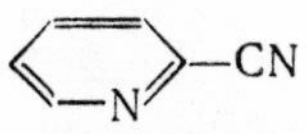	77	3.9476 3,0396	3.1298 2.8978	4.7182 3.9583	34.66 7.16	[32, 33, 59, 180, 181]
	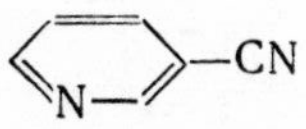	77	3.9007 3,0110	3.0504 2.8170	4.6338 3,8854	36.7 10.0	[33, 59, 180, 181]
	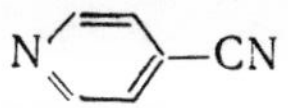	77	4.0997 2.9353	3.0576 2.9073	4.7715 3.8951	43.7 1.44	[32, 33, 59, 180, 181]
$C_6H_5NO_2$	$C_6H_5NO_2$	298	—	—	2.6	—	[18]. NMR 1.
C_6H_6BrN	4-$BrC_6H_4NH_2$	77 275	3.3417 3.2613	2.860 2.7764	4.135 4.025	23.1 24.1	[21, 23, 185]. cf. Br
C_6H_6ClN	4-$ClC_6H_4NH_2$	77 292	3.3382 3.2645	2.8365 2.7694	4.117 4.023	24.3 24.6	[21, 23, 185]. cf. Cl

C_6H_7N	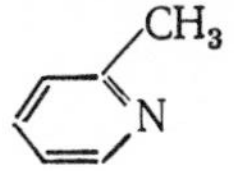	77	3.7135	2.9757	4.4594	33.1	[59, 180]
	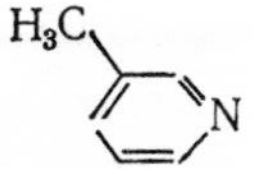	77	3,9205	3.0099	4.6203	39.42	[59, 180]
	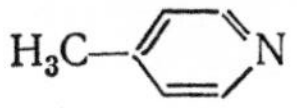	77 190 —	3.6884 3,6484 —	2.9326 2.9120 —	4.4140 4.3736 —4.82 (g)	34.24 33.7 58 (g)	[22, 27, 59, 64, 67]
$C_6H_8N_2$	$NC(CH_2)_4CN$	77	2.8615	2.8190	3.787	2.2	[33, 177]
	1,4-$(NH_2)_2C_6H_4$	77	3.2129 3.2109 3.1902	2.694 2.690 2.674	3.910	26.4	[21, 23, 185]
$C_6H_{11}N$	$C_5H_{11}CN$	77	2.8953	2,8038	3.7994	4.8	[33, 177]
	$(CH_3)_2CHCH_2CH_2CN$	77	2.9118	2.8332	3.8300	2.7	[33]
$C_6H_{12}N_2$	$N(CH_2CH_2)_3N$	77	3.6935	—	4.9247	0	[48]
$C_6H_{12}N_4$	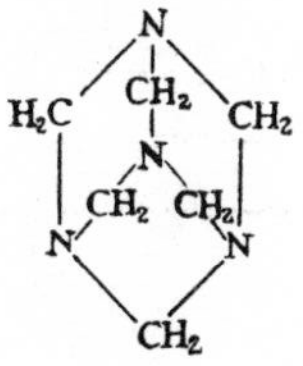	77 199 273 299.6	3.4076 3.3560 3.3199 3.3062	— — — —	4.543 4.475 4.426 4.408	0 0 0 0	[2, 68]
$C_6H_{14}N_2$	$H_2N(CH_2)_6NH_2$	77	3.394	2.687	4.054	34.88	[27]
$C_6H_{15}Cl_3CoN_3$	$[Co(en)_3]Cl_3$	Room	—	—	3.16	—	[5]. NMR 1.
C_7H_5N	C_6H_5CN	77	3.0133	2.8098	3.8854	10.73	[32]
C_7H_5NO	C_6H_5NCO	298	—	—	3.9	—	[18]. NMR 1..

Empirical formula	Structural formula	T, °K	ν_+, MHz	ν_-, MHz	e^2Qq, MHz	η, %	References, notes
C_7H_5NS	C_6H_5NCS	298	—	—	4.4	—	[18]. NMR 1.
$C_7H_7NO_2$	$2\text{-}CH_3C_6H_4NO_2$	298	—	—	2.1	—	[18]. NMR 1..
C_7H_9N	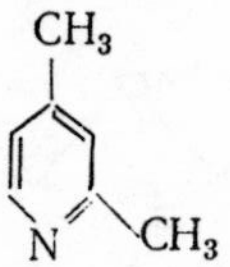	77	3.5492	2.9325	4.3211	28.54	[59]
	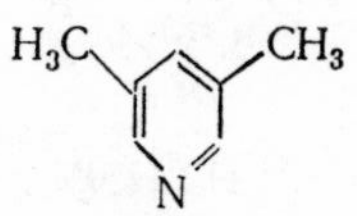	77	3.9513	3.0089	4 6401	40.62	[59]
$C_7H_9NO_2$	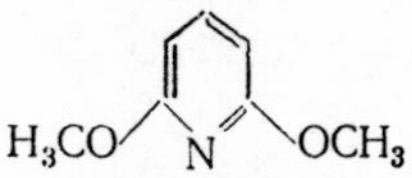	77	3.0022	2.7822	3.8563	11.4	[59]
$C_7H_{10}N_2$	$NC(CH_2)_5CN$	77	2.8814	2.8246	3.8040	2.98	[33, 177]
C_8H_7N	$C_6H_5CH_2CN$	77	2.9085	2.8811	3.8597	1.4	[33]
$C_8H_{11}N$	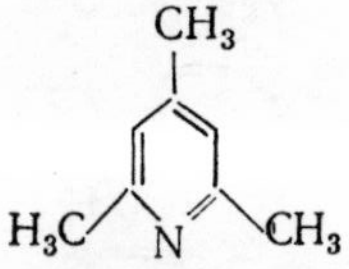	77	3.4448 3.4485	2.9353 2.9215 }	4.250	24.14	[59]
$C_8H_{12}N_2$	$NC(CH_2)_6CN$	77	3.6935		4.9247	0	[48]

	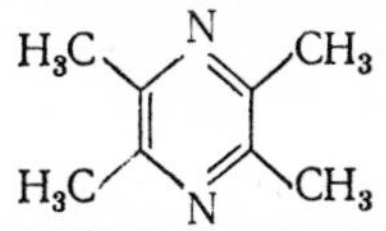	77			4.672	45.0	[54, 55]
$C_8H_{12}N_4$	$NCC(CH_3)_2N{=}NC(CH_3)_2CN$	77	3.0146	2.9468	3.9743	3.41	[33]
C_9H_9N	$C_6H_5CH_2CH_2CN$	77	2.8848	2.8267	3.8073	3.05	[33]
$C_{10}H_{16}N_2$	$NC(CH_2)_8CN$	77	2.8962	2.8063	3.8017	4.7	[33, 177]
$C_{12}H_8N_2$	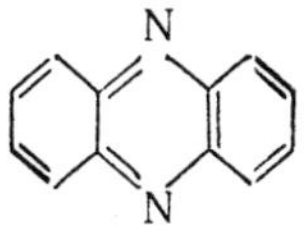	77			4.324	37.2	[54, 55]
$C_{12}H_9NS$	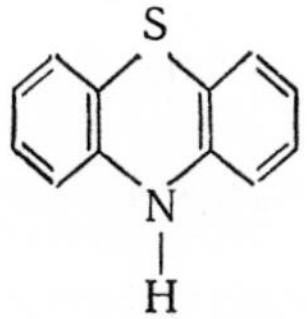	77			4.240	22.6	[69]
$Cl_2CoH_{12}N_5O_2$	$[Co(NH_3)_4NO_2]Cl_2$	Room	—	—	3.21	—	[5]. NMR 1. cf. Co
$Cl_3CoH_{16}N_4O_{14}$	$[Co(NH_3)_4(H_2O)_2]\cdot(ClO_4)_3$	»	—	—	2.85	—	The same
$Cl_3CoH_{17}N_5O_{13}$	$[Co(NH_3)_5H_2O]\,[ClO_4]_3$	»	—	—	2.70	—	[5]. NMR 1. cf. Co
$Cl_3CoH_{18}N_6$	$[Co(NH_3)_6]Cl_3$	»	—	—	3.62	—	»
$CoH_6N_6NaO_8$	$Na[Co(NH_3)_2(NO_2)_4]$	»	— —	— —	2.09 4.25	— —	»

Empirical formula	Structural formula	T, °K	ν_+, MHz	ν_-, MHz	e^2Qq, MHz	η, %	References, notes
$CoN_6Na_3O_{12}$	$Na_3[Co(NO_2)_6]$	Room	—	—	3.2	—	[5]. NMR 1.
			—	—	3.9	—	
FNO	NOF	—	—	—	—5.0	...	[70]. g
F_3N	NF_3	77	—	—	7.0681	11.2	[71, 72]
		—	—	—	—7.07 (g)	...	
F_3NS	NSF_3	—	—	—	1,19	...	[73]. g
HNO_2	HNO_2	—	—	—	—5.39	29	[74]. g
HNO_3	HNO_3	—	—	—	+0.93	...	[75]. g
					—0.82		
HN_3	HN_3	—	—	—	+4.85	...	[76, 77]. g
					<0.7		
					—1.35	...	
H_3N	NH_3	77	2.3706	—	3.1607	0	[58, 78—83, 172]
		—	—	—	4.084 (g)	0	
D_3N	ND_3	77	2.4231	—	3.2308	0	[58, 78—83, 172]
		—	—	—	4.20 (g)	0	
H_4N_2	NH_2NH_2	77	4.6667	2.6717	4.892	81.6	[84, 85, 186]
			4.5080	2.6170	4,750	79.6	
		100	4.6606	2.6692	4.886	81.5	
			4.5037	2.6157	4.746	79.6	
		120	4.6528	2.6660	4.879	81.4	
			4.4969	2.6137	4.740	79.5	
		140	4.6433	2,6625	4.870	81.3	
			4.4880	2.6108	4.732	79.3	

		160	4,6312 4,4780	2,6575 2,6070	4.859 4,723	81.2 79,2	
		180	4,6179 4.4648	2,6515 2,6018	4,846 4,711	81,15 79,1	
		200	4,6022 4,3616	2,6440 2.5060	4,831 4.578	81.1 81.06	
		220	4.5806 4.4371	2,6306 2,5898	4.807 4.685	81,12 78.9	
		240	4.5681 4.4216	2.6270 2,5825	4.857 4.669	79,9 78,8	
		244	4.5645 4.4185	2,6250 2,5810	4.793 4.666	80.9 78.7	
H_6NO_4P	$[NH_4]H_2PO_4$	Room	—	—	0.0246	0	[86, 187]. NMR cr.
$D_4H_2NO_4P$	$[ND_4]H_2PO_4$	»	—	—	0,0273	0	cf. D
KN_3	KN_3	»	— —	— —	1.79 1.028	4 3	[87]. NMR cr.
$NNaO_2$	$NaNO_2$	77 437	4,929	3.757	5,792 4,923	40.5 34,7	[88, 188, 189]
$NNaO_3$	$NaNO_3$	Room	—	—	0,720	. . .	[89]. NMR cr. cf. Na
NO	NO	—	—	—	—1.9	. . .	[90—93]. g
N_2	N_2	4,2	3.4873		4.650	0	[94, 95]
	N_2 (in hydroquinone clathrate)	4.2	3.5802 3.5818 3.5850 3.5868 3.5884 3.5923 3,5940	. . .	. . .	. . .	[96]. $[N_2]$ =0.2—0.8%
N_2O	N_2O	—	— —	— —	—0.792 —0,238	. . . . . .	[97—100]. g

Empirical formula	Structural formula	T, °K	ν, MHz	References, notes
		^{75}As		
$AgAsH_2O_4$	AgH_2AsO_4	77	39.19	*[101, 102]
			38.11	
		196	39.22	
			37.29	
		300	39.26	
			37.83	
Ag_3AsS_3	Ag_3AsS_3	77	67.34	[103]
$AsBr_3$	$AsBr_3$	83	63.569	[104—107]. cf. Br
		202	62,417	
		248	61.946	
		291	61.315	
$AsCl_3$	$AsCl_3$	77	78.950	*[105—109]. cf. Cl
		213	76.873	
		243	76.207	
		$e^2Qq = -173$ MHz		g
$AsCsH_2O_4$	CsH_2AsO_4	4.2	37.308	*[101, 102, 110, 111]
		77	37,236	
$AsCsD_2O_4$	CsD_2AsO_4	77	42.98	*[102,111]
		196	42.95	
AsCuS	CuAsS	77	69.23	[112]
		273	68.31	
		298	68.15	
		373	67.76	
AsF_3	AsF_3	$e^2Qq = -236.23$ MHz		[113,114]. g
$AsHNa_2O_4$	Na_2HAsO_4	77	38.37	*[101, 102]
$AsHNa_2O_4 + 7H_2O$	$Na_2HAsO_4 \cdot 7H_2O$	77	42.110	*[101, 102, 110, 115]
		196	42.752	
		300	43.70	
AsH_2KO_4	KH_2AsO_4	77	35.07	*[101, 102, 110, 111, 116]
		Room	3.53	
AsD_2KO_4	KD_2AsO_4	77	42.62	*[102, 111]
AsH_2LiO_4	LiH_2AsO_4	77	47.24	*[101, 102]
AsD_2LiO_4	LiD_2AsO_4	77	48.60	*[102]
AsH_2NaO_4	NaH_2AsO_4	77	54.50	*[101, 102]
		196	54.50	
		300	54.54	
AsD_2NaO_4	NaD_2AsO_4	77	55,78	*[102]
AsH_2O_4Rb	RbH_2AsO_4	77	35.70	*[101, 102, 110, 112]
AsD_2O_4Rb	RbD_2AsO_4	77	42,68	*[102, 111]
AsH_3	AsH_3	$e^2Qq = -160.1$ MHz		[117, 118]. g
AsH_2D	AsH_2D	$e^2Qq = -164$ MHz		[119]. g
AsD_3	AsD_3	$e^2Qq = -165.6$ MHz		[117, 118]. g
AsH_6NO_4	$[NH_4]H_2AsO_4$	77	35.18	*[101, 102, 110, 111]
		196	33,70	
		202	32.63	
AsD_6NO_4	$[ND_4]D_2AsO_4$	77	38.71	*[102, 111]
			39.45	
		196	38.78	
			39.45	
AsIS	AsSI	77	66.552	*[190, 191]. cf. I
AsISe	AsSeI	77	60,038	The same

Empirical formula	Structural formula	T, °K	ν, MHz	References, notes
AsI_3	AsI_3 Phase I	77 213 302 373	29.338 31.079 32.198 33.057	[105—108]. cf. I p.t. at 373°K
	Phase II	77 301	116.860 116.193	
$AsI_3 + 24S$	$AsI_3 \cdot 3S_8$	77 217 303	49.501 48.398 47.545	[108]. cf. I
AsS	$AsS(As_4S_4)$	77	89.22 91.45 92.05 92.96	*[121]
		300	88.38 90.65 91.02 91.72	
AsS_2Tl	$TlAsS_2$	77	81.144 84.120	*[190, 191]
$AsSe_2Tl$	$TlAsSe_2$	77	64.00 65.58	*[190, 191]
		300	61.996 62.96 64.02	
As_2O_3	$As_2O_3(As_4O_6)$ arsenolite	77 195 292	116.781 116.535 116.210	[122, 123]
	claudetite	77 195 305	110.909 109.898 108.601	[122, 124]
As_2S_3	As_2S_3	77	70.336 72.804	*[125]
		300	69.547 71.942	
As_2Se_2	As_2Se_2	77	79.80 81.63 82.02 82.38	*[120, 191]
		300	78.99 80.93 81.10 81.32	
As_2Se_3	As_2Se_3	77	55.96 60.12	*[190, 191]
CH_3AsF_2	$As(CH_3)F_2$	$e^2Qq = -$ 220 MHz		[126, 127]. g
C_3H_9As	$As(CH_3)_3$	$e^2Qq = -$ 203.15 MHz		[128]. g
$C_4H_{14}AsNO_3$	$[(CH_3)_4N]H_2AsO_4$	77 196	40.75 40.69	*[101, 102]
$C_4H_{12}D_2AsNO_4$	$[(CH_3)_4N]D_2AsO_4$	77	48.96 49.40	*[102]
$C_{18}H_{15}As$	$As(C_6H_5)_3$	77	98.500 98.900	*[105, 107]
		292,5	96.783 97.320	
$C_{18}H_{15}AsCl_2$	$(C_6H_5)_3AsCl_2$	77	119.86	*[193]. cf. Cl (supplement)

Empirical formula	Structural formula	Isotope	T, °K	ν, MHz 1/2—3/2	ν, MHz 3/2—5/2	ν, MHz 5/2—7/2	e^2Qq, MHz	η, %	References, notes
				Sb					
Ag_3S_3Sb	Ag_3SbS_3	121	77	49.84	99.70	—	332.3	0	[103]
		123	77	30.28	60.52	90.78	423.6	0	
Ag_5S_4Sb	Ag_5SbS_4	121	77	54.942	108.280	—	361.8	10.7	[128a]
			300	53.415	...	...	...	...	
		123	77	34.265	65.462	98.700	461.1	10.7	
			300	...	64.090	...	...	...	
BrSSb	SbSBr	121	77	51.44	100.64	—	336.4	13.9	*[129, 192]. cf. Br
		123	77	32.52	60.50	91.62	427.6	13.9	
Br_3Sb	$SbBr_3$ Phase I	121	77	50.273	99.215	—	343.95	10	[108, 130—133]. cf. Br
			273.4	48.659	96.441	—	321.93	8.39	
			289	48.480	...	—	320.80	8.3	
			304.4	48.356	95.872	—	320.01	8.24	
		123	77	31.186	60.107	...	422.99	9.5	
			273.4	30.048	58.399	...	410.39	8.36	
			289	29.933	58.106	...	408.96	8.3	
			304.4	29.843	58.057	...	407.93	8.22	
	Phase II	121	304	49.061	94.160	—	316.02	18.1	
		123	304	31.990	56.544	85.989	402.689	18.2	
$Br_3Sb + {}^1/_2C_6H_6$	$2SbBr_3 \cdot C_6H_6$	121	77	50.717	...	—	323.395	20.4	[131, 133—136]. cf. Br
			292	49.161	...	—	316.62	18.1	
		123	77	33.568	...	...	412.245	20.4	
			292	32.062	...	...	403.66	18.1	
$Br_3Sb + C_7H_8$	$SbBr_3 \cdot C_6H_5CH_3$	121	77	49.500	...	—	...	...	[134—136]
$Br_3Sb + C_7H_8O$	$SbBr_3 \cdot C_6H_5OCH_3$	121	77	50.194	...	—	330.44	10.8	[131, 132, 134, 135, 136]. cf. Br
		123	77	31.342	...	...	421.32	10.8	
$Br_3Sb + C_8H_{10}$	$SbBr_3 \cdot C_6H_5C_2H_5$	121	77	48.992	...	—	316.21	17.5	[131—136]. cf. Br
			295	47.256	...	—	...	...	

		123	77	31.817	...	...	403.07	17.5	
			292	30.126	...	...	...	...	
	$SbBr_3 \cdot 1,2\text{-}(CH_3)_2C_6H_4$	121	77	46.323	...	—	306.47	8,4	[131, 132, 134—136]. cf. Br
		123	77	28,627	...	...	390.89	8.4	
	$SbBr_3 \cdot 1,3\text{-}(CH_3)_2C_6H_4$	121	77	46.574	...	—	...	...	[131, 134—136]
	$SbBr_3 \cdot 1,4\text{-}(CH_3)_2C_6H_4$	121	77	49.963	...	—	318.015	20.9	[131, 132, 136]. cf. Br
				48.798	...	—	321,711	11,1	
			292	48.405	...	—	311.16	18.6	
				48.039	...	—	316.95	9,8	
		123	77	33.238	...	...	405.386	20.9	
				30.493	...	...	410.098	11,1	
			292	31,680	...	...	396.99	18.6	
				29.847	...	...	415.67	9.8	
$Br_3Sb + C_8H_{10}O$	$SbBr_3 \cdot C_6H_5OC_2H_5$	121	77	47,610	...	—	310.746	14	[131, 134—136]. cf. Br
		123	77	30.225	...	...	396.120	14	
$Br_3Sb + {}^1/_2C_9H_{12}$	$2SbBr_3 \cdot \cdot 1,3,5\text{-}(CH_3)_3C_6H_3$	121	77	51.279	...	—	...	...	[131, 134, 136]
				48.459	...	—	...	...	
$Br_3Sb + {}^1/_2C_{10}H_8$	$2SbBr_3 \cdot C_{10}H_8$	121	77	50.301	...	—	334,72	4.1	[131, 132, 134—136]. cf. Br
			292	49,017	...	—	326.38	3.3	
		123	77	30,654	...	...	426.686	4.1	
			292	29.835	...	...	416,04	3.3	
$Br_3Sb + {}^1/_2C_{12}H_{10}$	$2SbBr_3 \cdot C_6H_5C_6H_5$	121	77	47,207	...	—	314.41	3.0	[131, 132, 134, 136]. cf. Br
			292	45.906	...	—	302.40	0	
		123	77	28.717	...	...	400.79	3.0	
			292	27.856	...	...	388,91	0	
$Br_3Sb + {}^1/_2C_{13}H_{12}$	$2SbBr_3 \cdot (C_6H_5)_2CH_2$	121	77	49.434	...	—	323.91	12.6	[131, 132]. cf. Br
			292	46,503	...	—	308.68	6.3	
		123	77	31.140	...	...	412.90	12,6	
			292	28.506	...	...	393,45	6.3	
$C_3H_9Br_2Sb$	$(CH_3)_3SbBr_2$	121	...	...	...	...	631.127	...	[137]. cf. Br
$C_3H_9Cl_2Sb$	$(CH_3)_3SbCl_2$	121	...	...	...	...	663.093	...	[137[. cf. Cl
$C_3H_9I_2Sb$	$(CH_3)_3SbI_2$	121	...	...	...	...	550.650	...	[137]. cf. I

Structural formula	Empirical formula	Isotope	T, °K	ν, MHz $^{1}/_{2}$—$^{3}/_{2}$	ν, MHz $^{3}/_{2}$—$^{5}/_{2}$	ν, MHz $^{5}/_{2}$—$^{7}/_{2}$	e^2Qq, MHz	η, %	References, notes
$C_6H_6Cl_5Sb$	(trans-ClCH=CH)$_3$·$SbCl_2$	121	77	84,48	...	—	562,57	3.3	*[193]. cf. Cl
		123	77	51,40	102.36	...	716,94	3,3	
$C_9H_{21}O_3Sb$	$Sb(OC_3H_7)_3$	121	77	85.804	167.570	—	560.66	13,8	[138]
		123	77	...	101,111	...	714.7	13,8	
$C_{12}H_{27}Br_2Sb$	$(C_4H_9)_3SbBr_2$	121	77	101,84	...	—	676,74	5.6	*[193]. cf. Br
		123	77	62,28	...	...	862,66	5,6	
	(iso-$C_4H_9)_3SbBr_2$	121	77	100.28	...	—	666.64	5.2	*[193]. cf. Br
		123	77	60.84	120.36	...	843,79	5.3	
$C_{18}H_{15}Br_2Sb$	$(C_6H_5)_3SbBr_2$	121	77	83.68	165.93	—	553.83	8.2	*[193]. cf. Br
				89.22	178.43	—	594.00	3.6	
		123	77	51.68	100.52	...	706.38	8,2	
				54.28	108.16	...	757.49	3.6	
$C_{18}H_{15}Cl_2Sb$	$(C_6H_5)_3SbCl_2$	121	77	90.36	178.26	—	595,40	10,35	*[193]
		123	77	56.24	107.80	...	758.97	10,35	
$C_{18}H_{15}Sb$	$Sb(C_6H_5)_3$	121	77	76.912	152,471	—	509.00	8,0	[138, 139]
				78,274	156.200	—	520,84	2,6	
			294	75,550	150.040	—	500.68	7.0	
				76.659	153,020	—	510,37	2,6	
		123	77	47.545	92.267	138.859	648,56	8,3	
				47,841	94.741	142.174	664,18	1,8	
			294	...	90.950	...	635,20	7.0	
				46.65	92.915	...	650,73	2.2	
$C_{24}H_{45}B_{30}Sb$	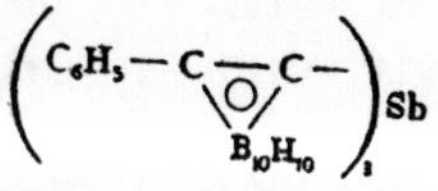 tri-(C-phenyl-o-boranyl)-antimony	121	77	72.26	144.31	—	481,28	3,4	*
		123	77	43.98	87.55	...	613,26	3.4	

Cl_3Sb	$SbCl_3$	121	77	59.724	...	—	383.6	18.8	*[130, 131, 138, 140—144]. cf. Cl
			83	59.709	114.300	—	383.69	18.9	
			273.8	58.374	112.875	—	378.27	16.36	
			292.4	58,193	112.644	—	377.3	16.2	
			304,1	58.078	112.498	—	376.90	15.92	
		123	77	39.118	68.641	...	488.8	18.8	
			83	39,094	68.648	104.460	489.34	18.7	
			273,8	37.626	67.925	...	482.20	16.35	
			292,4	37.445	67.802	102.828	480.6	16.2	
			304.1	37,333	67.726	...	480.45	15.92	
$Cl_3Sb + C_6H_4N_2O_4$	$SbCl_3 \cdot 1,3\text{-}(NO_2)_2C_6H_4$	121	77	59,532	...	—	395.44	5.7	[131]
		123	77	36,421	...	...	504.08	5,7	
$Cl_3Sb + C_6H_5Cl$	$SbCl_3 \cdot C_6H_5Cl$	121	77	60.952	...	—	...	...	[131, 134, 136]. cf. Cl
$Cl_3Sb + {}^1/_2C_6H_6$	$2SbCl_3 \cdot C_6H_6$	121	77	59,565	...	—	392.577	10,3	[131, 132, 134—136, 144]. cf. Cl
				59,955	...	—	388.849	16.1	
			292	58,720	...	—	382.192	15	
		123	77	37,096	...	...	500.448	10.3	
				38.554	...	...	495.319	16.1	
			292	37,541	...	...	487.410	15	
$Cl_3Sb + {}^1/_2C_6H_6O$	$2SbCl_3 \cdot C_6H_5OH$	121	77	61.988	...	—	394,20	21.3	[132, 134—136]. cf. Cl
				60,870	...	—	385.94	22.0	
				58.960	...	—	391.21	6.6	
				57.967	...	—	382.21	10.1	
			196	58,343	...	—	...	...	
				57,190	...	—	...	...	
			208	60.699	...	—	...	...	
				59,584	...	—	...	...	
			293	59,500	...	—	...	...	
				58,487	...	—	...	...	
		123	77	41,328	...	...	502.25	21.3	
				40.803	...	...	491.90	22.0	
				36,183	...	...	498.73	6.6	
				36,069	...	...	487.25	10.1	

Empirical formula	Structural formula	Isotope	T, °K	ν, MHz 1/2—3/2	ν, MHz 3/2—5/2	ν, MHz 5/2—7/2	e^2Qq, MHz	η, %	References, notes
$Cl_3Sb + {}^1/_2C_7H_8$	$2SbCl_3 \cdot C_6H_5CH_3$	121	77	59.998	...		399.99	0	[131, 134—136]
				59.307	...		395.38	0	
				58.640	...		390.93	0	
		123	77	36.033	...		509.88	0	
				35.804	...		504.01	0	
				35.405	...		498.34	0	
$Cl_3Sb + C_7H_8$	$SbCl_3 \cdot C_6H_5CH_3$	121	77	57.312	...	—	371.6	16.1	[131, 134—136]
		123	77	36.860	...	...	473.7	16.1	
$Cl_3Sb + {}^1/_2C_7H_8O$	$2SbCl_3 \cdot C_6H_5OCH_3$	121	77	63.114	...	—	411.914	14.1	[131, 132, 134—136]. cf. Cl
				61.200	...	—	401.317	12.4	
		123	77	40.110	72,432	...	525.022	14.1	
				38.521	...	...	511.548	12.4	
$Cl_3Sb + C_7H_8O$	$SbCl_3 \cdot C_6H_5OCH_3$	121	77	58.457	...	—	385.190	10.4	The same
		123	77	36.426	...	...	491.067	10.4	
$Cl_3Sb + C_8H_8O$	$SbCl_3 \cdot C_6H_5COCH_3$	121	77	62.042	...	—	392.16	22.7	[131, 132, 134, 136]
			292	59.310	...	—	379.04	20.1	
		123	77	41.801	...	...	499.81	22.7	
			292	39.230	...	...	483.29	20.1	
$Cl_3Sb + {}^1/_2C_8H_{10}$	$2SbCl_3 \cdot C_6H_5C_2H_5$	121	77	60.146	...	—	...	...	[131, 134—136]
				58.761	...	—	...	...	
$Cl_3Sb + C_8H_{10}$	$SbCl_3 \cdot C_6H_5C_2H_5$	121	77	58.688	113.158	—	379.41	17.0	[131, 132, 134—136]. cf. Cl
				58.89	...	—	...	...	
		123	77	37.988	...	...	483.70	17.0	
$Cl_3Sb + {}^1/_2C_8H_{10}$	$2SbCl_3 \cdot 1,2\text{-}(CH_3)_2C_6H_4$	121	77	58.887	...	—	390.8	6.4	[131, 132, 134—136]
		123	77	36.120	...	...	498.2	6.4	
$Cl_3Sb + C_8H_{10}$	$SbCl_3 \cdot 1,2\text{-}(CH_3)_2C_6H_4$	121	77	57.074	...	—	...	...	The same
$Cl_3Sb + {}^1/_2C_8H_{10}$	$2SbCl_3 \cdot 1,3\text{-}(CH_3)_2C_6H_4$	121	77	62.928	...	—	...	...	[131, 134—136]
				58.503	...	—	...	...	

$Cl_3Sb + C_8H_{10}$	$SbCl_3 \cdot 1,3\text{-}(CH_3)_2C_6H_4$	121	77	58.450	...	—	383.4	12.2	[131, 134—136]
		123	77	36.745	...	...	488,8	12.2	
$Cl_3Sb + {}^1/_2C_8H_{10}$	$2SbCl_3 \cdot 1,4\text{-}(CH_3)_2C_6H_4$	121	77	58.262	...	—	375.4	18	[131]. cf. Cl
			295	57,281	...	—	368.0	18.7	
		123	77	37.980	...	...	473.6	18	
			295	37.502	...	...	467.4	18.7	
$Cl_3Sb + C_8H_{10}$	$SbCl_3 \cdot 1,4\text{-}(CH_3)_2C_6H_4$	121	77	60.190	...	—	391.14	15.4	[131, 132, 134—136]. cf. Cl
			292	58.851	...	—	383.98	14.2	
		123	77	38.534	...	...	498.6	15.4	
			292	37.421	...	...	489.4	14.2	
$Cl_3Sb + C_8H_{10}O$	$SbCl_3 \cdot C_6H_5OC_2H_5$	121	77	56.063	...	—	368.29	11.7	The same
			196	55,284	...	—	364.44	10.2	
			292	54,621	...	—	360.82	9,2	
		123	77	35.167	...	...	469.59	11,7	
			196	34.410	...	...	464.53	10.2	
			292	33,842	...	...	458.88	9,2	
$Cl_3Sb + C_9H_{12}$	$SbCl_3 \cdot 1,3,5\text{-}(CH_3)_3C_6H_3$	121	77	56.374	...	—	...	...	[131, 135]
$Cl_3Sb + {}^1/_2C_{10}H_8$	$2SbCl_3 \cdot C_{10}H_8$	121	77	59.427	...	—	396.2	0	[131, 132, 134—136]. cf. Cl
			292	58,051	...	—	386,8	0	
		123	77	36,056	72,161	...	505,0	0	
			292	35,272	...	...	493.1	0	
$Cl_3Sb + {}^1/_2C_{12}H_{10}$	$2SbCl_3 \cdot C_6H_5C_6H_5$	121	77	58,491	...	—	385.8	9.8	[131, 134—136]. cf. Cl
				55,578	...	—	363.2	13.7	
			292	57,051	...	—	376.6	9,6	
				55.150	...	—	361.8	12.2	
		123	77	36,356	...	...	492.1	9.8	
				35.253	...	...	463.1	13,7	
			292	35.411	...	...	479.98	9.6	
				34.674	...	...	461.2	12.2	
$Cl_3Sb + {}^1/_2C_{13}H_{12}$	$2SbCl_3 \cdot (C_6H_5)_2CH_2$	121	77	57.510	...	—	377.2	12.2	[131, 132]. cf. Cl
		123	77	35.706	68,700	...	480.9	12.2	
$Cl_3Sb + C_{14}H_{10}$	$SbCl_3 \cdot C_{14}H_{10}$	121	77	59,350	...	—	384.7	16.1	The same
			295	57.678	...	—	376.9	13.7	
		123	77	38.160	...	...	490.4	16.1	
			295	36,564	...	...	480.3	13,7	

Empirical formula	Structural formula	Isotope	T, °K	ν, MHz $^1/_2$—$^3/_2$	ν, MHz $^3/_2$—$^5/_2$	ν, MHz $^5/_2$—$^7/_2$	e^2Qq, MHz	η, %	References, notes
Cl_5Sb	$SbCl_5$	121	210	12.70	25.40	—	84.67	0	[145]. cf. Cl
			228	...	25.40	—	...	...	
			249	12.67	25.39	—	84.63	0	
		123	210	7.72	15.43	23.15	97,53	0	
			228	...	...	23.14	...	...	
			249	7.72	15,43	23.11	97.48	0	
$Cl_5Sb + C_2H_3N$	$SbCl_5 \cdot CH_3CN$	121	77	32.47	64.87	—	216.27	3.0	[145]. cf. Cl
			195	32.25	64.36	—	214.61	4.0	
			300	32.02	63.93	—	213.16	4.0	
		123	77	19.70	39.43	59,06	275.68	3.0	
			195	19.57	39.10	58.63	273.56	4.0	
			300	19.46	38.63	58.20	271.71	4,0	
$Cl_5Sb + Cl_3OP$	$SbCl_5 \cdot POCl_3$	121	77	34.05	58.37	—	199.53	37.0	[145]. cf. Cl
			195	33.79	59.35	—	202,06	33.0	
		123	77	25.30	34.48	53.76	254.34	37.0	
			195	24.69	35.17	54.65	257.57	33.0	
$CuPbS_3Sb$	$PbCuSbS_3$	121	77	56.43	112.86	—	376.2	0.2	[146]
				57.44	107.85	—	363	22.8	
		123	77	34.25	68.51	102.76	479.6	0.2	
				38.68	64.49	98.69	463	22.8	
F_3Sb	SbF_3	121	...	160.96	...	—	1073	0.3	[147]
		123	...	97.62	...	...	1367	0.3	
H_3Sb	SbH_3	121	—	—	—	—	458.7	...	[148]. g
		123	—	—	—	—	586.0	...	
H_2DSb	SbH_2D	121	—	—	—	—	455	...	[119]. g
		123	—	—	—	—	575	...	
D_3Sb	SbD_3	121	—	—	—	—	465,4	...	[148]. g
		123	—	—	—	—	592.8	...	

ISSb	SbSI	121	77	51.50	92.00	—	312.34	31.0	*[129, 149, 192]. cf. I
		123	77	36.69	54.50	84.51	397.75	30.9	
I_3Sb	SbI_3	121	77	12.7	25.40	—	84.67	0.3	[108, 138, 147]. cf. I
		123	77	...	15.40	23.10	107.8	0.3	
$I_3Sb + 24S$	$SbI_3 \cdot 3S_8$	123	250	37.77	...	...	251.8	0	[108]. cf. I
			297	37.461	...	...	249.7	0	
O_3Sb_2	Sb_2O_3 senarmonite	121	77	83.21	166.45	—	554.8	0	[138, 139, 142]
			296	82.66	165.33	—	551.10	0	
		123	77	50.511	101.01	151.89	707.1	0	
			296	50.195	100.41	150.54	702.7	0	
	valentinite	121	77	91.97	158.52	—	541.4	36	[147]
		123	77	68.30	93.49	145.91	690.0	36	
S_3Sb_2	Sb_2S_3 (stibnite)	121	77	47.71	95.41	—	318.0	0.9	[130, 150]
				42.98	73.29	—	250.8	37.5	
			300	44.35	88.69	—	295.6	1.0	
				40.99	69.59	—	238.3	38.1	
		123	77	28.96	57.90	86.86	405.4	0.9	
				32.27	43.13	67.45	319.4	37.6	
			300	26.93	53.85	80.86	377.0	0.7	
				30.24	40.99	64.18	303.4	37.0	
		121	77	42.98	72.96	—	249.76	38.2	*[194]
				47.58	95.13	—	317.14	1.4	
		123	77	32.03	42.87	67.16	316.83	38.2	
				28.88	57.79	86.70	404.42	1.4	
Sb	Sb (met.)	121	4.2	11.5289	23.0611	—	76.85	0	[151]
		123	4.2	6.9996	13.9997	21.0000	97.98	0	
Sb_2Se_3	Sb_2Se_3	121	77	34.99	57.75	—	198.87	41.7	*[194]
				40.92	81.84	—	272.75	0.8	
		123	77	27.01	33.96	53.35	253.51	41.7	
				24.85	49.68	74.52	347.80	0.8	

Empirical formula	Structural formula	T, °K	ν, MHz 1/2—3/2	3/2—5/2	5/2—7/2	7/2—9/2	e^2Qq, MHz	η, %	References, notes
			^{209}Bi						
Bi	Bi (met.)	4,2	—	—	—	—	58.5	...	[152, 153]. NMR cr.
$BiBr_3$	$BiBr_3$	300	...	26.756	39.884	...	340.5	55.3	[154]
$BiCl_3$	$BiCl_3$	83	33.726	25.905	37.900	52.704	325.5	58.3	[155]. cf. Cl
		299	31.876	25.131	37,340	51.740	318.8	55.5	
$BiCl_3 + C_2H_3N$	$BiCl_3 \cdot CH_3CN$	77	...	28.78	45.30	61.07	373	30	*[195]
			...	29.39	45,96	61.71	372	25	
			...	29.75	46.24	62.17	373	21	
$BiCl_3 + C_3H_6O$	$BiCl_3 \cdot (CH_3)_2CO$	77	...	29,25	45.90	61.70	372	27	*[195]
$BiCl_3 + C_5H_5N$	$BiCl_3 \cdot Py$	77	33.81	25.98	37.88	52.68	326	58	*[195]
		295	31.93	25.07	37.38	51.78	318	56	
$BiCl_3 + C_7H_8O$	$BiCl_3 \cdot C_6H_5OCH_3$	77	33.85	25.90	37.88	52.75	325	59	*[195]
			25.08	28.24	44.55	60.10	363	32	
$BiCl_3 + C_8H_8O$	$BiCl_3 \cdot C_6H_5COCH_3$	77	40.47	31.23	45.81	63,71	393	58	*[195]. cf. Cl
		295	40.00	30.32	43.84	61.10	378	60	
BiIS	BiSI	77	25.06	32.16	50.30	67.65	408	26.1	*[129]. cf. I
Bi_2O_3	α-Bi_2O_3 (bismite)	77	25.18	45.77	69.47	92.71	556.6	10.0	*[147, 194]
			39.34	37.21	58.33	79.44	483.2	41.3	
	γ-Bi_2O_3 (sillenite)	77	56.0	40.8	60.9	86.2	495.3	66	[147]
Bi_2S_3	Bi_2S_3	77	14.25	28.29	42.45	56.63	339.8	1.7	*[194]
			21.58	14.65	16.32	23,74	153.1	88.9	
$Bi_4Ge_3O_{12}$	$Bi_4Ge_3O_{12}$	77	21.18	43.31	63.47	...	507.8	1.0	*[196]
$Bi_4O_{12}Ti_3$	$Bi_4Ti_3O_{12}$	77	32.24	31.16	48.92	66.49	404.3	40	*[196]
			...	19.67	30.6	40.86	245.9	20.3	
$Bi_8O_{14}Ti$	Bi_8TiO_{14}	77	33.84	55.67	85.29	114.08	685.5	15.1	*[196]
$Bi_{12}GeO_{20}$	$Bi_{12}GeO_{20}$	77	34.25	55.75	...	...	687.2	15.5	*[196]
$Bi_{12}O_{20}Si$	$Bi_{12}SiO_{20}$	77	34,29	56.02	85.75	114.65	686.6	15.4	*[196]
$C_{18}H_{15}Bi$	$Bi(C_6H_5)_3$	77	...	56.453	85.347	111.971	684.1	8,75	[156, 197]
		83	29.785	55.214	83.516	111.438	669.06	9	

Empirical formula	Structural formula	T, °K	e^2Qq, MHz	η, %	References, notes
		^{51}V			
Ga_5V_2	V_2Ga_5	77 300	0.332 0.334	60 60	[157]. NMR cr.
H_4NO_3V	$(NH_4)VO_3$	293	2.88	30	*[158. 198]. NMR cr.
KO_3V	KVO_3	293	4.36	75	[159, 198]. NMR cr.
NaO_3V	$NaVO_3$	293	3.65	60	[158, 198]. NMR cr.
$Na_3O_4V+14H_2O$	$Na_3VO_4 \cdot 14H_2O$	Room	3.1		[158]. NMR cr. cf. Na
$Na_4O_7V_2+16H_2O$	$Na_4V_2O_7 \cdot 16H_2O$	»	3,1		The same
O_5V_2	V_2O_5	77 Room	0.860 0.800	10 4	[158, 160]. NMR cr.
SiV_3	V_3Si	16.3 28.8 78—300	3.262 2.674 2.924 0.210	... ...	[161, 199]. NMR cr.
V	V (in corundum)	Room	0.021	...	[162]. NMR cr.

Empirical formula	Structural formula	T, °K	Transition	ν, MHz	e^2Qq, MHz	η, %	References, notes
			^{93}Nb				
Cl_5Nb	$NbCl_5$	77	$^1/_2—^3/_2$ $^3/_2—^5/_2$ $^5/_2—^7/_2$ $^7/_2—^9/_2$	5.7154 6.0516 9.5466 12.9132	78.3	34.9	[163, 164]. cf. Cl
		296.5	$^1/_2—^3/_2$ $^3/_2—^5/_2$ $^5/_2—^7/_2$ $^7/_2—^9/_2$ $^1/_2—^5/_2$ $^1/_2—^7/_2$	5.4228 6.0621 9.5612 12.9032 11.486 21,046	78.1	32.2	
$KNbO_3$	$KNbO_3$	77	$^1/_2—^3/_2$ $^3/_2—^5/_2$ $^5/_2—^7/_2$ $^7/_2—^9/_2$	... 1.335 2.004 2.672	16.0	0	[165, 166]. p.t. at 223°K
		298	$^1/_2—^3/_2$ $^3/_2—^5/_2$ $^5/_2—^7/_2$ $^7/_2—^9/_2$	3.030 2.085 2.527 3.648	23.12	80.6	
$LiNbO_3$	$LiNbO_3$	77	$^1/_2—^3/_2$ $^3/_2—^5/_2$ $^5/_2—^7/_2$ $^7/_2—^9/_2$	0.988 1.958 2.942 3.950	23.516	<3	[167, 200, 201]. cf. Li
		298	$^1/_2—^3/_2$ $^3/_2—^5/_2$ $^5/_2—^7/_2$ $^7/_2—^9/_2$	0.924 1.842 2.763 3.685	22.10	<3	
$NaNbO_3$	$NaNbO_3$	Room	—	—	19.7	82	[202]

BIBLIOGRAPHY

1. Townes, C., A. Holden, and F. Meritt. — Phys. Rev. 74 (1948), 1113.
2. Watkins, G. and R. V. Pound. — Phys. Rev. 85 (1952), 1062.
3. Casabella, P. and P. Bray. — J. Chem. Phys. 28 (1958), 1182.
4. Casabella, P. and P. Bray. — Bull. Am. Phys. Soc., ser. 2, 3 (1958), 21.
5. Hartmann, H. and H. Sillescu. — Theoret. chim. Acta 2 (1964), 371.
6. Negita, H., P. Casabella, and P. Bray. — J. Chem. Phys. 32 (1960), 314.
7. Simmons, J., W. Anderson, and W. Gordy. — Phys. Rev. 77 (1950), 77.
8. Simmons, J., W. Anderson, and W. Gordy. — Phys. Rev. 86 (1952), 1055.
9. Bhattachoria, B. and W. Gordy. — Phys. Rev. 119 (1960), 144.
10. Kern, C. and M. Karplus. — J. Chem. Phys. 42 (1965), 1062.
11. Shoolery, J., R. Shulman, and D. Jost. — J. Chem. Phys. 19 (1951), 250.
12. Dousmanis, G., T. Sanders, C. Townes, and H. Zeiger. — J. Chem. Phys. 21 (1953), 1416.
13. Ridgen, J. and R. Jackson. — J. Chem. Phys. 45 (1966), 3646.
14. Guibe, L. and E. Lucken. — Molec. Phys. 14 (1968), 73.
15. Kurland, R. and E. Wilson. — J. Chem. Phys. 27 (1957), 585.
16. Guibe, L. and E. Lucken. — Compt. rend. 263 (1966), 815.
17. Levine, I. — J. Molec. Spectrosc. 8 (1962), 276.
18. Monig, W. and H. Gutowsky. — J. Chem. Phys. 38 (1963), 1155.
19. Dixon, W. and E. Wilson. — J. Chem. Phys. 35 (1961), 191.
20. Sheridan, J. and A. Turner. — Proc. Chem. Soc., p. 21, 1960.
21. Minematsu, M. — J. Phys. Soc. Japan 14 (1959), 1030.
22. Guibe, L. — Arch. Sci. 13 (1960), 657.
23. Kojima, S., M. Minematsu, and M. Tanaka. — J. Chem. Phys. 31 (1959), 271.
24. Chiba, T., M. Toyama, and Y. Morino. — J. Phys. Soc. Japan 14 (1959), 379.
25. Guibe, L. — Compt. rend. 250 (1960), 1635.
26. Widman, R. — J. Chem. Phys. 43 (1965), 2922.
27. Guibe, L. — Ann. Phys. 7 (1962), 177.
28. Smith, D. — Bull. Am. Phys. Soc., ser. 2, 8 (1963), 350.
29. Smith, D. and R. Cotts. — J. Chem. Phys. 41 (1964), 2403.
30. Nishikawa, T. — J. Phys. Soc. Japan 17 (1962), 720.
31. Ikeda, R., D. Nakamura, and M. Kubo. — Bull. Chem. Soc. Japan 40 (1967), 701.
32. Negita, H. and P. Bray. — J. Chem. Phys. 33 (1960), 1876.
33. Colligiani, A., L. Guibe, P. Haigh, and E. Lucken. — Molec. Phys. 14 (1968), 89.
34. Sheridan, J. and W. Gordy. — J. Chem. Phys. 20 (1952), 735.
35. Casabella, P. and P. Bray. — J. Chem. Phys. 29 (1958), 1105.
36. Ring, H., H. Edwards, M. Kessler, and W. Gordy. — Phys. Rev. 72 (1947), 1262.
37. Coles, D., W. Good, and R. Hughes. — Phys. Rev. 79 (1950), 224.
38. Kessler, M., H. Ring, R. Trambarulo, and W. Gordy. — Phys. Rev. 79 (1950), 54.
39. Trambarulo, R. and W. Gordy. — Phys. Rev. 79 (1950), 224.
40. Kemp, M., J. Pochan, and W. Flygare. — J. Phys. Chem. 71 (1967), 765.
41. Curl, R., V. Rao, K. Sastry, and J. Hodgeson. — J. Chem. Phys. 39 (1963), 3335.
42. Lett, R. and W. Flygare. — J. Chem. Phys. 47 (1967), 4730.
43. Hess, H., A. Bauder, and H. Günthard. — J. Molec. Spectrosc. 22 (1967), 208.
44. Ikeda, R., D. Nakamura, and M. Kubo. — J. Phys. Chem. 70 (1966), 3626.
45. Schempp, E. and P. Bray. — Phys. Rev. Lett. 25A (1967), 414.
46. Abe, Y. and S. Kojima. — J. Phys. Soc. Japan 17 (1962), 720.
47. Abe, J. — J. Phys. Soc. Japan 18 (1963), 1804.
48. Haigh, P. and L. Guibe. — Compt. rend. 261 (1965), 2328.
49. Pierce, L., R. Nelson, and C. Thomas. — J. Chem. Phys. 43 (1965), 3423.
50. Kojima, S. and M. Minematsu. — J. Phys. Soc. Japan 15 (1960), 355.
51. Westenberg, A. and E. Wilson. — J. Am. Chem. Soc. 72 (1950), 199.
52. Sussman, A. and S. Alexander. — Solid State Communs. 5 (1967), 259.
53. Costein, C. and B. Stoicheff. — J. Chem. Phys. 30 (1959), 776.
54. Guibe, L. and E. Lucken. — Proc. XIII Colloque AMPERE, 1965.
55. Guibe, L. and E. Lucken. — Molec. Phys. 10 (1966), 273.

56. Laurie, V. — J. Chem. Phys. 31 (1959), 1500.
57. Mann, D. and D. Lide. — Bull. Am. Phys. Soc. 2 (1957), 212.
58. O'Konski, C. and R. Flautte. — J. Chem. Phys. 27 (1957), 815.
59. Guibe, L. and E. Lucken. — Molec. Phys. 14 (1968), 79.
60. Schempp, E. and P. Bray. — J. Chem. Phys. 46 (1967), 1186.
61. Schempp, E. and P. Bray. — J. Chem. Phys. 48 (1968), 2380.
62. Herberich, G. — Z. Naturf. 22a (1967), 543.
63. Tzalmona, A. — Phys. Rev. Lett. 20 (1966), 478.
64. Guibe, L. — Compt. rend. 250 (1960), 3014.
65. Sorensen, G. — J. Molec. Spectrosc. 22 (1967), 325.
66. Arnold, W., H. Dreizler, and H. Rudolph. — Z. Naturf. 23a (1968), 301.
67. Rudolph, H., H. Dreizler, and H. Seiler. — Z. Naturf. 22a (1967), 1738.
68. Matzkanin, G., T. O'Neal, and T. Scott. — J. Chem. Phys. 44 (1966), 4171.
69. Koo, J. and E. Hahn. — Bull. Am. Phys. Soc., ser. 2, 13 (1968), 356.
70. Guarnieri, A., G. Zuliani, and P. Favero. — Nuovo Cim. 45B (1966), 84.
71. Sheridan, J. and W. Gordy. — Phys. Rev. 79 (1950), 513.
72. Matzkanin, G., T. Scott, and P. Haigh. — J. Chem. Phys. 42 (1965), 1646.
73. Kirchholf, W. and E. Wilson. — J. Am. Chem. Soc. 84 (1962), 334.
74. Cox, A. and R. Kuczkowski. — J. Am. Chem. Soc. 88 (1966), 5071.
75. Millen, D. and J. Morton. — J. Chem. Soc., p. 1523, 1960.
76. Rogers, J. and D. Williams. — Phys. Rev. 88 (1952), 677.
77. Forman, R. and D. Lide. — J. Chem. Phys. 39 (1963), 1133.
78. Simmons, J. and W. Gordy. — Phys. Rev. 73 (1948), 713.
79. Simmons, J. and W. Gordy. — Phys. Rev. 74 (1948), 123.
80. Weiss, M. and M. Strandberg. — Phys. Rev. 83 (1951), 567.
81. Gunter-Mohr, G., R. White, A. Schawlow, W. Gordy, and D. Coles. — Phys. Rev. 94 (1954), 1184.
82. Erlandson, G. and W. Gordy. — Phys. Rev. 106 (1957), 513.
83. Lehrer, S. and C. O'Konski. — J. Chem. Phys. 43 (1965), 1941.
84. Abe, Y. and S. Kojima. — J. Phys. Soc. Japan 18 (1963), 1843.
85. Abe, Y., Y. Kamishina, and S. Kojima. — J. Phys. Soc. Japan 21 (1966), 2083.
86. Chiba, T. — Bull. Chem. Soc. Japan 38 (1965), 259, 490.
87. Forman, R. — J. Chem. Phys. 45 (1966), 1118.
88. Oja, T., R. Marino, and P. Bray. — Phys. Rev. Lett. 26A (1967), 11.
89. Gourdji, M. and L. Guibe. — Compt. rend. 260 (1965), 1131.
90. Beringer, R. and J. Castler. — Phys. Rev. 78 (1950), 340, 581.
91. Burkhard, D. — J. Chem. Phys. 21 (1953), 1531.
92. Mizushima, M. — Phys. Rev. 94 (1954), 569.
93. Henry, A., E. Rawson, and R. Beringer. — Phys. Rev. 94 (1954), 343.
94. Terman, F. and T. Scott. — Bull. Am. Phys. Soc., ser. 2, 3 (1958), 23.
95. Scott, T. — J. Chem. Phys. 36 (1962), 1459.
96. Meyer, H. and T. Scott. — J. Phys. Chem. Sol. 11 (1959), 215.
97. Coles, D., E. Elyash, and J. Gorman. — Phys. Rev. 72 (1947), 973.
98. Coles, D. and R. Hughes. — Phys. Rev. 76 (1949), 178.
99. Telenbaum, S. — Phys. Rev. 88 (1952), 772.
100. Sancho, M. and M. Harmony. — J. Chem. Phys. 45 (1966), 1812.
101. Zhukov, A. P., L. S. Golovchenko, and G. K. Semin. — Izv. AN SSSR, chem. ser., p. 1399, 1968.
102. Zhukov, A. R., L. S. Golovchenko, and G. K. Semin. — Chem. Communs., p. 854, 1968.
103. Pen'kov, I. N. and I. A. Safin. — Fiz. Tverd. Tela 6 (1964), 2467.
104. Kojima, S., K. Tsukada, S. Ogawa, A. Shimauchi, and Y. Abe. — J. Phys. Soc. Japan 9 (1954), 805.
105. Barnes, R. and P. Bray. — J. Chem. Phys. 23 (1955), 407.
106. Barnes, R. and P. Bray. — J. Chem. Phys. 23 (1955), 1178.
107. Bray, P. and R. Barnes. — J. Chem. Phys. 23 (1955), 1182.
108. Ogawa, S. — J. Phys. Soc. Japan 13 (1958), 618.
109. Kisliuk, P. and C. Townes. — Phys. Rev. 83 (1951), 210.
110. Zhukov, A. P. and G. K. Semin. Electrical Engineering, ser. 1, p. 176, 1968.

111. Zhukov, A. P., I. S. Rez, V. I. Pachomov, and G. K. Safin. — Physica Status Solidi 27 (1968), 129.
112. Pen'kov, I. N. and I. A. Safin. — Doklady AN SSSR 170 (1966), 626.
113. Dailey, B., K. Rusinov, R. Shulman, and C. Townes. — Phys. Rev. 74 (1948), 1243.
114. Kusliuk, P. and S. Geschwind. — J. Chem. Phys. 21 (1953), 828.
115. Jones, E. and E. Uehling. — J. Chem. Phys. 36 (1962), 1690.
116. Bjorkstam, J. — Bull. Am. Phys. Soc., ser. 2, 8 (1963), 464.
117. Tache, A., G. Blevins, and W. Gordy. — Phys. Rev. 95 (1954), 299.
118. Blevins, G. — Phys. Rev. 97 (1955), 684.
119. Loomis, C. and M. Strandberg. — Phys. Rev. 81 (1951), 798.
120. Pen'kov, I. N. and I. A. Safin. — Kristallografiya 13 (1968), 330.
121. Pen'kov, I. N. and I. A. Safin. — Doklady AN SSSR 153 (1963), 692.
122. Bray, P., G. O'Keefe, and R. Barnes. — J. Chem. Phys. 25 (1956), 792.
123. Krüger, H. and V. Meyer-Berkhout. — Z. Phys. 132 (1952), 221.
124. Bray, P., G. O'Keefe, and R. Barnes. — Bull. Am. Phys. Soc., ser. 2, 1 (1956), 11.
125. Pen'kov, I. N. and I. A. Safin. — Doklady AN SSSR 156 (1964), 139.
126. Nugent, L. and C. Cornwell. — Spectrochim. Acta 17 (1961), 1098.
127. Nugent, L. and C. Cornwell. — J. Chem. Phys. 37 (1962), 523.
128. Lide, D. — Spectrochim. Acta 15 (1959), 473.
128a. Pen'kov, I. N. and I. A. Safin. — Doklady AN SSSR 168 (1966), 1148.
129. Belyaev, L. M., V. A. Lyakhovitskaya, E. V. Bryukhova, T. A. Babushkina, A. P. Zhukov, A. S. Naumov, D. A. Tambovtsev, and G. K. Semin. — Reports of the All-Union Conference on Ferroelectricity, p. 36. Riga, 1968. (Russian)
130. Wang, T. — Phys. Rev. 99 (1955), 566.
131. Grechishkin, V. S. and I. A. Kyuntseli. — Trudy ENI Permsk. State Univ. 12, No. 1 (1966), 9.
132. Grechishkin, V. S. and A. D. Gordeev. — Trudy ENI Permsk. State Univ. 12, No. 4 (1966), 29.
133. Negita, H., T. Okuda, and T. Kashima. — J. Chem. Phys. 45 (1966), 1076.
134. Grechishkin, V. S. and I. A. Kyuntsel'. — Optika Spektrosk. 16 (1964), 161.
135. Grechishkin, V. S. and I. A. Kyuntsel'. — Zh. Strukt. Khim. 5 (1964), 53.
136. Grechishkin, V. S. and I. A. Kyuntsel'. — Optika Spektrosk. 15 (1963), 832.
137. Parker, D. — Diss. Abstr., Sect. B, 20 (1959), 2044.
138. Barnes, R. and P. Bray. — J. Chem. Phys. 23 (1955), 1177.
139. Barnes, R. and P. Bray. — J. Chem. Phys. 23 (1955), 1182.
140. Dehmelt, H. and H. Krüger. — Z. Phys. 130 (1951), 385.
141. Kantimati, B. — Indian J. Pure Appl. Phys. 4 (1966), 131.
142. Ogawa, S. — J. Phys. Soc. Japan 12 (1957), 1105.
143. Chihara, H., N. Nakamura, and H. Okuma. — J. Phys. Soc. Japan 24 (1968), 306.
144. Okuda, T., A. Nakoo, M. Shiroyama, and H. Negita. — Bull. Chem. Soc. Japan 41 (1968), 61.
145. Schneider, R. and J. Di Lorenzo. — J. Chem. Phys. 47 (1967), 2343.
146. Pen'kov, I. N. and I. A. Safin. — Doklady AN SSSR 161 (1965), 1404.
147. Safin, I. A. — Zh. Strukt. Khim. 4 (1963), 267.
148. Jache, A. et al. — Phys. Rev. 97 (1955), 680.
149. Krainik, N. N., S. N. Popov, and N. E. Myl'nikova. — Fiz. Tverd. Tela 8 (1966), 3664.
150. Safin, I. A. and I. N. Pen'kov. — Doklady AN SSSR 147 (1962), 410.
151. Hewitt, R. and B. Williams. — Phys. Rev. 129 (1963), 1188.
152. Hewitt, R. and B. Williams. — Phys. Rev. Lett. 12 (1964), 216.
153. Williams, B. and R. Hewitt. — Phys. Rev. 146 (1966), 286.
154. Swiger, E., P. Green, G. McKown, and J. Graybeal. — J. Phys. Chem. 69 (1965), 949.
155. Robinson, H. — Phys. Rev. 100 (1955), 1731.
156. Robinson, H., H. Dehmelt, and W. Gordy. — Phys. Rev. 89 (1953), 1305.
157. Lütgemeier, H. — Z. Naturf. 21a (1966), 541.
158. Saraswati, V. — J. Phys. Soc. Japan 23 (1967), 761.
159. Gornostansky, S. and C. Stoger. — J. Chem. Phys. 48 (1968), 1416.
160. Ioffe, V. A. and L. V. Dmitrieva. — Zh. Strukt. Khim. 7 (1966), 462.
161. Gossard, A. — Phys. Rev. 149 (1966), 246.
162. Lawrence, N. and J. Lambe. — Phys. Rev. 132 (1963), 1029.
163. Reddoch, A. — US Atomic Energy Communs., UCRL-8972, 1959.
164. Reddoch, A. — J. Chem. Phys. 35 (1961), 1085.
165. Cotts, R. and W. Knight. — Phys. Rev. 96 (1954), 1285.

166. Hewitt, R. — Phys. Rev. 121 (1961), 45.
167. Kind, R., H. Granicher, B. Derighetti, F. Waldner, and E. Brun. — Solid State Communs. 6 (1968), 439.
168. Kintzinger, J. and J. Lehn. — Molec. Phys. 17 (1969), 135.
169. Gur'evich, A. S. and L. I. Strelets. — Zh. Strukt. Khim. 10 (1969), 926.
170. O'Konski, C. and K. Torizuka. — J. Chem. Phys. 51 (1969), 461.
171. Lide, D. — J. Chem. Phys. 27 (1957), 353.
172. Osokin, D. Ya., L. A. Safin, and I. A. Nuretdinov. — Doklady AN SSSR 186 (1969), 1128.
173. Moniz, W. and C. Poranski. — J. Phys. Chem. 73 (1969), 4145.
174. Dreizler, H. and A. Mirri. — Z. Naturf. 23a (1968), 1313.
175. Zussman, A. and S. Alexander. — J. Chem. Phys. 49 (1968), 3792.
176. Li, G. and M. Harmony. — J. Chem. Phys. 50 (1969), 3674.
177. Onda, S., R. Ikeda, D. Nakamura, and M. Kubo. — Bull. Chem. Soc. Japan 42 (1969), 1771.
178. Osokin, D. Ya., L. A. Safin, and I. A. Nuretdinov. — Doklady AN SSSR 190 (1970), 357.
179. Tzalmona, A. — J. Chem. Phys. 50 (1969), 366.
180. Ikeda, R., S. Ouda, D. Nakamura, and M. Kubo. — J. Phys. Chem. 72 (1968), 2501.
181. Schempp, E. and P. Bray. — J. Chem. Phys. 49 (1968), 3450.
182. Marino, R., L. Guibe, and P. Bray. — J. Chem. Phys. 49 (1968), 5104.
183. Colligiani, A., R. Ambrosetti, and R. Angelone. — J. Chem. Phys. 52 (1970), 5022.
184. Ikeda, R., D. Nakamura, and M. Kubo. — J. Phys. Chem. 72 (1968), 2982.
185. Yim, C., M. Whitehead, and D. Lo. — Can. J. Chem. 46 (1968), 3595.
186. Zussman, A. and S. Alexander. — J. Chem. Phys. 49 (1968), 5179.
187. Hartland, A. — Proc. Phys. Soc., Solid State Phys., ser. 2, p. 264, 1969.
188. Bray, P. — Phys. Rev. Lett. 27a (1968), 263.
189. Ikeda, R., M. Mikami, D. Nakamura, and M. Kubo. — J. Magnet. Resonance 1 (1969), 211.
190. Kravchenko, E. A., S. A. Dembovskii, A. P. Chernov, and G. K. Semin. — Izv. AN SSSR, phys. ser. 33 (1969), 279.
191. Kravchenko, E. A., S. A. Dembovski, A. P. Chernov, and G. K. Semin. — Physica Status Solidi 31 (1969), K19.
192. Popov, S. N., N. N. Krainik, and N. E. Myl'nikova. — Izv. AN SSSR, phys. ser. 33 (1969), 271.
193. Svergun, V. I., A. E. Borisov, N. N. Novikova, E. V. Bryukhova, T. A. Babushkina, and G. K. Semin. — Izv. AN SSSR, chem. ser., p. 484, 1970.
194. Kravchenko, E. A. and G. K. Semin. — Izv. AN SSSR, ser. inorg. mater. 5 (1969), 1161.
195. Maksyutin, Yu. K., E. N. Gur'yanova, and G. K. Semin. — Izv. AN SSSR, chem. ser., p. 1632, 1970.
196. Buslaev, Yu., E. A. Kravchenko, V. I. Pachomov, V. M. Skorikov, and G. K. Semin. — Chem. Phys. Letters 5 (1969), 455.
197. Petrov, L. N., I. A. Kyuntsel', and V. S. Grechishkin. — Vest. Leningr. State Univ., p. 167, 1969.
198. Baugher, J., P. Taylor, T. Oja, and P. Bray. — J. Chem. Phys. 50 (1969), 4919.
199. Tret'yakov, B. N. and V. B. Kuritsyn. — Pis'ma Zh. Eksper. Teoret. Fiz. 9 (1969), 117.
200. Schempp, E., G. Peterson, and J. Carruthers. — J. Chem. Phys. 53 (1970), 306.
201. Peterson, G. and P. Bridenbaugh. — J. Chem. Phys. 48 (1968), 3402.
202. Wolf, F., D. Kline, and H. Story. — J. Chem. Phys. 53 (1970), 3538.

Group VI elements

Empirical formula	Structural formula	T, °K	ν, MHz	e^2Qq, MHz	References, notes
			^{17}O		
CO	CO	—	—	4.43	[1]. g
COS	COS	—	—	—1.32	[2, 3]. g cf. S
H_2O	H_2O	263	—	6.66	[4, 15—17]. NMR cr.
		—	—	9.82 (g)	η = 93.5%
		Room	—	7.7 (l)	η = 40.7%
			^{33}S		
CHNS	HCNS	—	—	—27.5	[5]. g cf. N
COS	COS	—	—	—29.130	[6—8]. g cf. O
		—	—	21.90*	
CS	CS	—	—	12.838	[9, 10, 12]. g
S	S (cryst.)	Room	22.801	...	[11]
			22.866	...	
			22.896	...	
			22.964	...	
			^{75}Se		
COSe	COSe	—	—	946	[12—14]. g
		—	—	752.09**	

* Data on ^{35}S.

** Data on ^{79}Se.

BIBLIOGRAPHY

1. Rosenblum, B. and A. Nethercot. — J. Chem. Phys. 27 (1957), 828.
2. Geschwind, S., G. Gunther-Mohr, and G. Silvey. — Phys. Rev. 83 (1951), 209.
3. Geschwind, S., G. Gunther-Mohr, and G. Silvey. — Phys. Rev. 85 (1952), 474.
4. Christ, H. and P. Diehl. — Proc. XI Colloque AMPERE, p. 224, 1963.
5. Dousmanis, G., T. Sandres, C. Townes, and H. Zeiger. — J. Chem. Phys. 21 (1953), 1416.
6. Cohen, V., W. Koski, and T. Wentink. — Phys. Rev. 76 (1949), 703.
7. Townes, C. and S. Geschwind. — Phys. Rev. 74 (1948), 626.
8. Eshbach, J., R. Hillger, and M. Strandberg. — Phys. Rev. 85 (1952), 532.
9. Bird, R. and R. Mockler. — Phys. Rev. 91 (1953), 222.
10. Mockler, R. — Phys. Rev. 98 (1955), 1837.
11. Dehmelt, H. — Phys. Rev. 91 (1953), 313.
12. Aamodt, L. — Phys. Rev. 98 (1955), 1224.
13. Aamodt, L., P. Fletcher, G. Silvey, and C. Townes. — Phys. Rev. 94 (1954), 789.
14. Hardy, W., G. Silvey, C. Townes, B. Burke, M. Strandberg, G. Parker, and V. Cohen. — Phys. Rev. 92 (1953), 1532.
15. Spiess, H., B. Garrett, R. Sheline, and S. Rabideau. — J. Chem. Phys. 51 (1969), 1201.
16. Verhoeven, J., A. Dymanus, and H. Bluyssen. — J. Chem. Phys. 50 (1969), 3330.
17. Garrett, B., A. Denison, and S. Rabideau. — J. Phys. Chem. 71 (1967), 2606.

Group VII elements. ^{35}Cl

The data were obtained at 77°K unless stated otherwise.

Empirical formula	Structural formula	ν, MHz	s/n	References, notes
	Group I element — chlorine bond			
ClH	HCl	26.469	...	[1—4]
ClD	DCl	27.310	...	[1, 2, 4]
ClT	TCl	$e^2Qq = 67$ MHz		[5]. g
$C_4H_{12}ClN + HCl$	$[N(CH_3)_4]Cl \cdot HCl$	20.21	...	[6, 6a]
$C_4H_{12}ClN + DCl$	$[N(CH_3)_4]Cl \cdot DCl$	21.12	...	[6a]
$C_8H_{20}ClN + HCl$	$[N(C_2H_5)_4]Cl \cdot HCl$	11.89	...	[7]. $T = 299$ °K
ClNa	NaCl	$e^2Qq = 67.0$ MHz		[5]. g cf. Na
ClK	KCl	$e^2Qq = 0.04$ MHz		[8, 9]. g cf. K
ClRb	RbCl	$e^2Qq = 0.774$ MHz		[10]. g cf. Rb
ClCs	CsCl	$e^2Qq \leqslant 3$ MHz		[11, 12]. g cf. Cs
$Cl_2Cu + 2H_2O$	$CuCl_2 \cdot 2H_2O$	9.05	...	[13]. $T = 4.3$ °K
$AuCl_3$	$AuCl_3$	36.116 33.340 23.285	...	[14]
$AuCl_3 + 2H_2O$	$AuCl_3 \cdot 2H_2O$	26.722 26.809 27.324 27.605	...	[14]. $T = 298$ °K
$AuCl_4H_4N$	$(NH_4)[AuCl_4]$	26.62 26.98 27.43 27.62	...	[15]. $T = 291$ °K
$AuCl_4K$	$K[AuCl_4]$	26.75 27.36	...	[15, 16]. $T = 291$ °K
$AuCl_4Na$	$Na[AuCl_4]$	28.87 29.47	...	[16]
$AuCl_4Na + 2H_2O$	$Na[AuCl_4] \cdot 2H_2O$	25.67 27.46 28.44 28.83	...	[15, 183]. $T = 273$ °K
$AuCl_4Na + 2D_2O$	$Na[AuCl_4] \cdot 2D_2O$	25.64 27.46 28.43 28.83	...	[183]. $T = 273$ °K

Empirical formula	Structural formula	ν, MHz	s/n	References, notes
	Group II element — chlorine bond			
CH_3ClHg	CH_3HgCl	14.75 $e^2Qq = -42$ MHz (g)	3	*[26]
C_2Cl_4Hg	$CCl_2{=}CClHgCl$	18.624	5	*cf. C—Cl
$C_2Cl_4Hg + C_2H_6OS$	$CCl_2{=}CClHgCl \cdot DMSO$	18.684	2	*cf. C—Cl
$C_2Cl_4Hg + C_3H_7NO$	$CCl_2{=}CClHgCl \cdot DMF$	18.261	...	*cf. C—Cl
C_6Cl_6Hg	C_6Cl_5HgCl	18.880	15	*[17]. cf. C—Cl
$C_2H_2Cl_2Hg$	trans- $ClCH{=}CHHgCl$	17.770 17,520	4 4	*[185]. cf. C—Cl
$C_2H_2Cl_2Hg + C_4H_8O$	trans-$CHCl{=}CHHgCl \cdot THF$	17.42	4	The same
$C_2H_2Cl_2Hg$	cis-$CHCl{=}CHHgCl$	16.590	2—2.5	»
$C_7H_6Cl_2Hg$	2-$ClC_6H_4CH_2HgCl$	16.188	3	*[184]. cf. C—Cl
Cl_2Hg	$HgCl_2$	22.522 22.874	5 5	*[18—25, 188]. cf. Hg
$Cl_2Hg + C_2H_6OS$	$HgCl_2 \cdot DMSO$	19.494 19.878	2—3 2—3	*[186, 187]
$Cl_2Hg + C_3H_7NO$	$HgCl_2 \cdot DMF$	20.208 20.456	4 3—4	*[186]
$Cl_2Hg + C_4H_8O$	$HgCl_2 \cdot THF$	19.272 20,584	7 7	*[186, 187]
$Cl_2Hg + C_4H_8O_2$	$HgCl_2 \cdot$ diox	21.196	4	*[22, 186—188]
$Cl_2Hg + C_4H_{10}O_2$	$HgCl_2 \cdot DME$	21.328	30	*[186]
	Group III element — chlorine bond			
BCl_3	BCl_3	21.582 21.578	...	[2, 27]. η = 24% (theoretical)
$B_3Cl_3H_3N_3$	$(HNBCl)_3$	19.937	...	[28, 29]
$C_2H_2B_{10}Cl_{10}$	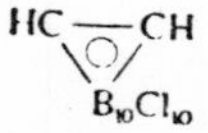 B-decachloro-o-borane †	25.720 25,125 24,950 24,695	70 7 18 5	*[31]

† Here and in what follows we use the sequence of atoms in the molecules of o- and m-boranes adopted in [208].

Empirical formula	Structural formula	ν, MHz	s/n	References, notes
$C_2H_2B_{10}Cl_{10}$	$HCB_{10}Cl_{10}CH$ B-decachloro-m-borane	24.660 25.050 25.195 25.315 25.540 25.605 25.640 25.965 26.060	5 10 5 5 5 5 5 5 5	*[29, 31]
$C_2H_4B_{10}Cl_8$	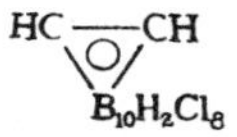 B-octachloro-o-borane	24.410 24.784 24.968 25.128 25.560 25.660	10 8 14 45 15 10	*[31]
$C_2H_8B_{10}Cl_4$	 B-9,10,11,12-tetrachloro-o-borane	23.74 23.91 24.39 24.62	4 4 3 2	*[31]
$C_2H_9B_{10}Cl_3$	$HCB_{10}H_7Cl_3CH$ B-3,9,10-trichloro-m-borane	23.86	4	*[31]. Nonsymmetrical line
$C_2H_{10}B_{10}Cl_2$	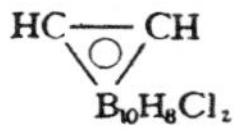 B-10,12-dichloro-o-borane	23.43	3	*[30—32]
	$HCB_{10}H_8Cl_2CH$ B-9,10-dichloro-m-borane Phase I Phase II	 23.41 24.04 23.66	 3 3 3	*[30—32]
$C_2H_{11}B_{10}Cl$	 B-10(12)-chloro-o-borane	23.06	...	*[30—31]
	$HCB_{10}H_9ClCH$ B-9(10)-chloro-m-borane	23.42 23.62	10 10	*[30—31]
$C_3H_{10}B_{10}Cl_4$	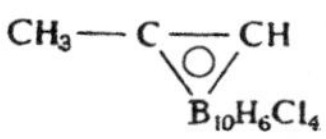 C-methyl-B-9,10,11,12-tetrachloro-o-borane	24.495 24.204 24.040	5 10 5	*[31]

Empirical formula	Structural formula	ν, MHz	s/n	References, notes
$C_3H_{11}B_{10}Cl_3$	HC—C—CH$_2$Cl B$_{10}$H$_8$Cl$_2$ C-chloromethyl-B-10, 12-dichloro-o-borane	23.34 23.52	2 3	*[31]. cf. C—Cl
$C_3H_{12}B_{10}Cl_2$	CH$_3$—C—CH B$_{10}$H$_8$Cl$_2$ C-methyl-B-10,12-dichloro-o-borane	23.44 23.48	5 5	*[31]
	HCB$_{10}$H$_8$Cl$_2$CCH$_3$ C-methyl-B-9,10-dichloro-m-borane	23.61	10	*[31]
$C_4H_{12}B_2Cl_4N_2$	[(CH$_3$)$_2$N·BCl$_2$]$_2$	21.008 21.156	1 2	[29]. $T = 273$ °K
$C_4H_{12}B_{10}Cl_2$	CH$_2$=CH—C—CH B$_{10}$H$_8$Cl$_2$ C-vinyl-B-10,12-dichloro-o-borane	23.61 23.39	8 15	*[31]
$C_{18}H_{15}B_3Cl_3N_3$	(C$_6$H$_5$—N·BCl)$_3$	20.87 20.97	1 2	[29]. $T = 273$ °K
$C_{20}H_{44}B_4Cl_4N_4$	(tert-C$_5$H$_{11}$N·BCl)$_4$	20.337 20.359	2 1	[29]. $T = 293$ °K
AlCl	AlCl	$e^2Qq=-8.8$MHz		[33]. g cf. Al
$AlCl_3 + C_4H_{10}O$	AlCl$_3$·O(C$_2$H$_5$)$_2$	11.30	2	[34]. $T = 83$ °K
$AlCl_4Ga$	Ga[AlCl$_4$]	10.296 10.382 10.460 10.312	···	[14, 35, 36]. $T = 297$ °K
$AlCl_4Na$	NaAlCl$_4$	11.583 11.385 11.272 11.009	···	[36]. $T = 297$ °K
$AlCl_6I$	ICl$_2$[AlCl]$_4$	11.42 11.40 11.31	···	[35]. $T = 297$ °K. cf. I—Cl
C_3AlCl_7		10.004	···	[36]. $T = 297$ °K. cf. C—Cl

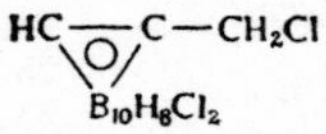

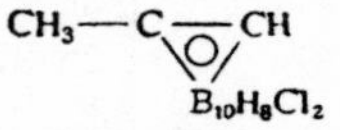

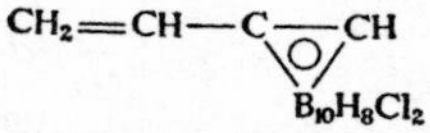

Empirical formula	Structural formula	ν, MHz	s/n	References, notes
ClGa	GaCl	$e^2Qq = -20$ MHz		[37]. g cf. Ga
Cl_3Ga	$GaCl_3$	15.135 19.542 21.016	...	*[38, 189]. cf. Ga
		At 306.2 °K $\eta_1 = 47.3\%$; $\eta_2 = 8.9\%$; $\eta_3 = 3.4\%$		
ClIn	InCl	$e^2Qq = -13$ MHz		[37, 39, 40]. g cf. In
$Cl_5H_{10}InN_2O$	$(NH_4)_2[InCl_5 \cdot H_2O]$	10.175 9.815 10.848 10.985	...	[41]. $T = 110$ °K. cf. In
ClTl	TlCl	$e^2Qq = -15.8$ MHz		[37]. g
Cl_3Tl	$TlCl_3$	15.080 14.925	...	[14, 19]
$CeCl_3$	$CeCl_3$	4.387	...	[42, 206]. $T = 4.2$ °K
Cl_3Gd	$GdCl_3$	5.29	...	[44, 206].
		At 4.2 °K $\eta = 42.5\%$		
Cl_3Nd	$NdCl_3$	4.729	...	[42, 206]. $T = 4.2$ °K
Cl_3Sm	$SmCl_3$	5.033	...	The same
Cl_3Pr	$PrCl_3$	4.561	...	[42, 43, 206]. $\eta = 58\%$
Cl_4Th	$ThCl_4$	5.9181	5	[45]
	Group IV element — chlorine bond			
$BrCl_3Si$	$BrSiCl_3$	20.47	...	[46]
CCl_6Si	Cl_3CSiCl_3	20.08	...	[46]
$CHCl_5Si$	$Cl_2HCSiCl_3$	19.75	...	*[46—49, 51, 52]. cf. C—Cl
CH_2Cl_4Si	ClH_2CSiCl_3	19.460 19.548	5 10	The same
$CH_2Cl_6Si_2$	$Cl_3SiCH_2SiCl_3$	19.163 19.131 19.050	...	[46, 47, 49, 50]
CH_3Cl_3Si	CH_3SiCl_3	18.955 19.020 19.155	10 6 10	*[47, 49, 51—54]

Empirical formula	Structural formula	ν, MHz	s/n	References, notes
CH_4Cl_2Si	CH_3SiHCl_2	18.236	...	[46]
$C_2H_2Cl_4Si$	$H_2C{=}CClSiCl_3$	19.401 20.361	3 3	[53]
$C_2H_2Cl_6Si_2$	$Cl_3SiCH{=}CHSiCl_3$	19.30	3	*[52]
$C_2H_3Br_2Cl_3Si$	$BrH_2CCHBrSiCl_3$	19.150 19.839	5 10	[53]
$C_2H_3Cl_3Si$	$H_2C{=}CHSiCl_3$	18.962 19.119 19.316	2.5 2 2	*[46, 47, 49, 51—53, 55]
$C_2H_4BrCl_3Si$	$BrCH_2CH_2SiCl_3$	19.397 19.034	1 2	[46]
$C_2H_4Cl_4Si$	$CH_3CHClSiCl_3$	19.53 19.33	1 2	[46]
	$Cl_2HC(CH_3)SiCl_2$	18.74	...	[46, 48, 49, 51]. cf. C—Cl
	$Cl(CH_2)_2SiCl_3$	19.030 19.368	10 5	*[46, 52]. cf. C—Cl
$C_2H_4Cl_6GeSi$	$Cl_3GeCH_2CH_2SiCl_3$	19.00	...	[46]. cf. Ge—Cl
$C_2H_4Cl_6Si_2$	$Cl_3Si(CH_2)_2SiCl_3$	18.914 19.030	10 20	*[46, 47, 49, 50]
$C_2H_5Cl_3Si$	$C_2H_5SiCl_3$			*[46, 47, 49, 53, 56—58]
	Phase I	18.756 18.840 18.864	20 20 20	
	Phase II	18.666 18.768 18.876	8 12 10	
	Phase III	18.621 18.639 18.755 18.887 18.933	5 5 10 5 5	
	$ClH_2CSiCl_2(CH_3)$	18.01 18.20	...	*[46—53]. cf. C—Cl
$C_2H_6Cl_2Si$	$(CH_3)_2SiCl_2$	17.756	...	*[46, 47, 49, 51, 53, 54, 59, 60]
	$C_2H_5SiHCl_2$	17.954 18.179	20 20	[53]

Empirical formula	Structural formula	ν, MHz	cf.	References, notes
C_2H_7ClSi	$(CH_3)_2SiHCl$	17.241 17.143 17.113	...	[46]
$C_3HCl_6F_3Si$	$CF_3CCl_2CHClSiCl_3$	20.484	3	*[52]. cf. C—Cl
$C_3H_2Cl_3F_3Si$	$CF_3CH{=}CHSiCl_3$	19.292 19.331 19.443	10 10 10	*[52]
$C_3H_2Cl_5F_3Si$	$CF_3CHClCHClSiCl_3$	19.512 20.088	5 5	*[201]. cf. C—Cl
$C_3H_4Cl_3F_3Si$	$CF_3CH_2CH_2SiCl_3$			*[46, 47, 49, 51, 52]
	Phase I	19.028 19.167 19.377	20 20 20	
	Phase II	18.876 18.984 19.224	...	
$C_3H_4Cl_3NSi$	$NCCH_2CH_2SiCl_3$	19.076 19.210	1 2	[46, 47, 49—51]
$C_3H_5Cl_5Si$	$ClCH_2C(CH_3)ClSiCl_3$	19.740 19.272	2 1	[46]
$C_3H_6BrCl_3Si$	$Br(CH_2)_3SiCl_3$	18.95	...	[46]
$C_3H_6Cl_2Si$	$CH_2{=}CHSiCl_2(CH_3)$	17.910	...	[46—49, 55]
	 $SiCl_2$	17.590 17.840	...	*[46, 61]
$C_3H_6Cl_4Si$	$Cl(CH_2)_3SiCl_3$	18.755 18.774 19.033	...	*[47—50, 52]
	$ClCH_2(CH_3)CHSiCl_3$	19.029	...	[46]
	$ClCH_2CHCl(CH_3)SiCl_2$	18.55 18.27	...	[46]
$C_3H_7Cl_3Si$	$C_3H_7SiCl_3$	18.770 18.890 18.945	15 15 15	*[52]
	iso-$C_3H_7CSiCl_3$	18.746 18.770 18.825 18.890 18.950	1 1 1 2 1	*[46, 47, 49, 50, 52]
$C_3H_8Cl_2Si$	$ClH_2CSiCl(CH_3)_2$	17.022 17.090	...	[46—51]. cf. C—Cl
	$CH_3(C_2H_5)SiCl_2$	17.73	...	[46]

Empirical formula	Structural formula	ν, MHz	s/n	References, notes
$C_3H_9ClO_3Si$	$(CH_3O)_3SiCl$	16.87	...	[46]
C_3H_9ClSi	$(CH_3)_3SiCl$	16.506	20	*[46, 47, 49, 51—55]
$C_4H_2Cl_{10}Si$	$(HCCl_2CCl_2)_2SiCl_2$	20.83	...	[46]
$C_4H_5Cl_2F_3Si$	$CF_3CH{=}CHSiCl_2(CH_3)$	18.087	10	*[52]
$C_4H_6Cl_3F_3Si$	$CF_3CH_2CHClSiCl_2(CH_3)$	18.165	2	*[52]. cf. C—Cl
$C_4H_6Cl_6Si$	$(ClCH_2CHCl)_2SiCl_2$	17.44	...	[46]
$C_4H_7Cl_2F_3Si$	$CF_3CH_2CH_2SiCl_2(CH_3)$	17.680	...	[46, 51]
$C_4H_7Cl_2N$	$NCCH_2CH_2SiCl_2(CH_3)$	17.70	...	[46, 51]
$C_4H_8Cl_2Si$	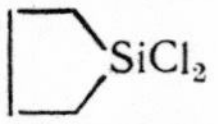	17.755	...	[46, 61]
$C_4H_8Cl_4Si$	$ClCH_2CH_2(CH_3)CHSiCl_3$	19.0	...	[46]
C_4H_9ClSi	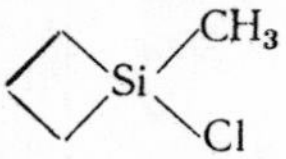	16.612	...	*[61]
	$CH_2{=}CHSiCl(CH_3)_2$	16.77	...	[46, 51]
$C_4H_9Cl_3Si$	$CH_3(CH_2)_3SiCl_3$	18.89	...	[46]
	$(CH_3CHCl)(C_2H_5)SiCl_2$	18.04	...	[46]
	$(ClCH_2CH_2)(C_2H_5)SiCl_2$	17.82	...	[46]
	$(ClCH_2CH_2CH_2)Si(CH_3)Cl_2$	17.75	...	[46]
$C_4H_{10}Cl_2O_2Si$	$(C_2H_5O)_2SiCl_2$	17.88	...	[46]
$C_4H_{10}Cl_2Si$	$(CH_3)(C_3H_7)SiCl_2$	17.713	...	[46]
	$(C_2H_5)_2SiCl_2$			*[46, 47, 49, 52—54]
	Phase I	17.515 17.740	5 10	
	Phase II	17.324 17.407 17.543 17.651	3 3 6 6	
$C_4H_{12}Cl_2OSi_2$	$Cl(CH_3)_2SiOSi(CH_3)_2Cl$	16.373 16.288	...	[46]
$C_5H_6Cl_2SSi$	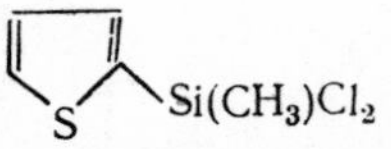	18.44 18.33 18.08	1 1 2	[46]
$C_5H_8Br_2ClF_3Si$	$CF_3CHBrCHBrSiCl(CH_3)_2$	17.010	2—3	*[201]
$C_5H_8Cl_5F_3GeSi$	$CF_3(CH_2)_2Si(Cl_2)(CH_2)_2GeCl_3$	18.335 18.862	10 10	*[201]. cf. Ge—Cl

Empirical formula	Structural formula	ν, MHz	s/n	References, notes
$C_5H_9Cl_3Si$	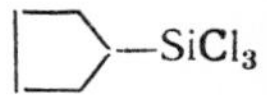	19.307 19.119 18.959	...	[46]
$C_5H_{10}ClF_3Si$	$(F_3CCH_2CH_2)(CH_3)_2SiCl$	16.531	...	[46]
$C_5H_{10}Cl_2Si$	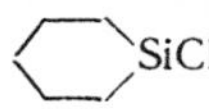	17.83	...	[46, 61]
$C_5H_{11}ClSi$	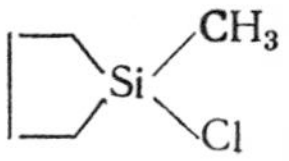			[46, 61]
	Phase I	16.775	...	
	Phase II	16.97		
$C_5H_{11}Cl_3Si$	$(ClCH_2CH_2CH_2)Si(C_2H_5)Cl_2$	17.72	...	[46]
	$C_5H_{11}SiCl_3$	18.90	...	[46]
$C_5H_{12}Cl_2Si$	$ClCH_2CH_2CH_2Si(CH_3)_2Cl$	16.68	...	[46]. cf. C—Cl
$C_6H_3Cl_5Si$	2,5-$Cl_2C_6H_3SiCl_3$	19.620	...	[46, 47, 49, 54]
$C_6H_4Cl_4Si$	3-$ClC_6H_4SiCl_3$			[46, 47, 49, 54]
	Phase I	19.145 19.187 19.295	...	
	Phase II	19.210		
$C_6H_5Cl_3Si$	$C_6H_5SiCl_3$	19.050 19.130 19.160	...	[47, 49, 51, 53, 54]
$C_6H_8Cl_2F_6Si$	$(CF_3CH_2CH_2)_2SiCl_2$	17.808	...	[46]
$C_6H_{10}Cl_2Si$	$(CH_2{=}CHCH_2)_2SiCl_2$	17.97	...	[46]
$C_6H_{13}Cl_3Si$	$C_6H_{13}SiCl_3$	18.89	...	[46]
$C_6H_{15}ClO_3Si$	$(C_2H_5O)_3SiCl$	16.451	...	[46]
$C_6H_{15}ClSi$	$(C_2H_5)_3SiCl$	16.460	2	*[52, 201]
$C_6H_{18}Cl_2O_2Si_3$	$[Cl(CH_3)_2SiO]_2Si(CH_3)_2$	16.28	...	[46]
$C_7H_2Cl_8Si$	$C_6Cl_5CH_2SiCl_3$	19.514 19.465 19.235	...	[46]
$C_7H_5Cl_5Si$	$C_6H_5CCl_2SiCl_3$	20.372 20.156 19.946 19.880 19.874 19.693	...	[46]. cf. C—Cl

Empirical formula	Structural formula	ν, MHz	s/n	References, notes
$C_7H_5Cl_5Si$ (contd.)	$C_6H_5SiCl_2(CCl_3)$	19.20	...	[46]. cf. C—Cl
$C_7H_6Cl_4Si$	$Cl_2CH(C_6H_5)SiCl_2$	18.82	...	The same
$C_7H_7Cl_3OSi$	$2\text{-}CH_3C_6H_4OSiCl_3$	19.46	...	[46]
$C_7H_7Cl_3Si$	$ClH_2CSiCl_2(C_6H_5)$	18.220	5	*[52]. cf. C—Cl
$C_7H_8Cl_2Si$	$C_6H_5SiCl_2(CH_3)$	18.118 18.137	2 1	[46, 51]
$C_7H_8Cl_3NSi$	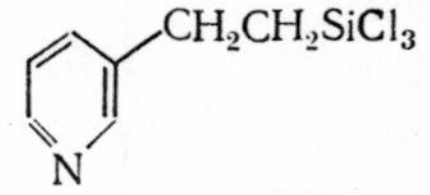	18.93	...	[46]
$C_7H_{15}Cl_3Si$	$C_7H_{15}SiCl_3$	18.870	...	[46]
$C_8H_5Cl_7Si$	$C_6H_5CCl_2(CCl_3)SiCl_2$	20.15	...	[46]
$C_8H_9Cl_3Si$	$C_6H_5CH_2CH_2SiCl_3$	18.97	...	[46]
$C_8H_{10}Cl_2Si$	$(C_6H_5CH_2)(CH_3)SiCl_2$	17.986	...	[46]
$C_8H_{18}Cl_2Si$	$(C_4H_9)_2SiCl_2$	17.73	...	[46]
$C_9H_{12}ClF_9Si$	$(CF_3CH_2CH_2)_3SiCl$	17.091	3	*[46, 201]
$C_9H_{19}Cl_3Si$	$C_9H_{19}SiCl_3$	18.88	...	[46]
$C_{10}H_8Cl_2SSi$	$(\alpha\text{-}C_4H_3S)(C_6H_5)SiCl_2$	18.35	...	[46]
$C_{10}H_8Cl_2Si$	$2\text{-}C_{10}H_7SiHCl_2$	18.507	...	[46]
$C_{10}H_{14}Cl_2Si$	$(C_6H_5CH_2CH_2CH_2)(CH_3)SiCl_2$	17.69	...	[46]
	$(C_6H_5CH_2CH_2)(C_2H_5)SiCl_2$	17.73	...	[46]
$C_{11}H_{16}Cl_2Si$	$[C_6H_5(CH_2)_4](CH_3)SiCl_2$	17.68	...	[46]
	$(CH_3)[C_6H_5CH_2CH_2(CH_3)CH]SiCl_2$	17.95	...	[46]
$C_{12}H_{26}Cl_2Si$	$(C_6H_{13})_2SiCl_2$	17.625	...	[46]
$C_{14}H_{14}Cl_2O_2Si$	$(2\text{-}CH_3C_6H_4O)_2SiCl_2$	19.04	...	[46]
$C_{14}H_{30}Cl_2Si$	$(C_7H_{15})_2SiCl_2$	17.843 17.704	...	[46]
$C_{18}H_{38}Cl_2Si$	$(C_9H_{19})_2SiCl_2$	17,71	...	[46]
ClH_3Si	H_3SiCl	$e^2Qq = -40$ MHz		[62, 63]. g
ClD_3Si	D_3SiCl	$e^2Qq = -39.4$ MHz		[63]. g
ClF_3Si	F_3SiCl	$e^2Qq = -43$ MHz		[64, 65]. g

Empirical formula	Structural formula	ν, MHz	s/n	References, notes
Cl_3HSi	$SiHCl_3$			*[2, 46, 47, 49, 50, 52, 53, 66]
	Phase I	18.882 18.930 19.025 19.082 19.132 19.292	...	
	Phase II	18.915 19.024 19.146		
		$e^2Qq = 42$ MHz (g)		
Cl_4Si	$SiCl_4$	20.273 20.408 20.415 20.464	...	*[2, 34, 47, 49, 52, 60, 190] $\eta = 45\%$
Cl_6OSi_2	$Cl_3SiOSiCl_3$			[46, 47, 49, 50]
	Phase I	19.300		
	Phase II	19.860 19.956	2 1	
Cl_6Si_2	Si_2Cl_6	19.319 19.287 19.271	3 3 3	[47, 49, 50, 53]
H_2Cl_2Si	H_2SiCl_2	18.571		[46]
$CHCl_5Ge$	$\mathbf{Cl}_2HCGeCl_3$	24.60 24.85	5 10	*[52]. cf. C—Cl
CH_2Cl_4Ge	$\mathbf{Cl}H_2CGeCl_3$			*[52]. cf. C—Cl
	Phase I	23.708 23.752 23.784	50 50 50	
	Phase II	23.040 23.728	20 50	
CH_3Cl_3Ge	$CH_3Ge\mathbf{Cl}_3$	22.941 22.707	2 1	*[46]
$C_2H_4BrCl_3Ge$	$BrCH_2CH_2GeCl_3$	23.627 22.924	1 2	[46]
$C_2H_4Cl_4Ge$	$ClCH_2CH_2GeCl_3$	23.540 22.948	20 40	*[52]. cf. C—Cl
»	$(ClCH_2)_2GeCl_2$	22.168 22.292	...	*[52]. cf. C—Cl

Empirical formula	Structural formula	ν, MHz	s/n	References, notes
$C_2H_4Cl_6GeSi$	$Cl_3SiCH_2CH_2GeCl_3$	22.85	...	[46]. cf. Si—Cl
$C_2H_5Cl_3Ge$	$C_2H_5GeCl_3$	22.164 22.308 22.788	30 30 30	*[52, 67, 68]
$C_2H_6Cl_2Ge$	$(CH_3)_2GeCl_2$	20.180 20.320	10 10	*[46, 52, 55, 67, 68]
$C_3H_2Cl_5F_3Ge$	$CF_3CH_2CCl_2GeCl_3$	24.26	2	*[52]. cf. C—Cl
$C_3H_4Cl_3F_3Ge$	$CF_3CH_2CH_2GeCl_3$	23.252 23.282 23.388	20 20 20	*[52, 67, 68]
$C_3H_4Cl_3GeO$	$ClCOCH_2CH_2GeCl_3$	22.75 22.67	...	[46]
$C_3H_5Cl_3GeO_2$	$CH_3COOCH_2GeCl_3$	23.819 23.556 22.993	...	[46]
$C_3H_6Cl_4Ge$	$Cl(CH_2)_3GeCl_3$	23.260 22.584	10 5	*[46, 52]. cf. C—Cl
$C_3H_7Cl_3Ge$	$C_3H_7GeCl_3$	22.592 22.608 22.736	10 10 8	*[52]
C_3H_9ClGe	$(CH_3)_3GeCl$	17.919	5	*[52, 67]
$C_4H_4Cl_2Ge$		21.619	...	[46]
$C_4H_9Cl_3Ge$	iso-$C_4H_9GeCl_3$	21.308 21.148	10 5	*[52]
$C_5H_8Cl_4Ge$	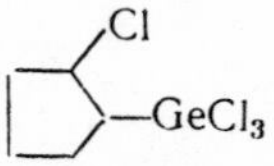	23.4 22.1	1 2	[46]. cf. C—Cl
$C_5H_8Cl_5F_3GeSi$	$CF_3(CH_2)_2SiCl_2(CH_2)_2GeCl_3$	22.162 22.392 23.524	10 10 10	*[201]. cf. Si—Cl
$C_5H_9Cl_3Ge$	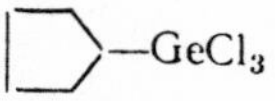	22.624 22.497 22.329	...	[46]
$C_6H_{15}ClGe$	$(C_2H_5)_3GeCl$	17.850	5	*[52, 67]
$C_7H_5Cl_3FeGeO_2$	$[C_5H_5Fe(CO)_2]GeCl_3$	20.096 20.248 20.346	5 5 5	*[201]
$C_8Cl_2Co_2GeO_8$	$Cl_2Ge[Co(CO)_4]_2$	19.692 20.400	5 5	*[202]. cf. Co

Empirical formula	Structural formula	ν, MHz	s/n	References, notes
$C_9H_{21}ClGe$	$(C_3H_7)_3GeCl$	18.096	10	*[52]
$ClGeH_3$	H_3GeCl	$e^2Qq = -36$ MHz		[69, 70]. g cf. Ge
Cl_3CsGe	$CsGeCl_3$	12.084	10	*[71, 72]
Cl_3GeH	$HGeCl_3$			*[46, 52, 58]
	Phase I	23.336	15	
	Phase II	23.328 23.076	10 5	
	Phase III	23.620 23.484 23.080	5 4 5	
$Cl_3GeH + 2C_4H_{10}O$	$HGeCl_3 \cdot 2(C_2H_5)_2O$	13.426 13.964 14.384	10 10 10	*[72]
Cl_4Ge	$GeCl_4$	25.450 25.715 25.735 25.745	50 50 50 50	*[2, 52, 67, 68], 190] $\eta = 35\%$
CH_2Cl_4Sn	$ClCH_2SnCl_3$	23.072 21.832	10 5	*[73]. cf. C—Cl
CH_3Cl_3Sn	CH_3SnCl_3	21.85 19.43	15 30	*
$C_2H_4Cl_4Sn$	$(ClCH_2)_2SnCl_2$	18.030	8	*[73]. cf. C—Cl
$C_2H_5Cl_3Sn$	$C_2H_5SnCl_3$	21.428 20.664 18.905	40 40 40	*
$C_2H_6Cl_2Sn$	$(CH_3)_2SnCl_2$	15.466	6	*[74]
$C_3H_6Cl_4Sn$	$(ClCH_2)_3SnCl$	11.63	3	*[73]. cf. C—Cl
C_3H_9ClSn	$(CH_3)_3SnCl$	11.73	5	*
$C_4H_9Cl_3Sn$	$C_4H_9SnCl_3$	20.212 20.448 20.521	3 3 3	[74, 75]
$C_4H_{10}Cl_2Sn$	$(C_2H_5)_2SnCl_2$	15.490 15.635 15.810	30 15 15	*[74]
$C_6H_5Cl_3Sn$	$C_6H_5SnCl_3$	20.112 20.632 21.279	6 3 6	[74, 75]
$C_6H_{14}Cl_2Sn$	$(C_3H_7)_2SnCl_2$	14.930	2	[74]

Empirical formula	Structural formula	ν, MHz	s/n	References, notes
$C_6H_{15}ClSn$	$(C_2H_5)_3SnCl$	11.952		*
$C_8H_{18}Cl_2Sn$	$(C_4H_9)_2SnCl_2$			[74, 75]
	Phase I	17.273	...	
	Phase II	15.820 15.948	2 2	$T = 200$ °K
$C_{12}H_{10}Cl_2Sn$	$(C_6H_5)_2SnCl_2$			[74, 75]
	Phase I	17.440 17.902 18.020 18.722	2 2 2 2	
	Phase II	16.947		
$C_{16}H_{34}Cl_2Sn$	$(C_8H_{17})_2SnCl_2$	15.136	2	[74]. $T = 303$ °K
$C_{18}H_{15}ClSn$	$(C_6H_5)_3SnCl$			*[74, 76, 201]
	Phase I	13.66	1.2	
	Phase II	16.750 16.985	2 2	$T = 303$ °K
Cl_4Sn	$SnCl_4$	23.720 24.140 24.226 24.296	50 50 50 50	*[2, 22, 34, 190] $\eta = 25\%$
$Cl_4Sn + CH_3NO_2$	$SnCl_4 \cdot CH_3NO_2$	20.232 24.008 24.264 24.384	1 3 4 3	*[77]
$Cl_4Sn + 2C_2H_5ClO$	$SnCl_4 \cdot 2CH_3OCH_2Cl$	20.058 21.648 21.912	6 3 3	*[203]. cf. C—Cl
$Cl_4Sn + 2C_4H_{10}O$	$SnCl_4 \cdot 2(C_2H_5)_2O$	19.438 19.468	6 6	*[22, 77]
$Cl_4Sn + C_4H_{10}O_2$	$SnCl_4 \cdot$ DME	19.620	5	*[77]
$Cl_4Sn + C_6H_4ClNO_2$	$SnCl_4 \cdot 4\text{-}ClC_6H_4NO_2$	23.024 24.350 24.784	10 10 10	*[77]. cf. C—Cl
$Cl_4Sn + C_6H_5NO_2$	$SnCl_4 \cdot C_6H_5NO_2$	20.838 20.958 23.160 23.208 23.336	8 8 4 4 8	*[22, 34, 77]
		23.888 23.916 24.064	1 2 1	$SnCl_4$ in a clathrate complex

Empirical formula	Structural formula	v, MHz	s/n	References, notes
$Cl_4Sn + 2C_2H_6O$	$SnCl_4 \cdot 2C_2H_5OH$	17.498 17.750 20.120 20.568	5 5 5 5	*[77]
$Cl_4Sn + C_7H_7Cl$	$SnCl_4 \cdot C_6H_5CH_2Cl$	24.090 23.644 19.014 19.146	2 2 2 2	*[203]
$Cl_4Sn + 2Cl_3OP$	$SnCl_4 \cdot 2POCl_3$	21.132 19.794 19.030	3 1.5 1.5	*[22, 77, 78]. cf. P—Cl
$Cl_6H_8N_2Sn$	$(NH_4)_2[SnCl_6]$	15.61 15.65	...	[79]
Cl_6K_2Sn	$K_2[SnCl_6]$	15.73	...	[79]. p.t. at 264.5°K
Cl_6Rb_2Sn	$Rb_2[SnCl_6]$	15.63	...	[79]
Cl_4Pb	$PbCl_4$	22.68	3	[80]
$Cl_6H_8N_2Pb$	$(NH_4)_2[PbCl_6]$	17.26	...	[79]
$C_5H_5Cl_3Ti$	$C_5H_5TiCl_3$	8.042 8.145	6 3	*
Cl_2Ti	$TiCl_2$	4.17	...	[81]. $T = 297$ °K
Cl_3Ti	$TiCl_3$	7.39	...	The same
Cl_4Ti	$TiCl_4$	5.9802 6.0380 6.0807 6.1118	...	[34, 45, 82]

Group V element — chlorine bond

C_2H_4ClN	$\begin{array}{c}CH_2\\ |\quad\rangle NCl\\ CH_2\end{array}$	45.604	100	*[191]
$C_4H_4ClNO_2$	$CH_2CH_2CONClCO$ (bracketed below)	54.100	...	[83]
$C_5H_6Cl_2N_2O_2$	$C(CH_3)_2CONClCONCl$ (bracketed below)	53.627 55.950	2 2	[53]
C_6H_4ClNO	$O=\langle\text{ring}\rangle=NCl$	44.992	...	[83]
$C_6H_4Cl_3NO_2S$	$4\text{-}ClC_6H_4SO_2NCl_2$	52.038	2.5	*[84]. cf. C—Cl
$C_7H_7Cl_2NSO_2$	$4\text{-}CH_3C_6H_4SO_2NCl_2$	51.600	1	[53]

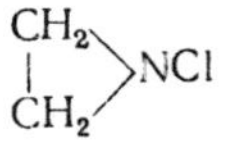

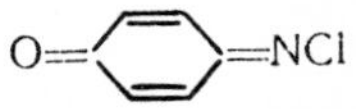

Empirical formula	Structural formula	ν, MHz	s/n	References, notes
$C_7H_8ClNSO_2$	4-$CH_3C_6H_4SO_2NHCl$	45.724	1.5	[53]
$C_9H_7Cl_3N_2O_2S$	4-$ClC_6H_4SO_2NCl$ (N–) $NCClHCCH_2$	51.680	2	*[84]. cf. C—Cl
$C_9H_9Cl_4NO_2S$	4-$ClC_6H_4SO_2NCl$ (N–) $ClH_2CClHCCH_2$	51.048	7	*[84]. cf. C—Cl
ClNO	NOCl	$e^2Qq =$ $= -57.2$ MHz $\eta = 32.5\%$		[85]. g
$ClNO_2$	NO_2Cl	$e^2Qq =$ $= -94.28$ MHz $\eta = 8.6\%$		[85]. g
CH_2Cl_3OP	$ClCH_2P(O)Cl_2$	26.860 27.175	50 50	*[86—88, 89a]. cf. C—Cl
CH_2Cl_3P	$ClCH_2PCl_2$	27.025 26.275	20 20	*[89a]. cf. C—Cl
CH_3ClFOP	$CH_3P(O)ClF$	25.835	7—10	*[87, 89a]
CH_3Cl_2OP	$CH_3P(O)Cl_2$	26.426 26.729	...	[86]
CH_3Cl_2OPS	$CH_3OP(S)Cl_2$	27.625 27.830	10 10	*[87, 89a]
	$CH_3SP(O)Cl_2$	27.100 27.083	...	*[89a]
$CH_3Cl_2O_2P$	$CH_3OP(O)Cl_2$	27.070 27.460	15 15	*[86—88, 89a]
CH_3Cl_2P	CH_3PCl_2	26.078 26.123	50 50	*[86, 89a]
CH_3Cl_2PS	$CH_3P(S)Cl_2$	27.6	...	*[87, 89a]
CH_3Cl_3NP	$CH_3N{=}PCl_3$	30.355 30.015	40 30	*
$C_2H_3Cl_2OP$	$CH_2{=}CHP(O)Cl_2$	26.248 26.270 26.428	7 10 7	*[88, 89a]
$C_2H_5Cl_2P$	$C_2H_5PCl_2$	25.940 25.805 25.740	8 15 8	*[89a]
$C_2H_5Cl_2OP$	$C_2H_5P(O)Cl_2$	26.200 26.325	100 100	*[86—89a]
$C_2H_5Cl_2O_2P$	$C_2H_5OP(O)Cl_2$	27.085 27.288	20 20	*[88, 89a]
C_2H_6ClOP	$(CH_3)_2P(O)Cl$	24.114	30	*[87, 89, 89a]

Empirical formula	Structural formula	ν, MHz	s/n	References, notes
C_2H_6ClOPS	$(CH_3)(CH_3O)P(S)Cl$	25.450	2	*[87, 89a]
$C_2H_6ClO_3P$	$(CH_3O)_2P(O)Cl$	25.420	20	*[89a]
C_2H_6ClPS	$(CH_3)_2P(S)Cl$	26.055	50	*[87, 89a]
$C_2H_6Cl_2NOP$	$(CH_3)_2NP(O)Cl_2$	25.785 27.073	30—50 30—50	*[87, 89a]
$C_2H_6Cl_2NPS$	$(CH_3)_2NP(S)Cl_2$	26.985 28.095	10 10	*[87, 89a]
$C_3H_7Cl_2OP$	$(C_3H_7)P(O)Cl_2$	26.040 26.273	50 50	*[88, 89a]
	$(iso\text{-}C_3H_7)P(O)Cl_2$	25.510 26.293	30 25	*[87—89a]
$C_3H_7Cl_2OPS$	$(CH_3)(ClCH_2CH_2S)P(O)Cl$	25.004	15	*[89a]. cf. C—Cl
$C_3H_7Cl_2O_2P$	$(C_3H_7O)P(O)Cl_2$	26.895 27.115	5 5	*[88, 89a]
$C_3H_7Cl_2P$	$iso\text{-}C_3H_7PCl_2$	25.580 26.095	30 30	*[89a]
C_3H_8ClOP	$CH_3(C_2H_5)P(O)Cl$	23.76	2—3	*[87, 89, 89a]
C_3H_8ClOPS	$CH_3(C_2H_5O)P(S)Cl$	25.535	3	*[87, 89a]
$C_3H_8ClO_2P$	$CH_3(C_2H_5O)P(O)Cl$	26.445 26.744	...	[86]
C_3H_8ClPS	$CH_3(C_2H_5)P(S)Cl$	25.85	1.5	*[89a]
$C_3H_8ClPS_2$	$CH_3(C_2H_5S)P(S)Cl$	26.230	4	*[87, 89a]
C_3H_9ClNOP	$CH_3[(CH_3)_2N]P(O)Cl$	22.384	...	*[87, 89a]
$C_4H_9Cl_2OP$	$C_4H_9P(O)Cl_2$	25.940 25.965 26.080	10 10 20	*[88, 89a]
	$iso\text{-}C_4H_9P(O)Cl_2$	26.173 26.285	50 50	*[88, 89a]
$C_4H_9Cl_2OPS$	$C_2H_5(ClCH_2CH_2S)P(O)Cl$	25.095	20	*[89a]. cf. C—Cl
$C_4H_9Cl_2O_2P$	$(C_4H_9O)P(O)Cl_2$	26.833 26.875	5 5	*[88, 89a]
$C_4H_{10}ClOP$	$CH_3(C_3H_7)P(O)Cl$	23.688	5	*[87, 89, 89a]
	$iso\text{-}C_3H_7(CH_3)P(O)Cl$	23.644	10	*[87, 89, 89a]
	$(C_2H_5)_2P(O)Cl$	23.220 23.240	5 5	*[87, 89, 89a]
$C_4H_{10}ClOPS$	$(C_3H_7)(CH_3S)P(O)Cl$	24.400	...	*[89a]

Empirical formula	Structural formula	ν, MHz	s/n	References, notes
$C_4H_{10}ClO_3P$	$(C_2H_5O)_2P(O)Cl$	25.100	20	*[89a]
$C_4H_{10}ClPS$	$(C_2H_5)_2P(S)Cl$	25.35	10	*[87, 89a]
$C_4H_{10}Cl_2NOP$	$[(C_2H_5)_2N]P(O)Cl_2$	26.097 27.034	. . .	*[86, 87, 89a]
$C_4H_{11}ClNOP$	$C_2H_5[(CH_3)_2N]P(O)Cl$	21.904 22.164	5 10	*[89a]
$C_4H_{12}ClN_2OP$	$[(CH_3)_2N]_2P(O)Cl$	22.224	30—50	*[87, 89a]
$C_5H_{11}ClOPS$	iso-$C_3H_7(ClCH_2CH_2S)P(O)Cl$	24.446	10	*[89a]. cf. C—Cl
$C_5H_{12}ClOP$	$C_4H_9(CH_3)P(O)Cl$	23.93	3	*[89, 89a]
	$C_2H_5(C_3H_7)P(O)Cl$	23.788	5	*[89, 89a]
$C_5H_{13}ClNOP$	$CH_3[(C_2H_5)_2N]P(O)Cl$	22.796	3—5	*[86, 87, 89a]
$C_6H_4Cl_3OP$	3-$ClC_6H_4P(O)Cl_2$	26.500	5	*[89a]. cf. C—Cl
	4-$ClC_6H_4P(O)Cl_2$	26.120 26.670	100 100	*[89a]. cf. C—Cl
$C_6H_4Cl_3P$	3-$ClC_6H_4PCl_2$	26.505 26.615	3 3	*[89a]. cf. C—Cl
»	4-$ClC_6H_4PCl_2$	26.578 26.550 26.413 26.335	5 5 5 5	*[89a]. cf. C—Cl
$C_6H_4Cl_3PS$	4-$ClC_6H_4P(S)Cl_2$	27.540 27.995	15 15	*[89a]
$C_6H_5Cl_2OP$	$C_6H_5P(O)Cl_2$	26.635 26.765	50 50	*[86—88, 89a]
$C_6H_5Cl_2O_2P$	$C_6H_5OP(O)Cl_2$	27.285 27.393 27.890 28.000	6 6 4 4	*[86, 88, 89a]
$C_6H_5Cl_2P$	$C_6H_5PCl_2$	26.240 26.540	10 10	*[86, 89a]
$C_6H_5Cl_2PS$	$C_6H_5P(S)Cl_2$	27.840	. . .	*[87, 89a]
$C_6H_{14}ClOP$	$(C_3H_7)_2P(O)Cl$	23.832	3—5	*[89, 89a]
	(iso-$C_3H_7)_2P(O)Cl$	23.220	20	*[89, 89a]
$C_6H_{14}ClOPS$	$C_3H_7(C_3H_7S)P(O)Cl$	24.264	. . .	*[89a]
	C_3H_7(iso-$C_3H_7S)P(O)Cl$	24.832 24.840	4 4	*[89a]
$C_6H_{14}ClO_2P$	C_3H_7(iso-$C_3H_7O)P(O)Cl$	24.104	2	*[89a]
	iso-C_3H_7(iso-$C_3H_7O)P(O)Cl$	23.880	10	*[89a]

Empirical formula	Structural formula	ν, MHz	s/n	References, notes
$C_6H_{14}ClO_3P$	$(iso\text{-}C_3H_7O)_2P(O)Cl$	25.042	50	*[89a]
$C_6H_{14}ClPS$	$(C_3H_7)_2P(S)Cl$	26.00	2	*[89a]
$C_7H_7Cl_2OP$	$4\text{-}CH_3C_6H_4P(O)Cl_2$	26.540 26.553	20 20	*[88, 89a]
$C_7H_7Cl_2O_2P$	$4\text{-}CH_3C_6H_4OP(O)Cl_2$	27.783 28.495	40 40	*[88, 89a]
$C_7H_8ClO_2P$	$CH_3(C_6H_5O)P(O)Cl$	24.88 25.00	2—3 2—3	*[89a]
C_7H_8ClPS	$CH_3(C_6H_5)P(S)Cl$	25.78	2	*[89a]
$C_7H_{16}ClO_2P$	$C_3H_7(C_4H_9O)P(O)Cl$	23.700	3	*[89a]
$C_8H_{10}ClOP$	$C_2H_5(C_6H_5)P(O)Cl$	23.996 24.526	30 30	*[89, 89a]
$C_8H_{10}Cl_2NP$	$4\text{-}(CH_3)_2NC_6H_4PCl_2$	25.250 25.588 25.735 25.935	20 20 15 15	*[89a]
$C_8H_{18}ClOP$	$(C_4H_9)_2P(O)Cl$	23.67	1.5	*[89a]
$C_8H_{20}Cl_6NP$	$[(C_2H_5)_4N]\,PCl_6$	29.32 30.06 30.34	15 25 13	[80]
$C_9H_{12}ClOPS$	$C_3H_7S(C_6H_5)P(O)Cl$	25.008	10	*[89a]
$C_{12}H_{10}Cl_4N_3P_3$	$P_3N_3Cl_4(C_6H_5)_2\text{-}1,1$	27.4 27.8	3 1	[90]
ClF_2OP	$P(O)F_2Cl$	23.320	50	*[87, 89a]
Cl_2FOP	$P(O)Cl_2F$	28.465 29.048	10—15 10—15	*[2, 87, 89a]
Cl_2F_3P	PCl_2F_3	31.49	...	[91]
Cl_3F_2P	PCl_3F_2	31.26 31.89	...	[91]
Cl_3OP	$POCl_3$	28.930 28.975	10 20	*[2, 22, 86, 87, 89a, 92—94, 192]
		e^2Qq = —55.4 MHz (g)		
$Cl_3OP + AsCl_3$	$POCl_3\cdot AsCl_3$	29.154 29.168 29.208	4 5 6	[22]. cf. As—Cl
$Cl_3OP + Cl_3Fe$	$POCl_3\cdot FeCl_3$	30.263	...	[78]
$Cl_3OP + Cl_5Nb$	$POCl_3\cdot NbCl_5$	30.366 30.234	5 10	*
$Cl_3OP + Cl_5Sb$	$POCl_3\cdot SbCl_5$	30.58 30.64	27 16	[78, 95]. cf. Sb, Cl

Empirical formula	Structural formula	ν, MHz	s/n	References, notes
$2Cl_3OP + Cl_4Sn$	$2POCl_3 \cdot SnCl_4$	30.117 30.213	2—3 5	*[22, 77, 78]. cf. Sn—Cl
$Cl_3OP + Cl_5Ta$	$POCl_3 \cdot TaCl_5$	30.420 30.540	20 10	*
$Cl_3OP + Cl_4Ti$	$(POCl_3 \cdot TiCl_4)_2$	29.987 30.112 30.313	...	[78]
Cl_3P	PCl_3	26.208 26.104	20 10	*[2, 86, 89a, 93, 94, 192]
		$e^2Qq = -53.4$ MHz [g]		
Cl_3PS	$PSCl_3$	29.675 29.800	5 10	*[87, 89a]
Cl_4FP	PCl_4F	28.99 32.54	10 30	[91]
$Cl_4F_6P_2$	$PCl_4[PF_6]$	32.46	15	[80]
Cl_5P	PCl_5 (ionic form, $[PCl_4]$ $[PCl_6]$)	28.400 29.715 30.070 30.475 30.570 32.274 32.388 32.424 32.598	15 15 25 20 20 50 50 30 30	*[80, 89a, 93, 96, 97]
	PCl_5 (mol.)	29.242 29.274 33.751	2 2 6	[97]
$Cl_6N_3P_3$	$(PNCl_2)_3$	28.316 28.327 28.547 28.656	1 2 1 2	[90, 93, 98—102]
$Cl_6N_3P_3 + ClHO_4$	$(PNCl_2)_3 \cdot HClO_4$	28.304 28.329 28.602 28.688	1 2 1 2	[90]
$Cl_8N_4P_4$	$(PNCl_2)_4$			[90, 98—103]
	Phase I	27.814 28.657 28.699 29.073	3 3 3 3	
	Phase II	28.611 28.635		

Empirical formula	Structural formula	ν, MHz	s/n	References, notes
$Cl_{10}N_5P_5$	$(PNCl_2)_5$	27.687	3	[90]
		27.841	4	
		27.927	3	
		27.942	2	
		28 222	3	
		28.247	3	
		28.440	2	
		28.485	3	
		28.744	1	
$Cl_{10}PSb$	$PCl_4[SbCl_6]$	32.275	...	[97]
		32.478		
		31.87	5	[80].
		32.35	12	cf. Sb—Cl
		32.51	20	
$Cl_{12}N_6P_6$	$(PNCl_2)_6$			[90].
	Phase I	28.4	2	Broad diffusion line in phase I
	Phase II	27.682		
		27.747		
		27.793		
		27.826		
		27.873		
		27.922		
		27.972		
		28.026		
		28.039		
		28.089		
		28.309		
		28.415		
$lAsCl_8$	$AsCl_4[AlCl_4]$	36.82	5	[80]
		37.03	7	
sCl_3	$AsCl_3$	24.960	...	[2, 22, 104, 193]. cf. As
		25.058		
		25.406		
$sCl_3 + Cl_3OP$	$AsCl_3 \cdot POCl_3$	24.799	2	[22].
		25.125	2	cf. P—Cl
		25.481	3	
$_3H_{20}AsCl_6N$	$[(C_2H_5)_4N]\,[AsCl_6]$	28.37	5	[105]
		29.15	5	
		30.07	5	
$_2H_{10}AsCl$	$(C_6H_5)_2AsCl$	24.312	20	*[194]. cf. As (supplement)
gCl_6Cs_2Sb	$Cs_2Ag[SbCl_6]$	12.344	...	
$_3H_9Cl_2Sb$	$(CH_3)_3SbCl_2$	$eQq = -28.091$ MHz		[106]. cf. Sb

Empirical formula	Structural formula	ν, MHz	s/n	References, notes
$C_8H_{20}Cl_6NSb$	$[(C_2H_5)_4N]\,[SbCl_6]$	24.01 24.21 24.67 24.86	4 6 8 5	[80]
Cl_3Sb	$SbCl_3$	19.306 20.914	20 10	*[2, 22, 104, 107—110]. cf. Sb
		At 300 °K $\eta_1 = 15.3\%$; $\eta_2 = 5.7\%$		
$Cl_3Sb + {}^1/_2C_6H_6$	$2SbCl_3 \cdot C_6H_6$	18.712 18.872 20.072 20.336 20.466 20.526	2 2 2 2 2 2	[22, 110—116]. cf. Sb
		At 294 °K $\eta_1 = 16.0\%$; $\eta_2 = 11.2\%$; $\eta_3 = 11.9\%$; $\eta_4 = 11.7\%$; $\eta_5 = 5.5\%$; $\eta_6 = 8.5\%$		
$Cl_2Sb + {}^1/_2C_6H_6O$	$2SbCl_3 \cdot C_6H_5OH$	19.132 19.295 19.457 19.520 19.676 19.768 19.798 19.920 20.110 20.525 20.619 20.676	1.5 1.5 1.5 1.5 1.5 1.5 1.5 1.5 2 2 2 2	[114]. cf. Sb
$Cl_3Sb + {}^1/_2C_7H_8O$	$2SbCl_3 \cdot C_6H_5OCH_3$	19.177 19.280 20.180 20.457 20.504 20.844	2 2.5 2 2.5 2.5 2	[114]. cf. Sb
$Cl_3Sb + C_7H_8O$	$SbCl_3 \cdot C_6H_5OCH_3$	19.195 19.705 20.316	3.5 3 3	[114]. cf. Sb
$Cl_3Sb + {}^1/_2C_8H_{10}$	$2SbCl_3 \cdot 1,4\text{-}(CH_3)_2C_6H_4$	18.361 19.609 20.631	2 2—3 3	[22]. cf. Sb
$Cl_3Sb + C_8H_{10}$	$SbCl_3 \cdot 1,4\text{-}(CH_3)_2C_6H_4$	18.500 20.402 20.956	1.5 2 1.5	[144]. cf. Sb

Empirical formula	Structural formula	ν, MHz	s/n	References, notes
$:l_3Sb + C_8H_{10}$	$SbCl_3 \cdot C_6H_5C_2H_5$	19.200 19.555 20,076	2.5 2.5 1.5	[111–116]. cf. Sb
$:l_3Sb + C_8H_{10}O$	$SbCl_3 \cdot C_6H_5OC_2H_5$	19,156 19,380 19,665	1.5 1.5 2	[114, 115]. cf. Sb
$:l_3Sb + {}^1/_2C_{10}H_8$	$2SbCl_3 \cdot C_{10}H_8$	19.260 19.966 20.800	2 2.5 2.5	[22, 114]. cf. Sb
$:l_3Sb + {}^1/_2C_{12}H_{10}$	$2SbCl_3 \cdot C_6H_5C_6H_5$	18.844 19.217 19.311 19.749 20.388 20.661	1.3 2 1.3 1.3 2 1.5	[22, 114]. cf. Sb
$l_3Sb + {}^1/_2C_{13}H_{12}$	$2SbCl_3 \cdot (C_6H_5)_2CH_2$	19.342 19.810 20,196	1.5 1.5 1.5	[114, 115]. cf. Sb
$l_3Sb + C_{13}H_{12}$	$SbCl_3 \cdot (C_6H_5)_2CH_2$	19.342 19.810 20.196	...	[115]. cf. Sb
$l_3Sh + C_{14}H_{10}$	$SbCl_3 \cdot C_{14}H_{10}$ (anthracene)	19.337 19.545 20.548	1.5 1.5 1.5	[114] cf. Sb
l_5Sb	$SbCl_5$			*[95, 96]. cf. Sb
	Phase I (mol.)	27.780 27.930 30.180	20 20 60	
	Phase II (dimeric)	18.76 27.42 27.85 28.01 29.80 29.98 30.42 30.50	8 11 11 16 10 10 10 10	
$l_5Sb + C_2H_3N$	$SbCl_5 \cdot CH_3CN$	24.99 26.37	6 22	[95]. cf. Sb
$l_5Sb + Cl_3OP$	$SbCl_5 \cdot POCl_3$	24.44 25.85 26.22 27.33	5 9 8 21	[78, 95]. cf. Sb, P–Cl
l_6Cs_2KSb	$Cs_2K[SbCl_6]$	13.476	...	*
l_6Cs_2NaSb	$Cs_2Na[SbCl_6[$	13.544	...	*

Empirical formula	Structural formula	ν, MHz	s/n	References, notes
Cl_6NOSb	$NO[SbCl_6]$	22.35 22.97 23.15 23.37 25.03 25.48	6 5 4 3 4 4	[80]
$Cl_{10}PSb$	$PCl_4[SbCl_6]$	24.03 24.96 22.80 23.02	... 2 2	[97] [80]. cf. P—Cl
$BiCl_3$	$BiCl_3$	15.387 19.531	...	[117].T=83°K cf. Bi
Cl_3V	VCl_3	9.40	...	[81]. T = 297 °K
Cl_3OV	$VOCl_3$	11.54	...	[34]. T = 83 °K. cf. ^{51}V (sup.)
Cl_5Nb	$NbCl_5$	13.280	...	[45, 118]. cf. Nb
Cl_5Ta	$TaCl_5$	12.30 (doublet) 13.6 (triplet)	... ...	*[45, 119, 207]
	Group VI element — chlorine bond			
$AgClO_2$	$AgClO_2$	54.08	...	[121]. T = 297 °K. η = 0.73%
$AgClO_3$	$AgClO_3$	29.421	...	[120, 122, 123]
$AlCl_3O_9 + 6H_2O$	$Al(ClO_3)_3 \cdot 6H_2O$	29.86	...	[120]
$BaCl_2O_6 + H_2O$	$Ba(ClO_3)_2 \cdot H_2O$	29,922	5	[120, 122, 124—128]
CH_3ClO	CH_3OCl	57.96 e^2Qq = η	50	*[129, 130] —114.0 MHz (g); = 4.0% (g)
$CaCl_2O_6$	$Ca(ClO_3)_2$	29.98	...	[126, 131]. T = 90 °K
$CaCl_2O_6 + 6H_2O$	$Ca(ClO_3)_2 \cdot 6H_2O$	30.11	...	[120]
$CdC_2O_6 + 2H_2O$	$Cd(ClO_3)_2 \cdot 2H_2O$	29.893	...	[120]
$ClCsO_3$	$CsClO_3$	28.37	...	[132]. T = 297 °K cf. Cs
ClH_4NO_3	$(NH_4)ClO_3$	28.46	...	[123]. T = 298 °K

Empirical formula	Structural formula	ν, MHz	s/n	References, notes
$ClKO_3$	$KClO_3$	28.954	...	*[100, 120, 122, 127, 132—137]. cf. K
$ClNaO_2 + xH_2O$	$NaClO_2 \cdot xH_2O$	53.33	...	[121]
$ClNaO_3$	$NaClO_3$	30.630	...	*[108, 120, 122, 132, 133, 138, 139]. cf. Na
ClO_3Rb	$RbClO_3$	28.81	...	[132]. $T = 297$ °K. cf. Rb
$Cl_2CoO_6 + 6H_2O$	$Co(ClO_3)_2 \cdot 6H_2O$	30.080	...	[120]
$Cl_2CuO_6 + 6H_2O$	$Cu(ClO_3)_2 \cdot 6H_2O$	30.064	13—16	[120, 140]
Cl_2MgO_6	$Mg(ClO_3)_2$	29.99	...	[126]. $T = 90$ °K
$Cl_2MgO_6 + 6H_2O$	$Mg(ClO_3)_2 \cdot 6H_2O$	29.885	3	[120, 140]
$Cl_2NiO_6 + 6H_2O$	$Ni(ClO_3)_2 \cdot 6H_2O$	30.572	...	[120, 122]
$Cl_2O_6Pb + H_2O$	$Pb(ClO_3)_2 \cdot H_2O$	29.712 29.233	...	[120]
Cl_2O_6Sr	$Sr(ClO_3)_2$	29.869	6	[120, 140]
$Cl_2O_6Zn + 4H_2O$	$Zn(ClO_3)_2 \cdot 4H_2O$	31.001	...	[120]
$Cl_2O_6Zn + xH_2O$	$Zn(ClO_3)_2 \cdot xH_2O$	30.026	...	[122]
$Cl_3GaO_9 + H_2O$	$Ga(ClO_3)_3 \cdot xH_2O$	29.832 29.388	...	[120]
$Cl_4O_{12}Pt$	$Pt(ClO_3)_4$	30.94	...	[120]
CCl_4O_2S	CCl_3SO_2Cl	36.294	4	*[87]. cf. C—Cl
$CH_2Cl_2O_2S$	$ClCH_2SO_2Cl$	34.632 34.128	20 15	* cf. C—Cl
CH_3ClO_2S	CH_3SO_2Cl	33.228	20	*[87]
CH_3ClO_3S	CH_3OSO_2Cl	35.538 35.766	20 20	*[87]
$C_2H_3ClO_2S$	$CH_2{=}CHSO_2Cl$	32.820	...	*[87]
$C_2H_4Cl_2O_2S$	$ClCH_2CH_2SO_2Cl$	33.672	5	* cf. C—Cl
$C_2H_5ClO_2S$	$C_2H_5SO_2Cl$	32.519	2	[141]
$C_2H_5ClO_3S$	$C_2H_5OSO_2Cl$	35.658	20	*[87]
$C_2H_6ClNO_2S$	$[(CH_3)_2N]SO_2Cl$	32.166	15	*[87]
$C_3H_7ClO_2S$	$C_3H_7SO_2Cl$	32.136	50	*[87]

Empirical formula	Structural formula	ν, MHz	s/n	References, notes
$C_3H_7ClO_2S$	iso-$C_3H_7SO_2Cl$	32.652	15	*[87]
$C_4H_3ClO_2S_2$	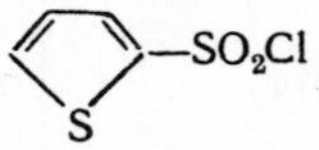	33.151	20	[141]
$C_4H_6Cl_2O_2S$	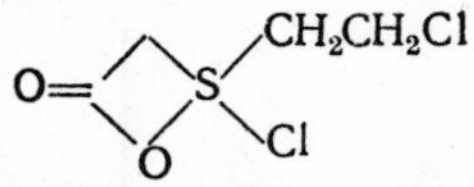	34.320	2	[53]
$C_4H_9ClO_2S$	$C_4H_9SO_2Cl$	32.759	3—4	[141]
$C_4H_{10}ClNO_2S$	$[(C_2H_5)_2N]SO_2Cl$	31.944	20	*[87]
$C_6ClF_5O_2S$	$C_6F_5SO_2Cl$	35.154	10	*[199]
$C_6Cl_6O_2S$	$Cl_5C_6SO_2Cl$	34.320	20	*[199]. cf.. C—Cl
$C_6HCl_5O_2S$	2,3,5,6-$Cl_4C_6HSO_2Cl$	34.408	10	The same
$C_6H_2Cl_2F_2O_2S$	2,4-F_2-5-$ClC_6H_2SO_2Cl$	33.858	2—3	»
$C_6H_2Cl_3FO_2S$	2-F-4,5-$Cl_2C_6H_2SO_2Cl$	34.644	20	»
	2,5-Cl_2-4-$FC_6H_2SO_2Cl$	34.266	20	»
$C_6H_2Cl_4O_2S$	2,4,5-$Cl_3C_6H_2SO_2Cl$	34.299	30	»
$C_6H_3ClF_2O_2S$	2,5-$F_2C_6H_3SO_2Cl$	33.446	10	*[199]
$C_6H_3ClN_2O_4S$	2,4-$(NO_2)_2C_6H_3SCl$	34.752	10	*[200]
$C_6H_3ClN_2O_4S$ + + Cl_5Sb	2,4-$(NO_2)_2C_6H_3SCl \cdot SbCl_5$	34.014	6	*[200]
$C_6H_3Cl_2NO_2S$	4-Cl-2-$NO_2C_6H_3SCl$	37.332	5	*[200]. cf. C—Cl
$C_6H_3Cl_2NO_4S$	3-NO_2-4-$ClC_6H_3SO_2Cl$	33.096	20	*[199]. cf. C—Cl
$C_6H_3Cl_3O_2S$	2,4-$Cl_2C_6H_3SO_2Cl$	33.828	30	* The same
	2,5-$Cl_2C_6H_3SO_2Cl$	34.350 34.449	8 10	*[141, 199] cf. C—Cl
$C_6H_4BrClO_2S$	4-$BrC_6H_4SO_2Cl$	32.895	3	[124]
C_6H_4BrClS	2-BrC_6H_4SCl	38.143	10	*[200]
$C_6H_4ClFO_2S$	4-$FC_6H_4SO_2Cl$	32.808	20	*[199]
$C_6H_4ClNO_2S$	2-$NO_2C_6H_4SCl$	35.966 36.330	10 10	*[200]
	4-$NO_2C_6H_4SCl$	38.738	3	*[200]
$C_6H_4ClNO_4S$	3-$NO_2C_6H_4SO_2Cl$	32.832 34.337	~1 3	[124]

Empirical formula	Structural formula	ν, MHz	s/n	References, notes
$C_6H_4ClNO_4S$	$4\text{-}NO_2C_6H_4SO_2Cl$	33.438	4	[124].
$C_6H_4Cl_2O_2S$	$4\text{-}ClC_6H_4SO_2Cl$	32.826	3	*[124], cf. C—Cl
$C_6H_4Cl_2O_4S_2$	$1,4\text{-}(SO_2Cl)_2C_6H_4$	34.452 33.912 33.708 32.568	10 10 10 10	*
$C_6H_4Cl_2S$	$2\text{-}ClC_6H_4SCl$	38.465	20	*[200]. cf. C—Cl
	$4\text{-}ClC_6H_4SCl$	37.947	10	The same
$C_6H_5ClO_2S$	$C_6H_5SO_2Cl$	32.892 32.538 32.470 32.538	15 15 10 20	[96, 141] *[199]
$C_6H_5ClO_3S$	$C_6H_5OSO_2Cl$	36.540	30	*[204]
C_6H_5ClS	C_6H_5SCl	37.016	50	*[200]
$C_7H_5ClO_4S$	$3\text{-}SO_2ClC_6H_4COOH$	33.672	1	[53]
$C_7H_7ClO_2S$	$4\text{-}CH_3C_6H_4SO_2Cl$	32.254	50	*[96, 199]
C_7H_7ClS	$4\text{-}CH_3C_6H_4SCl$	37.002 37.212	20 20	*[200]
$C_8H_4ClO_2MnS$	$(C_5H_4SO_2Cl)Mn(CO)_3$	33.56	5	*[182, 205]. cf. Mn
$C_8H_4ClO_2ReS$	$(C_5H_4SO_2Cl)Re(CO)_3$	33.84	4	*[182]. cf. Re
$C_8H_9ClO_2S$	$2,5\text{-}(CH_3)_2C_6H_3SO_2Cl$	32.7579 32.5269 32.3178 32.2698	15 2 6 8	[141]
$ClFO_2S$	SO_2FCl	39.354	...	*[87]
Cl_2OS	$SOCl_2$	32.088 31.884 31.584	30 25 50	*[2, 92, 93]
Cl_2O_2S	SO_2Cl_2	37.822 37.613	4—5 4—5	[93, 140]
Cl_2S_2	S_2Cl_2	35.605 35.994	10 10	[53]
$Cl_3N_3S_3$	$(SNCl)_3$	29.842	...	[93]. $T = 285$ °K

Empirical formula	Structural formula	ν, MHz	s/n	References, notes
Cl_4Se	$SeCl_4$	34.534 34.685 35.041 35.696 36.171 36.794	3 <2 2—3 2 2 <2	[140]
Cl_6K_2Se	K_2SeCl_6	20.576	...	[142]
$Cl_6H_8N_2Se$	$(NH_4)_2SeCl_6$	20.877	...	[142]
$C_2H_6Cl_2Te$	$(CH_3)_2TeCl_2$			[143, 195]
	Phase I	17.940	...	
	Phase II	18.127 15.685		
Cl_3Te	$TeCl_3$	14.756 14.812	...	[19]. $T = 297\ °K$
Cl_4Te	$TeCl_4$	27.304 27.370 27.471 27.926 28.110 28.461	...	[144].T=83°K
$Cl_6H_8N_2Te$	$(NH_4)_2TeCl_6$	15.137	...	[145]
Cl_2Cr	$CrCl_2$	8.52	...	[81]. $T = 297\ °K$
Cl_2CrO_2	CrO_2Cl_2	15.68	2	[34].T=83 °K
Cl_3Cr	$CrCl_3$	12.812 12.848	2 1	[81, 206]. $T = 297\ °K$
Cl_2MoO_2	MoO_2Cl_2	15.40	...	[34].T= 83 °K
Cl_6K_2W	K_2WCl_6	10.22	...	[146, 147]. $T = 295.5\ °K$
Cl_6W	WCl_6	10.172	...	[148]
	Group VII element — chlorine bond			
ClF	ClF	70.700	...	[2, 149, 150]. $T = 20\ °K$
		$e^2Qq = -145.99$ MHz (g)		
ClF_3	ClF_3	75.1295	...	[2, 151]
		$e^2Qq = -145$ MHz (g) $\eta = 11.7\%$ (g)		
Cl_2	Cl_2	54.247	...	[2, 46, 152, 153, 196]

Empirical formula	Structural formula	ν, MHz	s/n	References, notes
BrCl	BrCl	$e^2Qq =$ = —103.6 MHz		[154] g cf. Br
$AlCl_6I$	$ICl_2[AlCl_4]$	39.086 38.690	...	[35]. $T = 297$ °K. cf. Cl—Al
$C_4H_{12}Cl_2IN$	$[(CH_3)_4N]ICl_2$	19.35	...	[155]. $T = 295$ °K
$C_5H_6Cl_2IN$	$[C_5H_5NH]ICl_2$	17.62	...	[155]. $T = 293$ °K
$C_5H_6Cl_4IN$	$[C_5H_5NH]ICl_4$	22.38	...	[155]. $T = 294$ °K
$C_6H_5Cl_2I$	$C_6H_5ICl_2$	22.406 24.260	...	[156]
ClI	α-ICl	37.184	...	[157—159]. $T = 293$ °K. cf. I
		$e^2Qq =$ —82.5 MHz (g)		
	β-ICl	37.202	3—4	[140]
Cl_2CsI	$CsICl_2$	20.0895	25	[160]. $T = 76.9$ °K. cf. I
Cl_2H_4IN	$(NH_4)ICl_2$	26.14	...	[155]. $T = 294$ °K
Cl_2IK	$KICl_2$	19.627 19.002 18.764 18.408	14 14 14 14	[160]. $T = 298$ °K. cf. I
$Cl_2IK + H_2O$	$KICl_2 \cdot H_2O$	19.207	12	[155, 160]. cf. I
Cl_3I	ICl_3 $(ICl_3)_2$	35.680 33.916 ... 34.918 33.413 13.740	1,5 2,5	[35, 140, 161 162, 197]. cf. I $T = 297$ °K
Cl_4CsI	$CsICl_4$			*[160]. cf. I
	Phase I	22.612 22.892	7 7	
	Phase II	22.476 23.064	10 10	
Cl_4H_4IN	$(NH_4)ICl_4$	22.23	...	[155]. $T = 295$ °K

Empirical formula	Structural formula	ν, MHz	s/n	References, notes
$Cl_4H_4IN + H_2O$	$(NH_4)ICl_4 \cdot H_2O$	19.98 24.68 27.96	...	[155]. $T = 296$ °K
Cl_4IK	$KICl_4$	22.604	34	[155, 160]. cf. I
$Cl_4IK + H_2O$	$KICl_4 \cdot H_2O$	28.500 25.013 20.351	22 17 21	The same
$Cl_4INa + H_2O$	$NaICl_4 \cdot 2H_2O$	20.010 22.656 23.251 23.827	14 19 16 13	[160]. cf. I
Cl_4IRb	$RbICl_4$			*[160]. cf. I
	Phase I	22.540	10	
	Phase II	22.628 22.116	20 20	
Cl_6IP	$(PCl_4)ICl_2$	18.94	...	[155]. $T = 293$ °K
ClO_3Re	ReO_3Cl	$e^2Qq = -34$ MHz		[163—165]. g cf. Re
Cl_6Cs_2R	$Cs_2[ReCl_6]$	14.604	...	[166]
$Cl_6H_8N_2Re$	$(NH_4)_2[ReCl_6]$	14.125	...	[166]
Cl_6K_2Re	$K_2[ReCl_6]$	13.966 13.986	...	[146, 147, 167]
Cl_6RbRe	$Rh_2[ReCl_6]$	14.248	...	[166]
	Group VIII element — chlorine bond			
Cl_2Fe	$FeCl_2$	2.37	...	[168]
Cl_3Fe	$FeCl_3$	10.118	...	[169]
$C_7H_5ClFeO_2$	$C_5H_5Fe(CO)_2Cl$	18.39	...	*
$C_4H_{10}Cl_3CoN_2 +$ $+ HCl + 4H_2O$	trans-$[Co(en)_2Cl_2]Cl \cdot HCl \cdot 4H_2O$	16.058 2.858	...	[170]. Room temp. cf. Co
Cl_2Co	$CoCl_2$	2.62	...	[171, 172]. $T = 4{,}2$ °K
$Cl_2Co + 2H_2O$	$CoCl_2 \cdot 2H_2O$	$e^2Qq =$ $= 9.866$ MHz $\eta = 44\%$		[173]. $T = 76$ °K
$Cl_2Co + 6H_2O$	$CoCl_2 \cdot 6H_2O$	$e^2Qq = 5.508$ MHz $\eta = 8\%$		[174, 198]

Empirical formula	Structural formula	ν, MHz	s/n	References, notes
Cl_6K_2Pd	$K_2[PdCl_6]$	26.75	2	[175, 181]
Cl_6K_2Os	$K_2[OsCl_6]$	16.897	...	[142, 176]
Cl_6IrK_2	$K_2[IrCl_6]$	20.841	7	[142, 176]
Cl_4K_2Pt	$K_2[PtCl_4]$	18.13	...	*[177]
Cl_6Cs_2Pt	$Cs_2[PtCl_6]$	26.70	...	[178]. $T = 78$ °K
$Cl_6H_2Pt + 6H_2O$	$H_2[PtCl_6]\cdot 6H_2O$	26.55	...	[175, 181]. $T = 298$ °K
$Cl_6H_{12}N_4Pt_2$	$(NH_3)_4Pt_2Cl_6$	29.00 24.62 17.15	...	*
$Cl_6H_8N_2Pt$	$(NH_4)_2[PtCl_6]$	26.282	...	[79]. $T = 78$ °K
Cl_6K_2Pt	$K_2[PtCl_6]$	26.021	3	[179, 180]
$Cl_6Na_2Pt + 6H_2O$	$Na_2[PtCl_6]\cdot 6H_2O$	25.730 26.470 27.040	...	[175, 181]. $T = 296$ °K
Cl_6PtRb_2	$Rb_2[PtCl_6]$	26.44	...	[178]. $T = 78$ °K

BIBLIOGRAPHY

1. Allen, H. — J. Phys. Chem. 57 (1953), 501.
2. Livingston, R. — J. Phys. Chem. 57 (1953), 496.
3. Meal, H. and H. Allen. — Phys. Rev. 90 (1953), 348.
4. Okuma, H., N. Nakamura, and H. Chihara. — J. Phys. Soc. Japan 24 (1968), 452.
5. Burrus, C. — Phys. Rev. 97 (1955), 1661.
6. Evans, J. and G. Lo. — J. Phys. Chem. 70 (1966), 2702.
6a. Haas, T. and S. Welsh. — J. Phys. Chem. 71 (1967), 3363.
7. Evans, J. and G. Lo. — J. Phys. Chem. 71 (1967), 3697.
8. Lee, C., B. Fabrikand, R. Carlson, and I. Rabi. — Phys. Rev. 91 (1953), 1395.
9. Lee, C., B. Fabrikand, R. Carlson, and I. Rabi. — Phys. Rev. 86 (1952), 607.
10. Trischka, J. and R. Braunstein. — Phys. Rev. 96 (1954), 968.
11. Luce, R. and J. Trischka. — Phys. Rev. 82 (1951), 323.
12. Luce, R. and J. Trischka. — Phys. Rev. 83 (1951), 851.
13. Robinson, W., W. O'Sullivan, and W. Simmons. — Bull. Am. Phys. Soc., ser. 2, 6 (1961), 105.
14. Segel, S. and R. Barnes. Catalog of Nuclear Quadrupole Interactions and Resonance Frequencies in Solids. — US Atomic Energy Commission, 1962.
15. Caglioti, V., S. Sartori, and C. Furlani. — Proc. VIII Intern. Conf. on Coordinate Chemistry, p. 81, Vienna, 1964.
16. Caglioti, V., C. Furlani, F. Orestano, and F. Capece. — Atti Accad. naz. Lincei Rc. Classe scienze fiziche, matematiche e naturali 35 (1963), 10.
17. Bregadze, V. I., T. A. Babushkina, O. Yu. Okhlobystin, and G. K. Semin. — Teor. Eksperim. Khim. 3 (1967), 547.
18. Dehmelt, H., H. Robinson, and W. Gordy. — Phys. Rev. 93 (1954), 480.
19. Hultsch, H. and R. Barnes. — Phys. Rev. Lett. 1 (1958), 227.
20. Buyle-Bodin, M. and A. Monfils. — Compt. rend. 236 (1953), 1157.

21. Dehmelt, H., H. Robinson, and W. Gordy. — Phys. Rev. 93 (1954), 920.
22. Biedenkapp, D. and A. Weiss. — Z. Naturf. 19a (1964), 1518.
23. Dinesh and P. Narasimhan. — J. Chem. Phys. 45 (1966), 2170.
24. Negita, H., T. Tanaka, T. Okuda, and H. Shimada. — Inorg. Chem. 5 (1966), 2126.
25. Ramakrishna, J. — Indian J. Pure Appl. Phys. 5 (1967), 197.
26. Gordy, W. and J. Sheridan. — J. Chem. Phys. 22 (1954), 92.
27. Chiba, T. — J. Phys. Soc. Japan 13 (1958), 860.
28. Nakamura, D., H. Watanabe, and M. Kubo. — Bull. Chem. Soc. Japan 34 (1961), 142.
29. Smith, J. and D. Tong. — Chem. Communs., p. 3, 1965.
30. Stanko, V. I., Yu. T. Struchkov, A. I. Klimova, E. V. Bryukhova, and G. K. Semin. — Zh. Obshch. Khim. 36 (1966), 1707.
31. Bryukhova, E. V., V. I. Stanko, A. I. Klimova, N. S. Titova, and G. K. Semin. — Zh. Strukt. Khim. 9 (1968), 39.
32. Zakharkin, L. I., O. Yu. Okhlobystin, G. K. Semin, and T. A. Babushkina. — Izv. AN SSSR, chem. ser., p. 1913, 1965.
33. Lide, D. — J. Chem. Phys. 42 (1965), 1013.
34. Dehmelt, H. — J. Chem. Phys. 21 (1953), 380.
35. Evans, J. and G. Lo. — Inorg. Chem. 6 (1967), 836.
36. Lucken, E. and C. Mazeline. — J. Chem. Soc. (A), p. 153, 1968.
37. Barrett, A. and M. Mandel. — Phys. Rev. 109 (1958), 1572.
38. Dehmelt, H. — Phys. Rev. 92 (1953), 1240.
39. Hofft, J. — Z. Phys. 163 (1961), 262.
40. Delvigne, G. and H. de Wijn. — J. Chem. Phys. 45 (1966), 3318.
41. Carr, S., B. Garrett, and W. Moulton. — J. Chem. Phys. 47 (1967), 1170.
42. Mangum, B. and B. Utton. — Bull. Amer. Phys. Soc., ser. 2, 12 (1967), 1043.
43. Hughes, W., C. Montgomery, W. Moulton, and E. Carlson. — J. Chem. Phys. 41 (1964), 3470.
44. Carlson, E. — Bull. Am. Phys. Soc., ser. 2, 11 (1966), 377.
45. Reddoch, A. — US Atomic Energy Commission, UCRL-8972, 1959.
46. Biryukov, I. P., M. G. Voronkov, and I. A. Safin. — Izv. AN LatvSSR, chem. ser., No. 6 (1966), 638.
47. Biryukov, I. P., M. G. Voronkov, and I. A. Safin. — Teor. Eksperim. Khim. 1 (1965), 373.
48. Biryukov, I. P., M. G. Voronkov, G. V. Motsarev, V. R. Rozenberg, and I. A. Safin. — Doklady AN SSSR 162 (1965), 130.
49. Biryukov, I. P., M. G. Voronkov, and I. A. Safin. — In: "Radiospektroskopiya tverdogo tela," p. 252, Atomizdat, 1967.
50. Biryukov, I. P. and M. G. Voronkov. — Izv. AN LatvSSR, chem. ser., No. 2 (1965), 153.
51. Biryukov, I. P., M. G. Voronkov, B. N. Pavlov, and D. Ya. Shtern. — Izv. AN LatvSSR, chem. ser. No. 4 (1965), 501.
52. Semin, G. K., T. A. Babushkina, V. I. Robas, G. Ya. Zueva, M. A. Kadina, and V. I. Svergun. — In: "Radiospektroskopicheskie i kvantovokhimicheskie metody v strukturnykh issledovaniyakh," p. 225. "Nauka," 1967.
53. Hooper, H. and P. Bray. — J. Chem. Phys. 33 (1960), 334.
54. Biryukov, I. P. and M. G. Voronkov. — Izv. AN LatvSSR, chem. ser., No. 4 (1963), 495.
55. Biryukov, I. P., I. A. Safin, and M. G. Voronkov. — Izv. AN LatvSSR, chem. ser., No. 2 (1964), 181.
56. Babushkina, T. A., V. I. Robas, and G. K. Semin. — In: "Radiospektroskopiya tverdogo tela," p. 221. Atomizdat, 1967.
57. Williams, Q. and T. Weatherly. — J. Chem. Phys. 22 (1954), 572.
58. Babushkina, T. A. and G. K. Semin. — Kristallografiya 13 (1968), 529.
59. Biryukov, I. P. and M. G. Voronkov. — Izv. AN LatvSSR, chem. ser., No. 6 (1963), 689.
60. Biryukov, I. P., I. A. Safin, and M. G. Voronkov. — Izv. AN LatvSSR, chem. ser., No. 6 (1963), 695.
61. Biryukov, I. P., M. G. Voronkov, E. D. Babich, T. N. Arkhipova, V. M. Volovin, and N. S. Nametkin. — Doklady AN SSSR 161 (1965), 1336.
62. Sharbough, A. — Phys. Rev. 74 (1948), 1870.
63. Bak, B., J. Bruhn, and J. Rastrup-Anderson. — J. Chem. Phys. 21 (1953), 753.
64. Sheridan, J. and W. Gordy. — Phys. Rev. 77 (1950), 719.
65. Sheridan, J. and W. Gordy. — J. Chem. Phys. 19 (1951), 965.
66. Mitzlaff, M., R. Holm, and H. Hartmann. — Z. Naturf. 22a (1967), 288, 1415.
67. Semin, G. K. — In: "Radiospektroskopiya tverdogo tela," p. 205. Atomizdat, 1967.

Biryukov, I. P. and M. G. Voronkov. — Teor. Eksperim. Khim. 1 (1965), 124.
Townes, C., J. Mays, and B. Dailey. — Phys. Rev. 76 (1949), 137, 700.
Mays, J. and C. Townes. — Phys. Rev. 81 (1951), 940.
Graybeal, J. and M. Hsu. — Bull. Am. Phys. Soc., ser. 2, 12 (1967), 468.
Babushkina, T. A., G. K. Semin, S. P. Kolesnikov, O. M. Nefedov, and V. I. Svergun. — Izv. AN SSSR, chem. ser., p. 1055, 1969.
Semin, G. K., T. A. Babushkina, A. K. Prokof'ev, and R. G. Kostyanovskii. — Izv. AN SSSR, chem. ser., p. 1401, 1968.
Green, P. and J. Graybeal. — J. Am. Chem. Soc. 89 (1967), 4305.
Swiger, E. and J. Graybeal. — J. Am. Chem. Soc. 87 (1965), 1464.
Srivastava, T. — J. Organometall. Chem. 10 (1967), 373.
Kravchenko, E. A., Yu. K. Maksyutin, E. N. Gur'yanova, and G. K. Semin. — Izv. AN SSSR, chem. ser., p. 1271, 1968.
Rogers, M. and J. Ryan. — J. Phys. Chem. 72 (1968), 1340.
Nakamura, D. — Bull. Chem. Soc. Japan 36 (1963), 1662.
Di Lorenzo, J. and R. Schneider. — Inorg. Chem. 6 (1967), 766.
Barnes, R. and S. Segel. — Phys. Rev. Lett. 3 (1959), 462.
Barnes, R. and R. Engardt. — J. Chem. Phys. 21 (1958), 248.
Segel, S., R. Barnes, and P. Bray. — J. Chem. Phys. 25 (1956), 1286.
Freidlina, R. Kh., N. A. Rybakova, G. K. Semin, and E. A. Kravchenko. — Doklady AN SSSR, 176 (1967), 352.
Millen, D. and J. Pannell. — J. Chem. Soc., p. 1322, 1961.
Lucken, E. and M. Whitehead. — J. Chem. Soc., p. 2459, 1961.
Neimysheva, A. A., G. K. Semin, T. A. Babushkina, and I. L. Knunyants. — Doklady AN SSSR, 173 (1967), 585.
Tsvetkov, E. N., G. K. Semin, T. A. Babushkina, D. I. Lobanov, and M. I. Kabachnik, — Izv. AN SSSR, chem. ser., p. 2375, 1967.
Neimysheva, A. A., V. A. Pal'm, G. K. Semin, N. A. Loshadkin, and I. L. Knunyants. — Zh. Obshch. Khim. 37 (1967), 2255.
a. Semin, G. K. and T. A. Babushkina. — Teor. Eksperim. Khim. 4 (1968), 835.
Whitehead, M. — Can. J. Chem. 42 (1964), 1212.
Holmes, R., R. Carter, and G. Peterson. — Inorg. Chem. 3 (1964), 1748.
Livingston, R. — Phys. Rev. 82 (1951), 289.
Negita, H. and Z. Hirano. — Bull. Chem. Soc. Japan 31 (1958), 660.
Nave, C., T. Weatherly, and Q. Williams. — Bull. Am. Phys. Soc., ser. 2, 11 (1966), 312.
Schneider, R. and J. Di Lorenzo. — J. Chem. Phys. 47 (1967), 2343.
McCall, D. and H. Gutowsky. — J. Chem. Phys. 21 (1953), 1300.
Chihara, H., N. Nakamura, and S. Seki. — Bull. Chem. Soc. Japan 40 (1967), 50.
Kaplansky, M. and M. Whitehead. — Can. J. Chem. 45 (1967), 1669.
Torizuka, K. — J. Phys. Soc. Japan 11 (1956), 84.
Negita, H. and S. Satou. — Bull. Chem. Soc. Japan 29 (1956), 426.
Negita, H. and S. Satou. — J. Chem. Phys. 24 (1956), 621.
Negita, H. — J. Sci. Hiroshima Univ. A21 (1958), 261.
Dixon, M., H. Jenkins, J. Smith, and D. Tong. — Trans. Faraday Soc. 63 (1967), 2852.
Ogawa, S. — J. Phys. Soc. Japan 13 (1958), 618.
Di Lorenzo, J. and R. Schneider. — J. Phys. Chem. 72 (1968), 761.
Parker, D. — Diss. Abstr. Sect. B, 20 (1959), 2044.
Dehmelt, H. and H. Krüger. — Z. Phys. 130 (1951), 385.
Wang, T. — Phys. Rev. 99 (1955), 566.
Chihara, H., N. Nakamura, and H. Okuma. — J. Phys. Soc. Japan 24 (1968), 306.
Okuda, T., A. Nakao, M. Shiroyama, and H. Negita. — Bull. Chem. Soc. Japan 41 (1968), 61.
Grechishkin, V. S. and I. A. Kyuntsel'. — Optika Spektrosk. 16 (1964), 161.
Grechishkin, V. S. and I. A. Kyuntsel'. — Zh. Strukt. Khim. 5 (1964), 53.
Grechishkin, V. S. and I. A. Kyuntsel'. — Trudy ENI Permsk. State Univ. 11, No. 2 (1964), 119.
Grechishkin, V. S. and I. A. Kyuntsel'. — Trudy ENI Permsk. State Univ. 12, No. 1 (1966), 9.
Grechishkin, V. S. and I. A. Kyuntsel'. — Trudy ENI Permsk. State Univ. 12, No. 4 (1966), 29.
Grechishkin, V. S. and I. A. Kyuntsel'. — Optika Spektrosk. 15 (1963), 832.

117. Robinson, H. — Phys. Rev. 100 (1955), 1731.
118. Reddoch, A. — J. Chem. Phys. 35 (1961), 1085.
119. Duchesne, J., A. Monfils, and J. Garson. — Physica 22 (1956), 816.
120. Moshier, R. — Inorg. Chem. 3 (1964), 199.
121. Ragle, J. — J. Chem. Phys. 32 (1960), 403.
122. Livingston, R. — Rec. Chem. Prog. 20 (1959), 173.
123. Anderson, L. and W. Goodwin. — Bull. Am. Phys. Soc., ser. 2, 5 (1960), 412.
124. Bray, P. and R. Ring. — J. Chem. Phys. 21 (1953), 2226.
125. Torizuka, K. — J. Phys. Soc. Japan 9 (1954), 645.
126. Grechishkin, V. S. and F. I. Skripov. — Doklady AN SSSR 126 (1959), 1229.
127. Zeldes, H. and R. Livingston. — J. Chem. Phys. 26 (1957), 1102.
128. Nakamura, D., K. Ito, and M. Kubo. — Inorg. Chem. 1 (1962), 592.
129. Rigden, J. — Spectrochim. Acta 17 (1961), 1096.
130. Rigden and Butcher. — J. Chem. Phys. 40 (1964), 2109.
131. Grechishkin, V. S. — Zh. Eksper. Teoret. Fiz. 36 (1959), 630.
132. Emshwiller, M., E. Hahn, and D. Kaplan. — Phys. Rev. 118 (1960), 414.
133. Wang, T., C. Townes, A. Schawlow, and A. Holden. — Phys. Rev. 86 (1952), 809.
134. Kushida, T., N. Benedek, and N. Bloembergen. — Phys. Rev. 104 (1956), 1364.
135. Utton, D. — J. Chem. Phys. 47 (1967), 371.
136. Brodskii, A. D. and V. I. Solov'ev. — Izmerit. Tekh., No. 9 (1962), 39.
137. Kaplan, D. and E. Hahn. — Bull. Am. Phys. Soc., ser. 2, 2 (1957), 384.
138. Ting, Y., E. Manring, and D. Williams. — Phys. Rev. 96 (1954), 408.
139. Partlow, W. and W. Moulton. — J. Chem. Phys. 39 (1963), 2381.
140. Bray, P. — J. Chem. Phys. 23 (1955), 703.
141. Bray, P. and D. Esteva. — J. Chem. Phys. 22 (1954), 570.
142. Nakamura, D., K. Ito, and M. Kubo. — International Symposium on Molecular Structure and Spectroscopy, Tokyo, 1962; Inorg. Chem. 2 (1963), 61.
143. Zeil, W. — Angew. Chem. 76 (1964), 654.
144. Schmitt, A. and W. Zeil. — Z. Naturf. 18a (1963), 428.
145. Nakamura, D., K. Ito, and M. Kubo. — J. Am. Chem. Soc. 84 (1962), 163.
146. Ikeda, R., D. Nakamura, and M. Kubo. — J. Phys. Chem. 69 (1965), 2101.
147. Nakamura, D. and M. Kubo. — Proc. VIII International Conference on Coordinate Chemistry, Vienna, 1964.
148. Hamlen, R. and W. Koski. — J. Chem. Phys. 25 (1956), 360.
149. Gilbert, D., A. Roberts, and P. Griswold. — Phys. Rev. 76 (1949), 1723.
150. Gilbert, D., A. Roberts, and P. Griswold. — Phys. Rev. 77 (1950), 742.
151. Smith, D. — J. Chem. Phys. 21 (1959), 609.
152. Chihara, H. and N. Nakamura. — J. Phys. Soc. Japan 21 (1966), 202; 22 (1967), 202.
153. Livingston, R. — J. Chem. Phys. 19 (1951), 803.
154. Smith, D., M. Tidwell, and D. Williams. — Phys. Rev. 79 (1950), 1007.
155. Kurita, Y., D. Nakamura, and N. Nayakawa. — J. Chem. Soc. Japan, Pure Chemistry Section 79 (1958), 1093.
156. Evans, J. and G. Lo. — J. Phys. Chem. 71 (1967), 2730.
157. Townes, C., F. Merritt, and B. Wright. — Phys. Rev. 73 (1948), 1334.
158. Townes, C., F. Merritt, and B. Wright. — Ibid., p. 1249.
159. Kojima, S., K. Tsukada, S. Ogawa, and A. Shimauchi. — J. Chem. Phys. 23 (1955), 1963.
160. Cornwell, C. and R. Yamasaki. — J. Chem. Phys. 27 (1957), 1060.
161. Yamasaki, R. and C. Cornwell. — J. Chem. Phys. 30 (1959), 1265.
162. Hagiwara, S., K. Kato, Y. Abe, and M. Minematsu. — J. Phys. Soc. Japan 12 (1957), 1166.
163. Javan, A., G. Silvey, C. Townes, and A. Grosse. — Phys. Rev. 91 (1953), 222.
164. Javan, A. and A. Engelbrecht. — Phys. Rev. 96 (1954), 649.
165. Lotspeich, J. — J. Chem. Phys. 31 (1959), 633, 643.
166. Ikeda, R., A. Sasane, D. Nakamura, and M. Kubo. — J. Phys. Chem. 70 (1966), 2926.
167. Ikeda, R., D. Nakamura, and M. Kubo. — Bull. Chem. Soc. Japan 36 (1963), 1056.
168. Jones, W. and S. Segel. — Phys. Rev. Lett. 13 (1964), 528.
169. Narath, A. — J. Chem. Phys. 40 (1964), 1169.
170. Hartmann, H., M. Fleissner, and H. Sillescu. — Naturwissenschaften 50 (1963), 591.

. Jones, W. and K. Brog. — Bull. Am. Phys. Soc., ser. 2, 13 (1968), 473.
2. Barnes, R., R. Lecander, and S. Segel. — Bull. Am. Phys. Soc., ser. 2, 8 (1963), 260.
3. Narath, A. — Phys. Rev. **140** (1965), 552.
4. Schumacher, R. and H. Hartmann. — Bull. Am. Phys. Soc., ser. 2, 10 (1965), 316.
5. Ito, K., D. Nakamura, Y. Kurita, Ko. Ito, and M. Kubo. — J. Am. Chem. Soc. 83 (1961), 4526.
6. Ito, Ko., D. Nakamura, K. Ito, and M. Kubo. — Inorg. Chem. 2 (1963), 690.
7. Marram, E., E. McNiff, and J. Ragle. — J. Phys. Chem. 67 (1963), 1719.
8. Nakamura, D., K. Ito, and M. Kubo. — J. Phys. Chem. **68** (1964), 2986.
9. Nakamura, D., Y. Kurita, K. Ito, and M. Kubo. — J. Am. Chem. Soc. 82 (1960), 5783.
0. Jeffrey, K. and K. Armstrong. — Bull. Am. Phys. Soc., ser. 2, 13 (1968), 45.
1. Ito, Ko, K. Ito, D. Nakamura, and M. Kubo. — Bull. Chem. Soc. Japan **35** (1962), 518.
2. Nesmejanov, A. N., G. K. Semin, E. V. Bruchova, T. A. Babuschkina, K. N. Anisimov, N. E. Kolobova, and Yu. V. Makarov. — Tetrahedron Lett., p. 3987, 1968.
3. Fryer, C. and J. Smith. — J. Chem. Soc. (A), p. 1029, 1970.
4. Nemeyanov, A. N., O. Yu. Okhlobystin, E. V. Bryukhova, V. I. Bregadze, D. N. Kravtsov, B. A. Faingor, L. S. Golovchenko, and G. K. Semin. — Izv. AN SSSR, chem. ser., p. 1928, 1969.
5. Bryukhova, E. V., T. A. Babushkina, M. V. Kashutina, O. Yu. Okhlobystin, and G. K. Semin. — Doklady AN SSSR **183** (1968), 827.
6. Maksyutin, Yu. K., E. N. Gur'yanova, and G. K. Semin. — Uspekhi Khimii **39** (1970), 727.
7. Brill, T. and Z. Hugus. — Inorg. Chem. **9** (1970), 984.
8. Brill, T. — J. Inorg. Nucl. Chem. 32 (1970), 1869.
9. Peterson, G. and P. Bridenbaugh. — J. Chem. Phys. 51 (1969), 238.
0. Graybeal, J. and P. Green. — J. Phys. Chem. 73 (1969), 2948.
1. Osokin, D. Ya., I. A. Safin, I. A. Nuretdinov, and. — Doklady AN SSSR **190** (1970), 357.
2. Nave, C., T. Weatherly, and Q. Williams. — J. Chem. Phys. **49** (1968), 1413.
3. Biedenkapp, D. and A. Weiss. — Z. Naturf. 23b (1968), 174.
4. Svergun, V. I., T. A. Babushkina, G. N. Shvedova, L. V. Kudryavtseva, and G. K. Semin. — Izv. AN SSSR, chem. ser., p. 479, 1970.
5. Kondo, S., E. Kakiuchi, and T. Shimizu. — Bull. Chem. Soc. Japan **42** (1969), 2050.
6. Grechishkin, V. S., A. D. Gordeev, and Yu. A. Galishevskii. — Zh. Strukt. Khim. **10** (1969), 743.
7. Goto, N. — Res. Bull. Fac. Education, Oito Univ. 3 (1969), 23.
8. Legrandt, J. P. and J. P. Renardt. — Compt. rend. **266** (1968), 1165.
9. Semin, G. K., T. A. Babushkina, V. M. Vlasov, and G. G. Yakobson. — Izv. Sib. Otdel. AN SSSR, ser. chem. sci., No. 12, Issue 5 (1969), 99.
0. Babushkina, T. A. and M. I. Kalinkin. — Izv. AN SSSR, chem. ser., p. 157, 1969.
1. Bryukhova, E. V., T. A. Babushkina, V. I. Svergun, and G. K. Semin. — Uchen. Zap. Mosk. Oblast. Pedag. Inst. im. N. K. Krupskoi 222 (1969), 64.
2. Nesmeyanov, A. N., G. K. Semin, E. V. Bryukhova, K. N. Anisimov, N. E. Kolobova, and V. N. Khandozhko. — Izv. AN SSSR, chem. ser., p. 1936, 1969.
3. Maksyutin, Yu. K., V. P. Makridin, E. N. Gur'yanova, and G. K. Semin. — Izv. AN SSSR, chem. ser., p. 1634, 1970.
4. Semin, G. K., A. A. Neimysheva, and T. A. Babushkina. — Izv. AN SSSR, chem. ser., p. 486, 1970.
5. Nesmeyanov, A. N., G. K. Semin, E. V. Bryukhova, T. A. Babushkina, K. N. Anisimov, N. E. Kolobova, and Yu. V. Makarov. — Izv. AN SSSR, chem. ser., p. 1953, 1968.
6. Carlson, E. and H. Adams. — J. Chem. Phys. 51 (1969), 390.
7. Buslaev, Yu. A., E. A. Kravchenko, S. M. Sinitsyna, and G. K. Semin. — Izv. AN SSSR, chem. ser., p. 2816, 1969.
8. Stanko, V. I., Yu. T. Struchkov, and A. I. Klimova. — Zh. Strukt. Khim. **7** (1966), 629.

Group VII elements. C—Cl bond

The data were obtained at 77°K unless stated otherwise.

Empirical formula	Structural formula	ν, MHz	s/n	References, notes
$CBrClF_2$	CF_2BrCl	38.342	...	[1]
$CBrCl_3$	$CBrCl_3$	40.46	200	*[2]
$CBrCl_3Hg$	CCl_3HgBr	37.590 37.898 38.794 39.032	6 6 3 3	*[3,4]. cf. Hg—Br
$CBrCl_3Hg + C_4H_8O_2$	$CCl_3HgBr \cdot diox$	36.554 36.906 36.995 37.814 38.010 38.339	3 3 3 6 6 6	*[236]
$CBrCl_3Hg +$ $+ nC_4H_{10}O_2$	$CCl_3HgBr \cdot nDME$	37.410 37.898	10 20	*[236]
CBr_2Cl_2	CCl_2Br_2	40.34	2	*[2]
$CClFN_2O_4$	$CClF(NO_2)_2$	40.880	10	*[237]
$CClF_3$	CF_3Cl	38.089	...	[1]
$CClN$	$ClCN$	41.736	...	[6—12]. cf. N
		$e^2Qq = -83.33$ MHz (g) $\eta = 0$		
$CClN_3O_6$	$CCl(NO_2)_3$	42.9486	4	*[13]
CCl_2F_2	CF_2Cl_2	38.450	...	[1]
$CCl_2N_2O_4$	$CCl_2(NO_2)_2$ Phase I Phase II Phase III	 41.601 41.909 42.133 42.287	 30 30 15 15	*[13,14]
CCl_2O	$COCl_2$	35.081 36.225	...	[1, 15—17]. $\eta = 25\%$
CCl_3F	$CFCl_3$	39.161 39.517 39.703	...	[5,18,19]. $e^2Qq =$ $= -78.05$ MHz (g)
CCl_3NO_2	CCl_3NO_2	41.546 41.235 41.003	10 10 10	[20]
CCl_4	CCl_4	40.4560 40.5080 40.5275 40.5340 40.5600	...	*[1,15,18, 21,22,209]. $\eta < 10\%$

Empirical formula	Structural formula	ν, MHz	s/n	References, notes
CCl_4 (contd.)	CCl_4 (contd.)	40.5730 40.5990 40.6250 40.6328 40.6455 40.6900 40.7095 40.7745 40.7940 40.8070		
$CCl_4 + C_3H_6O$	$CCl_4 \cdot (CH_3)_2CO$	40.280 40.380 40.696 40.970	2 2 2 2	[23]
$CCl_4 + C_3H_7NO$	$CCl_4 \cdot$DMF	39.984 40.271 40.334 40.772 40.978 41.059 41.111	30—40 10 10 7 10 10 10	*[24]
$CCl_4 + C_4H_{10}O$	$CCl_4 \cdot (C_2H_5)_2O$	40.131 40.184 40.320 40.502 40.926 41.041 41.108 41.164	. . .	*[23,24]
$CCl_4 + 2C_4H_{10}O$	$CCl_4 \cdot 2(C_2H_5)_2O$	40.495 40.738	1.5 1.5	[23]
$CCl_4 + {}^1/_3C_6H_6$	$3CCl_4 \cdot C_6H_6$	40.306 40.496 40.648 40.800	. . .	[23]
$CCl_4 + {}^1/_2C_6H_6$	$2CCl_4 \cdot C_6H_6$	40.280 40.576 40.904	2 2 2	[23]
$CCl_4 + C_6H_6$	$CCl_4 \cdot C_6H_6$	40.288 40.368 40.400	. . .	[23]
$CCl_4 + C_8H_{10}$	$CCl_4 \cdot 1,4\text{-}(CH_3)_2C_6H_4$	40.472 40.689	. . .	[25]
CCl_4Hg	CCl_3HgCl	37.604 37.968 39.032 39.172	10 10 5 5	*[3,4]

Empirical formula	Structural formula	ν, MHz	s/n	References, notes
CCl_4O_2S	CCl_3SO_2Cl	40.957	4	*[238]. cf. Cl—S
		41.097	4	
		41.214	4	
CCl_4S	CCl_3SCl	39.166	3	[20]
		39.827	2,5	
		40.117	5	
CCl_6Si	CCl_3SiCl_3	39.02	...	[26]. cf. Cl—Si
$CHBr_2Cl$	CBr_2ClH	38.24	10	*[2]
$CHClF_2$	CHF_2Cl	35.249	...	[18, 27—29]. $T = 20\ ^\circ K$
		$e^2Qq = -71.71$ MHz (g) $\eta = 1.2\%$		
$CHClN_2O_4$	$CHCl(NO_2)_2$	41.180	50	*[237]
$CHClN_2O_4 + C_6H_{14}O_3$	$CHCl(NO_2)_2 \cdot DG$	40.460	30	*[237]
$CHCl_2F$	$CHCl_2F$	36.479	...	[18, 27, 30, 31]
		36.508		
		$e^2Qq = -76.75$ MHz (g)		
$CHCl_3$	$CHCl_3$	38.254	100	*[1, 18, 19, 21, 32—34]
		38.308	50	
		$e^2Qq = -82,35$ MHz (g)		
$CHCl_3 + CH_4O$	$CHCl_3 \cdot CH_3OH$	37.507	2,5	[23, 210]
		37.804	2	
		37.968	2	
		37.290	2	
$CHCl_3 + C_2H_8N_2$	$CHCl_3 \cdot H_2NCH_2CH_2NH_2$	38.32	...	[34]
		38.27		
$CHCl_3 + C_3H_6O$	$CHCl_3 \cdot (CH_3)_2CO$	37.310	3	[23, 34, 210]
		37.650	3	
		37.752	3	
$CHCl_3 + {}^1/_2C_4H_{10}O$	$2CHCl_3 \cdot (C_2H_5)_2O$	37.170	4	[23, 210]
		37.420	8	
		37.704	4	
		37.842	8	
$CHCl_3 + C_4H_{10}O$	$CHCl_3 \cdot (C_2H_5)_2O$	37.618	50	*[23, 24, 210]
		37.891	50	
$CHCl_3 + 2C_4H_{10}O$	$CHCl_3 \cdot 2(C_2H_5)_2O$	37.282	20	*[23, 24]
		37.947	25	
		38.185	30	
$CHCl_3 + 3C_4H_{10}O$	$CHCl_3 \cdot 3(C_2H_5)_2O$	37.282	10	*[23, 210]
		37.947	12	
		38.185	15	

Empirical formula	Structural formula	ν, MHz	s/n	References, notes
$CHCl_3 + C_4H_{11}N$	$CHCl_3 \cdot (C_2H_5)_2NH$	37.46 37.17 36.92	...	[34]
$CHCl_3 + C_5H_5N$	$CHCl_3 \cdot Py$	37.825 37.771 37.706	5 5 5	[34, 210]
$CHCl_3 + C_5H_{11}N$	$CHCl_3 \cdot C_5H_{11}N$	37.44 37.16 36.98	...	[34]
$CHCl_3 + C_6H_6$	$CHCl_3 \cdot C_6H_6$	38.32	...	[34]. Several unresolved lines
$CHCl_3 + C_6H_{15}N$	$CHCl_3 \cdot (C_2H_5)_3N$	37.146 37.104 36.974	3 3 3	[34, 210]
$CHCl_3 + C_8H_{10}$	$CHCl_3 \cdot 1,2\text{-}(CH_3)_2C_6H_4$	38.32 38.27 37.67 37.32	...	[23, 34, 210]
	$CHCl_3 \cdot 1,3\text{-}(CH_3)_2C_6H_4$	37.814 37.842	1.5 1.5	[23, 34, 210
	$CHCl_3 \cdot 1,4\text{-}(CH_3)_2C_6H_4$	38.32 38.27 37.73 37.61	...	[34]
$CHCl_3 + C_9H_{12}$	$CHCl_3 \cdot 1,3,5\text{-}(CH_3)_3C_6H_3$	37.933 37.723 37.586	4 4 4	[23, 34, 210]
$CHCl_3 + C_{10}H_{14}$	$CHCl_3 \cdot 1,2,4,5\text{-}(CH_3)_4C_6H_2$	38.32 38.27	...	[34]
$CHCl_5Ge$	$CHCl_2GeCl_3$	39.375	15	*[35]. cf. Cl—Ge
$CHCl_5Si$	$CHCl_2SiCl_3$	38.171 37.240	7 7	*[35, 36]. cf. Cl—Si
CH_2BrCl	CH_2BrCl	36.14	50	*[2, 37]
CH_2ClF	CH_2FCl	33.799	...	[1]. $T=20$ °K
CH_2ClI	CH_2ClI	36.498	10	*[2, 37]. cf. I
CH_2ClNO_2	$CH_2Cl(NO_2)$	37.636	50	*[237]

Empirical formula	Structural formula	ν, MHz	s/n	References, notes
CH_2Cl_2	CH_2Cl_2	35.991	100	*[1, 18, 21, 32, 37—40]
		$e^2Qq = -76.2$ MHz (g)		
		$\eta = 4.76\%$ (g)		
$CH_2Cl_2 + C_4H_{10}O$	$CH_2Cl_2 \cdot (C_2H_5)_2O$	34.656 35.166 35.274 35.424 35.694 35.892	10 30 10 10 10 30	*[24]
$CH_2Cl_2 + 2C_4H_{10}O$	$CH_2Cl_2 \cdot 2(C_2H_5)_2O$	35.130 35.328	30 30	*[24]
$CH_2Cl_2O_2S$	$ClCH_2SO_2Cl$	38.612 39.039	10 10	*[238]. cf. Cl—S
CH_2Cl_3OP	$(CH_2Cl)P(O)Cl_2$	38.224 38.286	15 15	*[41, 42a]. cf. Cl—P
CH_2Cl_3P	$(CH_2Cl)PCl_2$	36.09	10	*[42a]. cf. Cl—P
CH_2Cl_3PS	$(CH_2Cl)P(S)Cl_2$	38.584	10	The same
CH_2Cl_4Ge	$(CH_2Cl)GeCl_3$	37.926	5	*cf. Cl—Ge
CH_2Cl_4Si	$ClCH_2SiCl_3$	36.786	10	*[26, 35, 42]. cf. Cl—Si
CH_2Cl_4Sn	$ClCH_2SnCl_3$	37.870	15	*[43]. cf. Cl—Sn
$CH_2Cl_4Sn + C_6H_6$	$ClCH_2SnCl_3 \cdot C_6H_6$	37.366	30	*cf. Cl—Sn
CH_3Cl	CH_3Cl	34.029	...	[1, 9, 10, 18, 44—48]
		$e^2Qq = -74.74$ MHz (g)		
CD_3Cl	CD_3Cl	$e^2Qq = -74.41$ MHz		[47]. g
CH_4ClO_3P	$(CH_2Cl)P(O)(OH)_2$	37.244	10	*
$CH_4ClO_3P + C_6H_7N$	$(CH_2Cl)P(O)(OH)_2 \cdot C_6H_5NH_2$	35.208	3	*
$CH_4ClO_3P + 2C_6H_7N$	$(CH_2Cl)P(O)(OH)_2 \cdot 2C_6H_5NH_2$	34.500	2	*
CH_5ClSi	H_3SiCH_2Cl	$e^2Qq = -68.7$ MHz $\eta = 4.8\%$		[49]. g
$C_2AgCl_3O_2$	CCl_3COOAg	37.961 39.445 39.795	10 10 10	*[3]

Empirical formula	Structural formula	ν, MHz	s/n	References, notes
$C_2BBr_2Cl_3 + C_5H_5N$	$CCl_2{=}CClBBr_2 \cdot Py$	36.379 37.408 37.527	3 3 3	*
C_2BrCl_3Hg	$CCl_2{=}CClHgBr$	35.546 36.180 37.153 37.625 37.944 38.077	10 10 10 10 10 10	*[4]. cf. Br
C_2ClF_3	$CF_2{=}CFCl$	38.017 38.171	10 10	[50, 211]
		$e^2Qq = -89.14$ MHz (g) $\eta = 13.08$ (g)		
C_2ClF_3O	CF_3COCl	33.432	20	*
$C_2ClN_3O_4$	$CCl(NO_2)_2(CN)$	42.854 42.998	15 15	*[237]
$C_2Cl_2N_4O_8$	$Cl(NO_2)_2CC(NO_2)_2Cl$	43.014 43.672	4 4	*[13]
$C_2Cl_2O_2$	ClOCCOCl	33.621	12—15	*[52, 53, 212]
$C_2Cl_3CuO_2$	$Cl_3CCOOCu$	39.995 40.075 40.523	4 4 4	*
$C_2Cl_3LiO_2$	CCl_3COOLi	39.114 40.181	1—2 1—2	[54]
$C_2Cl_3LiO_2 + H_2O$	$CCl_3COOLi \cdot H_2O$	38.366 38.394 38.439 38.466 39.105 39.118 39.213 39.83— —39.85	2 2 1—2 1—2 1—2 1—2 2—3 1—2	[54]
C_2Cl_3N	CCl_3CN	41.7296 41.6660 41.5340	. . .	[55]. cf. N
$C_2Cl_3NaO_2$	CCl_3COONa	38.332 38.892 39.452	20 20 20	*[3, 54]
$C_2Cl_3NaO_2 + CH_4O$	$CCl_3COONa \cdot CH_3OH$	37.730 38.220 38.332 38.409 38.570	5 5 10 5 5	*[3]

Empirical formula	Structural formula	ν, MHz	s/n	References, notes
$C_2Cl_3NaO_2+CH_4O$ (contd.)	$CCl_3COONa\cdot CH_3OH$ (contd.)	38.640	5	
		38.885	5	
		38.969	5	
		39.242	5	
		39.452	10	
$C_2Cl_3NaO_2 + 3H_2O$	$CCl_3COONa\cdot 3H_2O$	37.982	8	*[3, 54]
		38.542	8	
		38.903	8	
		39.302	8	
		39.323	8	
		39.382	8	
C_2Cl_4	$CCl_2{=}CCl_2$	38.374	50	*[20]
		38.535	50	
		38.684	50	
		38.794	50	
C_2Cl_4Hg	$Cl_2C{=}CClHgCl$	35.424	10	*[3, 4]. cf. Cl—Hg
		37.132	10	
		37.562	10	
$C_2Cl_4Hg + C_2H_6OS$	$CCl_2{=}CClHgCl\cdot DMSO$	35.894	10	* cf. Cl—Hg
		37.260	10	
		37.290	10	
$C_2Cl_4Hg + C_3H_7NO$	$CCl_2{=}CClHgCl\cdot DMF$	35.568	10	* cf. Cl—Hg
		36.912	10	
		37.518	10	
C_2Cl_4O	CCl_3COCl	33.721	30	*[56]
		40.208	10	
		40.473	10	
		40.613	10	
C_2Cl_6	Cl_3CCCl_3	40.548	20	*[1, 57—59]
		40.681	30	
		40.712	20	
		40.758	35	
$C_2Cl_6 + 3CH_4NS$	$CCl_3CCl_3\cdot 3(NH_2)_2CS$	40.222	5	*
		40.527	10	
		40.590	15	
C_2Cl_6Hg	$(Cl_3C)_2Hg$	37.037	50	*[3, 4]
		37.681	50	
		38.675	50	
$C_2Cl_6Hg + C_4H_8O$	$(Cl_3C)_2Hg\cdot THF$	36.780	10	*[236]
		36.212	12	
		37.212	8	
		37.320	8	
		37.398	8	
		37.500	5	
		37.560	8	
		37.650	10	
		37.848	8	

Empirical formula	Structural formula	ν, MHz	s/n	References, notes
$C_2Cl_6Hg + C_4H_{10}O_2$	$(Cl_3C)_2Hg \cdot DME$	36.978 37.230 37.194 37.596 37.884 38.010	15 15 15 15 15 15	*[236]
C_2HCl	HC≡CCl	$e^2Qq =$ = —79.7 MHz		[60, 61]. g
C_2DCl	DC≡CCl	$e^2Qq =$ = —79.6 MHz		[62]. g
C_2HClF_2	CF_2=CHCl	$e^2Qq =$ = —84.3 MHz η = 20.6%		[63]. g
C_2HClF_2O	$F_2HCCOCl$	32.184	10	*
$C_2HClF_2O_2$	ClF_2CCOOH	37.485	...	[64]
$C_2HCl_2FO_2$	$Cl_2FCCOOH$	39.594	...	[64]
$C_2HCl_2F_3$	Cl_2HCCF_3	38.694	...	[64]
C_2HCl_2N	Cl_2HCCN	39.4434 39.7537 39.9233 40.0840	5 7 5 4	[55]. cf. N
C_2HCl_3	ClHC=CCl_2	36.351 36.533 37.415 37.730 37.975	7 7 15 5 5	*
C_2HCl_3O	$Cl_2HCCOCl$	38.353 38.521 39.189 39 386 32.147 32.962	2—3 2—3 3 3 2—3 2—3	[52]
	Cl_3CCOH	38.794 39.172 39.291	...	*[50]
$C_2HCl_3O_2$	CCl_3COOH	39.964 40.160 40.236	20 20 20	*[54, 56, 65]
$C_2HCl_3O_2$ + + $C_2Cl_3KO_2$	$CCl_3COOH \cdot CCl_3COOK$	39.113 39.147 39.747	3 3 3	[54, 65a]
$C_2HCl_3O_2$ + + $^1/_2C_2Cl_3NaO_2$ + + H_2O	$2CCl_3COOH \cdot CCl_3COONa \cdot 2H_2O$	39.423 39.684 39.703	3 3 3	[54]

Empirical formula	Structural formula	ν, MHz	s/n	References, notes
$C_2HCl_3O_2$ + + $^1/_2C_2Cl_3NaO_2$ + + H_2O (contd.)	$2CCl_3COOH \cdot CCl_3COONa \cdot 2H_2O$ (contd.)	39.733	3	
		39.930	3	
		40.151	3	
		40.245	3	
		40.442	3	
		40.549	3	
$C_2HCl_3O_2$ + + $C_2Cl_3O_2Rb$	$CCl_3COOH \cdot CCl_3COORb$	39.121	3	[54]
		39.226	1—2	
		40.187	2	
$C_2HCl_3O_2$ + + $C_2Cl_3O_2Tl$	$CCl_3COOH \cdot CCl_3COOTl$	39.044	6	[54]
		39.099	6	
		39.824	6	
$C_2HCl_3O_2$ + + $C_2H_4Cl_3NO_2$	$CCl_3COOH \cdot CCl_3COONH_4$	39.291	3	[54, 65a]
		39.398	3	
		39.766	3	
$C_2HCl_3O_2$ + C_3H_6O	$CCl_3COOH \cdot (CH_3)_2CO$	39.565	4	[54]
		39.610	4	
		39.644	3—4	
		39.776	4	
		39.983	3	
		40.085	4	
$C_2HCl_3O_2$ + $C_3H_8O_2$	$CCl_3COOH \cdot CH_3COOC_2H_5$	39.604	2	[54]
		39.710	2	
		39.744	2	
		39.789	2	
		39.911	2	
		40.227	2	
$C_2HCl_3O_2$ + + $^1/_2C_2H_8O_2$	$2CCl_3COOH \cdot$ diox	39.325	2	[54]
		39.459	2	
		39.524	3—4	
		39.906	4	
		40.301	3	
		40.320	3	
$C_2HCl_3O_2$ + $C_4H_{10}O$	$CCl_3COOH \cdot (CH_3)_3COH$	39.686	4	[54]
		39.845	4	
		40.056	4	
$C_2HCl_3O_2$ + + $C_6H_4Cl_2O$	$CCl_3COOH \cdot 2,4\text{-}Cl_2C_6H_3OH$	35.458	4	[54]
		36.006	4	
		39.930	3—4	
		40.063	3—4	
		40.357	3—4	
$C_2HCl_3O_2$ + + C_6H_5ClO	$CCl_3COOH \cdot 4\text{-}ClC_6H_4OH$	34.874	1—2	[54]
		34.903	2	
		39.533	2—3	
		39.709	2—3	
		40.015	2—3	

Empirical formula	Structural formula	ν, MHz	s/n	References, notes
$C_2HCl_3O_2$ + $+C_6H_5ClO$ (contd.)	$CCl_3COOH \cdot 4\text{-}ClC_6H_4OH$(contd.)	40.102 40.370 40.620	2—3 2 2	
$C_2HCl_3O_2 + C_6H_6O$	$CCl_3COOH \cdot C_6H_5OH$	39.770 39.818 40.413	5 4 4—5	[54]
$C_2HCl_3O_2$ + $+ {}^1/_2C_6H_{10}O_4$	$2CCl_3COOH \cdot (COOC_2H_5)_2$	39.489 39.550 39.727 39.739 39.839 39.893	3 3 3 3 3 3	[54]
$C_2HCl_3O_2 + C_7H_6O$	$CCl_3COOH \cdot C_6H_5CHO$	39.438 39.599 40.357	1—2 1—2 1—2	[54]
$C_2HCl_3O_2 + C_7H_6O_2$	$CCl_3COOH \cdot C_6H_5COOH$	39.268 39.645 39.716 40.214 40.274 40.488	1—2 1—2 1—2 1—2 1—2 1—2	[54]
$C_2HCl_3O_2 + C_7H_8O$	$CCl_3COOH \cdot 2\text{-}CH_3C_6H_4OH$	39.753 39.916 40.262	2 1—2 2	[54]
	$CCl_3COOH \cdot 3\text{-}CH_3C_6H_4OH$	39.635 40.242 40.310	3 3 3	[54]
	$CCl_3COOH \cdot 4\text{-}CH_3C_6H_4OH$	39.687 40.176 40.236	3—4 3—4 2	[54]
$C_2HCl_3O_2$ + $+ {}^1/_2C_7H_8O_2$	$2CCl_3COOH \cdot$ 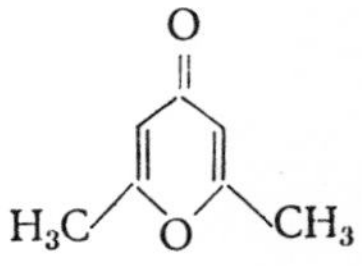	38.594 38.630 39.135 39.230 39.606 39.922	5 5 5 5 3—4 5	[54]
$C_2HCl_3O_2 + C_7H_8O_2$	$CCl_3COOH \cdot$ 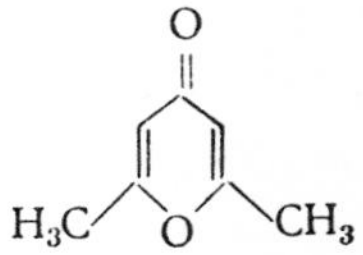	39.358 39.367 39.734 39.748 39.947 39.974	3 3 2 2 1—2 1—2	[54]
$C_2HCl_3O_2$ + $+ {}^1/_2C_8H_6O_3$	$2CCl_3COOH \cdot$ 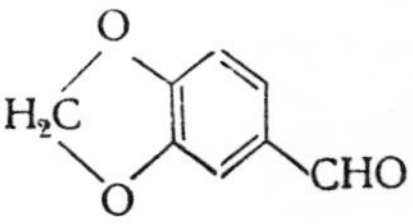	39.488 39.498 39.696 39.740	2 2 2 2	[54]

Empirical formula	Structural formula	ν, MHz	s/n	References, notes
$C_2HCl_3O_2$ + $^1/_2C_8H_6O_3$ (contd.)	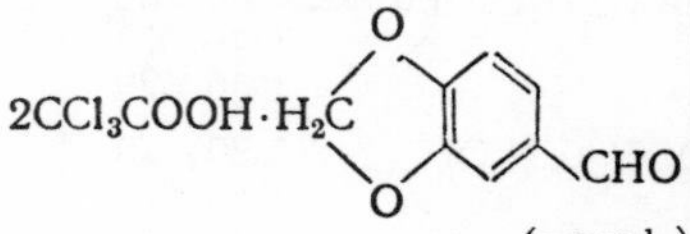 (contd.)	39.887 39.918 39.938 39.991 40.009 40.109 40.506 40.661	2 2 2 2 2 2 2 2	
$C_2HCl_3O_2 + C_8H_6O_3$	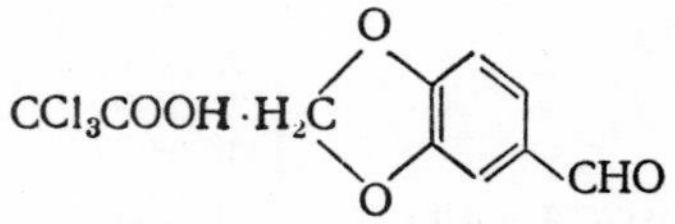	39.387 39.555 40.326	4 4 4	[54]
$C_2HCl_3O_2 + C_8H_8O$	$CCl_3COOH \cdot C_6H_5COCH_3$	39.373 39.729 40.148	2—3 4 4	[54]
$C_2HCl_3O_2 + C_8H_8O_2$	$CCl_3COOH \cdot 2\text{-}CH_3C_6H_4COOH$	39.822 40.170	2 3—4	[54]
	$CCl_3COOH \cdot 3\text{-}CH_3C_6H_4COOH$	39.930 40.089 40.268	1—2 1 1	[54]
	$CCl_3COOH \cdot 4\text{-}H_3COC_6H_4CHO$	39.766 39.793	3 1—2	[54]
$C_2HCl_3O_2$ + + $^1/_2C_8H_8O_3$	$2CCl_3COOH \cdot$ $\cdot\, 2\text{-}CH_3O\text{-}4\text{-}CHOC_6H_3OH$	39.476 39.635 39.911 39.964 40.159 40.236	1—2 1—2 1—2 1—2 1—2 1—2	[54]
$C_2HCl_3O_2 + C_8H_8O_3$	$CCl_3COOH \cdot$ $\cdot\, 2\text{-}CH_3O\text{-}4\text{-}CHOC_6H_3OH$	39.353 39.781 40.098 40.122 40.230	1—2 1—2 1—2 1—2 1—2	[54]
$C_2HCl_3O_2 + C_6H_{10}O$	$CCl_3COOH \cdot 2,5\text{-}(CH_3)_2C_6H_3OH$	39.697 40.160 40.185	5 5 5	[54]
	$CCl_3COOH \cdot 2,6\text{-}(CH_3)_2C_6H_3OH$	39.665 39.914 39.948 40.143 40.623	2 2 2 3—4 2	[54]
	$CCl_3COOH \cdot 3,4\text{-}(CH_3)_2C_6H_3OH$	39.998 40.151 40.222	3 2—3 1—2	[54]

Empirical formula	Structural formula	ν, MHz	s/n	References, notes
$C_2HCl_3O_2 + C_8H_{10}O$	$CCl_3COOH\cdot 3,5\text{-}(CH_3)_2C_6H_3OH$	39.593 39.791 40.292	5 4 5	[54]
$C_2HCl_3O_2 +$ $+ {}^1/_4C_{10}H_{10}O_4$	$4CCl_3COOH\cdot C_6H_4(COOCH_3)_2\text{-}1,4$	39.534 39.788 39.955	3 3—4 3—4	[54]
$C_2HCl_3O_2 +$ $+ C_{14}H_{12}O_2$	$CCl_3COOH\cdot C_6H_5COOCH_2C_6H_5$	39.892 40.051 40.071	2 2 2	[54]
$C_2HCl_3O_2 +$ $+ C_{14}H_{14}O$	$CCl_3COOH\cdot C_6H_5CH_2OCH_2C_6H_5$	39.348 39.380 39.462 39.702 39.715 39.855	1—2 1—2 1—2 1—2 1—2 1—2	[54]
$C_2HCl_3O_2 + H_2O$	$CCl_3COOH\cdot H_2O$	39.484 39.785 39.823 39.974 40.079 40.191 40.261 40.294 40.632	10 10 10 10 10 10 10 10 10	*[3, 54]
C_2HCl_5	Cl_2HCCCl_3	38.696 38.745 39.752 39.832	10 10 10 20	*[1]
C_2H_2ClF	$CH_2{=}CClF$	$e^2Qq =$ $= -74.4$ MHz		[66]. g
	cis-$CHCl{=}CHF$	$e^2Qq =$ $= -73.7$ MHz		[67]. g
$C_2H_2ClFN_2O_4$	$ClCH_2CF(NO_2)_2$	38.546	40	*[237]
$C_2H_2ClF_3$	CH_2ClCF_3	37.387	20	*[102]
C_2H_2ClN	CH_2ClCN	38.122	100	*[55, 68, 69]. cf. N
		$e^2Qq = -80$ MHz (g) $\eta = 0$ (g)		
$C_2H_2ClN_3O_6$	$C(NO_2)_3CH_2Cl$	39.564	20	*[237]
$C_2H_2ClNaO_2$	$CH_2ClCOONa$	34.794	...	*[56, 65]

Empirical formula	Structural formula	ν, MHz	s/n	References, notes
$C_2H_2Cl_2$	cis-CHCl=CHCl	34.968 35.029	...	[1, 70—72]. $T = 20\ °K$
		$e^2Qq = -72.3$ MHz (g); $\eta = 11.76\%$ (g)		
	trans-CHCl=CHCl	35.496 $\eta = 26.5\%$	...	[1, 48, 70, 73, 74, 213]
		$e^2Qq = -78.7$ MHz (g);		
	$CH_2{=}CCl_2$	36.258 36.524 36.873	8 15 8	*[50, 75]
$C_2H_2Cl_2FI$	CCl_2ICH_2F	38.812 38.966	50 60	* cf. I
$C_2H_2Cl_2Hg$	cis-CHCl=CHHgCl	33.132	10	*[239]. cf. Cl—Hg
	trans-CHCl=CHHgCl	31.746 32.408 33.462	5 5 5	*[3, 4, 239]
	cis-CHCl=CHHgCl · trans-CHCl=CHHgCl	33.75	...	*[239]
$C_2H_2Cl_2Hg$ + + C_3H_7NO	trans-CHCl=CHHgCl·DMF	32.730 33.324	8 4	*[239]
$C_2H_2Cl_2Hg + C_4H_8O$	trans-CHCl=CHHgCl·2THF	33.036	20	*[239]. cf. Cl—Hg
$C_2H_2Cl_2Hg + C_5H_5N$	trans-CHCl=CHHgCl·Py	31.926	10	*[239]
$C_2H_2Cl_2N_2O_4$	$(NO_2)_2CClCH_2Cl$	40.936 40.898 38.045 37.923	30 30 20 20	*[237]
$C_2H_2Cl_2O$	$CH_2ClCOCl$	37.517 30.437	2—3 2	[52]
$C_2H_2Cl_2O_2$	$CHCl_2COOH$	37.979 38.807	...	[56]
$C_2H_2Cl_3NO$	CCl_3CONH_2	38.857 39.480 39.599 39.662 39.809 39.998	...	*[56, 77]
$C_2H_2Cl_4$	$Cl_2CHCHCl_2$	37.331 37.583 37.793 38.066	10 15 10 15	*[20, 76]

Empirical formula	Structural formula	ν, MHz	s/n	References, notes
$C_2H_2Cl_4$ (contd.)	$ClCH_2CCl_3$	36.40 39.02	3 10	*[76, 78]
$C_2H_2Cl_6Ge$	$(CHCl_2)_2GeCl_2$	36.946 37.051 37.233 37.338 38.444 38.913	10 5 5 5 10 5	*[35]
C_2H_3Cl	$H_2C{=}CHCl$	33.411	50	*[16, 70, 79, 80]. $\eta = 7\%$
		$e^2Qq = -70.16$ MHz (g) $\eta = 14.22\%$ (g)		
$C_2H_3ClN_2O_4$	$CCl(NO_2)_2(CH_3)$	40.142	100	*[13]
	$(NO_2)_2CHCH_2Cl$	37.038 37.194	20 20	[*237]
C_2H_3ClO	CH_3COCl	28.855 29.070	...	*[81]
		$e^2Qq = -59.2$ MHz (g); $\eta = 27\%$ (g)		
C_2D_3ClO	CD_3COCl	$e^2Qq =$ $= -57.7$MHz (g) $\eta = 25$ 6% (g)		[81]
$C_2H_3ClO_2$	CH_3OCOCl	34.224	10	*[22, 56, 59, 65, 82]
	$CH_2ClCOOH$ Phase I	36.131 36.429	...	
	Phase II	36.45		
	Phase III	36.14		
$C_2H_2DClO_2$	$CH_2ClCOOD$	36.15	...	[56]
$C_2H_3Cl_2NO$	$CHCl_2CONH_2$	37.238 37.750	...	[77, 83]
$C_2H_3Cl_2NO_2$	$CCl_2(NO_2)(CH_3)$	38.990 39.074	15—20 15—20	*[237]
$C_2H_3Cl_3$	CH_3CCl_3	37.829 38.052	...	[21, 57]
$C_2H_3Cl_3O_2$	$CCl_3CH(OH)_2$ Phase I	39.806 39.769 39.450 39.302 39.184 39.131	...	[56, 65, 84]

Empirical formula	Structural formula	ν, MHz	s/n	References, notes
$C_2H_3Cl_3O_2$ (contd.)	$CCl_3CH(OH)_2$ Phase I (contd.)	38.979		
		38.946		
		38.697		
		38.530		
		38.508		
		38.498		
	Phase II	39.515	...	
		39.429		
		38.190		
$C_2HD_2Cl_3O_2$	$CCl_3CH(OD)_2$			[84]
	Phase I	39.816	...	
		39.783		
		39.465		
		39.306		
		39.194		
		39.141		
		38.991		
		38.956		
		38.711		
		38.550		
		38.538		
		38.519		
	Phase II	39.527	...	
		39.441		
		38.209		
$C_2H_3Cl_5Si$	$ClCH_2CHClSiCl_3$	36.9	...	[26]
		35.6		
C_2H_4ClF	CH_2FCH_2Cl	33.948	1.5—2	*
C_2H_4ClNO	$CH_2ClCONH_2$	34.882	...	[56, 65, 77]
$C_2H_4ClNO_2$	$CH_2ClCH_2(NO_2)$	34.164	15—20	*[102]
	$CClH(NO_2)CH_3$	37.038	15—20	*[237]
$C_2H_4Cl_2$	CH_3CHCl_2	35.412	...	[27, 57, 85]
		$e^2Qq = -75.5$ MHz (g)		
	$ClCH_2CH_2Cl$	34.361	...	*[1, 38, 57, 86—89].
		29.92	...	$T = 238$ °K
$C_2D_4Cl_2$	$ClCD_2CD_2Cl$	34.3343	...	*[89, 90]
$C_2H_4Cl_2O$	$ClCH_2OCH_2Cl$	32.381	...	[1]
		32.587		
$C_2H_4Cl_2O_2S$	$ClCH_2CH_2SO_2Cl$	35.334	3—5	*[238]. cf. Cl—S
$C_2H_4Cl_2S$	$ClCH_2SCH_2Cl$	34.526	...	[64]
		34.749		

Empirical formula	Structural formula	ν, MHz	s/n	References, notes
$C_2H_4Cl_3NO$	$CCl_3CH(OH)NH_2$	38.220 38.518 38.603	...	[59, 91]
$C_2H_4Cl_4Ge$	$(ClCH_2)_2GeCl_2$	38.374	3	*[35]
	$ClCH_2CH_2GeCl_3$	34.386	10	*[35]. cf. Cl—Ge
$C_2H_4Cl_4Si$	$CH_3CHClSiCl_3$	35.01	...	[26]
	$ClCH_2CH_2SiCl_3$	34.008	20	*[35]. cf. Cl—Si
	$CHCl_2SiCl_2(CH_3)$	36.76	...	[26, 42]. cf. Cl—Si
$C_2H_4Cl_4Sn$	$(ClCH_2)_2SnCl_2$	36.708	50	*[43]. cf. Cl—Sn
C_2H_5Cl	$ClCH_2CH_3$	32.646 32.759	50 50	*[1, 20, 27, 41, 57, 92, 93, 94]
		$e^2Qq = -68.80$ MHz (g) $\eta = 3.5\%$ (g)		
C_2H_5ClO	$ClCH_2OCH_3$	29.817 30.206	40 16	[20, 64]
$C_2H_5ClO + {}^1/_2SnCl_4$	$2ClCH_2OCH_3 \cdot SnCl_4$	34.230	10	*[240]. cf. Cl—Sn
C_2H_5ClO	$ClCH_2CH_2OH$	33.003	10—15	*[102]
$C_2H_5ClO_2S$	$CH_3SO_2CH_2Cl$	38.217	...	*[102]
C_2H_5ClS	$(CH_2Cl)S(CH_3)$	33.104	...	[64]
$C_2H_5Cl_3Si$	$(CH_2Cl)SiCl_2(CH_3)$	36.11	7	*[26, 35, 36, 42]. cf. Cl—Si
$C_2H_{10}B_{10}Cl_2$	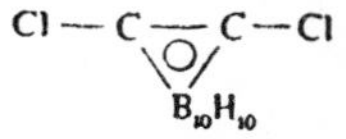 C,C-dichloro-o-borane	41.755	10	*[95]
C_3AlCl_7	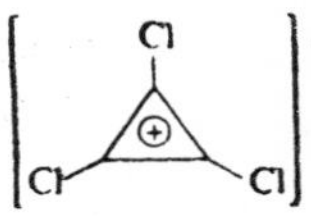 $AlCl_4$	39.960 40.166 40.171	...	[96]. At 293 °K $\eta = 35\%$. cf. Cl—Al

Empirical formula	Structural formula	ν, MHz	s/n	References, notes
$C_3Cl_3N_3$	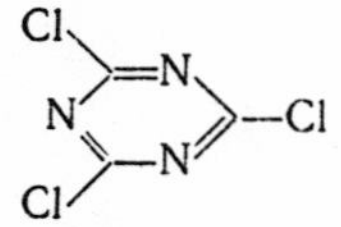	36.7708 36.7384	40 20	*[15.97–101] cf. N
		At 299 °K $\eta_1 = 26\%$; $\eta_2 = 23\%$		
C_3Cl_4	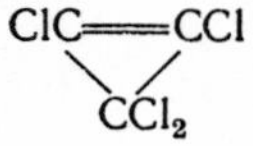	38.556 38.507 38.328 38.304 36.862 36.736 36.407 36.281	15 5 5 15 10 5 5 7	*
		36.41 36.73 38.241 38.260 38.513 38.743	. . .	[96, 214]
$C_3Cl_4F_2O$	$FCl_2CC(=O)CCl_2F$	39.104 39.312	. . .	[64]
C_3Cl_6	$CCl_3CCl{=}CCl_2$	38.588 38.595 38.728 38.980 39.029 39.715 40.114 40.576 40.660 40.936	10 10 10 10 10 10 10 20 10 10	*
$C_3HClN_3O_2$	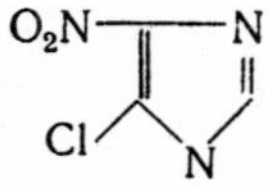	39.151	12	*
C_3HCl_5	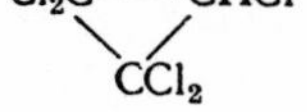	38.358 39.683 39.711 39.787 39.931	. . .	[96]
$C_3HCl_6F_3Si$	$CF_3CCl_2CHClSiCl_3$	38.815	20	*[241]. cf. Cl—Si
C_3HCl_7	$Cl_3CCCl_2CCl_2H$	38.267 39.140 39.754 39.900 40.440	30 20 20 20 20	[20]

Empirical formula	Structural formula	ν, MHz	s/n	References, notes
C_3HCl_7 (contd.)	$Cl_3CCCl_2CCl_2H$ (contd.)	41.186 41.276	20 20	
	$Cl_3CCHClCCl_3$	40.200 (av.)	...	[20]
$C_3H_2ClF_3O$	CF_3CH_2COCl	31.500	10	*
$C_3H_2Cl_2N_4O_8$	$ClC(NO_2)_2CH_2CCl(NO_2)_2$	41.0648	3	*[13]
$C_3H_2Cl_3NO_2$	$CCl_3CH{=}CHNO_2$	38.801 39.179 39.981	50 50 50	*
$C_3H_2Cl_5F_3Ge$	$CF_3CH_2CCl_2GeCl_3$	38.647 38.815	10 10	*[241]. cf. Cl—Ge
$C_3H_2Cl_5F_3Si$	$CF_3CHClCHClSiCl_3$	38.444 37.653	15 15	*[241]. cf. Cl—Si
C_3H_3Cl C_3D_3Cl	$CH_3C{\equiv}CCl$ $CD_3C{\equiv}CCl$	$e^2Qq =$ = —79.6 MHz		[103]. g
C_3H_3Cl	$HC{\equiv}CCH_2Cl$	35.8123		[20, 64, 104—106]. $\eta = 1{,}4\%$
		$e^2Qq =$ —75.8 MHz (g) $\eta = 0$ (g)		
C_3H_3ClO	$CH_2{=}CHCOCl$	29.950 30.130	15 17	*
$C_3H_3Cl_3$	$CCl_3CH{=}CH_2$	38.150 38.272 38.476	10 10 10	*[50, 75]
	$CH_2ClCH{=}CCl_2$	34.578 34.650 36.792 36.888 37.224 37.308	10 10 10 10 10 10	*
$C_3H_3Cl_3O_2$	CCl_3COOCH_3	39.676 39.760 39.858 39.872 40.152 40.191	10 10 10 10 10 10	*[20]
$C_3H_3Cl_5$	$CCl_3CHClCH_2Cl$	36.687 37.002 39.068 39.729	20 20 40 20	*
	$CHCl_2CHClCHCl_2$	37.186 37.196 37.629 38.109	3 3 3 2	[20]

Empirical formula	Structural formula	ν, MHz	s/n	References, notes
C_3H_4ClF	trans-$(CH_3)FC{=}CClH$	$e^2Qq =$ $= -73.49$MHz; $\eta = 8.44\%$		[107] g
C_3H_4ClN	CH_2ClCH_2CN	34.110	50	*[102]
$C_3H_4Cl_2$	$CH_2ClCCl{=}CH_2$	33.882	8	*
		34.224	60	
		34.896	5	
		35.064	50	
	$CHCl_2CH{=}CH_2$	35.928	50	*
		36.072	50	
	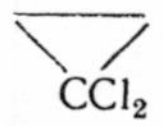	36.610	...	[108, 109]
		$e^2Qq = -76.4$ MHz (g) $\eta = 3.21\%$ (g)		
$C_3H_4Cl_2N_2O_4$	$(ClCH_2)_2C(NO_2)_2$	37.296	20	*[237]
		37.518	20	
$C_3H_4Cl_2O$	CH_2ClCH_2COCl	34.014	15	*
		29.475	15	
	$CH_2ClCOCH_2Cl$	35.943	...	[56]
	$CHCl_2COCH_3$	34.504	...	[110]. $T = 190$ °K
		34.981		
$C_3H_4Cl_2O_2$	$CHCl_2COOCH_3$	37.306	3	[20]
		37.351	3	
		37.385	3	
		37.621	2	
		37.828	2	
		38.334	3	
		38.724	3	
$C_3H_4Cl_4$	$CH_2ClCH_2CCl_3$	34.629	10	*[78, 111]
		38.150	10	
		38.304	10	
		38.825	10	
	$CH_2ClCCl_2CH_2Cl$	35.940	15	*
		36.750	15	
		37.254	30	
$C_3H_5BrCl_2$	$CH_2BrCH_2CCl_2H$	35.90	...	*
		36.10		
C_3H_5Cl		34.063	...	[108, 112, 113]
		$e^2Qq = -71.7$ MHz (g) $\eta = 15.6\%$ (g)		
	$CH_2{=}CHCH_2Cl$	33.455	5	*[20, 64, 115]

Empirical formula	Structural formula	ν, MHz	s/n	References, notes
C_3H_5Cl (contd.)	$CH_3CH{=}CHCl$	33.420	...	*[114, 115]
		$e^2Qq = -71.2$ MHz (g); $\eta = 11.3\%$ (g)		
	$CH_3CCl{=}CH_2$	32.629	...	*[115, 116]
		$e^2Qq = -68.52$ MHz (g); $\eta = 9.4\%$ (g)		
$C_3H_5ClN_2O_4$	$ClC(NO_2)_2C_2H_5$	39.9484	2—3	*[13]
	$CH_3C(NO_2)_2CH_2Cl$	37.002	40	*[237]
C_3H_5ClO	C_2H_5COCl	28.875 28.660	10 10	*[81a]
	$CH_2ClCOCH_3$	35.075 35.484	...	[1, 20, 56]
$C_3H_5ClO_2$	$CH_2ClCOOCH_3$	36.099 36.132	50 50	*
	$ClCH_2CH_2COOH$	33.948 33.978	30 30	*[20, 38]
	C_2H_5OCOCl	33.858	...	*[81a]
$C_3H_5Cl_2NO_2$	$C_2H_5CCl_2(NO_2)$	39.088 38.976	40 40	*[237]
$C_3H_5Cl_3$	$CH_3CCl_2CH_2Cl$	35.778 35.832 35.910 36.024 36.168	15 15 60 20 15	*
	$CH_2ClCHClCH_2Cl$	34.584 34.782 34.872 34.938 35.010	15 15 8 40 8	*[57]
$C_3H_5Cl_3O_2$	$CCl_3CH(OH)(OCH_3)$	38.59 38.74 38.83	...	[118]. $T = 90$ °K
$C_3H_5Cl_5Si$	$ClCH_2(CH_3)CClSiCl_3$	35.790	...	[26]. cf. Cl—Si
C_3H_6ClNO	$(CH_3)_2NCOCl$	31.8	...	[83]
$C_3H_6ClNO_2$	$CCl(CH_3)_2(NO_2)$	36.936	15—20	*[237]
$C_3H_6Cl_2$	$ClH_2CCHClCH_3$	32.51 33.11 34.42	2 4 8	*[102]
	$(CH_3)_2CCl_2$	34.883	...	*[1]

Empirical formula	Structural formula	ν, MHz	s/n	References, notes
$C_3H_6Cl_2$ (contd.)	$ClCH_2CH_2CH_2Cl$	32.868 32.967	30 30	*[20]
$C_3H_6Cl_3P$	$P(CH_2Cl)_3$	34.473 35.403 35.685	30 30 30	*
$C_3H_6Cl_3OP$	$(CH_2Cl)_3PO$	37.070 37.074 37.236 37.243 37.257 37.275	5 5 5 5 5 5	*
$C_3H_6Cl_4Ge$	$ClCH_2CH_2CH_2GeCl_3$	32.868	20	*[25, 35]. cf. Cl—Ge
$C_3H_6Cl_4Si$	$ClCH_2CH_2CH_2SiCl_3$	33.090	10	*[26, 35]. cf. Cl—Si
	$ClCH_2CHClSiCl_2(CH_3)$	35.462 34.970	. . .	[26]
	$ClCH_2(CH_3)CHSiCl_3$	34.420	. . .	[26]. cf. Cl—Si
$C_3H_6Cl_4Sn$	$(ClCH_2)_3SnCl$	35.166 35.772 36.171 36.708	16 40 25 40	*[43]
C_3H_7Cl	C_3H_7Cl	32.968		*[1, 26, 64]
	iso-C_3H_7Cl	31.939	5	*[20, 26, 27, 119]
		$e^2Qq = -67.82$ MHz (g); $\eta = 2.75\%$ (g)		
C_3H_7ClO	$CH_2ClCH_2OCH_3$	33.453	. . .	[64]
C_3H_7ClS	$CH_3SCH_2CH_2Cl$	32.634	50	*[81a]
$C_3H_7Cl_2OP$	$(CH_2Cl)_2P(O)(CH_3)$	36.241 36.950	10 10	*
$C_3H_7Cl_2OPS$	$Cl(CH_3)P(O)(SCH_2CH_2Cl)$	33.666	25	*[42a]. cf. Cl—P
$C_3H_8ClO_3P$	$(ClCH_2)P(O)(OCH_3)_2$	35.956 36.677	5 10	*
$C_3H_8Cl_2Si$	$ClCH_2SiCl(CH_3)_2$	34.80 34.85		[26, 42]. cf. Cl—Si
$C_3H_8Cl_2Sn$	$Cl(CH_3)_2SnCH_2Cl$	34.80	1.5—2	*[43]

Empirical formula	Structural formula	ν, MHz	s/n	References, notes
$C_3H_{13}B_{10}Cl$	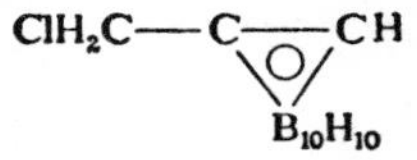 C-chloromethyl-o-borane	36.044	...	*[120, 215]
C_4BrClF_6O	$(CF_3)_2CBrCOCl$	34.164	3	*
$C_4CaCl_6O_4 + 4H_2O$	$(CCl_3COO)_2Ca \cdot 4H_2O$	38.812 38.860 39.017 39.188 39.266 39.408	1 2 1 1 2 1	[54]
C_4ClF_7	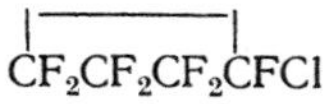	39.074	25	*
C_4ClF_7O	$(CF_3)_2CFCOCl$	33.558	5	*
$C_4Cl_2F_4$	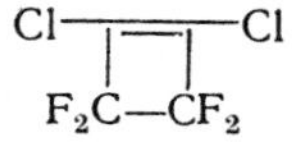	37.0005 37.1134	50 50	*[121]
$C_4Cl_2F_6$	$CF_3CCl{=}CClCF_3$	39.491	...	[64]
$C_4Cl_2O_2$	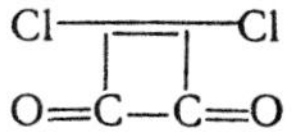	36.3 35.534 35.544	... 6 6	[121] } $T = 300$ °K
$C_4Cl_2O_3$	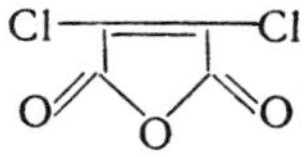	37.945 38.013	...	[122]. $T = 86$ °K
$C_4Cl_3N_3O_2$	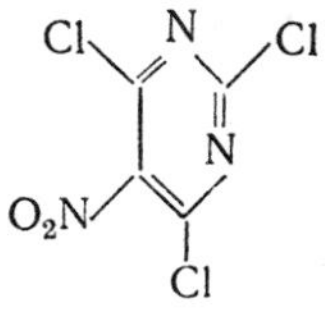	37.009 37.149 37.219 37.492 37.590 37.653 37.828 38.094 38.122	...	*
$C_4Cl_4N_2$	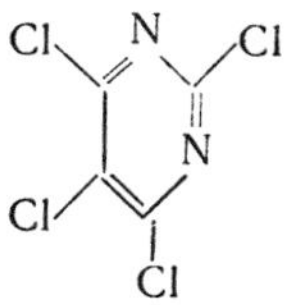	36.330 36.344 36.617 36.967 36.981 37.065 37.184 37.226 38.073 38.094 38.395 38.437	...	*[97, 123]

Empirical formula	Structural formula	ν, MHz	s/n	References, notes
C_4Cl_4O	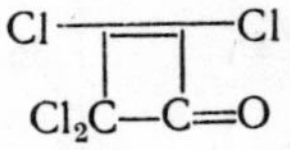	36.705 37.295 37.645 37.777	50 50 50 50	[121]
C_4Cl_4S	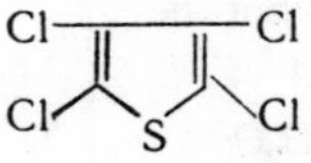	37.517 38.500	2 1	[122]. $T = 86\ °K$
C_4Cl_6	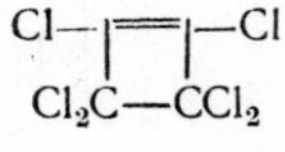	36.479 36.647 38.229 38.351 38.409	30 30 10 10 25	[121]
$C_4Cl_6CoO_4$	$(Cl_3CCOO)_2Co$	38.997 39.221	3 6	*
$C_4Cl_6CoO_4 +$ $+ 4C_5H_5N$	$(CCl_3COO)_2Co \cdot 4Py$	38.275 38.400 38.817	1—2 1—2	[54]
$C_4Cl_6CoO_4 + 4H_2O$	$(CCl_3COO)_2Co \cdot 4H_2O$	38.197 38.476 39.404 39.489 39.862 40.071	1—2 2 1—2 2 1—2 2	[54]
$C_4Cl_6CuO_4$	$(CCl_3COO)_2Cu$	39.564 39.865	3 6	*
$C_4Cl_6CuO_4 +$ $+ 4C_5H_5N + 2H_2O$	$(CCl_3COO)_2Cu \cdot 4Py \cdot 2H_2O$	37.891 38.012 38.378	2 2 2	[54]
C_4Cl_6Hg	$(CCl_2{=}CCl)_2Hg$	35.420 35.595 36.148 36.890 37.296	6 6 7 8 14	*
$C_4Cl_6HgO_4 + C_4H_8O$	$(CCl_3COO)_2Hg \cdot THF$	38.766 39.332 39.557 39.627 39.697 39.890 39.956 39.977 40.117	5 5 5 5 5 5 5 5 5	*[4, 124]

Empirical formula	Structural formula	ν, MHz	s/n	References, notes
$C_4Cl_6HgO_4$ + + $C_4H_{10}O_2$	$(CCl_3COO)_2Hg \cdot DME$	39.151 39.214 39.277 39.396 39.564 39.655 39.795	10 10 10 10 5 10 5	*[4, 124]
$C_4Cl_6Hg_2O_4$	$(CCl_3COO)_2Hg_2$	38.416 39.494 39.767 39.935 40.341	3—4 3—4 3—4 3—4 6—8	*[3, 4]
$C_4Cl_6MgO_4 + 6H_2O$	$(CCl_3COO)_2Mg \cdot 6H_2O$	38.695 38.952 39.072 39.084 39.229 39.964	1—2 2—3 2 2—3 2 2	[54]
$C_4Cl_6NiO_4$ + + $4C_5H_5N$	$(CCl_3COO)_2Ni \cdot 4Py$	38.228 38.372 38.801	1—2 1—2 1—2	[54]
$C_4HClF_2N_2$	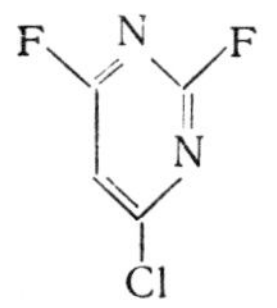	36.042	20	*
C_4HClF_6O	$(CF_3)_2CHCOCl$	32.736	10	*
C_4HClO_3	O=CCH=CClC=O |—O—|	37.187	7	[20, 122]
$C_4HCl_3N_2$	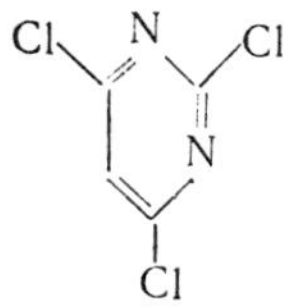	35.676 36.150 36.270	10 10 10	*[99, 123]

Empirical formula	Structural formula	ν, MHz	s/n	References, notes
$C_4HCl_3N_2$ (contd.)	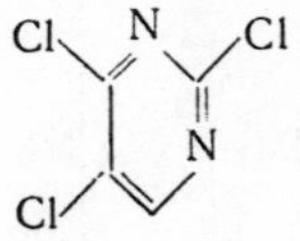	35.532 35.802 36.210 36.486 37.596 37.680	10 10 10 10 10 10	*
	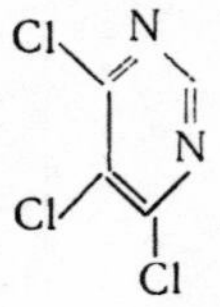	35.850 36.060 37.996	40 35 45	
$C_4H_2Cl_2N_2$	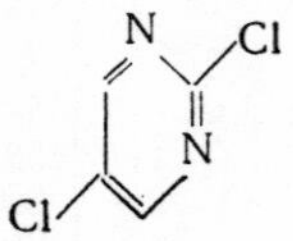	35.088 35.398	10 10	*
	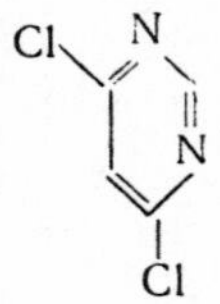	35.088 35.136 35.310 35.424	20 20 20 20	*[97, 123]
	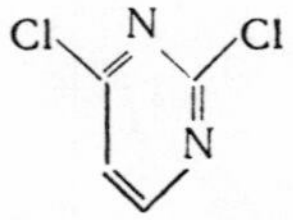	35.203 35.304 35.310 } 35.321 }	15 15 ...	* [97, 123]
	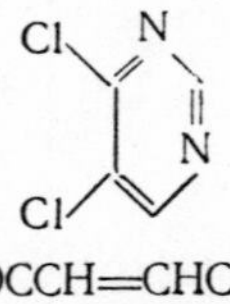	35.724 35.946 37.242 37.620	10 10 10 10	*
$C_4H_2Cl_2O_2$	ClOCCH=CHCOCl	30.380 30.970	5—6 5—6	[52]
$C_4H_2Cl_2O_3$	HOOCCCl=CClCHO	36.576 37.344	1.5 3	[20]
$C_4H_2Cl_2S$	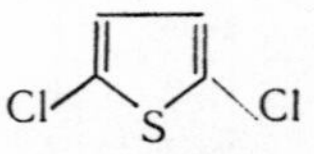	37.018 37.071	2 5	[20, 122]. p.t. at 83°K
$C_4H_2Cl_{10}Si$	$(CHCl_2CCl_2)_2SiCl_2$	39.34	...	[26]. cf. Cl—Si

Empirical formula	Structural formula	ν, MHz	s/n	References, notes
$C_4H_3ClN_2$	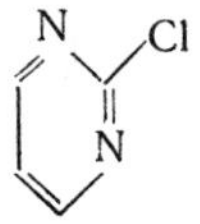	34.434 34.506	...	*[99]
$C_4H_3ClN_2O_2$	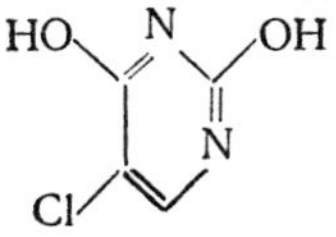	37,947	...	*
	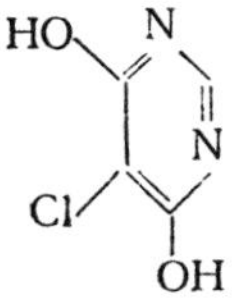	36.288	...	*
	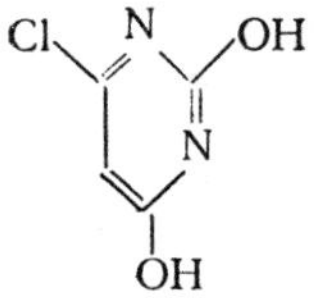	37.248	...	*
$C_4H_3Cl_3N_2$	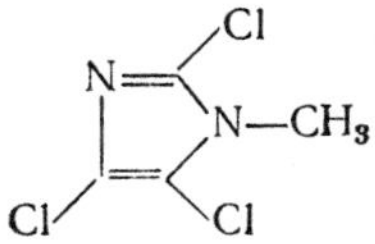	36.420 36.912 37.260 37.860	...	*[125]
$C_4H_4ClN_3O_2$	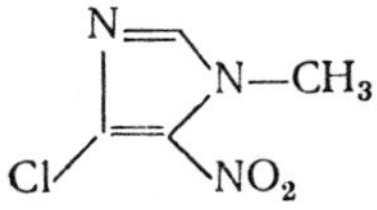	37.359 37.482	8 8	*
$C_4H_4Cl_2$	$ClCH_2C{\equiv}CCH_2Cl$	36.3728	...	[105, 216]
$C_4H_4Cl_2Hg$	trans-$(CHCl{=}CH)_2Hg$	31.590 31.698 31.932	2 4 2	*[239]
$C_4H_4Cl_2N_2$	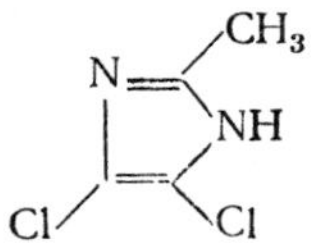	36.070 37.394	10 10	*[125]

Empirical formula	Structural formula	ν, MHz	s/n	References, notes
$C_4H_4Cl_2N_2$ (contd.)	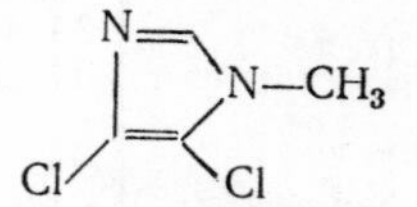	35.840 37.639 37.758	30 15 15	*[125]
$C_4H_4Cl_2O_2$	$ClOCCH_2CH_2COCl$	30.217	12—15	*[52]
$C_4H_4Cl_3NO_2$	$CCl_3CH{=}C(NO_2)CH_3$	39.480 39.655 39.711	30 25 30	*
$C_4H_4Cl_4$	$CCl_2{=}CClCH_2CH_2Cl$	38.003 37.646 35.966 33.732	5 5 5 3	*
$C_4H_4Cl_4O$	CHClCHClCHClCHClO (ring)	34.824	...	[83]
$C_4H_4Cl_4O_2$	CHClCHClOCHClCHClO (ring)	36.968	...	[83]
C_4H_5Cl	$CH_3C{\equiv}CCH_2Cl$	35.3618	...	[105]
$C_4H_5ClN_2$	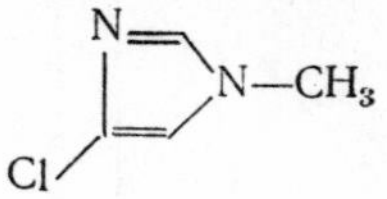	35.034	...	*[125]
	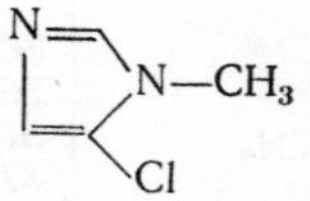	36.720 36.924	...	*[125]
$C_4H_5ClN_4$	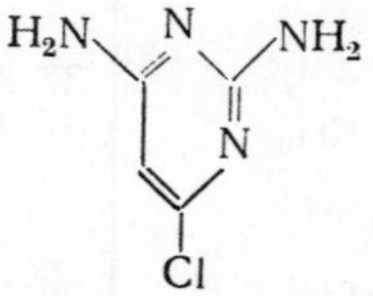	34.304	...	[123]
$C_4H_5ClN_4O_8$	$Cl(NO_2)_2CCH_2$ $H_3C(NO_2)_2C$	40.824	20	*[13]
C_4H_5ClOS	$OCSCH_2CClCH_3$ (ring)	35.06	10	*
$C_4H_5Cl_3OS$	$CCl_3COSC_2H_5$	39.162 39.435 39.834	40 40 40	*
$C_4H_5Cl_3O_2$	$CCl_3COOC_2H_5$	40.200 40.339	...	[117]

Empirical formula	Structural formula	ν, MHz	s/n	References, notes
$C_4H_6Cl_2$	CH$_2$CH$_2$CH$_2$CCl$_2$ |_____|	35.358 34.710	15 15	*
	ClCH$_2$CH=CHCH$_2$Cl	34.109	3	[20]
$C_4H_6Cl_2OS$	Cl$_2$CHCOSC$_2$H$_5$	37.559 37.849	40 40	*
$C_4H_6Cl_3F_3Si$	CF$_3$CH$_2$CHClSiCl$_2$(CH$_3$)	35.220 35.514	5 5	*[241]. cf. Cl—Si
$C_4H_6Cl_3NO_2$	(ClCH$_2$)$_3$CNO$_2$	35.706 35.928 36.096	10 10 10	*[237]
$C_4H_6Cl_4$	CCl$_3$CH$_2$CHClCH$_3$	33.744 38.836 38.360 38.178	10 12 10 10	*
C_4H_7Cl	CH$_2$=C(CH$_3$)CH$_2$Cl	33.777	...	[64]
$C_4H_7ClN_2O_4$	C$_3$H$_7$CCl(NO$_2$)$_2$	40.432	50	*[237]
C_4H_7ClO	C$_3$H$_7$COCl	29.005	8	*[81a]
	(CH$_3$)$_2$CHCOCl	28.94	5	*[81a]
C_4H_7ClOS	ClCH$_2$COSC$_2$H$_5$	35.315	50	*
$C_4H_7ClO_2$	ClCH$_2$COOC$_2$H$_5$	35.962	...	[56]
	CH$_3$COOCH$_2$CH$_2$Cl	34.104	6	[20]
$C_4H_7Cl_2NO_2$	(ClCH$_2$)$_2$C(CH$_3$)NO$_2$	35.568 35.742 35.874	5 10 5	*[237]
$C_4H_7Cl_3O_2$	CH$_3$CHClCCl$_2$CH(OH)$_2$	35.017 36.954 37.425	3 4 5	[20]
	CCl$_3$CH(OH)OC$_2$H$_5$	38.516 38.765 39.140	3 2.5 4	[20, 118, 126, 127]
$C_4H_8Cl_2$	Cl$_2$CHCH$_2$CH$_2$CH$_3$	35.117 35.210	5 5	[20]
	ClCH$_2$CH$_2$CH$_2$CH$_2$Cl	32.850	...	[20, 38]
	ClCH$_2$CH$_2$CHClCH$_3$	31.949 33.485	11 13	[20]
	H$_3$CCCl$_2$C$_2$H$_5$	34.722 34.793	5 5	[20]
$C_4H_8Cl_2O$	H$_3$CCHClOCHClCH$_3$	33.392 33.483	...	[64]
	ClCH$_2$CH$_2$OCH$_2$CH$_2$Cl	33.391 33.997	3 1.2	[20]

Empirical formula	Structural formula	ν, MHz	s/n	References, notes
$C_4H_8Cl_2O_3$	$ClCH_2CH_2OCOOCH_2CH_2Cl$	33.755 34.311	2 2	[20]
$C_4H_8Cl_4Si$	$ClCH_2CH(CH_3)CH_2SiCl_3$	33.525	...	[26]
$C_4H_8Cl_4Sn$	$(ClCH_2)_4Sn$	34.512 35.067	150 150	*[43]
$C_4H_8Cl_5P$	$ClP(CH_2Cl)_4$	38.927 39.263 39.445 39.480	5 5 5 5	*
C_4H_9Cl	$CH_3CHClCH_2CH_3$	31.875	...	*
	$(CH_3)_3CCl$	31.065	...	[20, 21, 26, 128]
		$e^2Qq = -66.9$ MHz (g)		
	$(CH_3)_2CHCH_2Cl$	32.96	...	[26]
	C_4H_9Cl	33.255	2.5	*[20, 102]
C_4H_9ClO	$C_2H_5OCH_2CH_2Cl$	33.708	100	*
$C_4H_9ClO_2$	$CH_2ClCH(OCH_3)_2$	34.608	...	[20]
$C_4H_9Cl_2OPS$	$(C_2H_5)P(O)Cl(SCH_2CH_2Cl)$	33.546	5	*[42a]. cf. Cl—P
$C_4H_9Cl_3Si$	$ClCH_2CH_2CH_2SiCl_2(CH_3)$	33.19	...	[26]. cf. Cl—Si
$C_4H_9Cl_4P$	$(ClCH_2)_3P(CH_3)Cl$	38.245 38.486 38.780	5 5 5	*
$C_4H_{10}ClN$	$(CH_3)_2NCH_2CH_2Cl$	33.266	10	[20]
$C_4H_{10}Cl_2Sn$	$(CH_3)_2Sn(CH_2Cl)_2$	33.840 33.960 34.080 34.248	2 2 2—3 2—3	*[43]
$C_4H_{10}Cl_3N$	$[(ClCH_2CH_2)_2NH_2]Cl$	35.412 34.244	20 25	[20]
$C_4H_{10}Cl_3P$	$(ClCH_2)_2P(CH_3)_2Cl$	37.436 37.590 37.814	5 10 5	*
$C_4H_{11}ClGe$	$(CH_3)_3GeCH_2Cl$	35.227	...	[26]
$C_4H_{11}ClSi$	$(CH_3)_3SiCH_2Cl$	35.39	50	*[26, 35]
$C_4H_{11}ClSn$	$(CH_3)_3SnCH_2Cl$	33.936	5	*[43]
$C_4H_{11}Cl_2N$	$[(CH_3)_3N(CH_2Cl)]Cl$	36.039	20	*
	$[(CH_3)_2NH(CH_2CH_2Cl)]Cl$	35.065	12	[20]

Empirical formula	Structural formula	ν, MHz	s/n	References, notes
$C_4H_{14}B_{10}Cl_2$	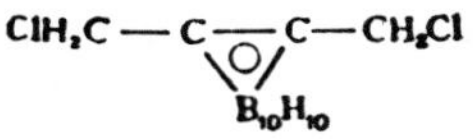 C,C'-di(chloromethyl)-o-borane	36.664 37.009 37.342 37.405	...	*[120, 215]
$C_5Cl_2F_6$	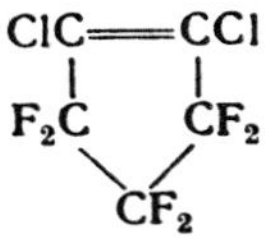	38.039	...	[64]
$C_5Cl_4O_2$	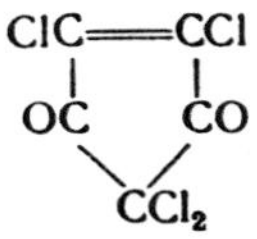	35.574	...	[121]
C_5Cl_6	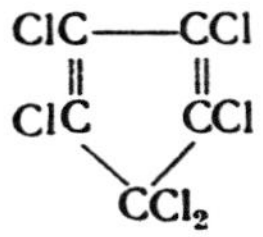			*[20, 129, 130]
	Phase I	36.953 37.279 37.457 38.822 39.099	10 10 20 10 10	
	Phase II	36.060 37.198 37.303 37.359 37.380 37.653 39.102 39.473 39.585		
C_5Cl_8	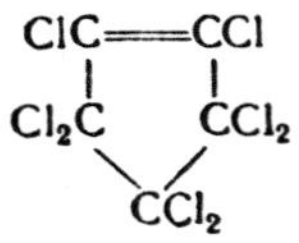	37.426 37.891 39.368 39.606 39.771 39.998 40.254 41.176	20 20 20 20 20 20 20 20	*[129]

Empirical formula	Structural formula	ν, MHz	s/n	References, notes
$C_5HCl_3N_2O$	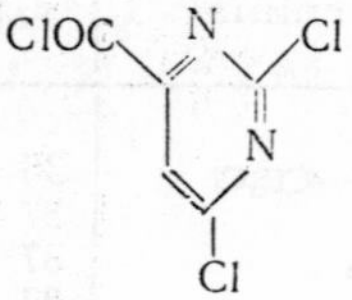	35.754 35.982 32.460	5 5 3	*
$C_5H_2Cl_2N_2O_2$	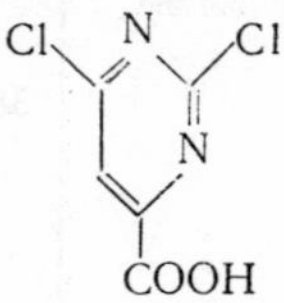	35.946 36.198	7 7	*
$C_5H_3BrCl_2N_2$	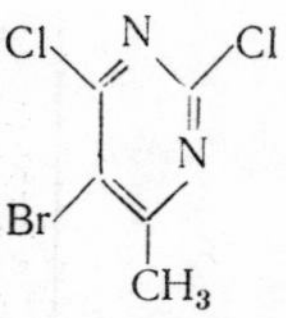	36.378	2	*
$C_5H_3ClF_6O$	$(CF_3)_2C(CH_3)COCl$	32.352 32.634	8 10	*
$C_5H_3ClO_2$	OCH=CHCH=CCOCl (ring closed between O and C)	30.726	3—4	[52]
$C_5H_3Cl_2IN_2$	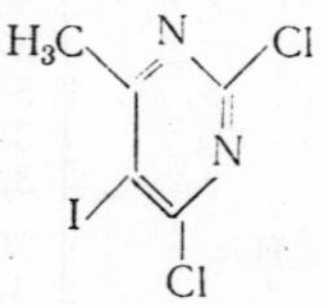	35.640 36.294	10 10	*
$C_5H_3Cl_2N$	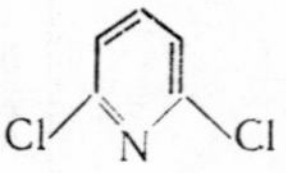	33.850 η = 11.8%	...	[131, 132]. T = 299 °K. cf. N
	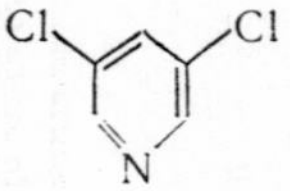	35.601 At 299 °K η = 8.6%	...	[97, 132]. cf. N
$C_5H_3Cl_2NO$	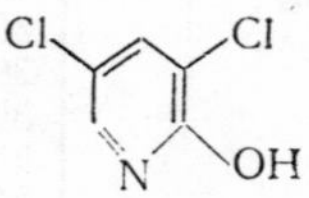	36.683 36.988	15 15	*

Empirical formula	Structural formula	ν, MHz	s/n	References, notes
$C_5H_3Cl_2NO$ (contd.)	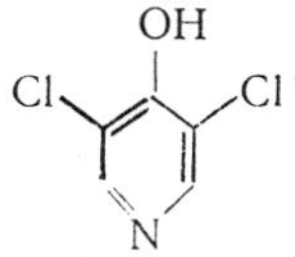	36.386 35.980	10 10	*
$C_5H_3Cl_3N_2$	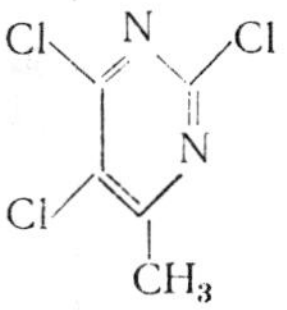	36.198 37.368	15 10	*
$C_5H_3Cl_3N_2OS$		38.731 39,081	3 6	*
C_5H_4ClN	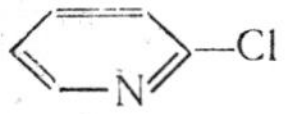	34.194	...	[99, 133]. cf. N
$C_5H_4ClN + HCl$	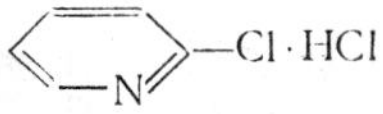	37.559	...	[133]
C_5H_4ClN	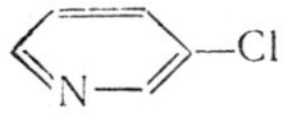	35.238	...	[97]. cf. N
	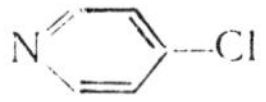	34.739 34.748 35.031 35.042	...	[97]. cf. N
$C_5H_4Cl_2N_2$	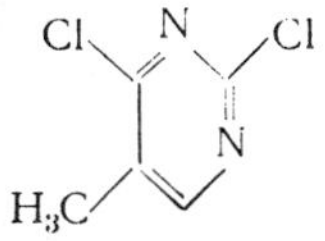	34.893 35.338	...	[123]
	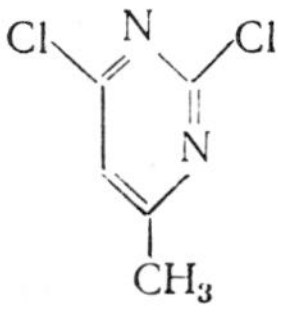	35.256	...	[97, 123]

Empirical formula	Structural formula	ν, MHz	s/n	References, notes
$C_5H_4Cl_2N_2$ (contd.)	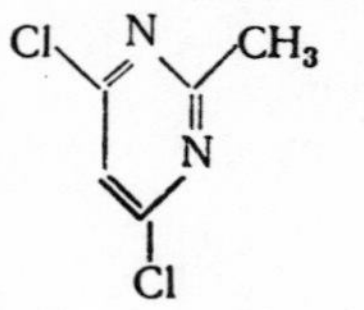	35.156	...	[99]. $T = 86\ °K$
	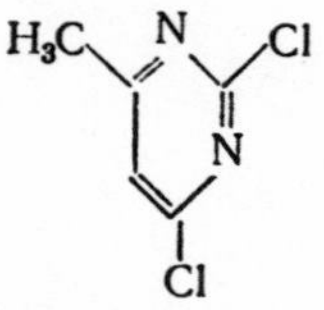	35.136 35.274	5 5	*
$C_5H_4Cl_2N_2S$	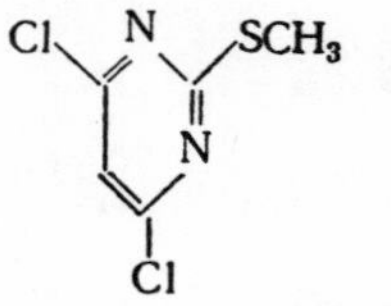	34.895 34.986 35.084 35.532 35.700	20 20 20 40 20	*
$C_5H_5BCl_6O_2$	$\left[(CCl_2{=}CCl)_2B{<}^{OCH_3}_{OH}\right]H$	34.860 35.770 37.030	7 7 7	*
$C_5H_5ClN_2$	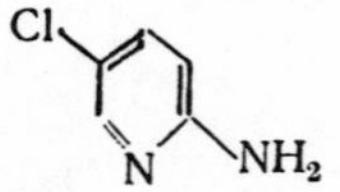	35.630	...	[97]
	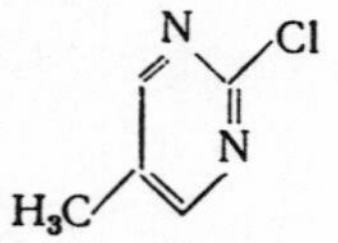	35.136 35.274	20 20	*
$C_5H_5ClN_2O_2$	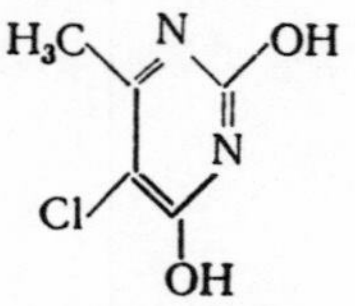	36.960	20	*
$C_5H_5Cl_2N_3$	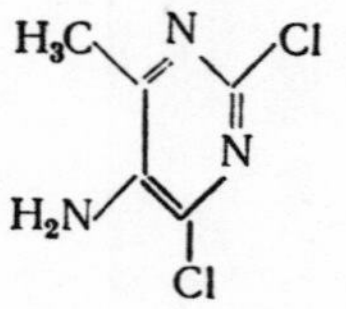	34.914 35.010 35.286	5 5 10	*

Empirical formula	Structural formula	ν, MHz	s/n	References, notes
C_5H_6ClNO	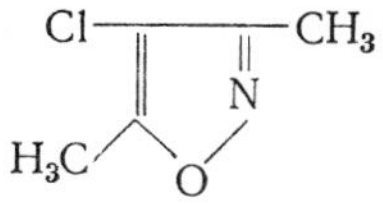	36.55	6	*
$C_5H_6ClN_3S$	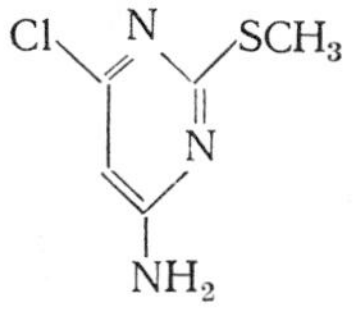	34.993	2	[97, 123]
$C_5H_6Cl_6$	$CCl_3CCl_2CH_2CH_2CH_2Cl$	33.474 38.087 40.075 40.194 40.313	15 30 15 15 15	*
C_5H_7Cl	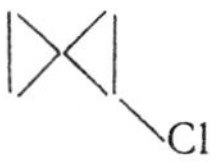	$e^2Qq =$ $= -73.45$ MHz $\eta = 0$		[134]. g
$C_5H_7ClN_2$	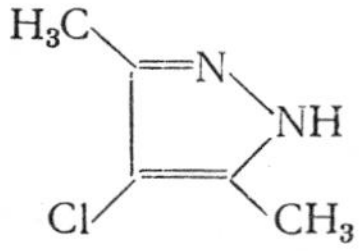	36.336	50	*
$C_5H_7ClN_4O_8$	$(NO_2)_2CClCH_2C(NO_2)_2C_2H_5$	42.032	50	*[13]
$C_5H_7ClO_2$	$CH_3C(OH)=CClCOCH_3$	36.106	10	*
$C_5H_7Cl_2NO_2$	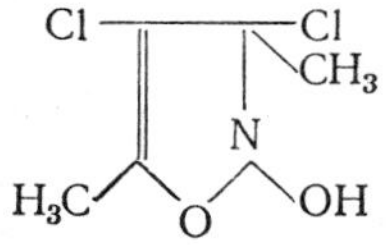	37.972 38.843	10 10	*[235]
$C_5H_7Cl_3O_2$	$CCl_3CH(OH)CH_2COCH_3$	38.234 38.293 38.295	...	[59, 91]
$C_5H_7Cl_4F_3Si$	$CF_3CH_2CH_2SiCl_2$ | ClH_2CCHCl	35.058 35.541	5—7 5—7	*[241]. cf. Cl—Si
$C_5H_7Cl_5$	$CCl_3CH_2CHClCH_2CH_2Cl$	39.092 39.060 37.572 34.332	5 10 4 4	*[242]
$C_5H_8Br_2Cl_2$	$CH_2BrCH_2CCl_2CH_2CH_2Br$	35.540	20	*cf. Br

Empirical formula	Structural formula	ν, MHz	s/n	References, notes
$C_5H_8Br_3Cl$	$CH_2BrCH_2CBrClCH_2CH_2Br$	35.654	8	*
C_5H_8ClN	$ClCH_2(CH_2)_3CN$	32.844	2	*
$C_5H_8Cl_2$	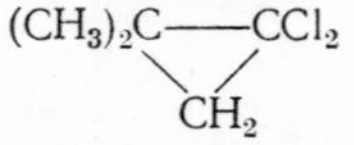	36.797	. . .	[108]
	$CH_2CH_2CH_2CH_2CCl_2$ (ring)	34.984 35.060	. . .	[135]
$C_5H_3Cl_4$	$ClCH_2(CH_2)_3CCl_3$	33.345 37.832 38.021 38.560	50 50 50 50	*[20, 78]
$C_5H_8Cl_4Ge$	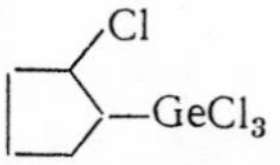	33.496	. . .	[26]. cf. Cl—Ge
$C_5H_9BrCl_2$	$BrCH_2(CH_2)_3CHCl_2$	35.418	2	*
C_5H_9ClGe	$(CH_3)_3GeC{\equiv}CCl$	38.20	. . .	[136]
C_5H_9ClSi	$(CH_3)_3SiC{\equiv}CCl$	38.15	. . .	[136]
$C_5H_{10}ClNO$	$(C_2H_5)_2NCOCl$	31.877	. . .	[83]
$C_5H_{10}Cl_2$	$ClCH_2(CH_2)_3CH_2Cl$	33.222	10	[20]
	$CH_3CHClCH_2CHClCH_3$	32.333 32.686	8 7	[20]
$C_5H_{10}Cl_2O_2$	$ClCH_2CH_2OCH_2OCH_2CH_2Cl$	33.249 33.724 33.932	~1 ~1 ~1	[20]
$C_5H_{11}Cl$	$C_5H_{11}Cl$	33.103	7	[20]
	$C_5H_{11}Cl$-2	30.878	1.5	[20]
	$CH_3CH(CH_3)CH_2CH_2Cl$	32.720	1.2	[20]
$C_5H_{11}ClO$	$H_9C_4OCH_2Cl$	29.920 30.695	20 8	*
$C_5H_{11}Cl_2OPS$	(iso-C_3H_7)—P(O)Cl | SCH_2CH_2C	33.657	12	*[42a]. cf. Cl—P
$C_5H_{12}ClO_3P$	$(ClCH_2)P(O)(OC_2H_5)_2$	36.607	100	*
$C_5H_{12}Cl_2Si$	$ClCH_2CH_2CH_2Si(CH_3)_2Cl$	33.00	. . .	[26]. cf. Cl—Si
$C_5H_{12}Cl_3N$	$[(ClCH_2CH_2)_2NH(CH_3)]Cl$	35.075	20	[20]
$C_5H_{13}ClSi$	$(CH_3)_3SiCHClCH_3$	32.425	. . .	[26]
$C_5H_{13}Cl_2N$	$[(ClCH_2CH_2CH_2)NH(CH_3)_2]\cdot Cl$	33.239	20	[20]

Empirical formula	Structural formula	ν, MHz	s/n	References, notes
$C_5H_{14}ClN_2OP$	$(ClCH_2)P(O)[N(CH_3)_2]_2$	35.502	100	*
$C_5H_{17}B_{10}Cl$	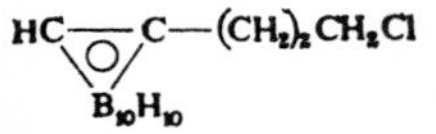 C-chloropropyl-o-borane	33.654	...	*[120]
C_6BrCl_4F	1,3-$FBrC_6Cl_4$	38.633 38.479 38.325 38.165	~1.5 ~1.5 ~1.5 ~1.5	*[231]
C_6BrCl_5	C_6Cl_5Br	38.30 38.43	10 40	*[2, 137]
$C_6Br_2Cl_4$	1,2-$Br_2C_6Cl_4$	38.36	10	*[2]
	1,3-$Br_2C_6Cl_4$	38.37	15	*[2]
	1,4-$Br_2C_6Cl_4$	38.26 38.39	30 50	*[2]
$C_6Br_3Cl_3$	1,2,3-$Cl_3C_6Br_3$	38.27 38.37	8 4	*[2]
	1,2,4-$Cl_3C_6Br_3$	38.26 38.36	20 10	*[2]
C_6Br_4ClF	1,2-$FClC_6Br_4$	38.60	1.5	*[2]
	1,3-$FClC_6Br_4$	38.19	1.5	*[2]
$C_6Br_4Cl_2$	1,2-$Cl_2C_6Br_4$	38.26	20	*[2]
	1,3-$Cl_2C_6Br_4$	38.20	10	*[2]
	1,4-$Cl_2C_6Br_4$	38.21 38.32	20 5	*[2]
C_6Br_5Cl	C_6Br_5Cl	38.20	10	*[2, 137]
C_6ClF_5	C_6ClF_5	39.410	50	*[2, 138, 139]
$C_6ClF_5 + C_6H_5Br$	$C_6F_5Cl \cdot C_6H_5Br$	39.543	10	*
$C_6ClF_5 + C_6H_5Cl$	$C_6F_5Cl \cdot C_6H_5Cl$	34.68 39.36	7 7	*[138, 139]
$C_6ClF_5 + C_6H_5I$	$C_6F_5Cl \cdot C_6H_5I$	39.435	8	*
$C_6ClF_5 + C_6H_6$	$C_6F_5Cl \cdot C_6H_6$	38.717	10	*[138, 139]
$C_6ClF_5 + C_6H_7N$	$C_6F_5Cl \cdot C_6H_5NH_2$	38.640	2	*[138, 139]
$C_6ClF_5 + C_6H_{15}N$	$C_6F_5Cl \cdot (C_2H_5)_3N$	39.284	10—15	*[138, 139]
$C_6ClF_5 + C_7H_8$	$C_6F_5Cl \cdot C_6H_5CH_3$	39.16	5	*[138, 139]

Empirical formula	Structural formula	ν, MHz	s/n	References, notes
$C_6ClF_5 + C_9H_{12}$	$C_6F_5Cl \cdot 1,3,5\text{-}(CH_3)_3C_6H_3$	38.684	5	*[138, 139]
$C_6Cl_2F_4$	$1,3\text{-}Cl_2C_6F_4$	39.46	2	*[231]
$C_6Cl_3F_3$	$1,3,5\text{-}Cl_3C_6F_3$	39.312	20	*[2, 138, 139]
$C_6Cl_3F_3 + C_9H_{12}$	$1,3,5\text{-}Cl_3C_6F_3 \cdot 1,3,5\text{-}(CH_3)_3C_6H_3$	38.850	7	[*138, 139]
$C_6Cl_3F_3 + C_{12}H_{18}$	$1,3,5\text{-}Cl_3C_6F_3 \cdot C_6(CH_3)_6$	38.83	4	*[138, 139]
$C_6Cl_4F_2$	$1,3\text{-}F_2C_6Cl_4$	37.923 38.810 39.144	5 10 5	*[2]
$C_6Cl_4N_2O_4$	$1,4\text{-}(NO_2)_2C_6Cl_4$	39.113	50	*[140]
$C_6Cl_4O_2$	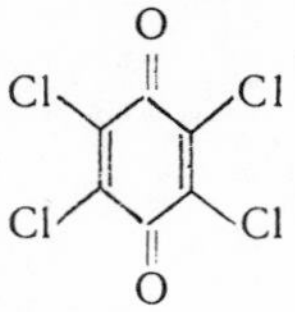	37.4417 37.4698 37.5148 37.5851	...	[83, 97, 141—143]. p.t. at 100°K
		At 77 °K $\eta = 19.6\%$ [213]; At 273 °K $\eta_1 = 21.4\%$, $\eta_2 = 21.1\%$		
$C_6Cl_4O_2 + C_9H_{12}$	$1,4\text{-}O_2C_6Cl_4 \cdot 1,3,5\text{-}(CH_3)_3C_6H_3$	37.387 37.639 37.646	2 3 2	[23]
$C_6Cl_4O_2 + C_{12}H_{18}$	$1,4\text{-}O_2C_6Cl_4 \cdot C_6(CH_3)_6$	37.5042 37.7161	...	[142]
C_6Cl_5F	C_6FCl_5	38.43 38.78 39.05	100 100 10	*[2, 144—146]
$C_6Cl_5NO_2$	$C_6Cl_5NO_2$	38.33	5	*[2]
C_6Cl_5NaO	C_6Cl_5ONa	35.442 35.772 37.236 37.788	20 20 40 20	*[147]
C_6Cl_6	C_6Cl_6	38.493 38.452 38.381	100 100 100	*[145, 148—150]. $\eta \approx 16\%$
C_6Cl_6Hg	C_6Cl_5HgCl	36.495 36.953 38.059 38.164 38.290	12 15 20 20 20	*[147]. cf. Cl—Hg

Empirical formula	Structural formula	ν, MHz	s/n	References, notes
C_6Cl_6O	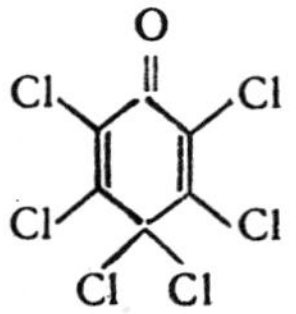	37.553 37.865	...	[126]
	C_6Cl_5OCl	37.055 38.157 39.997 40.100	~1 5 ~1.5 ~1	[126]
$C_6Cl_6O_2S$	$C_6Cl_5SO_2Cl$	39.648 39.354 38.801 38.731 38.444	30 30 30 30 30	*[225]. cf. S—Cl
$C_6HCl_3F_2$	2,4,6-Cl_3-3,5-F_2C_6H	38.03 38.30	2—3 4—5	*[231]
$C_6HCl_3F_2O_2S$	2,3,5-Cl_3-6-FC_6HSO_2F	37.492 37.732 37.982 38.136 38.689	3 6 3 3 3	*[225]
$C_6HCl_3N_2O_4$	2,4,5-Cl_3-1,3-$(NO_2)_2C_6H$	38.245 38.741 39.916	10 10 10	*[140]
	2,4,6-Cl_3-1,3-$(NO_2)_2C_6H$	37.835 38.010 39.165	10 10 10	*[140, 217]
C_6HCl_4F	2,4,5,6-Cl_4-3-FC_6H	37.163 37.716 38.028 38.703	2 8 3 4	*[231]
$C_6HCl_4FO_2S$	2,3,5,6-$Cl_4C_6HSO_2F$	37.625 37.800 38.619 39.340	10 10 10 10	*[225]
$C_6HCl_4NO_2$	2,3,5,6-$Cl_4C_6HNO_2$	37.458 37.811 37.950 38.069 38.464	10 10 10 10 10	*[140]

Empirical formula	Structural formula	ν, MHz	s/n	References, notes
$C_6HCl_4NO_2$ (contd.)	2,3,5,6-$Cl_4C_6HNO_2$ (contd.)	38.501 38.567 38.676	10 10 10	
C_6HCl_5	C_6Cl_5H	37.464 37.503 37.720 37.957 38.152	20 20 20 20 20	*[70, 145, 148, 162]
C_6HCl_5O	C_6Cl_5OH	37.125 37.371 38.086 38.188 38.577	10 10 10 10 10	*[17, 145, 151—153]. At 300 °K $\eta_1 = 10\%$; $\eta_2 = 16.3\%$; $\eta_3 = 16.4\%$; $\eta_4 = 18.5\%$; $\eta_5 = 16.7\%$
$C_6HCl_5O_2S$	2,3,5,6-$Cl_4C_6HSO_2Cl$	37.618 37.905 38.696 39.424	10 10 15 10	*[225]. cf. S—Cl
C_6HCl_5S	C_6Cl_5SH	37.141 37.267 38.238 38.296	3 3 8 3	*[145]
$C_6H_2BrCl_2F$	3-F-4,6-$Cl_2C_6H_2Br$	36.64 37.13	4 4	*[231]
$C_6H_2BrCl_3$	3,4,6-$Cl_3C_6H_2Br$	36.66 36.78	7 15	*[2]
$C_6H_2Br_2ClNO_2$	3,5-Br_2-4-Cl-$C_6H_2NO_2$	37.590	5	*[140]
$C_6H_2ClFN_2O_4$	4,6-$(NO_2)_2$-3-ClC_6H_2F	38.283	100	*
$C_6H_2ClF_3O_2S$	2,4-F_2-5-$ClC_6H_2SO_2F$	38.451	20	*[225]
$C_6H_2ClN_3O_6$	2,4,6-$(NO_2)_3C_6H_2Cl$	39.367	10	*[140, 154]
$C_6H_2ClN_3O_6$ + + C_6H_5Br	2,4,6-$(NO_2)_3C_6H_2Cl \cdot C_6H_5Br$	39.998	5	*[226, 227]
$C_6H_2ClN_3O_6$ + + C_6H_5Cl	2,4,6-$(NO_2)_3C_6H_2Cl \cdot C_6H_5Cl$	39.977 34.380	3 3	*[226, 227]
$C_6H_2ClN_3O_6$ + + C_6H_5I	2,4,6-$(NO_2)_3C_6H_2Cl \cdot C_6H_5I$	40.243	20	*[226, 227]
$C_6H_2ClN_3O_6$ + + C_6H_5F	2,4,6-$(NO_2)_3C_6H_2Cl \cdot C_6H_5F$	39.683	...	*[226, 227]
$C_6H_2ClN_3O_6$ + + $C_6H_5NO_2$	2,4,6-$(NO_2)_3C_6H_2Cl \cdot C_6H_5NO_2$	39.466	2	*[226, 227]

Empirical formula	Structural formula	ν, MHz	s/n	References, notes
$C_6H_2ClN_3O_6$ + + C_6H_6O	2,4,6-$(NO_2)_3C_6H_2Cl \cdot C_6H_5OH$	40.544	...	*[226, 227]
$C_6H_2ClN_3O_6$ + C_6H_6	2,4,6-$(NO_2)_3C_6H_2Cl \cdot C_6H_6$	39.578	15	*[226, 227]
$C_6H_2ClN_3O_6$ + + C_7H_7Br	2,4,6-$(NO_2)_3C_6H_2Cl \cdot$ 4-$BrC_6H_4CH_3$	39.606		*[226, 227]
$C_6H_2ClN_3O_6$ + + C_7H_7Cl	2,4,6-$(NO_2)_3C_6H_2Cl \cdot$ 4-$CH_3C_6H_4Cl$	34.704 34.836 39.907 40.152	5 5 5 5	*[226, 227]
$C_6H_2ClN_3O_6$ + + C_7H_7F	2,4,6-$(NO_2)_3C_6H_2Cl \cdot$ 4-$FC_6H_4CH_3$	40.040	...	*[226, 227]
$C_6H_2ClN_3O_6$ + + C_7H_7I	2,4,6-$(NO_2)_3C_6H_2Cl \cdot$ 4-$IC_6H_4CH_3$	39.431	...	*[226, 227]
$C_6H_2ClN_3O_6$ + C_7H_8	2,4,6-$(NO_2)_3C_6H_2Cl \cdot C_6H_5CH_3$	40.075 39.622	...	*[226, 227]
$C_6H_2ClN_3O_6$ + + C_7H_8O	2,4,6-$(NO_2)_3C_6H_2Cl \cdot$ 4-$HOC_6H_4CH_3$	39.501	...	*[226, 227]
$C_6H_2ClN_3O_6$ + + C_8H_{10}	2,4,6-$(NO_2)_3C_6H_2Cl \cdot$ · 1,4-$(CH_3)_2C_6H_4$	39.425	...	*[226, 227]
$C_6H_2ClN_3O_6$ + + C_9H_{12}	2,4,6-$(NO_2)_3C_6H_2Cl \cdot$ · 1,3,5-$(CH_3)_3C_6H_3$	40.387	15	*[226, 227]
$C_6H_2ClN_3O_6$ + + $C_{10}H_8$	2,4,6-$(NO_2)_3C_6H_2Cl \cdot C_{10}H_8$	39.9315	...	*[155]
$C_6H_2ClN_3O_6$ + + $C_{10}H_{14}$	2,4,6-$(NO_2)_3C_6H_2Cl \cdot$ · 1,2,4,5-$(CH_3)_4C_6H_2$	39.970	...	*[226, 227]
$C_6H_2ClN_3O_6$ + + $C_{11}H_{16}$	2,4,6-$(NO_2)_3C_6H_2Cl \cdot (CH_3)_5C_6H$	40.264	...	*[226, 227]
$C_6H_2ClN_3O_6$ + + $C_{14}H_{10}$	2,4,6-$(NO_2)_3C_6H_2Cl \cdot$ anthracene	39.802	...	*[226, 227]
	2,4,6-$(NO_2)_3C_6H_2Cl \cdot$ phenanthrene	39.914	...	*[226, 227]
$C_6H_2Cl_2FNO_2$	2,4-Cl_2-5-$FC_6H_2NO_2$	37.416 38.487	5 5	*[140]
$C_6H_2Cl_2F_2$	2,6-Cl_2-3,5-$F_2C_6H_2$	37.44	20	*[231]
$C_6H_2Cl_2F_2O_2S$	2,4-F_2-5-$ClC_6H_2SO_2Cl$	37.898	3	*[225]. cf. Cl—S
	3,4-Cl_2-6-$FC_6H_2SO_2F$	37.308 37.638	10 10	*[225]
$C_6H_2Cl_2KNO_3$	2,6-Cl_2-4-$NO_2C_6H_2OK$	34.992 35.100	10 10	*[3]

Empirical formula	Structural formula	ν, MHz	s/n	References, notes
$C_6H_2Cl_2N_2O_4$	1,3-$(NO_2)_2$-4,6-$Cl_2C_6H_2$	38.171 38.497	40 50	*[232]
$C_6H_2Cl_2O_2$	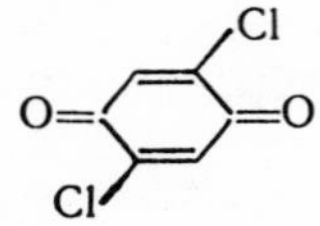	36.321	...	[97, 141]
	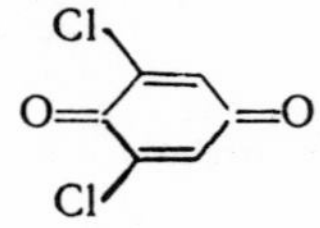	36.266 36.361	...	[97, 141]
$C_6H_2Cl_2O_4$	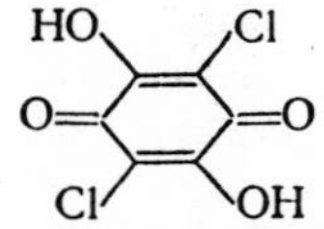	37.148	...	[97]
$C_6H_2Cl_3FO_2S$	2,4,5-$Cl_3C_6H_2SO_2F$	36.930 37.368 37.650	5 5 5	*[225]
	2,5-Cl_2-4-$FC_6H_2SO_2Cl$	37.611 37.961	50 50	*[225]. cf. S—Cl
	4,5-Cl_2-2-$FC_6H_2SO_2Cl$	37.134 37.146	40 40	The same
$C_6H_2Cl_3NO$	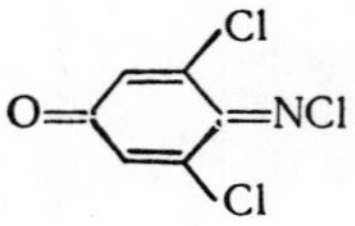	36.013 36.239 36.324 36.614 36.666 36.932	7 7 7 7 7 7	*
$C_6H_2Cl_3NO_2$	2,3,4-$Cl_3C_6H_2NO_2$	37.170 38.269 38.780	5 5 5	*[232]
	3,4,5-$Cl_3C_6H_2NO_2$	37.359 37.720	10 5	*[140, 218]
	2,4,5-$Cl_3C_6H_2NO_2$	36.585 36.717 37.338 37.435 38.289 38.441	15 15 15 15 15 15	*[140]
$C_6H_2Cl_4$	1,2,3,4-$Cl_4C_6H_2$	37.013 37.455 37.557	10 5 5	[70, 148]

Empirical formula	Structural formula	ν, MHz	s/n	References, notes
$C_6H_2Cl_4$ (contd.)	1,2,3,5-$Cl_4C_6H_2$	36.183 36.386 36.813 36.872 37.020 37.292 37.488 37.667	10 10 10 10 10 10 10 10	*[70, 148]
	1,2,4,5-$Cl_4C_6H_2$	36.898 36.843 36.738 36.708	...	*[148, 156—159, 161]. p.t. at 187 °K.
		At 294 °K $\eta = 12.5\%$		
$C_6H_2Cl_4O_2$	1,4-$(OH)_2C_6Cl_4$	37.205 37.373	...	[17, 154, 160]
		At 300 K° $\eta_1 = 9\%$; $\eta_2 = 13\%$		
$C_6H_2Cl_4O_2S$	2,4,5-$Cl_3C_6H_2SO_2Cl$	37.069 37.604 38.112	30 30 30	*[225]. cf. Cl—S
$C_6H_2Cl_5N$	$C_6Cl_5NH_2$	37.94 36.76	13 5	*[2, 145, 233]
$C_6H_3BrClNO_2$	2-Br-5-$ClC_6H_3NO_2$	35.931	10	*
$C_6H_3BrCl_2O$	2,4-Cl_2-6-BrC_6H_2OH	35.370 36.096	3 1.5	* Room temp.
	2,6-Cl_2-4-BrC_6H_2OH	35.826 35.907 36.480 36.540	6 6 6 6	*[220]
$C_6H_3BrCl_2O$ + + C_6H_6ClN	2,6-Cl_2-4-BrC_6H_2OH· ·1/2(4-$ClC_6H_4NH_2$)	36.302 36.162 35.455 35.259	20 10 20 10	* cf. Br
$C_6H_3BrCl_2O$+ C_6H_6IN	2,6-Cl_2-4-BrC_6H_2OH· ·1/2(4-$IC_6H_4NH_2$)	35 328 35.520 36.060	5 5 10	* cf. I
$C_6H_3BrCl_3N$	2,4,5-Cl_3-6-BrC_6HNH_2	35.916 36.678 36.912	4 4 8	*
$C_6H_3Br_2ClO$ + + C_6H_6ClN	2,6-Br_2-4-ClC_6H_2OH·4-$ClC_6H_4NH_2$	35.777	50	* cf. Br
$C_6H_3Br_2ClO$+ +C_6H_7N	2,6-Br_2-4ClC_6H_2OH·$C_6H_5NH_2$	35.676	20	* cf. Br.

Empirical formula	Structural formula	ν, MHz	s/n	References, notes
$C_6H_3ClFNO_4S$	3-NO_2-4-$ClC_6H_3SO_2F$	38.290	50	*[225]
$C_6H_3ClKNO_5S$	4-Cl-3-$NO_2C_6H_3SO_3K$	37.957	...	[156]
$C_6H_3ClNNaO_5S$	2-Cl-5-$NO_2C_6H_3SO_3Na$	36.627	...	[154]
$C_6H_3ClN_2O_4$	2,4-$(NO_2)_2C_6H_3Cl$	37.796	...	[118, 126, 164, 165]
	2,6-$(NO_2)_2C_6H_3Cl$	39.383	10	*[140]
$C_6H_3ClN_2S$	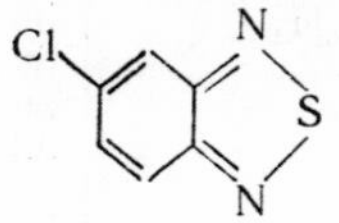	35.040	4	*
$C_6H_3ClN_2Se$	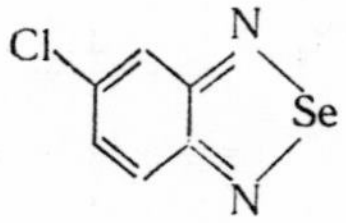	34.704	8	*
$C_6H_3ClO_2$	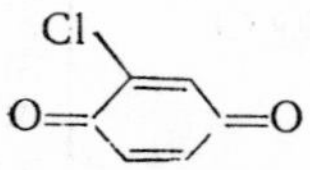	35.468	...	[141]. $T = 294$ °K
$C_6H_3Cl_2F$	2,4-$Cl_2C_6H_3F$	35.562	20	*[231]
		35.778	10	
		36.474	10	
		36.840	20	
	2,5-$Cl_2C_6H_3F$	35.514	20	*[231]
		36.516	20	
	2,6-$Cl_2C_6H_3F$	36.324	10	*
	3,4-$Cl_2C_6H_3F$	36.192	5	*[231]
		36.342	10	
		36.510	5	
$C_6H_3Cl_2FO$	2,6-Cl_2-4-FC_6H_2OH	35.610	8	*[220]
		35.685	8	
		36.420	8	
		36.750	8	
$C_6H_3Cl_2FO_2S$	2,4-$Cl_2C_6H_3SO_2F$	35.767	20	*[225]
		37.338	20	
$C_6H_3Cl_2IO$	2,6-Cl_2-4-IC_6H_2OH	35.796	15	*[220]
		36.390	15	
$C_6H_3Cl_2NO_2$	2,3-$Cl_2C_6H_3NO_2$	36.02	...	[165, 166]. $T = 301$ °K
		38.32		
	3,4-$Cl_2C_6H_3NO_2$	36.488	...	[151, 165]
		37.055		
	2,5-$Cl_2C_6H_3NO_2$	35.921	...	[151, 165, 167—169]
		37.874		

Empirical formula	Structural formula	ν, MHz	s/n	References, notes
$C_6H_3Cl_2NO_2S$	2-NO_2-4-ClC_6H_3SCl	35.826	5	*[228]. cf. Cl—S
$C_6H_3Cl_2NO_3$	2,6-Cl_2-4-$NO_2C_6H_2OH$	36.096 36.306 37.104 37.371	5 2 2 8—10	*[3, 219, 220]
$C_6H_3Cl_2NO_4S$	3-NO_2-4-$ClC_6H_3SO_2Cl$	37.206	20	*[225]. cf. Cl—S
$C_6H_3Cl_2NaO_2S$	2,5-$Cl_2C_6H_3SO_2Na$	35.2117 36.5040	6 9—10	[152]
$C_6H_3Cl_3$	1,2,3-$Cl_3C_6H_3$	36.214 36.238 36.268 36.523 36.973 37.031	8 8 8 8 8 8	*[70, 148, 154, 170]
	1,2,4-$Cl_3C_6H_3$	35.189 35.617 36.173 36.380 36.400 36.623	. . .	[70, 148, 171]
	1,3,5-$Cl_3C_6H_3$	35.545 35.894 36.115	. . .	*[70, 100, 148, 162].
		At 299 °K $\eta_1 = 9.3\%$; $\eta_2 = 11.1\%$; $\eta_3 = 12.9\%$		
$C_6H_3Cl_3O$	2,4,6-$Cl_3C_6H_2OH$	35.298 35.400 36.774	15 15 15	*[154, 172, 217, 220]
$C_6H_3Cl_3O$ + + C_6H_6BrN	2,4,6-$Cl_3C_6H_2OH$ 1/2(4-$BrC_6H_4NH_2$)	35.232 35.514 35.824 36.198 36.336 36.466	20 20 20 20 20 20	* cf. Br
$C_6H_3Cl_3O$ + + C_6H_6ClN	2,4,6-$Cl_3C_6H_2OH$ 1/2(4-$ClC_6H_4NH_2$)	36.474 36.204 35.814 35.496 35.262	20 10 10 10 10	*
$C_6H_3Cl_3O_2S$	2,4-$Cl_2C_6H_3SO_2Cl$	36.012 37.380	. . .	*[225]. cf. S—Cl

Empirical formula	Structural formula	ν, MHz	s/n	References, notes
$C_6H_3Cl_3O_2S$ (contd.)	2,5-$Cl_2C_6H_3SO_2Cl$	36.366 36.390 37.254 37.320	15 15 15 20	*[152, 225] cf. S—Cl
$C_6H_3Cl_4N$	2,3,4,5-$Cl_4C_6HNH_2$	36.378 36.684 37.098 37.825	8 8 8 8	*[140]
	2,3,5,6-$Cl_4C_6HNH_2$	35.950 36.120 36.217 36.362 36.475 36.565 36.677	7 7 7 7 7 7 15	*[140, 218]
	2,4,5,6-$Cl_4C_6HNH_2$	35.616 36.461 36.961 37.708	10 10 10 10	*[140]
$C_6H_4BClF_4N_2$	[4-$ClC_6H_4N_2$]·BF_4	35.44 35.68	1,5 1,5	*[230]
C_6H_4BClO	4-ClC_6H_4BO	34.438	20	*
C_6H_4BrCl	3-ClC_6H_4Br	34.85 34.99	4 8	*[231]
	4-ClC_6H_4Br	34.81	20	*[2]
$C_6H_4BrClHg$	4-ClC_6H_4HgBr	35.610	10	*[229]. cf. Br
$C_6H_4Br_2ClN$	2,6-Br_2-4-$ClC_6H_2NH_2$	35.124	10	* Room temps.
	2,4-Br_2-6-$ClC_6H_2NH_2$	34.620	10	The same
C_6H_4ClF	2-ClC_6H_4F	36,294	10	*[173, 231]
		$e^2Qq_{aa} = -74.0$ MHz (g)		
	3-ClC_6H_4F	34.97 35.05	5 5	*[231]
	4-ClC_6H_4F	34.818 35.226 35.286	10 10 10	*[2, 217]
$C_6H_4ClF + C_9H_{12}$	4-ClC_6H_4F·1,3,5-$(CH_3)_3C_6H_3$	35.126	10	*[138]
$C_6H_4ClFO_2S$	4-$ClC_6H_4SO_2F$	35.613	15	*[225]
C_6H_4ClHgI	4-ClC_6H_4HgI	35.175	2	*[229]. cf. I

Empirical formula	Structural formula	ν, MHz	s/n	References, notes
C_6H_4ClI	4-ClC_6H_4I	34.77	3—4	*[2]
$C_6H_4ClNO_2$	2-$ClC_6H_4NO_2$	37.260	. . .	[174]
	3-$ClC_6H_4NO_2$	35.457	. . .	[174]
	4-$ClC_6H_4NO_2$	34.88	7	*[2, 175]. $\Delta\nu \approx 400$ KHz
$C_6H_4ClNO_2$ + + $C_6H_4BrNO_2$	4-$ClC_6H_4NO_2 \cdot$ 4-$BrC_6H_4NO_2$	35.421	20	*[175]
$C_6H_4ClNO_2$ + + $C_6H_4INO_2$	4-$ClC_6H_4NO_2 \cdot$ 4-$IC_6H_4NO_2$	35.412	10	*[175]
$C_6H_4ClNO_2$ + + C_7H_7Br	4-$ClC_6H_4NO_2 \cdot$ 4-$CH_3C_6H_4Br$	34.86 35.04	2 3—4	*$\Delta\nu$ = = 200 KHz $\Delta\nu$ = 100 KHz
$C_6H_4ClNO_2$ + + C_7H_7Cl	4-$ClC_6H_4NO_2 \cdot$ 4-$CH_3C_6H_4Cl$	34.97 34.44	1.5 1.5	*
$C_6H_4ClNO_2$ + + $C_{10}H_8$	4-$ClC_6H_4NO_2 \cdot C_{10}H_8$	34.82	5	*$\Delta\nu$ = 50 KHz
$C_6H_4ClNO_3$	2-Cl-4-$NO_2C_6H_3OH$	36.175	20	*[154, 220]
	2-NO_2-4-ClC_6H_3OH	35.580	3	*[220]
$C_6H_4ClNO_5S$	2-Cl-5-$NO_2C_6H_3SO_3H$	36.751	. . .	[154]
$C_6H_4ClN_3O_4$	4-Cl-2,6-$(NO_2)_2C_6H_2NH_2$	37.280	. . .	[154]
$C_6H_4ClNaO_2S$	4-$ClC_6H_4SO_2Na$	35.1366	3—4	[152]
$C_6H_4Cl_2$	1,2-$Cl_2C_6H_4$	35.580 35.755 35.824	4 8 4	[70, 148, 174, 176]
	1,3-$Cl_2C_6H_4$	34.809 34.875 35.030	5 5 10	The same
	1,4-$Cl_2C_6H_4$			*[48, 77, 148, 167, 174, 177—187]
	Phase I	34.780	. . .	At 299 °K η = 7.6%
	Phase II	34.765	. . .	At 300.4 °K η = 8.3%
	Phase III	35.208	. . .	
$C_6H_4Cl_2$ + + $C_6H_2ClN_3O_6$	1,4-$Cl_2C_6H_4 \cdot$ 2,4,6-$(NO_2)_3C_6H_2Cl$	34.783	3	*
$C_6H_4Cl_2$ + C_8H_{10}	1,4-$Cl_2C_6H_4 \cdot$ 1,4-$(CH_3)_2C_6H_4$	34.836	10	*
$C_6H_4Cl_2Hg$	4-ClC_6H_4HgCl	35.60 35.70	2 2	*[229]

Empirical formula	Structural formula	ν, MHz	s/n	References, notes
$C_6H_4Cl_2N_2O_2$	2,6-Cl_2-4-$NO_2C_6H_2NH_2$	35,920	...	*[140]
$C_6H_4Cl_2O$	2,4-$Cl_2C_6H_3OH$	34,992 35,028 35.214 35,421 35,682 35,892	...	*[140, 217]
	2,5-$Cl_2C_6H_3OH$	34,647 35,277	...	[189]. $T = 301$ °K
	2,6-$Cl_2C_6H_3OH$	35,304 35,856	10 10	*[219, 220]
	3,5-$Cl_2C_6H_3OH$	34,99 35,28	...	[166]. Room temp.
$C_6H_4Cl_2O_2S$	4-$ClC_6H_4SO_2Cl$	35,967	30	*[126, 225]. cf. Cl—S
$C_6H_4Cl_2O_2S + H_2O$	2,5-$Cl_2C_6H_3SO_2H \cdot 2H_2O$	35,6222 36,1600 36,4500	2—3 7 3	[152]
$C_6H_4Cl_2S$	2-ClC_6H_4SCl	33,864	20	*[228]. cf. Cl—S
	4-ClC_6H_4SCl	34,979	20	The same
$C_6H_4Cl_3N$	2,4,5-$Cl_3C_6H_2NH_2$	36,030 35,967 35,408	5 5 5	*[233]
	2,4,6-$Cl_3C_6H_2NH_2$	34,925 34,976 35.177 35.591 35.780 35.885	...	[156, 165]
	3,4,5-$Cl_3C_6H_2NH_2$	35.988 36,509 37.380	30 30 30	*[140]
$C_6H_4Cl_3NO_2S$	4-$ClC_6H_4SO_2NCl_2$	35,532	5	*[190]. cf. Cl—N
$C_6H_4Cl_3OP$	3-$ClC_6H_4P(O)Cl_2$	35,076	5	*[42a]. cf. Cl—P
	4-$ClC_6H_4P(O)Cl_2$	35.310	100	The same
$C_6H_4Cl_3P$	3-$ClC_6H_4PCl_2$	35,091	15	*[42a]. cf. Cl—P

Empirical formula	Structural formula	ν, MHz	s/n	References, notes
$C_6H_4Cl_3P$ (contd.)	4-$ClC_6H_4PCl_2$	34.930 35.104	5 5	*[42a]. cf. Cl—P
$C_6H_4Cl_3PS$	4-$ClC_6H_4P(S)Cl_2$	35.269	10	The same
$C_6H_4Cl_4N_2$	1,4-$(NH_2)_2C_6Cl_4$	36.171 36.659	10 10	*[140]
$C_6H_4Cl_5FeN_2$	[4-$ClC_6H_4N_2$]$FeCl_4$	36.148 36.384	5 5	*[230]
$C_6H_4Cl_5N_2Sb$	[4-$ClC_6H_4N_2$]$SbCl_4$	36.281	4	*[230]
$C_6H_4Cl_7N_2Sb$	[4-$ClC_6H_4N_2$]$SbCl_6$	36.168	10	*[230]
C_6H_5BrClN	2-Br-5-$ClC_6H_3NH_2$	34.884	50	*
C_6H_5Cl	C_6H_5Cl	34.621	...	*[1, 32, 88, 148, 174, 191, 192]
		e^2Qq = —71,10 MHz (g) At 77 °K η = 10% [221]		
$C_6H_5Cl + Cl_3Sb$	$C_6H_5Cl \cdot SbCl_3$	34.574′	...	[193—195]. cf. Sb, Cl—Sb
$C_6H_5Cl + C_9H_{12}$	$C_6H_5Cl \cdot$ 1,3,5-$(CH_3)_3C_6H_3$	34.265	1,5	[23]
C_6H_5ClFN	3-Cl-4-$FC_6H_3NH_2$	36.127	...	[189, 222]. T = 90 °K
C_6H_5ClHgO	4-ClC_6H_4HgOH	35.124	4	*[229]
C_6H_5ClO	2-ClC_6H_4OH	35.268 35.364 35.546	15 15 15	*[219, 220]
	3-ClC_6H_4OH	34.766 34.825	...	[154]
	4-ClC_6H_4OH	34.700 34.945	...	[174, 191, 196, 223]
		At 293 °K η_1 = 7.1%; η_2 = 9.6%		
$C_6H_5ClO_2$	2-$ClC_6H_3(OH)_2$-1,4	35.765	50	*[140]
$C_6H_5Cl_2N$	2,4-$Cl_2C_6H_3NH_2$	34.734 34.863	5 5	*[140, 154, 165, 222]
	2,5-$Cl_2C_6H_3NH_2$	34.413 34.530	...	[17, 126, 167]
		At 292 °K η_1 = 7.9%; η_2 = 5.9%		
	2,6-$Cl_2C_6H_3NH_2$	34.398 35.010	10 10	*[189, 222 233]

Empirical formula	Structural formula	ν, MHz	s/n	References, notes
$C_6H_5Cl_2N$	3,4-$Cl_2C_6H_3NH_2$	35.673 35.872	...	[154, 165]
	3,5-$Cl_2C_6H_3NH_2$	34.909 35.133	10 10	*[140, 218]
$C_6H_5Cl_2NO$	2,6-Cl_2-4-$NH_2C_6H_2OH$	35.568 35.952	20 20	*[20, 220]
	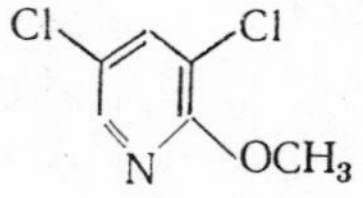	36.222	...	* Doublet
	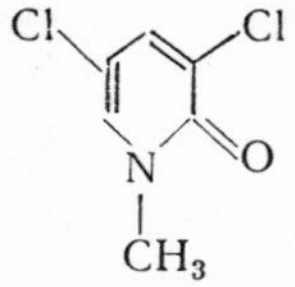	36.015 36.960	5 5	*
$C_6H_5Cl_3N_2$	2,4,6-$Cl_3C_6H_2NHNH_2$	35.569 35.810 36.235	...	*[156, 233]
$C_6H_6BClO_2$	4-$ClC_6H_4BO_2H_2$	34.875	100	*
$C_6H_6ClFN_2S$	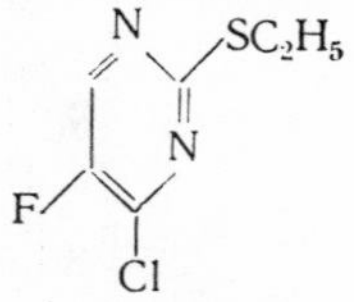	36.00	...	*
C_6H_6ClN	2-$ClC_6H_4NH_2$	33.953	10	*[233]
	3-$ClC_6H_4NH_2$	34.388 34.468	...	[154]
	4-$ClC_6H_4NH_2$	34.146	15	*[17, 174, 191]. cf. N
		At 299 °K $\eta = 4.89\%$		
$C_6H_6ClN + BrH$	4-$ClC_6H_4NH_2 \cdot HBr$	34.752	...	[122]. T=86°K
C_6H_6ClN + + $C_6H_3BrCl_2O$	$^1/_2$(4-$ClC_6H_4NH_2$)· ·2,4-Cl_2-6-BrC_6H_2OH	33.822	3	* cf. Br
	4-$ClC_6H_4NH_2$· 4-Br-2,6-$Cl_2C_6H_2OH$	34.800	10	* cf. Br
C_6H_6ClN + + $C_6H_3Br_2ClO$	4-$ClC_6H_4NH_2$· 4-Cl-2,6-$Br_2C_6H_2OH$	33.894	50	* cf. Br
C_6H_6ClN + + $C_6H_3Br_3O$	$^1/_2$(4-$ClC_6H_4NH_2$)· ·2,4,6-$Br_3C_6H_2OH$	34.032	20	* cf. Br

Empirical formula	Structural formula	ν, MHz	s/n	References, notes
C_6H_6ClN + + $C_6H_3Cl_3O$	½(4-$ClC_6H_4NH_2$)· ·2,4,6-$Cl_3C_6H_2OH$	34.950	10	*
C_6H_6ClN + ClH	4-$ClC_6H_4NH_2$·HCl	35.448	...	[122]. $T = 86$ °K
$C_6H_6ClO_2P$	4-$ClC_6H_4P(O)H(OH)$	35.017	10	*
$C_6H_6ClO_3P$	4-$ClC_6H_4P(O)(OH)_2$	34.904	5	*
C_6H_6ClP	3-$ClC_6H_4PH_2$	34.800	10	*
	4-$ClC_6H_4PH_2$	34.650	2	*
$C_6H_6Cl_4$	$Cl_2C{=}CHCH_2CCl_2CH{=}CH_2$	36.134 36.708	50 50	*
	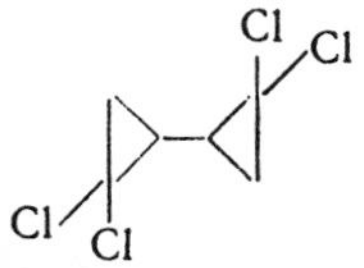	37.021 37.385		[108]
$C_6H_6Cl_6$	α-$C_6H_6Cl_6$	35.600 35.688 36.226 36.231 36.474 36.816	...	[197]. Room temp.
	β-$C_6H_6Cl_6$	37.099	...	[197, 198]
	γ-$C_6H_6Cl_6$	36.133 36.365 36.819 36.892 37.341	2,5 2,5 3 4 3	[20, 197, 198]
	δ-$C_6H_6Cl_6$	36.341 36.689 36.752 36.861 37.240 37.932	...	[197—199]
		At 293 °K, $\eta_1 = 1.5\%$, $\eta_2 = 2.3\%$, $\eta_3 = 3.7\%$, $\eta_4 = 2.0\%$, $\eta_5 = 3.4\%$, $\eta_6 = 0.0\%$		

Empirical formula	Structural formula	ν, MHz	s/n	References, notes
$C_6H_7ClN_2$	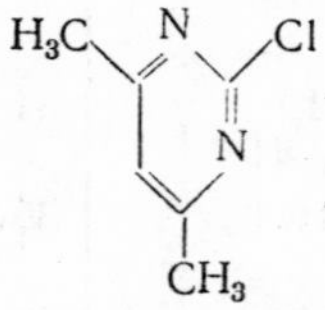	34.335	20	*
$C_6H_7ClN_2OS$	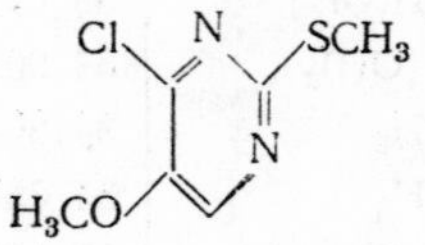	35.520 35.730	3 2—3	*
$C_6H_7ClN_2O_2$	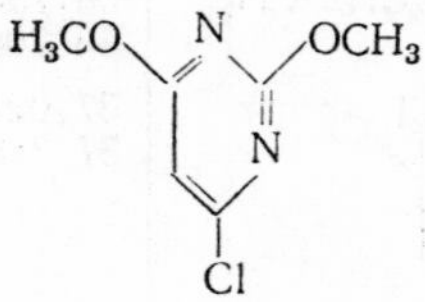	34.626	10	*[97, 123]
	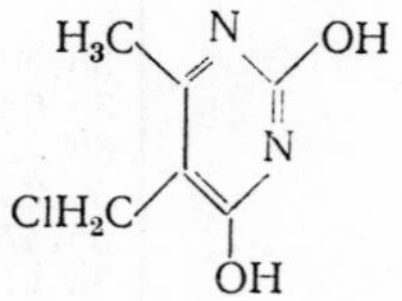	33.948	3	*
$C_6H_7Cl_2N_3$	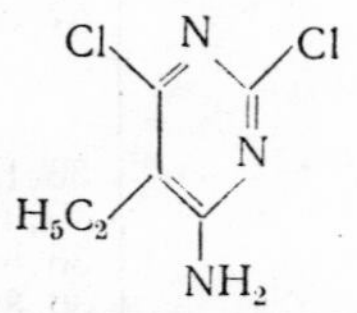	34.879 35.044	...	[99]. $T = 86$ °K
$C_6H_8ClNO_2$	$ClH_2CCH(OCOCH_3)CH_2CN$	34.674	50	*
$C_6H_8Cl_2O$	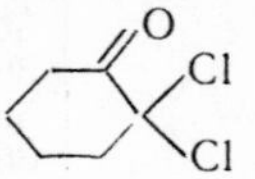	36.475 36.703	5 5	[200]
	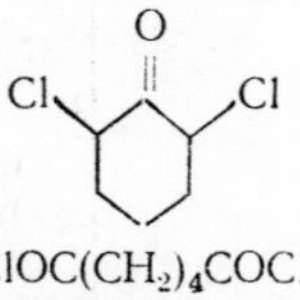	35.501	15	[200]
$C_6H_8Cl_2O_2$	$ClOC(CH_2)_4COCl$	28.978	20	[52]
$C_6H_8Cl_4$	$Cl_2C{=}CCl(CH_2)_3CH_2Cl$	37.70 35.67	7 3	*Δν = = 470 kHz

Empirical formula	Structural formula	ν, MHz	s/n	References, notes
$C_6H_8Cl_6$	$(-Cl_2CCH_2CH_2Cl)_2$	34.428 37.730 37.870	...	*
C_6H_9Cl	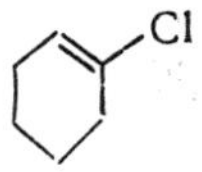	32.698	...	[105]
	$(CH_3)_3CC\equiv CCl$	39.15	...	[136]
$C_6H_9Cl_3O_3$	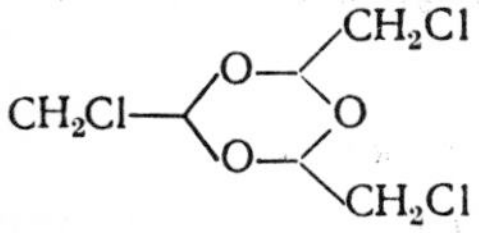	35.363 35.601	...	[56]
$C_6H_{10}Cl_2$	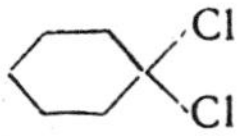	34.904 34.989	...	[105]
$C_6H_{10}Cl_3NO$	$Cl_3CCON(C_2H_5)_2$	38.969 39.466 39.956	40 40 30	*
$C_6H_{10}Cl_6O_2Si$	$(CCl_3)_2Si(OC_2H_5)_2$ Phase I	38.80 38.16	...	[26]
	Phase II	38.7954 38.5748 38.1842 38.1548 37.0054 37.8996		
$C_6H_{11}ClO_2$	$CH_2ClCOOC_4H_9$	36.136 36.363	10 4	[20]
$C_6H_{11}Cl_2NO$	$Cl_2HCCON(C_2H_5)_2$	37.611 37.016	20 20	*
$C_6H_{12}ClNO$	$ClCH_2CON(C_2H_5)_2$	35.357	5	*
$C_6H_{12}Cl_2$	$CH_3CHCl(CH_2)_2CHClCH_3$	31.819	1.2	[20]
	$CH_2Cl(CH_2)_4CH_2Cl$	33.008	17	[20]
$C_6H_{12}Cl_2N_2$	$[(CH_3)_2ClCN{=}]_2$	33.108	15	*
$C_6H_{13}Cl$	$C_6H_{13}Cl$	32.958	6	[20]
$C_6H_{13}ClO_2Si$	$ClCH_2CH_2COOSi(CH_3)_3$	33.768 33.585 33.450	4 4 ~1	[26]
$C_6H_{13}Cl_4N$	$[(ClCH_2CH_2)_3NH]Cl$	35.124 35.558	2.5 1.3	[20]

Empirical formula	Structural formula	ν, MHz	s/n	References, notes
$C_6H_{15}ClGe$	$(CH_3)_3GeCH_2CH_2CH_2Cl$	35.227	...	[26]
$C_6H_{15}ClSi$	$(CH_3)_3SiCH_2CH_2CH_2Cl$	32.958	5	*[26]
	$ClCH_2Si(CH_3)(C_2H_5)_2$	34.592	...	[26]
$C_6H_{15}Cl_2N$	$[(ClCH_2CH_2)NH(C_2H_5)_2]Cl$	34.660	20	[20]
$C_6H_{19}B_{10}Cl$	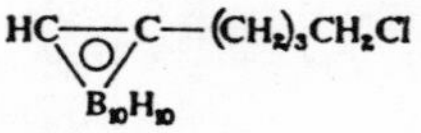 C-chlorobutyl-o-borane	33.273	...	*[120]
C_7ClF_5O	$3\text{-}ClC_6F_4COF$	40.184	3—5	*
$C_7Cl_2F_4O$	$3,5\text{-}Cl_2\text{-}2,4,6\text{-}F_3C_6COF$	39.648	50	*
		39.900	50	
C_7Cl_6O	Cl_5C_6COCl	30.798	...	*[140, 145]
		30.888	...	
		37.919	10	
		37.954	10	
		38.024	10	
		38.094	10	
		38.115	10	
		38.185	10	
		38.234	10	
		38.322	10	
		38.462	10	
		38.598	10	
$C_7HCl_2F_5$	$C_6F_5CHCl_2$	38.192	8	*
		38.367	8	
$C_7HCl_5O_2$	C_6Cl_5COOH	37.863	20	*[140, 145]
		37.951	20	
		38.213	20	
		38.325	20	
		38.430	20	
$C_7H_2ClF_5$	$C_6F_5CH_2Cl$	35.756	20	*
$C_7H_2Cl_8Si$	$C_6Cl_5CH_2SiCl_3$	38.133	...	[26]. cf. Cl—Si
		38.069		
		37.966		
		37.221		
		37.037		
$C_7H_3Cl_2NO$	$2,6\text{-}Cl_2\text{-}4\text{-}(CN)C_6H_2OH$	36.148	6	*[220]
		36.176	6	
		36.981	6	
		37.128	6	

Empirical formula	Structural formula	ν, MHz	s/n	References, notes
$C_7H_3Cl_3O$	2,4-$Cl_2C_6H_3COCl$	31.039 31.086 35.834 35.923 37.038 37.145	1.5 1.5 2 2 3 3	*[52]
$C_7H_3Cl_5$	$C_6Cl_5CH_3$	37.20 37.60	15 10	*[234]
	2,4-Cl_2-$C_6H_3CCl_3$	34.87 36.40 38.76	1 1 3	[166]. Room temp. $\eta_1 = 11\%$; $\eta_2 = 10\%$
$C_7H_3Cl_5Hg$	$C_6Cl_5HgCH_3$	35.940 36.012 37.368 37.536 37.668	5 5 5 5 5	*[147]
$C_7H_3Cl_5O$	$C_6Cl_5OCH_3$	37.904 37.926 37.985 38.133 38.367	50 50 50 50 50	*[140, 145]
$C_7H_3Cl_5S$	$C_6Cl_5SCH_3$	37.583 37.723 37.786 37.933 38.052 38.186 38.220 38.311	5 10 5 10 5 5 5 5	*[140, 145]
C_7H_4BrClO	4-BrC_6H_4COCl	30.105	1.5	*
C_7H_4ClFO	2-ClC_6H_4COF	36.400	10	*
$C_7H_4ClF_3$	2-$ClC_6H_4CF_3$	35.633	...	[174]
	3-$ClC_6H_4CF_3$	35.073	...	[174]
C_7H_4ClHgN	4-ClC_6H_4HgCN	34.644	15	*[229]
C_7H_4ClN	2-ClC_6H_4CN	35.410 35.453 35.500 35.541	...	[156]
	4-ClC_6H_4CN	35.150	10	*[102]
C_7H_4ClNO	2-ClC_6H_4NCO	34.635	2	[20, 174]
	3-ClC_6H_4NCO	34.653	...	[174]. $T = 196$ °K

Empirical formula	Structural formula	ν, MHz	s/n	References, notes
C_7H_4ClNO	4-ClC_6H_4NCO	35.259	...	[154]
$C_7H_4ClNO_3$	2-$NO_2C_6H_4COCl$	30.860	5	*
	3-$NO_2C_6H_4COCl$	31.185 31.395	10 10	*
	4-$NO_2C_6H_4COCl$	30.113 30.575	5 5	*
$C_7H_4ClNO_4$	2-Cl-5-$NO_2C_6H_3COOH$	37.337	...	[154]
	4-Cl-3-$NO_2C_6H_3COOH$	37.843	...	[154]
C_7H_4ClNS	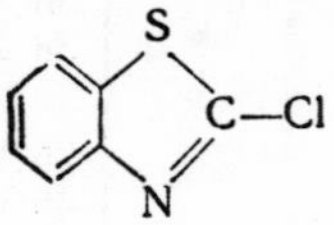	36.210	1.5	[20]
$C_7H_4Cl_2N_2$	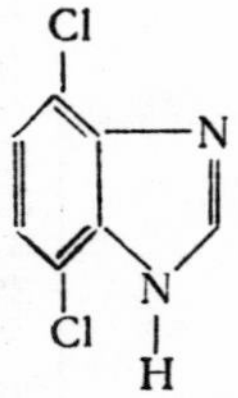	34.944 35.214	2 4	*
$C_7H_4Cl_2O$	4-ClC_6H_4COCl	30.28 34.86	3 5	*
	2,4-$Cl_2C_6H_3CHO$	35.461 35.986	10 10	[20]
	3,4-$Cl_2C_6H_3CHO$	35.685 36.762	3 3	[20]
$C_7H_4Cl_2O_2$	2,4-$Cl_2C_6H_3COOH$	35.528 37.432	...	[154]
	2,6-$Cl_2C_6H_3COOH$	35.622 35.682 36.018 36.066	15 15 15 15	*
	3,4-$Cl_2C_6H_3COOH$	36.576 37.298	...	[154]
	3,5-Cl_2-4-HOC_6H_2CHO	35.646 36.660	4 4	* [220]
$C_7H_4Cl_3F$	4-$FC_6H_4CCl_3$	39.056 38.772 36.968	...	[83]
$C_7H_4Cl_4$	4-$ClC_6H_4CCl_3$	34.564	...	[201]

Empirical formula	Structural formula	ν, MHz	s/n	References, notes
$C_7H_4Cl_4N_2O_2$	$4\text{-}NO_2C_6Cl_4NHCH_3$	37.338 38.115 38.528 38.829	20 20 20 20	*[233]
$C_7H_4Cl_4O$	$2,3,5,6\text{-}Cl_4C_6HOCH_3$	36.722 36.928 37.203 37.278 37.363 37.749	30 50 50 30 30 30	*[140]
$C_7H_4Cl_5N$	$C_6Cl_5NHCH_3$	36.709 37.587 37.695 37.968 37.981	15 15 15 15 15	*[140, 145]
$C_7H_5ClN_2$	$4\text{-}ClC_6H_4NHCN$	35.247 35.502	5 5	*
C_7H_5ClO	C_6H_5COCl	29.918	50	*[83]
	$2\text{-}ClC_6H_4CHO$	34.740	3	*
	$4\text{-}ClC_6H_4CHO$	34.607 34.623	...	[174]
$C_7H_5ClO_2$	$2\text{-}ClC_6H_4COOH$	36.3049	7—8	[152]
	$3\text{-}ClC_6H_4COOH$	35.232	20	*[174]
	$4\text{-}ClC_6H_4COOH$	34.673	...	[174]
	$2\text{-}HO\text{-}5\text{-}ClC_6H_3CHO$	34.968	...	[154]
$C_7H_5Cl_2NO_3$	$2,6\text{-}Cl_2\text{-}4\text{-}NO_2C_6H_2OCH_3$	36.966 37.152	8 8	*[220]
$C_7H_5Cl_3$	$C_6H_5CCl_3$	38.898 38.824 38.786 38.713 38.702 38.288	...	[26, 83]
	$2,4\text{-}Cl_2C_6H_3CH_2Cl$	35.505	1	[20]
	$3,4\text{-}Cl_2C_6H_3CH_2Cl$	35.854 35.673 36.436 36.232	12 6 6 13	[20]
$C_7H_5Cl_4N$	$2,3,5,6\text{-}Cl_4C_6HNHCH_3$	35.748 36.018 36.144 36.384	5 5 5 5	*[233]

Empirical formula	Structural formula	ν, MHz	s/n	References, notes
$C_7H_5Cl_4N$ (contd.)	2,3,5,6-$Cl_4C_6HNHCN_3$	36.468 36.882 37.056 37.374	5 5 5 5	
$C_7H_5Cl_4NO$	4-$CH_3OC_6Cl_4NH_2$	36.250 36.719 37.584	10 10 20	*
$C_7H_5Cl_5Si$	$C_6H_5CCl_2SiCl_3$	38.086 37.750 37.468 37.345	. . .	[26]. cf. Cl—Si
	$CCl_3SiCl_2(C_6H_5)$	38.78 38.10	4 2	[26, 42]. cf. Cl—Si
C_7H_6ClNO	2-ClC_6H_4NHCOH	34.860	8	*
	3-ClC_6H_4NHCOH	34.050 34.707	6 7	*
	4-ClC_6H_4NHCOH	34.590	12	*
	4-$ClC_6H_4CONH_2$	34.728 34.950 35.133	5 5 10	*
$C_7H_6ClNO_2$	2-Cl-6-$NO_2C_6H_3CH_3$	35.219	. . .	[154, 165, 168]
	2-Cl-4-$NO_2C_6H_3CH_3$	35.037	. . .	[172, 217]
	2-NO_2-4-$ClC_6H_3CH_3$	35.413 35.252	1.7 1.7	[172, 217]
	2-Cl-5-$NO_2C_6H_3CH_3$	34.50	. . .	[172] $T = 303.5$ °K
	3-$NO_2C_6H_4CH_2Cl$	34.216	3	[20]
	4-$NO_2C_6H_4CH_2Cl$	34.311	3	[126, 165]
$C_7H_6ClNO_3$	2-NO_2-3-CH_3-4-ClC_6H_2OH	35.660	. . .	[154]
$C_7H_6ClNO_4S$	3-NO_2-4-$ClC_6H_3SO_2CH_3$	37.653	10	*
$C_7H_6ClNaO_3S$	3-Cl-4-$CH_3C_6H_3SO_3Na$	34.91	1	[20]
$C_7H_6Cl_2$	$C_6H_5CHCl_2$	36.373 35.770	. . .	[26]
	2-$ClC_6H_4CH_2Cl$	34.560 34.795 35.225	9 10 4.5	[20]
	3-$ClC_6H_4CH_2Cl$	34.426 34.622 34.683 34.875 35.050	. . .	[201]

Empirical formula	Structural formula	ν, MHz	s/n	References, notes
$C_7H_6Cl_2$	4-$ClC_6H_4CH_2Cl$	33.754 34.567	. . .	[20, 174, 201]. $\eta = 7\%$
	2,6-Cl_2-$C_6H_3CH_3$	34.755 34.788	2 1.5	[20]
$C_7H_6Cl_2Hg$	2-$ClC_6H_4CH_2HgCl$	33.648	15	*[229]. cf. Hg—Cl
$C_7H_6Cl_2O$	2,4-$Cl_2C_6H_3OCH_3$	35.734 36.256	. . .	[154]
	2,6-$Cl_2C_6H_3OCH_3$	35.586 35.826	. . .	*[220]
	2,6-Cl_2-4-$CH_3C_6H_2OH$	35.301 35.742	10 10	*[220]
	2,4-Cl_2-5-$CH_3C_6H_2OH$	34.749 35.450	. . .	[154]
$C_7H_6Cl_3NO$	2,4,6-Cl_3-3-$CH_3OC_6HNH_2$	35.916 36.018 36.450	4 4 4	*
$C_7H_6Cl_4N_2$	4-$H_2NC_6Cl_4NHCH_3$	36.396 36.426 36.558 36.690	20 20 20 20	*[233]
$C_7H_6Cl_4Si$	$(C_6H_5)(CHCl_2)SiCl_2$	37.06	. . .	[26]. cf. Cl—Si
C_7H_7Cl	$C_6H_5CH_2Cl$	33.627	. . .	[1, 26, 174]
	2-$ClC_6H_4CH_3$	34.19	5	[2]
	3-$ClC_6H_4CH_3$	34.36	10	*[234]
	4-$ClC_6H_4CH_3$	34.53	7	*[2, 202]
		$e^2Qq = -71.7$ MHz (g)		
C_7H_7ClO	3-CH_3-4-ClC_6H_3OH	34.535 34.887	. . .	[154]
	2-Cl-5-$CH_3C_6H_3OH$	34.13 34.53	. . .	[172]. $T = 303.5$ °K
	4-$ClC_6H_4OCH_3$	34.753	. . .	[174]
$C_7H_7Cl_2NO$	3,5-Cl_2-4-$CH_3OC_6H_2NH_2$	35.787 35.913	10 10	*
$C_7H_7Cl_2OP$	$(ClCH_2)P(O)(C_6H_5)Cl$	36.900	4	*
$C_7H_7Cl_3Si$	$(ClCH_2)(C_6H_5)SiCl_2$	36.024	5	*[241]. cf. Cl—Si
C_7H_8ClNO	2-CH_3O-5-$ClC_6H_3NH_2$	34.220	. . .	[154]

Empirical formula	Structural formula	ν, MHz	s/n	References, notes
$C_7H_{10}Cl_6$	$CCl_3CH_2CHClCH_2$ | ClH_2CH_2CClHC	38.967	2	*[242]
		38.717	4	
		38.605	12	
		38.017	4	
		37.860	2	
		33.648	4	
		33.294	2	
		33.156	4	
		32.998	2	
$C_7H_{12}ClNO_2$		34.734	5	*
		34.818	5	
		35.076	5	
		35.304	5	
$C_7H_{12}Cl_2$	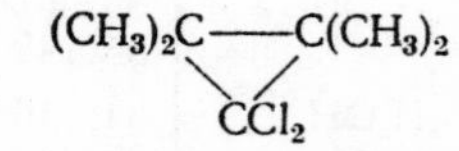	36.407	. . .	*
$C_7H_{12}Cl_4$	$CCl_3CH_2CHCH_3$ | $H_3CClHCH_2C$	31.830	6	*
		32.190	8	
		37.350	6	
		37.402	5	
		38.325	20	
$C_7H_{14}Cl_2$	$ClCH_2(CH_2)_5CH_2Cl$	32.604	10	*[87]
		32.988	10	
$C_7H_{15}Cl$	$C_7H_{15}Cl$	33.106	2	[20]
$C_7H_{15}ClN_2$	$(CH_3)_2ClCN{=}NC(CH_3)_3$	33.108	15	*
$C_7H_{15}Cl_3Si$	$(C_2H_5)_3SiCCl_3$			
	Phase I	37.430	. . .	[26]
	Phase II	37.659		
$C_7H_{16}Cl_2O_3Si$	$Cl_2CHSi(OC_2H_5)_3$	36.232	1	[26]
		36.114	3	
		35.784	2	
$C_7H_{16}Cl_2Si$	$(CH_3)_3SiCH_2CH(CH_2Cl)_2$	33.36	. . .	[26]
$C_7H_{17}ClO_3Si$	$ClCH_2Si(OC_2H_5)_3$	35.122	. . .	[26]
$C_8Cl_4O_3$	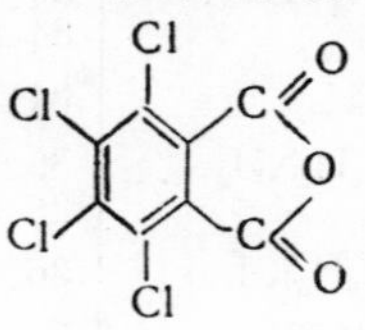	38.249	1	*[20]
		38.413	1.1	

Empirical formula	Structural formula	ν, MHz	s/n	References notes
$C_8Cl_6O_2$	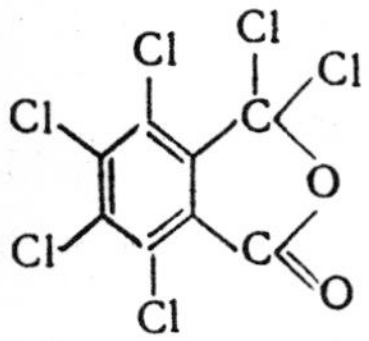	37.597 38.045 38.220 38.465 38.591 39.662	10 10 10 10 10 10	*
	$C_6Cl_4(COCl)_2$-1.4	31.524 31.608 37.874	15 15 15	*[140]
$C_8H_2Cl_2O_3$	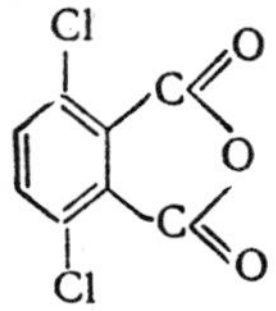	36.732 37.062	5 10	*
$C_8H_2Cl_4O_4$	$C_6Cl_4(COOH)_2$-1.2	38.450 38.401 38.230 38.148	...	[20, 217]
	$C_6Cl_4(COOH)_2$-1,4	37.872	10	*[140]
	$C_6H_4(COCl)_2$-1,2	29.861	...	[33]. T=196°K
$C_8H_4Cl_2O_2$	$C_6H_4(COCl)_2$-1,3	30.276 31.074	3 3	*
	$C_6H_4(COCl)_2$-1,4	31.134	10	*
$C_8H_4Cl_6$	1,4-$(CCl_3)_2C_6H_4$	38.294 38.712 38.904 39.192 39.276 39.342 39.408 39.630	...	[20, 224]
$C_8H_5Cl_3HgO_2$	$Cl_3CCOOHgC_6H_5$	39.123 39.298 39.375	5 5 5	*[4]
$C_8H_5Cl_5$	$C_6Cl_5C_2H_5$	36.848 36.946 37.765 37.835 37.954	20 20 20 20 20	*[234]
C_8H_6BrClO	4-$ClC_6H_4COCH_2Br$	34.8224	12	[152]. cf. Br

Empirical formula	Structural formula	ν, MHz	s/n	References, notes
C_8H_6ClN	4-$ClC_6H_4CH_2CN$	34,634 34,593	...	[154]
$C_8H_6Cl_2N_2O_3$	2,5-Cl_2-4-$NO_2C_6H_2NHCOCH_3$	36,318 37.996	100 100	*
$C_8H_6Cl_2O$	2,5-$Cl_2C_6H_3COCH_3$	35.285 35.335	1 1	[20]
	$C_6H_5OCH{=}CCl_2$	37.142 37.548	10 10	*
$C_8H_6Cl_2O_3$	2,4-$Cl_2C_6H_3OCH_2COOH$	35.146 36.226	...	[154, 172]
$C_8H_6Cl_3NO$	$C_6H_5NHCOCCl_3$	39.452 39.466 39.998	20 20 20	*
$C_8H_6Cl_4$	1,3-$(CH_3)_2C_6Cl_4$	35.69 36.83 37.57	20 50 20	*[2, 234]
	1,4-$(CH_3)_2C_6Cl_4$	36.60	50	*[2, 234]
$C_8H_6Cl_5N$	$C_6Cl_5N(CH_3)_2$	37.268 37.713 37,804 37.839 37.968	10 10 10 10 10	*[140, 145]
C_8H_7Cl	4-$ClC_6H_4CH{=}CH_2$	34.745	4	*
$C_8H_7ClHgO_2$	4-$ClC_6H_4HgOCOCH_3$	34.527	30	*[229]
C_8H_7ClO	$C_6H_5COCH_2Cl$	35.757	30	*[140]
	4-$ClC_6H_4COCH_3$	34.618	...	[174]
	$C_6H_5CH_2COCl$	29.943	30	*
$C_8H_7ClO_2$	4-$ClC_6H_4OCOCH_3$	34.932	40	*[217]
	$C_6H_5OCOCH_2Cl$	36.172 36.386 36.430 36.897	2 2 3 2	[20]
$C_8H_7ClO_3$	4-$ClC_6H_4OCH_2COOH$	34.212	...	*[154, 222]
$C_8H_7Cl_4N$	2,3,5,6-$Cl_4C_6HN(CH_3)_2$	36.600 36.810	30 30	*[233]
	3,4,5,6-$Cl_4C_6HN(CH_3)_2$	36.378 36.636 37.152 37.200 37.464 37,614 37.728 37,800	5	*[233]

Empirical formula	Structural formula	ν, MHz	s/n	References, notes
$C_8H_7Cl_5Si$	$C_6H_5CH_2SiCl_2(CCl_3)$	39.66 38.10	...	[26]
C_8H_8ClNO	$2\text{-}ClC_6H_4NHCOCH_3$	35.150	...	[156]
	$3\text{-}ClC_6H_4NHCOCH_3$	34.735	...	[154]
	$4\text{-}ClC_6H_4NHCOCH_3$	34.792	...	[154]
$C_8H_8Cl_2$	$1,4\text{-}(ClCH_2)_2C_6H_4$	33.863	3	[20]
$C_8H_8Cl_2HgO$	$4\text{-}CH_3OC_6H_4HgCHCl_2$	34.296 34.404	5 5	*
$C_8H_8Cl_2O$	$2,4\text{-}Cl_2C_6H_3OC_2H_5$	35.406 36.108	...	[154]
$C_8H_9ClN_2O$	$4\text{-}ClC_6H_4NHNHCOCH_3$	34.680	4	*
C_8H_9ClO	$4\text{-}ClC_6H_4OC_2H_5$	34.381	...	[174]
	$C_6H_5OCH_2CH_2Cl$	34.316	5	[20]
	$4\text{-}Cl\text{-}3,5\text{-}(CH_3)_2C_6H_2OH$	34.348 34.415	...	[156, 164]
$C_8H_{10}ClN$	$3\text{-}Cl\text{-}C_6H_4N(CH_3)_2$	34.164	20	*[233]
	$4\text{-}Cl\text{-}C_6H_4N(CH_3)_2$	33.939	20	*[233]
$C_8H_{10}ClOP$	$4\text{-}ClC_6H_4P(O)(CH_3)_2$	35.034 35.142	4 4	*
$C_8H_{10}ClP$	$4\text{-}ClC_6H_4P(CH_3)_2$	34.560	50	*
$C_8H_{10}Cl_4$	$[H_2C{=}C(CH_3)CCl_2{-}]_2$	36.350	50	*
$C_8H_{10}Cl_6N_6O_8$	$[NO_2CH_2CHCCl_3N(NO_2)CH_2{-}]_2$	38.927	3—5	*
$C_8H_{12}Cl_4$	$CCl_2{=}CCl(CH_2)_5CH_2Cl$	35.922 37.604	5 2.5	*$\Delta\nu \approx$ ≈ 500 kHz
$C_8H_{15}B_{10}Cl$	$C_6H_5{-}C{-}\overset{\diagup\!\diagup}{\underset{B_{10}H_{10}}{\diagdown O\diagup}}{-}C{-}Cl$ C-chloro-C-phenyl-o-borane	41.640	20	*[95]
	$2\text{-}ClC_6H_4{-}C\overset{\diagup\!\diagup}{\underset{B_{10}H_{10}}{\diagdown O\diagup}}CH$ C-(2-chlorophenyl)-o-borane	38.500	3	*
	$3\text{-}ClC_6H_4{-}C\overset{\diagup\!\diagup}{\underset{B_{10}H_{10}}{\diagdown O\diagup}}CH$ C-(3-chlorophenyl)-o-borane	35.406 35.112	2 2	*[120]

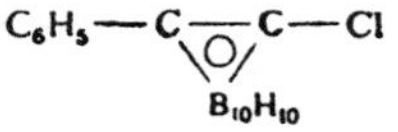

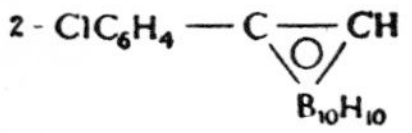

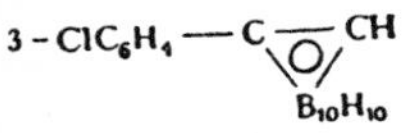

Empirical formula	Structural formula	ν, MHz	s/n	References, notes
$C_8H_{15}B_{10}Cl$	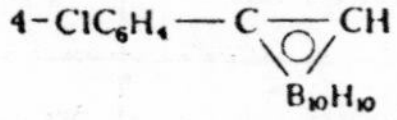 C-(4-chlorophenyl)-o-borane	35.442	...	*[120]
$C_8H_{15}Cl_5O_3Si$	$CCl_3CCl_2Si(OC_2H_5)_3$	40.095 39.752 39.396 38.867 37.486	...	[26]
$C_8H_{16}Cl_2$	$ClCH_2(CH_2)_6CH_2Cl$	33.084	10	*[87]
$C_8H_{16}Cl_4GeO_4$	$Ge(OCH_2CH_2Cl)_4$	33.732	10	*
$C_8H_{17}Cl$	$C_8H_{17}Cl$	33.009	5	[20]
	$CH_3CHCl(CH_2)_5CH_3$	32.302	1.5	[20]
$C_8H_{19}ClGeO_3$	$ClCH_2CH_2Ge(OC_2H_5)_3$	33.042	10	*
$C_9H_4ClO_4Mn$	$C_5H_4COClMn(CO)_3$	30.43	3	*[203]. cf. Mn
$C_9H_4ClO_4Re$	$C_5H_4COClRe(CO)_3$	30.66	3	*[203]. cf. Re
$C_9H_4Cl_2O_2$	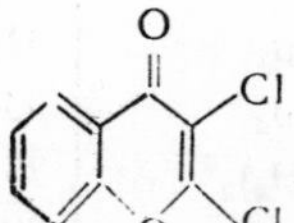	36.915 33.372	10 4	*
$C_9H_5ClO_2$	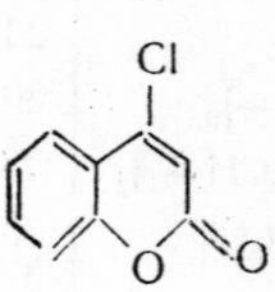	35.455	10	*
	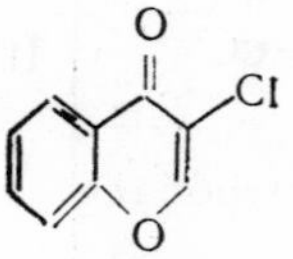	37.072	10	*
$C_9H_5Cl_2N$	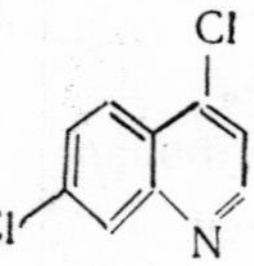	34.764 35.179 35.591	...	[133]

Empirical formula	Structural formula	ν, MHz	s/n	References, notes
$C_9H_5Cl_2NO$	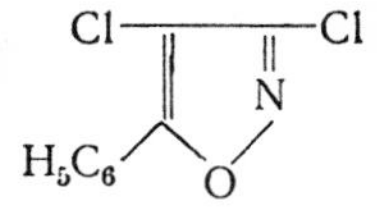	38.220	30—50	*
C_9H_6ClMnO	$ClCH_2C_5H_4Mn(CO)_3$	34.316	30	*[203]. cf. Mn
C_9H_6ClN	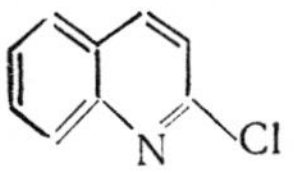	33.271	...	[99, 133]
C_6H_9ClN	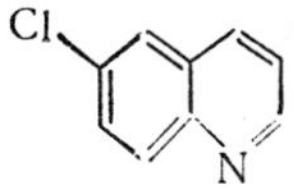	34.628	...	[99, 133]
	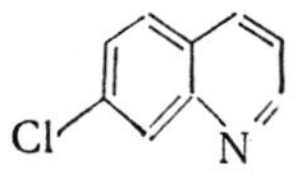	34.681	...	[99]. $T = 86\,°K$
$C_9H_6ClN + ClH$	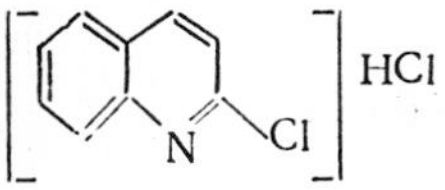	37.145	...	[133]
C_9H_6ClNO	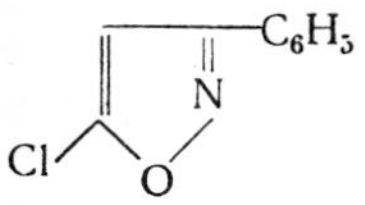	37.230 37.260	20 20	*
	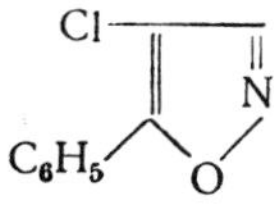	37.518	8	*
	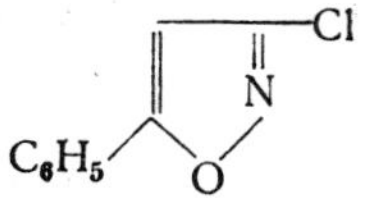	37.188	8	*
C_9H_6ClNO	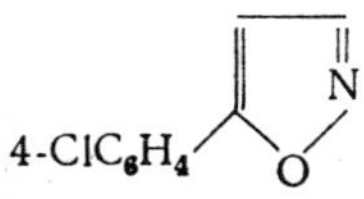	34.824	5	*

Empirical formula	Structural formula	ν, MHz	s/n	References, notes
$C_9H_6Cl_2HgN_2$	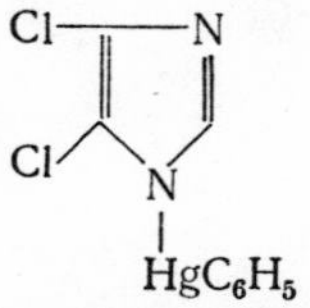	35.637 35.959	2 4	*[125]
C_9H_7ClO	trans - $C_6H_5CH{=}CHCOCl$	29.100 30.190	4 4	*
C_9H_7ClO	2-$ClC_6H_4OCH_2C{\equiv}CH$	35.262 35.616	5 5	*
	$C_6H_5COCH{=}CHCl$	33.927	7—10	*
$C_9H_7ClO_2$	4-$ClC_6H_4CH{=}CHCOOH$	34.227	—	[174]. $T = 196$ °K
$C_9H_7Cl_3N_2O_2S$	4-$ClC_6H_4SO_2NCl$ | $NCClHCH_2C$	35.196 39.130	3 3	*[190]. cf. Cl—N
$C_9H_7Cl_3O_3$	2-$HOC_6H_4CCl_2CHClCOOH$	38.220 38.006 35.760	3 3 2	*
$C_9H_7Cl_4NO_2$	4-$CH_3OC_6Cl_4NHCOCH_3$	37.398 37.527 37.691 39.893	15 15 15 15	*
$C_9H_8Cl_2$	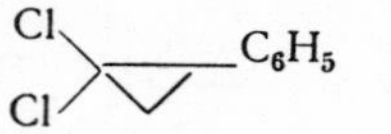	36.89	—	[26]
C_9H_9Cl	$C_6H_5CH{=}CHCH_2Cl$	33.612	1.2	[20]
C_9H_9ClO	4-$ClC_6H_4COC_2H_5$	34.702	—	[154]
	$C_6H_5COCH_2CH_2Cl$	33.789	20—30	*
$C_9H_9Cl_3$	1,3,5-$Cl_3C_6(CH_3)_3$	35.088 35.112 35.286	10 10 10	*[140]
$C_9H_9Cl_4NO_2S$	4-$ClC_6H_4SO_2NCl$ | $ClH_2CClHCH_2C$	33.516 35.418 35.724	. . .	*[190]. cf. Cl—N
$C_9H_{10}Cl_2FN_3O_3$	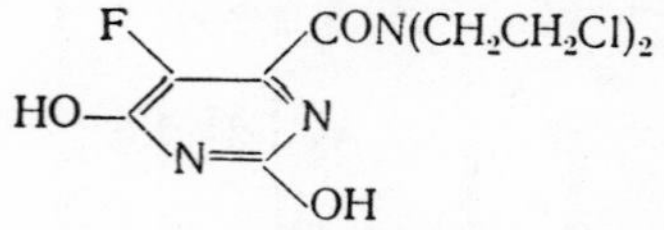	33.960	. . .	*
$C_9H_{10}Cl_3NO_2S$	4-$ClC_6H_4SO_2NH$ | $ClH_2CClHCH_2C$	33.816 34.836 35.184	2 4 4	*[190]

Empirical formula	Structural formula	ν, MHz	s/n	References, notes
$C_9H_{10}Cl_3O_2P$	$C_6H_5P(O)(CCl_3)(OC_2H_5)$	38.651 38.724 39.242 39.487 39.589 39.743	7 5 5 7 5 7	*
$C_9H_{11}Cl$	2,4,6-$(CH_3)_3C_6H_2Cl$	33.71	4	*[234]
$C_9H_{11}ClO_3S$	4-$CH_3C_6H_4SO_2CH_2CH_2Cl$	34.590	3	[20]
$C_9H_{11}Cl_2N_3O_3$	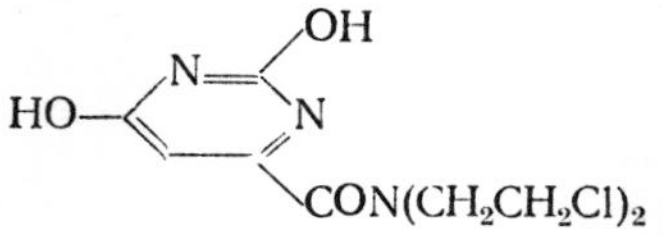	33.840	. . .	*
$C_9H_{11}Cl_4N_3$	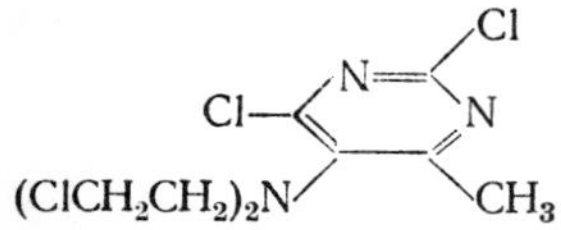	35.328 35.484	. . .	*
$C_9H_{12}Cl_2FN_3O$	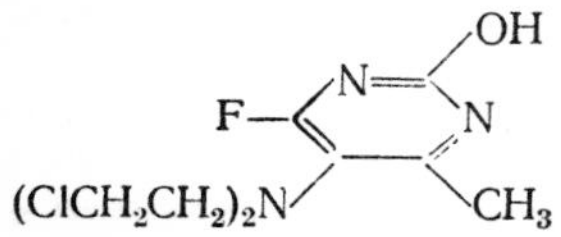	33.720 33.744	. . .	*
$C_9H_{13}ClIN$	$[2\text{-}ClC_6H_4N(CH_3)_3]\cdot I$	36.348	10	*[233]
	$[3\text{-}ClC_6H_4N(CH_3)_3]\cdot I$	34.806	10	*[233]
	$[4\text{-}ClC_6H_4N(CH_3)_3]\cdot I$	35.406	20	*[233]
$C_9H_{13}ClIP$	$[4\text{-}ClC_6H_4P(CH_3)_3]\cdot I$	34.932	70	*
$C_9H_{13}ClSi$	2-$ClC_6H_4Si(CH_3)_3$	34.16	. . .	[26]
	3-$ClC_6H_4Si(CH_3)_3$	34.438 34.217	2 1	[26]
	4-$ClC_6H_4Si(CH_3)_3$	34.52	. . .	[26]
$C_9H_{13}Cl_2N_3O_2$	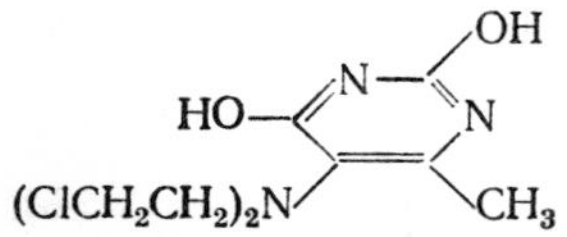	33.984	. . .	*
$C_9H_{16}ClN$	$ClCH_2(CH_2)_6CH_2CN$	33.198	2–3	*
$C_9H_{16}Cl_4$	$ClCH_2(CH_2)_7CCl_3$	38.528 38.308 37.384 33.384	50 50 50 50	*[78]

Empirical formula	Structural formula	ν, MHz	s/n	References, notes
$C_9H_{18}Cl_2$	$ClCH_2(CH_2)_7CH_2Cl$	33.000 32.550	8 5	*[87]
$C_9H_{19}Cl$	$C_9H_{19}Cl$	33.15	4	*
$C_9H_{22}ClN_2OP$	$ClCH_2P(O)\,[N(C_2H_5)_2]_2$	37.044 36.678	5 5	*
$C_{10}Cl_6O_2$	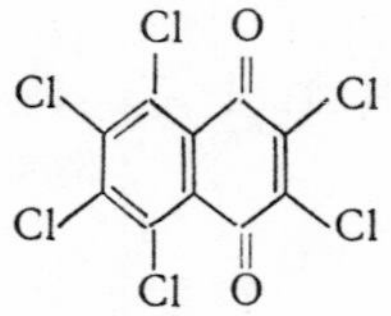	37.541 37.618 38.274 38.696 38.871 38.962	10 10 10 10 10 10	*
$C_{10}Cl_7F$	$1\text{-}FC_{10}Cl_7$	38.416 39.277	...	*
$C_{10}Cl_8$	$C_{10}Cl_8$	38.2053 38.2944 38.3508 38.3616 38.6590 38.7180 38.7385 38.7461	...	*[129, 204—206]
$C_{10}Cl_{10}$	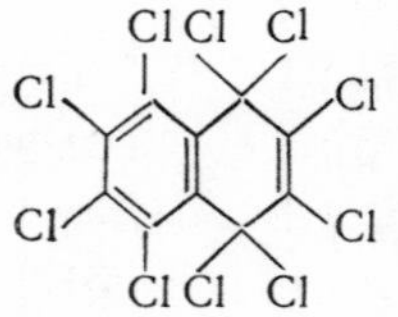	38.406 38.483 38.626 38.693 38.738 38.819 39.288 39.750 39.802 39.855	10 10 10 10 10 10 10 10 10 10	*[129]
$C_{10}H_2Cl_6$	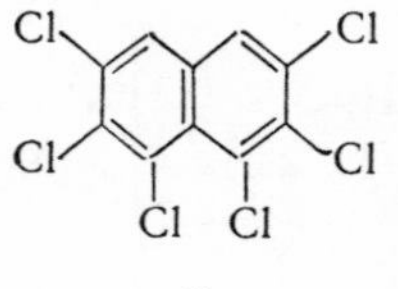	36.542 36.958 37.019 37.158 37.182 38.172	...	[205]
$C_{10}H_4Cl_2O_2$	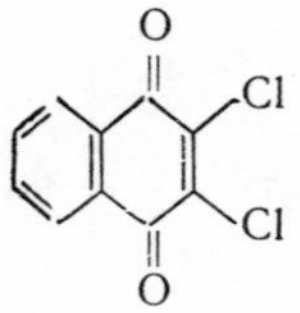	37.114	...	[97]

Empirical formula	Structural formula	ν, MHz	s/n	References, notes
$C_{10}H_4Cl_4$	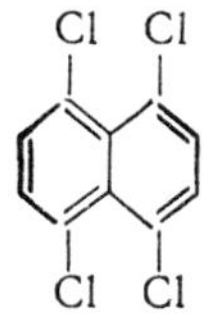	35.238 36.768	...	*[205, 207]
$C_{10}H_5ClN_2O_4$	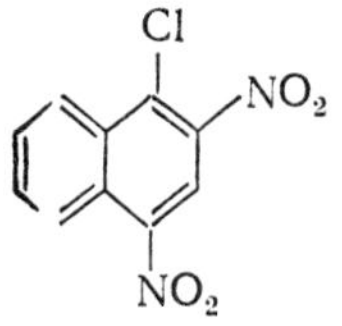	38.003	5	*
$C_{10}H_6ClF$	1,2-$ClC_{10}H_6F$	35.971	...	*[88, 129]
$C_{10}H_6Cl_2$	1,4-$Cl_2C_{10}H_6$	34.818	...	[205, 207]
	1,5-$Cl_2C_{10}H_6$	34.650	...	[205]
	1,8-$Cl_2C_{10}H_6$	35.125	...	[205, 207]
	2,6-$Cl_2C_{10}H_6$	36.420 36.300	5 5	*[129]
$C_{10}H_6Cl_2O$	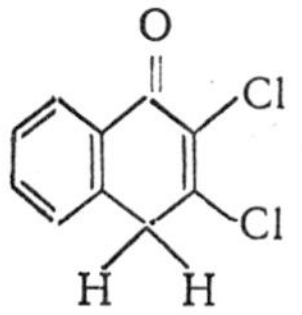	36.667 37.111	...	[20, 97]
$C_{10}H_6Cl_4O_4$	$C_6Cl_4(COOCH_3)_2$-1,2	37.751 37.881 37.986 38.157 38.210 38.217 38.290 38.493	6 6 6 6 6 6 6 6	*
	$C_6Cl_4(COOCH_3)_2$-1,4	37.471 37.520 37.576 37.786	10 10 10 10	*
$C_{10}H_7Cl$	1-$ClC_{10}H_7$	34.606	3	[20, 129]
	2-$ClC_{10}H_7$	34.69	10	*[208]

Empirical formula	Structural formula	ν, MHz	s/n	References, notes
$C_{10}H_7ClN_2$	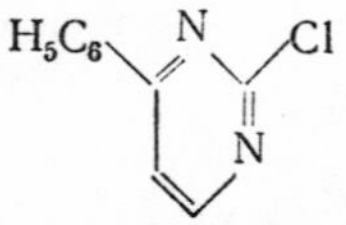	34.368 34.608	. . .	*
$C_{10}H_7Cl_2MnO_3$	$(ClCH_2)_2C_5H_3Mn(CO)_3$	34.584 33.786	20 20	*[203]. cf. Mn
$C_{10}H_7Cl_2N$	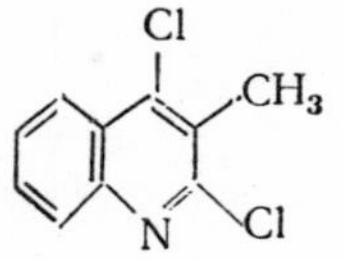	34.268 35.627	. . .	[99]. $T = 86$ °K
$C_{10}H_8ClN$	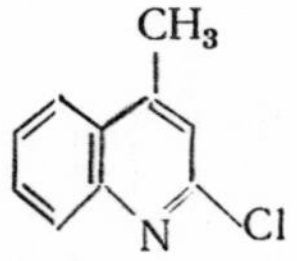	33.665	. . .	[97, 99]
$C_{10}H_8Cl_2Fe$	$(C_5H_4Cl)_2Fe$	35.469	8	*
$C_{10}H_8Cl_2HgN_2$	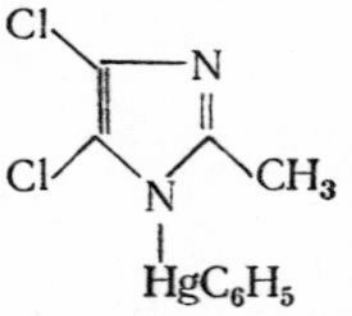	35.637 35.959	2 4	*[125]
$C_{10}H_8Cl_4$	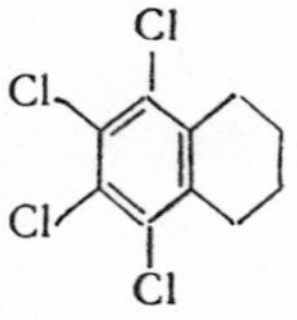	36.323 36.435 37.555 37.597	10 10 10 10	*
	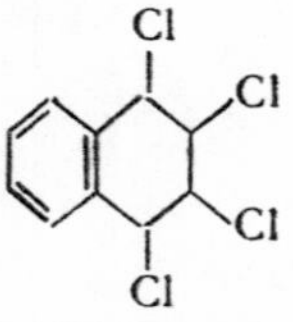	37.055 36.775 36.254 36.100	10 10 10 10	* [20 126]
$C_{10}H_9ClFe$	$C_5H_5FeC_5H_4Cl$	35.91	2	*

Empirical formula	Structural formula	ν, MHz	s/n	References, notes
$H_9Cl_2NO_2$	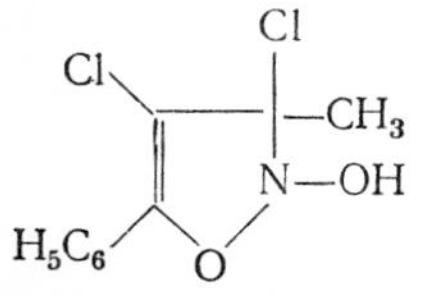	37.863 38.031 38.703 38.926	10 10 10 10	*[235]
	$2,5\text{-}Cl_2C_6H_3NHCOCH_2$ | H_3COC	35.000 35.232	. . .	[154]
$H_{10}BClFeO_2$	$(C_5H_4Cl)Fe[C_5H_4B(OH)_2]$	35.665	4	*
$H_{10}ClNO_2$	$2\text{-}ClC_6H_4NHCOCH_2COCH_3$	35.049	. . .	[156]
$H_{10}Cl_2N_2O_3$	$2\text{-}(Cl_2CHOOC)C_6H_4NH$ | H_3CHNOC	37.800 38.178	5 5	*
$H_{11}ClN_2O_3$	$C_6H_5N\langle^{CONHCH_3}_{OCH_2COCl}$	29.45	4	*
	$C_6H_5N\langle^{CONHCH_3}_{OCOCH_2Cl}$	37.275	10	*
$H_{11}ClO$	$2,4\text{-}(CH_3)_2C_6H_3COCH_2Cl$	35.602	3	[20]
$H_{12}Cl_4N_2$	$1,3\text{-}[(CH_3)_2N]_2C_6Cl_4$	36.894 37.104 37.434 37.542	10 10 10 10	*[233]
	$1,4\text{-}[(CH_3)_2N]_2C_6Cl_4$	37.002 37.278	20 18	*[233]
$H_{14}ClN$	$4\text{-}ClC_6H_4N(C_2H_5)_2$	34.079 34.161	. . .	[154]
$H_{14}ClO_2P$	$4\text{-}ClC_6H_4P(OC_2H_5)_2$	34.467	50	*
$H_{14}ClO_2PS$	$4\text{-}ClC_6H_4P(S)(OC_2H_5)_2$	34.824	20	*
$H_{14}ClO_3P$	$4\text{-}ClC_6H_4P(O)(OC_2H_5)_2$	34.698	5	*
$H_{16}ClN_2P$	$3\text{-}ClC_6H_4P[N(CH_3)_2]_2$	34.581	5	*
	$4\text{-}ClC_6H_4P[N(CH_3)_2]_2$	34.854	6	*
$H_{16}ClN_2PS$	$4\text{-}ClC_6H_4P(S)[N(CH_3)_2]_2$	34.908	10	*
$H_{16}Cl_2O_2$	$ClCO(CH_2)_8COCl$	29.118	2	[52]

Empirical formula	Structural formula	ν, MHz	s/n	References, notes
$C_{10}H_{16}Cl_4$	$[ClCH_2(CH_2)_3CCl{=}]_2$	33.108 34.833	30 30	*
$C_{10}H_{16}Cl_4I_2$	$ICH_2(CH_2)_3CCl_2$ | $ICH_2(CH_2)_3CCl_2$	37.627 37.140	3—4 3—4	*
$C_{10}H_{16}Cl_6$	$ClCH_2(CH_2)_3CCl_2$ | $ClCH_2(CH_2)_3CCl_2$	37.278 37.314 37.500 33.048 32.976	7 7 15 10 10	*
$C_{10}H_{18}Cl_4$	$C_4H_9CCl_2CCl_2C_4H_9$	37.422	100	*
	$Cl_3C(CH_2CHCH_3)_2CH_2$ | CH_3ClHC	38.136 37.793 37.660	6 1.5 1.5	*[243]
$C_{10}H_{20}Cl_2$	$ClCH_2(CH_2)_8CH_2Cl$	33.120	20—30	*[87]
$C_{10}H_{21}Cl$	$C_{10}H_{21}Cl$	33.024	3	[20]
$C_{11}H_4ClMnO_5$	$4\text{-}ClC_6H_4Mn(CO)_5$	34.572 34.770	5 5	*
$C_{11}H_7Cl_2HgNO$	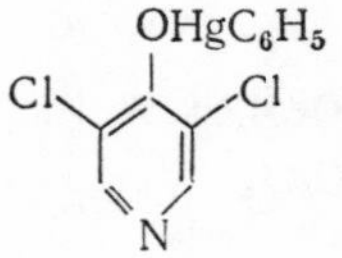	35.400 35.724	...	*
$C_{11}H_8Cl_2N_2$	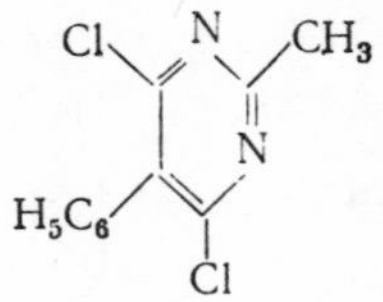	35.248	...	[99]. $T = 86\ °K$
$C_{11}H_8Cl_2O_3$	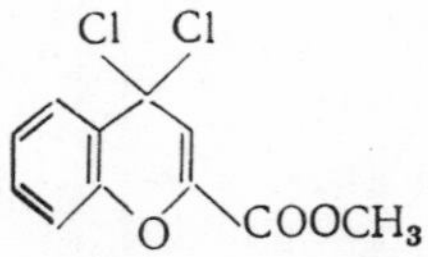	36.043	3	*
$C_{11}H_8Cl_4O_3$	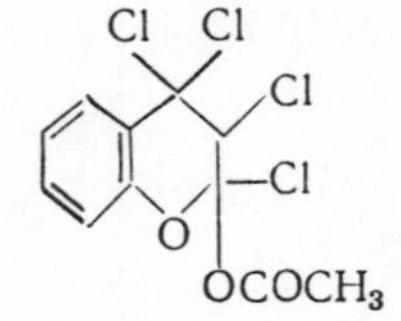	39.081 38.465 37.401	10 5 5	*

Empirical formula	Structural formula	ν, MHz	s/n	References, notes
$_{11}H_9ClF_7FeP$	$[(3\text{-}FC_6H_4Cl)Fe(C_5H_5)]PF_6$	37.51	2	*
	$[(4\text{-}FC_6H_4Cl)Fe(C_5H_5)]PF_6$	37.64	2	*
$_{11}H_9ClN_2O_2S$		35.33 35.51	10 10	*
$_{11}H_9Cl_2F_6FeP$	$[(1,4\text{-}Cl_2C_6H_4)Fe(C_5H_5)]PF_6$	37.23 37.74	2 2	*
$_1H_{10}BClF_4Fe$	$[(C_6H_6)Fe(C_5H_4Cl)]BF_4$	36.98	4	*
	$[(C_6H_5Cl)Fe(C_5H_5)]BF_4$	36.45	5	*
$_1H_{10}ClF_6FeP$	$[(C_6H_5Cl)Fe(C_5H_5)]PF_6$	37.016 37.450	3 3	*
$_1H_{11}ClN_2O_2$	$4\text{-}ClC_6H_4N{=}NCH(COCH_3)_2$	34.695	5	*
$_1H_{15}Cl_3O_2Si$	$C_6H_5(C_2H_5O)_2SiCCl_3$	37.5692 37.4754 37.0838	. . .	[26]
$_1H_{17}Cl_2NOSn$	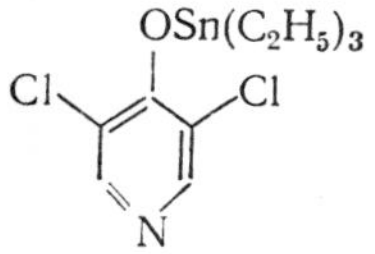	35.778 36.042	. . .	*
$H_{20}Cl_4$	$Cl_3C(CH_2)_9CH_2Cl$	33.312 37.275 38.332 38.574	10 10 10 10	*[78]
Cl_{10}	$Cl_5C_6C_6Cl_5$	37.10 37.58 38.03	5 10 10	*[129]
$Cl_{10}Hg$	$Cl_5C_6HgC_6Cl_5$	38.220 38.094 37.828 36.421	6 3 3 3	*[147]
$Cl_{10}O$	$Cl_5C_6OC_6Cl_5$	38.206 38.274 38.451 38.703	4 8—10 4 4	*[129]
H_4ClMnO_6	$4\text{-}ClC_6H_4COMn(CO)_5$	35.070	5—6	*
H_4ClO_6Re	$4\text{-}ClC_6H_4CORe(CO)_5$	34.860	5	* cf. Re
H_5Cl_5Hg	$Cl_5C_6HgC_6H_5$	35.868 37.758 38.122	4 4 2	*[147]

Empirical formula	Structural formula	ν, MHz	s/n	References, notes
$C_{12}H_7Cl_2HgNO_3$	2,6-$Cl_2C_6H_3OHgC_6H_4NO_2$-4	34.692	...	*
$C_{12}H_7Cl_5$	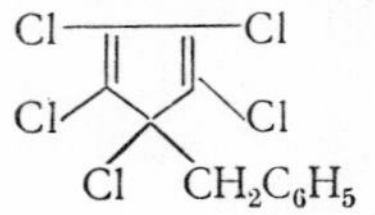	36.435 36.572 36.890 37.170 37.709	5 5 5 5 5	*
$C_{12}H_8Br_4CdCl_2N_4$	[4-$ClC_6H_4N_2$]$_2CdBr_4$	35.676 35.946	3 3	*[230]. cf. Br
$C_{12}H_8Br_4Cl_2CoN_4$	[4-$ClC_6H_4N_2$]$_2CoBr_4$	35.559 35.934	10 10	*[230]
$C_{12}H_8Br_4Cl_2HgN_4$	[4-$ClC_6H_4N_2$]$_2HgBr_4$	35.735	5	*[230]
$C_{12}H_8Br_4Cl_2N_4Zn$	[4-$ClC_6H_4N_2$]$_2ZnBr_4$	35.553 35.916	10 10	*[230]
$C_{12}H_8CdCl_2I_4N_4$	[4-$ClC_6H_4N_2$]$_2CdI_4$	36.258 36.982	2 2	*[230] cf. I
$C_{12}H_8CdCl_6N_4$	[4-$ClC_6H_4N_2$]$_2CdCl_4$	35.568 35.871	4 4	*[230]
$C_{12}H_8Cl_2$	4-$ClC_6H_4C_6H_4Cl$-4′	34.368 34.653 34.818 34.836	...	*[129]
$C_{12}H_8Cl_2Hg$	4-$ClC_6H_4HgC_6H_4Cl$-4′	34.105	100	*
$C_{12}H_8Cl_2HgN_2O_2$	2,6-Cl_2-4-$NO_2C_6H_2NH$ | H_5C_6Hg	34.599 35.328	3 2	*
$C_{12}H_8Cl_2HgO$	2,6-$Cl_2C_6H_3OHgC_6H_5$	34.056 34.932	8 8	*
$C_{12}H_8Cl_2O_2S$	4-$ClC_6H_4SO_2C_6H_4Cl$-4′	34.828	...	[154]
$C_{12}H_8Cl_3NO$	α-$C_{10}H_7NHCOCCl_3$	39.260 39.557 39.771	10 10 10	*
$C_{12}H_8Cl_6CoN_4$	[4-$ClC_6H_4N_2$]$_2CoCl_4$	35.34	2	*[230]
$C_{12}H_8Cl_6HgN_4$	[4-$ClC_6H_4N_2$]$_2HgCl_4$	35.400 35.688	3 3	*[230]
$C_{12}H_8Cl_6N_4Zn$	[4-$ClC_6H_4N_2$]$_2ZnCl_4$	35.346 35.816	1.5 1.5	*[230]
$C_{12}H_8Cl_8N_4Sn$	[2-$ClC_6H_4N_2$]$_2SnCl_6$	37.485	20	*[230]
	[3-$ClC_6H_4N_2$]$_2SnCl_6$	36.974	10	*[230]
	[4-$ClC_6H_4N_2$]$_2SnCl_6$	35.945	5	*[230]

Empirical formula	Structural formula	ν, MHz	s/n	References, notes
$_2H_9BrClN_3$	$2\text{-}ClC_6H_4N{=}NNHC_6H_4Br\text{-}4'$	34.860	15	*
	$4\text{-}ClC_6H_4N{=}NNHC_6H_4Br\text{-}4'$	35.28	10	*
$_2H_9ClHgO$	$2\text{-}ClC_6H_4OHgC_6H_5$	32.808 32.838	3 3	*
$_2H_9ClN_2O$	$4\text{-}ClC_6H_4N{=}NC_6H_4OH\text{-}4'$	34.632	50	*
$_2H_9ClN_4O_2$	$2\text{-}ClC_6H_4N{=}NNHC_6H_4NO_2\text{-}4'$	35.220 35.340	5 5	*
$_2H_9ClO_3S$	$4\text{-}ClC_6H_4SO_3C_6H_5$	35.347	. . .	[154]
$_2H_9ClO_4$	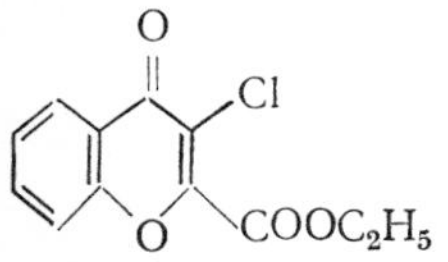	38.241	5	*
$_2H_9Cl_2N_3$	$2\text{-}ClC_6H_4N{=}NNHC_6H_4Cl\text{-}2'$	34.452 35.496	5 6	*
	$3\text{-}ClC_6H_4N{=}NNHC_6H_4Cl\text{-}3'$	34.893	8	*
	$4\text{-}ClC_6H_4N{=}NNHC_6H_4Cl\text{-}4'$	34.770 34.896 35.322	4 4 8	*
$_2H_{10}ClF_6FeO_2P$	$[(3\text{-}ClC_6H_4COOH)Fe(C_5H_5)]PF_6$	38.423	2	*
	$[(4\text{-}ClC_6H_4COOH)Fe(C_5H_5)]PF_6$	37.55	5	*
	$[(C_6H_5Cl)Fe(C_5H_4COOH)]PF_6$	37.268	2	*
$_2H_{10}ClN_3$	$4\text{-}ClC_6H_4N{=}NNHC_6H_5$	35.16	10	*
$_2H_{10}Cl_2N_2$	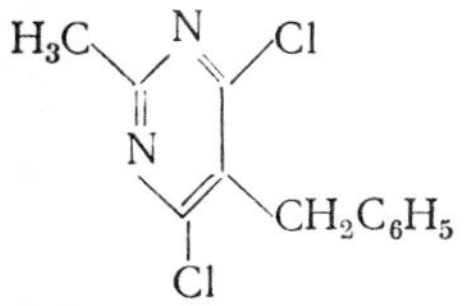	35.248	. . .	[99]. $T = 86\ °K$
$H_{10}ClO_2S$	$2\text{-}ClC_6H_4NHSO_2C_6H_5$	35.307	15	*
$H_{10}ClNO_2S$	$3\text{-}ClC_6H_4NHSO_2C_6H_5$	35.259	6	*
	$4\text{-}ClC_6H_4NHSO_2C_6H_5$	35.038	20	*
	$4\text{-}ClC_6H_4SO_2NHC_6H_5$	35.090 35.237	8 8	*
$H_{12}ClF_6FeP$	$[(3\text{-}ClC_6H_4CH_3)Fe(C_5H_5)]PF_6$	37.058	5	*
	$[(4\text{-}ClC_6H_4CH_3)Fe(C_5H_5)]PF_6$	36.624 37.240 37.373	2 2 4	*

Empirical formula	Structural formula	ν, MHz	s/n	References, notes
$C_{12}H_{13}ClN_2O_3$	$4\text{-}ClC_6H_4N{=}NCHCOCH_3$ (with $COOC_2H_5$ on CH)	34.884	10	*
$C_{12}H_{15}ClN_2O_3$	$C_6H_5N(OCOCH_2Cl)(CONHCH(CH_3)_2)$	37.051 36.862	9 3	*
$C_{12}H_{15}Cl_2NOS$	$C_6H_5NHCOCCl(CH_3)$ (with $CH_2SCH_2CH_2Cl$ on C)	33.561 33.408	5 5	*
$C_{12}H_{17}Cl$	$C_6(CH_3)_5CH_2Cl$	33.612	3	*
$C_{12}H_{18}Cl_2OSn$	$2,6\text{-}Cl_2C_6H_3OSn(C_2H_5)_3$	34.791 35.238	2 2	*
$C_{12}H_{18}Cl_4O_4$	$CCl_2(CH_2)_4COOH$ – $CCl_2(CH_2)_4COOH$	37.110 37.506	10 10	*
$C_{12}H_{20}Cl_4O_4Sn$	$(Cl_2HCCOO)_2Sn(C_4H_9)_2$	37.050 37.152	10 10	*
$C_{12}H_{22}Cl_2O_4Sn$	$(ClCH_2COO)_2Sn(C_4H_9)_2$	35.886	50	*
$C_{13}HClF_{10}$	$C_6F_5CHClC_6F_5$	38.738	20	*
$C_{13}H_6ClF_5$	$C_6H_5CHClC_6F_5$	38.838	20	*
$C_{13}H_8Cl_2O$	$4\text{-}ClC_6H_4COC_6H_4Cl\text{-}4'$	35.237	...	[154]
	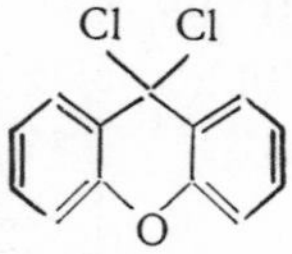	36.316 36.421	5 5	*
$C_{13}H_{10}ClHgN$	$2\text{-}ClC_6H_4N(CHO)\ (HgC_6H_5)$	34.220	5	*
$C_{13}H_{10}ClNO_5P$	$(4\text{-}NO_2C_6H_4)_2P(O)CH_2Cl$	37.152	3	*
$C_{13}H_{10}Cl_2$	$(C_6H_5)_2CCl_2$	36.883	50	*[26]
$C_{13}H_{10}Cl_2HgO$	$2,6\text{-}Cl_2C_6H_3OHgCH_2C_6H_5$	33.942 34.002 35.010	4 4 8	*
$C_{13}H_{10}Cl_2O$	$2,6\text{-}Cl_2C_6H_3OCH_2C_6H_5$	35.376 35.940	10 10	*
$C_{13}H_{11}Cl$	$ClCH(C_6H_5)_2$	34.21	3	*[26]
$C_{13}H_{11}Cl_2N_3$	$2\text{-}Cl\text{-}4\text{-}CH_3C_6H_3N{=}NNHC_6H_4Cl\text{-}2'$	34.251 34.959	6 6	*
$C_{13}H_{12}ClOP$	$(C_6H_5)_2P(O)CH_2Cl$	37.023	50	*

Empirical formula	Structural formula	ν, MHz	s/n	References, notes
$_{13}H_{12}ClO_3P$	$ClCH_2P(O)(OC_6H_5)_2$	37.013	100	*
$_{13}H_{14}BClF_4Fe$	$[(C_6H_5Cl)Fe(C_5H_4C_2H_5)]BF_4$	36.568	7	*
$_{13}H_{14}ClF_6FeP$	$[(C_6H_5Cl)Fe(C_5H_4C_2H_5)]PF_6$	36.554	3	*
$_{13}H_{14}Cl_2$	$-C_6H_5$ CCl_2	37.09	...	[26]
$_{13}H_{17}ClN_2O_3$	C_6H_5N $CONHC(CH_3)_3$ $OCOCH_2Cl$	36.890 37.072	10 10	*
$_{13}H_{24}Cl_4$	$ClCH_2(CH_2)_{11}CCl_3$	33.306 37.261 38.395 38.605	4 4 4 4	*[78]
$_{13}H_{25}ClO_2$	$ClH_2C(CH_2)_{11}COOH$	33.642	10	*
$_{14}H_4Cl_4O_2$		39.186 39.032 38.934 38.724 38.584 38.192 38.073 38.038 37.968	2 10 2 10 2 10 2 10 2	*
$_{14}H_6Cl_2O_2$		37.996 36.428	3 3	*
		36.777	3	[20]
		36.885	6	[20]

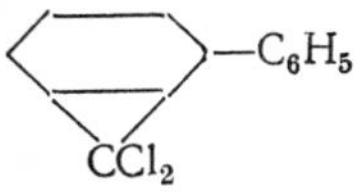

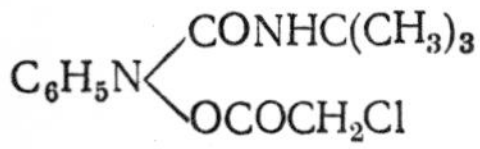

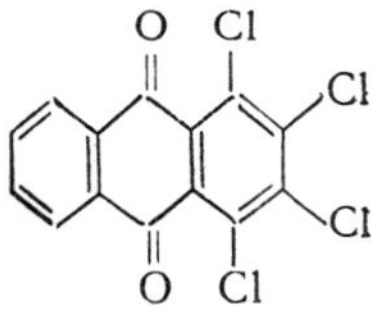

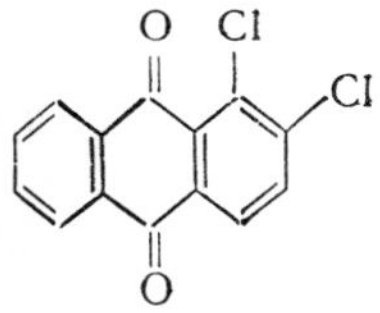

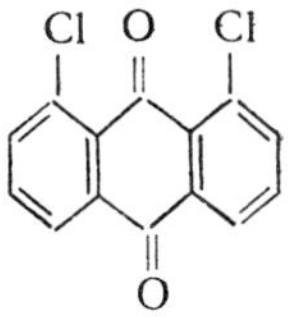

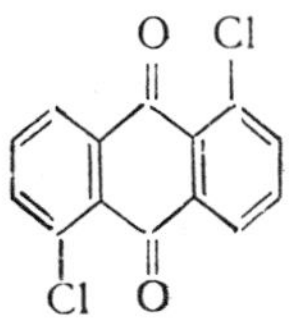

Empirical formula	Structural formula	ν, MHz	s/n	References, notes
$C_{14}H_6Cl_2O_2$	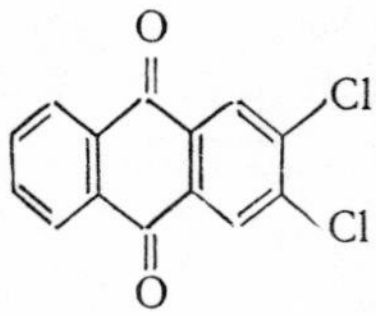	36.667 37.111	2 2	[20]
$C_{14}H_7ClO_2$	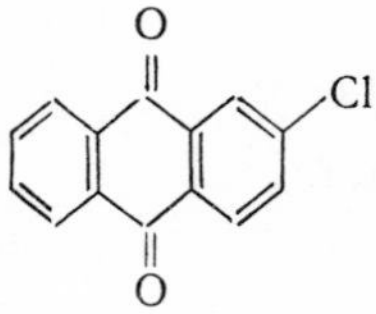	35.390	7	[20]
	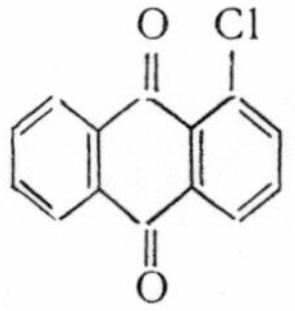	37.529	5	[20]
$C_{14}H_8Cl_2$	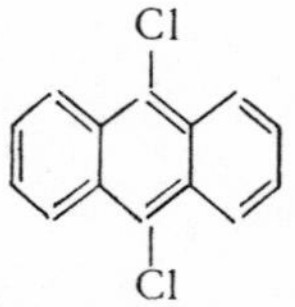	34.899	3.5	[20]
$C_{14}H_8Cl_2O_2$	2-$ClOCC_6H_4C_6H_4COCl$-2′	29.885	10	*
$C_{14}H_8Cl_2O_3$	4-$ClOCC_6H_4OC_6H_4COCl$-4′	30.120 29.870	5 5	*
$C_{14}H_9Cl$	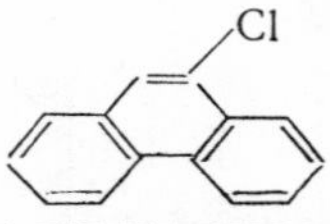	34.837 35.167	10 5	*[129]
$C_{14}H_9Cl_5$	(4-$ClC_6H_4)_2CHCCl_3$	34.868 34.977 38.487 38.814 39.039	2 2 2—3 2—3 2—3	[52]
$C_{14}H_{13}Cl_2HgNO$	2,6-$Cl_2C_6H_3OHgC_6H_4N(CH_3)_2$-4′	33.663 34.899	7 7	*
$C_{14}H_{13}Cl_2N_3$	2-Cl-4-$CH_3C_6H_3N{=}NNH$ (on N: 2′-Cl-4′-$CH_3C_6H_3$)	34.053 34.905 35.052	3 3 4	*

Empirical formula	Structural formula	ν, MHz	s/n	References, notes
$C_{14}H_{22}ClP$	$4\text{-}ClC_6H_4P(C_4H_9)_2$	34.749	5	*
$C_{14}H_{22}ClPS$	$4\text{-}ClC_6H_4P(S)\,(C_4H_9)_2$	35.142	10	*
$C_{14}H_{22}Cl_4O_4$	$[H_3COOC(CH_2)_4CCl_2\text{—}]_2$	37.247 37.548	5 5	*
$C_{14}H_{24}Cl_6$	$[ClCH_2(CH_2)_5CCl_2\text{—}]_2$	33.072 37.198 37.394	20 20 20	*
$C_{14}H_{26}Cl_4O_4$	$[HOOC(CH_2)_6CCl_2\text{—}]_2$	37.219 37.485	3 3	*
$C_{15}H_9Cl_3O_2$	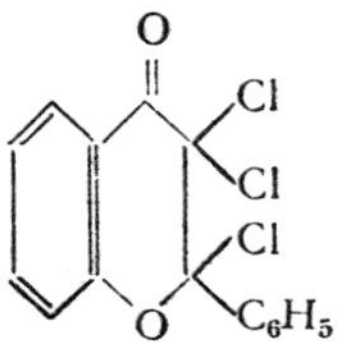	34.962 38.255 39.056	6 8 8	*
$C_{15}H_9Cl_5O$	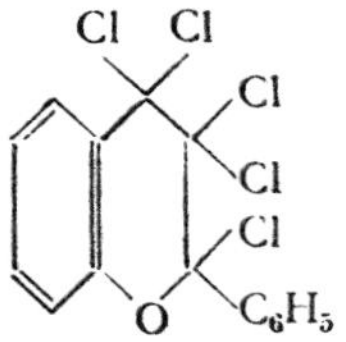	40.243 39.466 39.445 39.375 36.708	5 5 5 5 5	*
$C_{15}H_{10}ClNO$	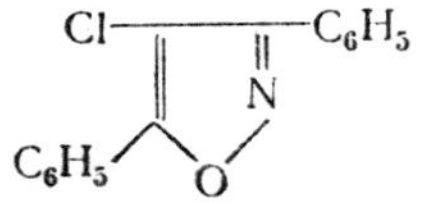	38.372	20	*
$C_{15}H_{11}ClO_2$	$(C_6H_5CO)_2CHCl$	37.954	50	*
$C_{15}H_{11}Cl_2NO$	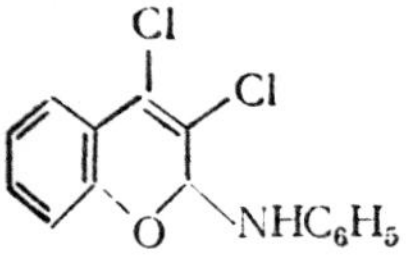	35.820 36.039	10 10	*
$C_{15}H_{11}Cl_2NO_2$	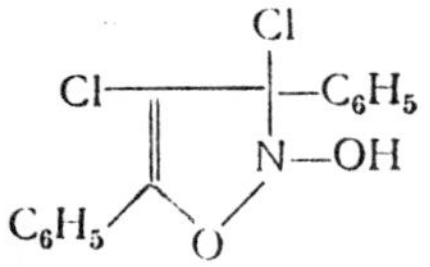	38.301 39.043	10 10	*[235]

Empirical formula	Structural formula	ν, MHz	s/n	References, notes
$C_{15}H_{12}ClNO_5$	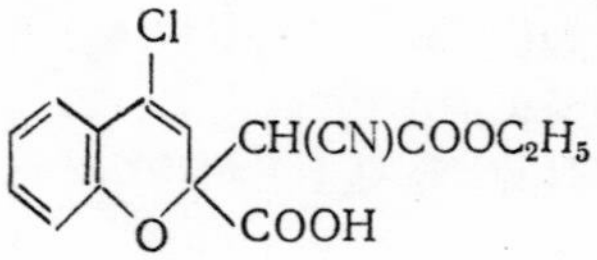	35.814	5	*
$C_{15}H_{12}Cl_2$	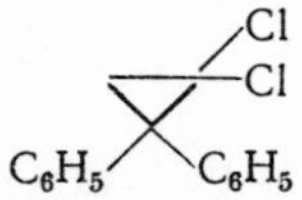	37.242 37.353	—	[108]
$C_{15}H_{14}Cl_2HgO$	$2,6\text{-}Cl_2C_6H_3OHg$ $2',4',6'\text{-}(CH_3)_3H_2C_6$	33.810 33.840 34.758 34.800	3 3 3 3	*
$C_{15}H_{16}ClO_3P$	$ClCH_2P(O)(OC_6H_4CH_3\text{-}4)_2$	37.520	50	*
$C_{16}Cl_{10}$	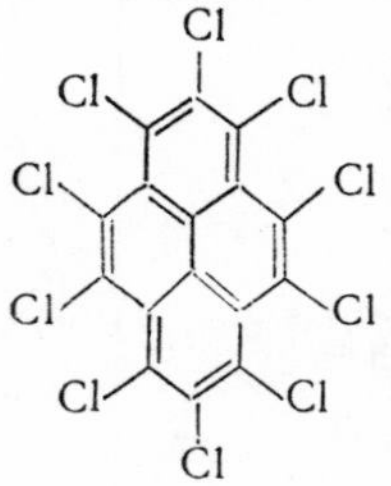	38.276 38.381 38.539 38.668 38.843	10 10 5 8 8	*[129]
$C_{16}H_{10}Cl_4$	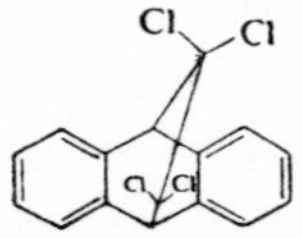	38.332 38.521 } 38.556 } 38.955	3 6 3	*[129]
$C_{16}H_{11}ClN_2$	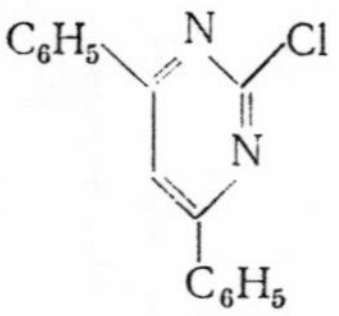	33.432	4	*
$C_{16}H_{11}ClN_2O$	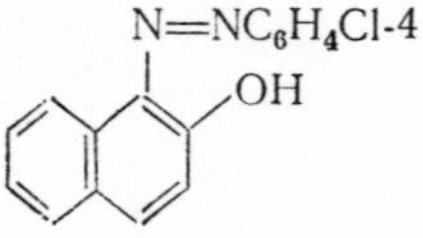	34.625	2	*

Empirical formula	Structural formula	ν, MHz	s/n	References, notes
$C_{16}H_{12}ClN_3$	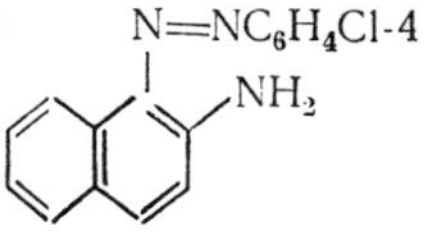	35.11	3	*
	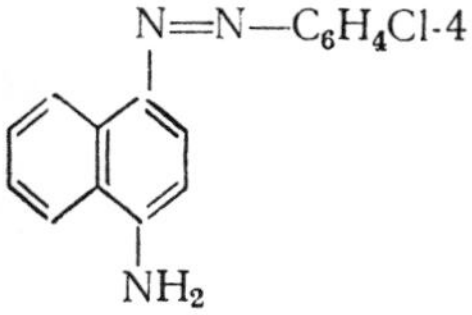	34.06	5	*
$C_{16}H_{12}Cl_2$	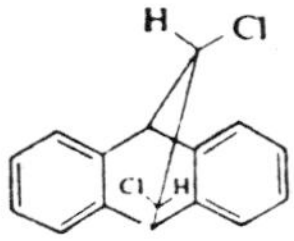	33.996 35.184	5 5	*[129]
$C_{16}H_{13}Cl_2NO_2$	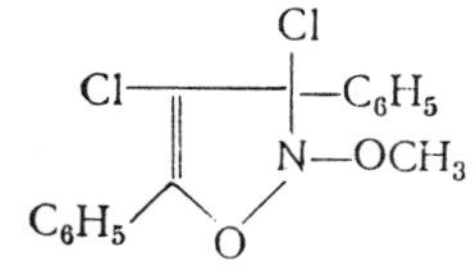	38.241 39.130	20 20	*
$C_{16}H_{26}Cl_4O_4$	$CCl_2(CH_2)_6COOH$ | $CCl_2(CH_2)_6COOH$	37.219 37.485	3 3	*
$C_{17}H_{13}ClN_2$	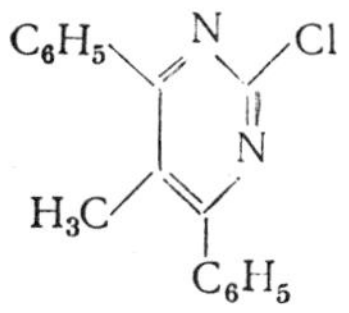	34.230	...	*
$C_{17}H_{13}ClO_4$	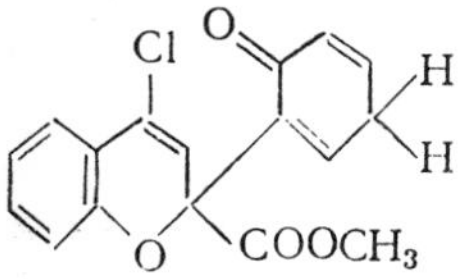	35.502	5	*
$C_{17}H_{13}Cl_2NO_3$	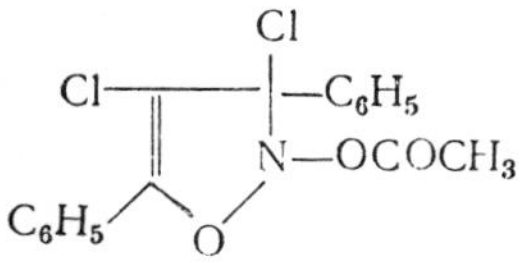	38.682 38.948 39.788	10 5 5	*

Empirical formula	Structural formula	ν, MHz	s/n	References, notes
$C_{17}H_{16}ClNO_2$	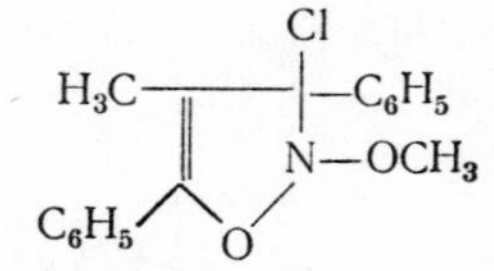	35.070	10	*
$C_{17}H_{16}ClNO_5$	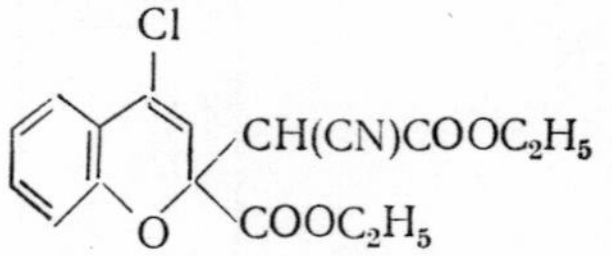	35.763	10	*
$C_{17}H_{32}Cl_4$	$Cl_3C(CH_2)_{15}CH_2Cl$	37.247 38.402 38.605	5 5 5	*
$C_{18}H_{12}Cl_2O_2$	$(2\text{-}ClC_6H_4{-}OCH_2C{\equiv}C{-})_2$	35.187 35.325	7 7	*
$C_{18}H_{13}BrClHgN_3$	$2\text{-}ClC_6H_4N{=}NNC_6H_4Br\text{-}4'$ (N–HgC_6H_5)	35.10	. . .	*
$C_{18}H_{13}ClHgN_4O_2$	$2\text{-}ClC_6H_4N{=}NNC_6H_4NO_2\text{-}4'$ (N–HgC_6H_5)	35.430	3	*
$C_{18}H_{13}Cl_2HgN_3$	$2\text{-}ClC_6H_4N{=}NNC_6H_4Cl\text{-}2'$ (N–HgC_6H_5)	33.783 35.616	6 6	*
	$3\text{-}ClC_6H_4N{=}NNC_6H_4Cl\text{-}3'$ (N–HgC_6H_5)	35.040	8	*
	$4\text{-}ClC_6H_4N{=}NNC_6H_4Cl\text{-}4'$ (N–HgC_6H_5)	33.666 34.902	6 5	*
$C_{18}H_{14}ClHgN_3$	$2\text{-}ClC_6H_4N{=}NNC_6H_5$ (N–HgC_6H_5)	34.90	. . .	*
$C_{18}H_{14}ClHgNO_2S$	$3\text{-}ClC_6H_4N(SO_2C_6H_5)\ (HgC_6H_5)$	35.004	. . .	*
	$4\text{-}ClC_6H_4N(SO_2C_6H_5)\ (HgC_6H_5)$	34.682 34.691	8 8	*
$C_{18}H_{14}ClN$	$4\text{-}ClC_6H_4N(C_6H_5)_2$	34.500	10	*
$C_{18}H_{14}ClOP$	$3\text{-}ClC_6H_4P(O)\ (C_6H_5)_2$	34.71	2	*
	$4\text{-}ClC_6H_4P(O)\ (C_6H_5)_2$	34.818	4	*
$C_{18}H_{14}ClP$	$4\text{-}ClC_6H_4P(C_6H_5)_2$	34.866	3	*
$C_{18}H_{14}ClPS$	$3\text{-}ClC_6H_4P(S)\ (C_6H_5)_2$	35.010	5	*
	$4\text{-}ClC_6H_4P(S)\ (C_6H_5)_2$	34.856	7	*

Empirical formula	Structural formula	ν, MHz	s/n	References, notes
$C_{18}H_{32}Cl_6$	$CCl_2(CH_2)_7CH_2Cl$ | $CCl_2(CH_2)_7CH_2Cl$	33.00 36.99 37.07 37.63	1 1.5 1.5 3	*
$C_{18}H_{34}Cl_4$	$CCl_2(CH_2)_7CH_3$ | $CCl_2(CH_2)_7CH_3$	37.352 37.394	20 20	*
$C_{19}H_{15}Cl$	$ClC(C_6H_5)_3$	34.332 34.551 34.842 35.046 35.058	10 10 10 10 10	*
$C_{19}H_{17}ClIP$	$[3\text{-}ClC_6H_4P(CH_3)(C_6H_5)_2]I$	35.670	2	*
	$[4\text{-}ClC_6H_4P(CH_3)(C_6H_5)_2]I$	35.076	3	*
$C_{19}H_{36}Cl_4$	$Cl_3C(CH_2)_{17}CH_2Cl$	37.191 38.451	3 6	*
$C_{20}H_{15}ClF_4NO_4P$	$CF_3CFClNP(O)(OC_6H_5)_3$	36.834	10	*
$C_{20}H_{18}Cl_2HgN_4$	$4\text{-}ClC_6H_4N{=}NNC_6H_4Cl\text{-}4'$ | $4'\text{-}(CH_3)_2NH_4C_6Hg$	34.43 34.56 34.64	6 6 4	*
$C_{20}H_{34}Cl_4O_4$	$CCl_2(CH_2)_6COOC_2H_5$ | $CCl_2(CH_2)_6COOC_2H_5$	37.156 37.639	...	*
$C_{21}H_{26}Cl_2HgO$	$2,6\text{-}Cl_2C_6H_3OHg$ | $2',4',6'\text{-}(iso\text{-}C_3H_7)_3H_2C_6$	34.182 35.256	3 3	*
$C_{23}H_{17}Cl_2NOPb$	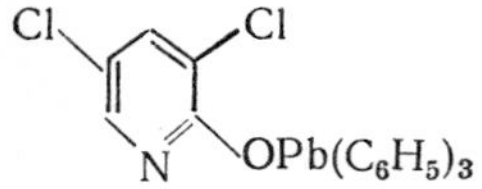	35.328 35.598 35.790 36.621	2 2 2 2	*
	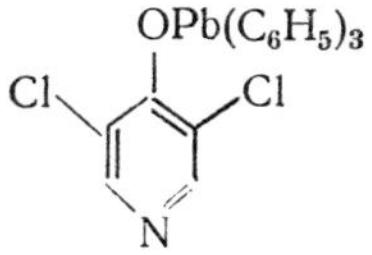	34.968 35.604	3 3	*
$C_{23}H_{17}Cl_2NOSn$	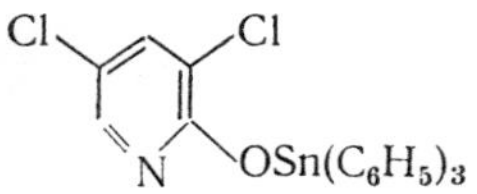	35.934 36.360	5 5	*
$C_{24}H_{18}Cl_2HgN_2O_4S_2$	$(4\text{-}ClC_6H_4NSO_2C_6H_5)_2Hg$	34.342 34.760	10 10	*
$C_{24}H_{18}Cl_2OPb$	$2,6\text{-}Cl_2C_6H_3OPb(C_6H_5)_3$	34.806 34.938	2 2	*

Empirical formula	Structural formula	ν, MHz	s/n	References, notes
$C_{24}H_{18}Cl_2OSn$	2,6-$Cl_2C_6H_3OSn(C_6H_5)_3$	34.800 35.244 35.448	2 4 2	*
$C_{25}H_{18}Cl_4O_2P_2$	$CH_2[P(O)(C_6H_4Cl\text{-}4)_2]_2$	34.644 35.376 35.574	4 2 2	*
$C_{25}H_{21}ClO_2P_2$	$CHCl[P(O)(C_6H_5)_2]_2$	38.066	10	*
$C_{32}H_{18}Cl_4N_4NiO_2$	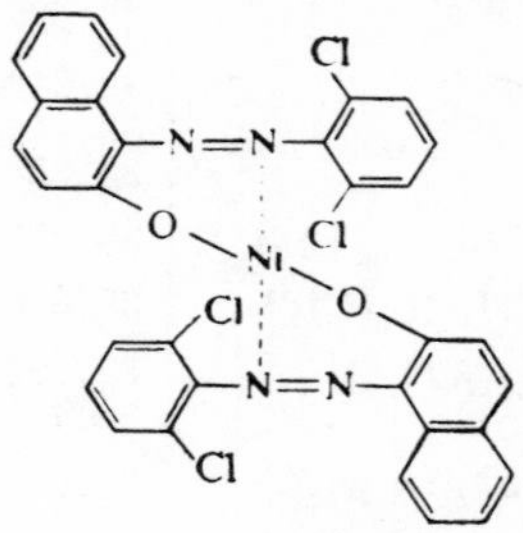	35.820 36.450	10 10	*
	Polymers			
$(C_2ClF_3)_n$	$[-CF_2CFCl-]_n$	39.94	6	*[178]
		Δν ≈ 650 kHz		
$(C_2HCl_3O)_n$	$[-CH(CCl_3)-O-]_n$	39.77	15	*[178]
		Δν ≈ 400 kHz		
$(C_2H_2Cl_2)_n$	$[-CH_2CCl_2-]_n$	36.6 37.25 38.04	3 6 15	*[75, 178]
$(C_3H_3Cl_3)_n$	$[-CH_2CHClCCl_2-]_n$	37.01	3	*[75, 178]
		Δν ≈ 800 kHz		
$(C_3H_5Cl)_n$	$[-CHClCH_2{=}CH_2-]_n$	32.75	10	*
		Δν ≈ 60 kHz		
$(C_{12}H_4Si_2Cl_2O_3)_n$	$[(3\text{-}C_6H_4ClSiO)_2O]_n$	34.99	10	*
		Δν ≈ 540 kHz		

BIBLIOGRAPHY

1. Livingston, R. — J. Phys. Chem. **57** (1953), 496.
2. Khotsyanova, T. L., V. I. Robas, and G. K. Semin. — In: "Radiospektroskopiya tverdogo tela," p. 233. Moscow, Atomizdat, 1967.
3. Semin, G. K. and V. I. Robas. — In: "Radiospektroskopiya tverdogo tela," p. 229. Atomizdat, 1967.
4. Babushkina, T. A., E. V. Bryukhova, F. K. Velichko, V. I. Pakhomov, and G. K. Semin. — Zh. Strukt. Khim. **9** (1968), 207.
5. Coles, D. and R. Hughes. — Phys. Rev. **76** (1949), 858.
6. Smith, A., H. Ring, W. Smith, and W. Gordy. — Phys. Rev. **74** (1948), 370.

7. Smith, A., H. Ring, W. Smith, and W. Gordy. — Ibid., p. 123.
8. Townes, C., A. Holden, and F. Meritt. — Phys. Rev. 74 (1948), 1113.
9. Geschwind, S., R. Günther-Mohr, and C. Townes. — Phys. Rev. 81 (1951), 288L.
10. Geschwind, S., R. Günther-Mohr, and C. Townes. — Phys. Rev. 82 (1951), 343.
11. Casabella, P. and P. Bray. — J. Chem. Phys. 28 (1958), 1182.
12. Casabella, P. and P. Bray. — Bull. Am. Phys. Soc., ser. 2, 3 (1958), 21.
13. Semin, G. K. and A. A. Fainzil'berg. — Zh. Strukt. Khim. 6 (1965), 213.
14. Babushkina, T. A., G. K. Semin, V. I. Robas, and I. A. Safin. — Fiz. Tverd. Tela 7 (1965), 924.
15. Negita, H. and Z. Hirano. — Bull. Chem. Soc. Japan 31 (1958), 660.
16. Bersohn, R. — J. Chem. Phys. 22 (1954), 2078.
17. Rehn, V. — J. Chem. Phys. 38 (1963), 749.
18. Livingston, R. — J. Chem. Phys. 19 (1951), 1434.
19. Long, M., Q. Williams, and T. Weatherly. — J. Chem. Phys. 33 (1960), 508.
20. Hooper, H. and P. Bray. — J. Chem. Phys. 33 (1960), 334.
21. Gutowsky, H. and D. McCall. — J. Chem. Phys. 32 (1960), 548.
22. Grechishkin, V. S. — Zh. Fiz. Khim. 35 (1961), 1803.
22a. Hartmann, H., H. Silescu, and K. Funke. — Z. Naturf. 18a (1963), 974.
23. Grechishkin, V. S. and I. A. Kyuntsel'. — Zh. Strukt. Khim. 7 (1966), 119.
24. Babushkina, T. A., G. K. Semin, S. P. Kolesnikov, O. M. Nefedov, and V. I. Svergun. — Izv. AN SSSR, chem. ser., p. 1055, 1969.
25. Hooper, H. — J. Chem. Phys. 41 (1964), 599.
26. Biryukov, I. P., M. G. Voronkov, and I. A. Safin. — Izv. AN LatvSSR, chem. ser., No. 6 (1966), 638.
27. Livingston, R. — Rec. Chem. Prog. 20 (1959), 173.
28. McLay, D. and C. Mann. — Can. J. Phys. 40 (1962), 61.
29. Beeson, E., T. Weatherly, and Q. Williams. — J. Chem. Phys. 37 (1962), 2926.
30. McLay, D. — Spectrochim. Acta 18 (1962), 1380.
31. McLay, D. — Can. J. Phys. 42 (1964), 720.
32. Livingston, R. — Phys. Rev. 82 (1951), 289.
33. Wolfe, P. — J. Chem. Phys. 25 (1956), 976.
34. Bennett, R. and H. Hooper. — J. Chem. Phys. 47 (1967), 4855.
35. Semin, G. K., T. A. Babushkina, V. I. Robas, G. Ya. Zueva, M. A. Kadina, and V. I. Svergun. — In: "Radiospektroskopicheskie i kvantovokhimicheskie metody v strukturnoi khimii," p. 225. "Nauka," 1967.
36. Biryukov, I. P., M. G. Voronkov, and I. A. Safin. — In: "Radiospektroskopiya tverdogo tela," p. 252. Atomizdat, 1967.
37. Babushkina, T. A. and G. K. Semin. — Zh. Strukt. Khim. 7 (1966), 114.
38. Akon, C. and T. Iredale. — J. Chem. Phys. 34 (1961), 340.
39. Meyers, R. and W. Gwinn. — J. Chem. Phys. 20 (1950), 1420.
40. Flygare, W. and W. Gwinn. — J. Chem. Phys. 36 (1962), 787.
41. Lucken, E. and M. Whitehead. — J. Chem. Soc., p. 2459, 1961.
42. Biryukov, I. P., M. G. Voronkov, G. V. Motsarev, V. R. Rozenberg, and I. A. Safin. — Doklady AN SSSR 162 (1965), 130.
42a. Semin, G. K. and T. A. Babushkina. — Teor. Eksperim. Khim. 4 (1968), 835.
43. Semin, G. K., T. A. Babushkina, A. V. Prokof'ev, and R. G. Kostyanovskii. — Izv. AN SSSR, chem. ser., p. 1401, 1968.
44. Gordy, W., J. Simmons, and A. Smith. — Phys. Rev. 74 (1948), 243.
45. Gordy, W., J. Simmons, and A. Smith. — Ibid., p. 1246.
46. Karplus, R. and A. Sharbough. — Phys. Rev. 75 (1949), 889.
47. Simmons, J. and J. Goldstein. — J. Chem. Phys. 20 (1952), 122.
48. Wang, T. — Phys. Rev. 99 (1955), 566.
49. Schwendeman, R. and G. Jacobs. — J. Chem. Phys. 36 (1962), 1251.
50. Semin, G. K., T. A. Babushkina, and V. I. Robas. — In: "Radiospektroskopiya tverdogo tela," p. 218. Atomizdat, 1967.
51. Mizzi, A., A. Guarneri, P. Fabero, and G. Zuliani. — Nuovo Cim. 25 (1962), 265.
52. Bray, P. — J. Chem. Phys. 23 (1955), 703.
53. Kinastowsky, S. and J. Pietrzak. — Bull. Acad. pol. Sci. sér. sci. chim. 16 (1968), 155.
54. Biedenkapp, D. and A. Weiss. — Ber. Bunsengesellsch. phys. Chem. 70 (1966), 788.
55. Graybeal, J. and C. Cornwell. — J. Phys. Chem. 62 (1958), 483.

56. Allen, H. — J. Phys. Chem. 57 (1953), 501.
57. Livingston, R. — J. Chem. Phys. 20 (1952), 1170.
58. Tatsuzaki, I. and Y. Yokozawa. — J. Phys. Soc. Japan 12 (1957), 802.
59. Grechishkin, V. S., G. B. Soifer, and Yu. G. Svetlov. — Izv. vuzov, Fizika, No. 5 (1963), 32.
60. Westenberg, A., J. Goldstein, and E. Wilson. — J. Chem. Phys. 17 (1949), 1319.
61. Westenberg, A., J. Goldstein, and E. Wilson. — Phys. Rev. 76 (1949), 472.
62. White, R. — Phys. Rev. 94 (1954), 789.
63. Jenkins, D. and T. Sugden. — Trans. Faraday Soc. 55 (1959), 1473.
64. Lucken, E. — J. Chem. Soc., p. 2954, 1959.
65. Allen, H. — J. Am. Chem. Soc. 74 (1952), 6074.
65a. Blinč, R., M. Mali, and Z. Trontelij. — Phys. Rev. Lett. 25A (1967), 289.
66. Bragg, J., T. Madison, and A. Sharbough. — Phys. Rev. 77 (1950), 148.
67. Howe, J. — J. Chem. Phys. 34 (1961), 1247.
68. Graybeal, J. — J. Chem. Phys. 32 (1960), 1258.
69. Wada, K., G. Kikushi, C. Matsumura, E. Hirota, and Y. Morino. — Bull. Chem. Soc. Japan 34 (1961), 337.
70. Bray, P., P. Barnes, and R. Bersohn. — J. Chem. Phys. 25 (1956), 813.
71. Shimizu, T. and H. Takuma. — J. Phys. Soc. Japan 15 (1960), 646.
72. Flygare, W. and J. Howe. — J. Chem. Phys. 36 (1962), 440.
73. Dehmelt, H. and H. Krüger. — Z. Phys. 130 (1951), 401.
74. Solimene, N. — J. Chem. Phys. 26 (1957), 1593.
75. Babushkina, T. A., V. I. Robas, and G. K. Semin. — Doklady AN SSSR 159 (1964), 164.
76. Semin, G. K. — Zh. Strukt. Khim. 9 (1965), 916.
77. Grechishkin, V. S. and G. B. Soifer. — Zh. Strukt. Khim. 3 (1962), 337.
78. Semin, G. K. and V. I. Robas. — Zh. Strukt. Khim. 7 (1966), 117.
79. Goldstein, J. and R. Livingston. — J. Chem. Phys. 19 (1951), 1613.
80. Kivelson, D., E. Wilson, and D. Lide. — J. Chem. Phys. 32 (1960), 205.
81. Sinott, K. — J. Chem. Phys. 34 (1961), 851.
81a. Neimysheva, A. A., G. K. Semin, T. A. Babushkina, and I. L. Knunyants. — Doklady AN SSSR 173 (1967), 585.
82. Negita, H. — J. Chem. Phys. 23 (1955), 214.
83. McCall, D. and H. Gutowsky. — J. Chem. Phys. 21 (1953), 1300.
84. Biedenkapp, D. and A. Weiss. — Z. Naturf. 22a (1967), 1124.
85. Flygare, W. — J. Molec. Spectrosc. 14 (1964), 145.
86. Ragle, J. — J. Phys. Chem. 63 (1959), 1395.
87. Semin, G. K. — In: "Radiospektroskopiya tverdogo tela," p. 205. Atomizdat, 1967.
88. Baisa, D. F. and A. F. Prokhot'ko. — Ukr. Fiz. Zh. 9 (1964), 573.
89. Tokuhiro, T. — J. Chem. Phys. 47 (1967), 2353.
90. Marram, E. and J. Ragle. — J. Chem. Phys. 37 (1962), 3015.
91. Grechishkin, V. S. and G. B. Soifer. — Trudy ENI Perms. State Univ. 11, No. 2 (1964), 125.
92. Wagner, R. and B. Dailey. — J. Chem. Phys. 26 (1957), 1588.
93. Schwendeman, R. and G. Jacobs. — J. Chem. Phys. 36 (1962), 1245.
94. Veselaro, V. G. and N. A. Irisova. — Radiotekhnika i Elektronika 2 (1957), 484.
95. Bryukhova, E. V., V. I. Stanko, A. I. Klimova, N. S. Titova, and G. K. Semin. — Zh. Strukt. Khim. 9 (1968), 39.
96. Lucken, E. and C. Marzeline. — J. Chem. Soc. (A), p. 153, 1968.
97. Bray, P., S. Moskowitz, H. Hooper, R. Barnes, and S. Segel. — J. Chem. Phys. 28 (1958), 99.
98. Morino, Y., T. Chiba, K. Shimozawa, and M. Toyama. — J. Phys. Soc. Japan 13 (1958), 869.
99. Dewar, M. and E. Lucken. — J. Chem. Soc., p. 2653, 1958.
100. Morino, Y. and M. Toyama. — J. Phys. Soc. Japan 15 (1960), 288.
101. Kojima, S. — J. Jap. Chem., Kagaku-noryoiki 13 (1962), 619.
102. Tsvetko, E. N., G. K. Semin, D. I. Lobanov, and M. I. Kabachnik. — Teor. Eksperim. Khim. 4 (1968), 452.
103. Costain, C. — J. Chem. Phys. 23 (1955), 2037.
104. Hirota, E., T. Oka, and Y. Morino. — J. Chem. Phys. 29 (1958), 444.
105. Shimozawa, T., T. Chiba, and Y. Morino. — Bull. Chem. Soc. Japan 35 (1962), 1041.
106. Hirota, E. and Y. Morino. — Bull. Chem. Soc. Japan 34 (1961), 341.

107. Stone, R., S. Srivastava, and W. Flygare. — J. Chem. Phys. 48 (1968), 1890.
108. Todd, J., M. Whitehead, and K. Weber. — J. Chem. Phys. 39 (1963), 404.
109. Flygare, W., A. Narath, and W. Gwinn. — J. Chem. Phys. 36 (1962), 200.
110. Saito, S. — J. Phys. Soc. Japan 16 (1961), 2361.
111. Semin, G. K. — Doklady AN SSSR 158 (1964), 1169.
112. Friend, J., K. Schneider, and B. Dailey. — J. Chem. Phys. 23 (1955), 1557.
113. Friend, J. and B. Dailey. — J. Chem. Phys. 29 (1958), 577.
114. Beaudet, R. — J. Chem. Phys. 37 (1962), 2398.
115. Koltenbah, D. and A. Silvidi. — Bull. Am. Phys. Soc., ser. 2, 12 (1967), 936.
116. Unland, T., V. Weiss, and W. Flygare. — J. Chem. Phys. 42 (1965), 2138.
117. Weatherly, T. and Q. Williams. — J. Chem. Phys. 22 (1954), 958.
118. Torizuka, K. — J. Phys. Soc. Japan 9 (1954), 645.
119. Tobiason, F. and R. Schwendeman. — J. Chem. Phys. 40 (1964), 1014.
120. Semin, G. K., V. I. Robas, V. I. Stanko, and V. A. Brattsev. — Zh. Strukt. Khim. 6 (1965), 305.
121. Brevard, C. and J. Lehn. — J. Chim. phys. 65 (1968), 727.
122. Dewar, M. and E. Lucken. — J. Chem. Soc., p. 426, 1959.
123. Hooper, H. and P. Bray. — J. Chem. Phys. 30 (1959), 957.
124. Velichko, F. K., L. A. Nikonova, and G. K. Semin. — Izv. AN SSSR, chem. ser., p. 84, 1967.
125. Nesmeyanov, A. N., D. N. Kravtsov, A. P. Zhukov, P. M. Kochergin, and G. K. Semin. — Doklady AN SSSR 179 (1968), 102.
126. Bray, P. and P. Ring. — J. Chem. Phys. 21 (1953), 2226.
127. Grechishkin, V. S. and G. B. Soifer. — Fiz. Tverd. Tela 4 (1962), 2268.
128. Linde, D. and M. Jen. — J. Chem. Phys. 38 (1963), 1504.
129. Semin, G. K., T. A. Babushkina, and V. I. Robas. — Zh. Fiz. Khim. 40 (1966), 2564.
130. Hayek, M., D. Gill, I. Agranat, and M. Rabinovitz. — J. Chem. Phys. 47 (1967), 3680.
131. Lucken, E. and C. Mazeline. — Proc. XIII Colloque AMPERE, Louvain, 1964.
132. Bucci, P., P. Cecchi, and A. Colligiani. — Chimica ind. 47 (1965), 213.
133. Segel, S., R. Barnes, and P. Bray. — J. Chem. Phys. 25 (1956), 1286.
134. Woerner, L. and M. Harmony. — J. Chem. Phys. 45 (1966), 2339.
135. Tokuhiro, T. — Bull. Tokyo Inst. Technol., No. 57 (1964), 31.
136. Zeil, W. — Angew. Chem. 76 (1964), 654.
137. Khotsyanova, T. A., T. A. Babushkina, and G. K. Semin. — Zh. Strukt. Khim. 7 (1966), 634.
138. Semin, G. K., V. I. Robas, G. G. Yakobson, and V. D. Shteingarts. — In: "Radiospektroskopiya tverdogo tela," p. 239. Atomizdat, 1967.
139. Semin, G. K., V. I. Robas, G. G. Yakobson, and V. D. Shteingarts. — Zh. Strukt. Khim. 6 (1965), 160.
140. Babushkina, T. A., V. I. Robas, and G. K. Semin. — In: "Radiospektroskopiya tverdogo tela," p. 221. Atomizdat. 1967.
141. Bucci, P. and R. Viani. — Gazz. chim. ital. 89 (1959), 1066.
142. Douglass, D. — J. Chem. Phys. 32 (1960), 1882.
143. Richardson, C. — J. Chem. Phys. 38 (1963), 510. Richardson, C. — Acta crystallogr. 16 (1963), 1063.
144. Khotsyanova, T. L., V. I. Robas, and G. K. Semin. — Zh. Strukt. Khim. 5 (1964), 644.
145. Semin, G. K., V. I. Robas, L. S. Kobrina, and G. G. Yakobson. — Zh. Strukt. Khim. 5 (1964), 915.
146. Babushkina, T. A., V. M. Kozhin, V. I. Robas, I. A. Safin, G. K. Semin, and T. L. Khotsyanova. — Kristallografiya 12 (1967), 143.
147. Bregadze, V. I., T. A. Babushkina, O. Yu. Okhlobystin, and G. K. Semin. — Teor. Eksperim. Khim. 3 (1967), 547.
148. Kojima, S. — J. Jap. Chem. (Kagaku-no-ryoiki) 13, No. 10, 1962.
149. Weatherly, T., E. Davidson, and Q. Williams. — J. Chem. Phys. 21 (1953), 761.
150. Duchesne, J. and A. Monfils. — J. Chem. Phys. 22 (1954), 562.
151. Weatherly, T., E. Davidson, and Q. Williams. — J. Chem. Phys. 21 (1953), 2073.
152. Bray, P. and D. Esteva. — J. Chem. Phys. 22 (1954), 570.
153. Sakurai, T. — Acta crystallogr. 15 (1962), 1164.
154. Bray, P. and R. Barnes. — J. Chem. Phys. 27 (1957), 551.
155. Fedin, E. I. and A. I. Kitaigorodskii. — Zh. Strukt. Khim. 2 (1961), 216.
156. Bray, P. — J. Chem. Phys. 23 (1955), 220.
157. Monfils, A. — Compt. rend. 241 (1955), 561.

158. Rehn, V. and C. Dean. — Bull. Am. Phys. Soc., ser. 2, 5 (1960), 345.
159. Dean, C., M. Pollak, B. Graven, and G. Jeffery. — Acta crystallogr. 11 (1958), 710.
160. Sakurai, T. — Acta crystallogr. 15 (1962), 443.
161. Johnson, B. — Nature 178 (1956), 590.
162. Morino, Y., M. Toyama, K. Ito, and S. Kyono. — Bull. Chem. Soc. Japan 35 (1962), 1667.
163. Chu, S., G. Jeffrey, and T. Sakurai. — Acta crystallogr. 15 (1962), 661.
164. Williams, Q. and T. Weatherly. — J. Chem. Phys. 22 (1954), 572.
165. Nagarajan, V. — Curr. Sci. 31 (1962), 233.
166. Sasikala, D. and C. Murty. — J. Phys. Soc. Japan 23 (1967), 139.
167. Negita, H., H. Yamamura, and H. Shiba. — Bull. Chem. Soc. Japan 28 (1955), 271.
168. Ramakrishna, J. — Indian J. Pure Appl. Phys. 4 (1966), 450.
169. Nagarajan, V. and C. Murty. — Curr. Sci. 32 (1962), 254.
170. Fedin, E. I. — Zh. Strukt. Khim. 1 (1960), 124.
171. Monfils, A. and J. Duchesne. — J. Chem. Phys. 22 (1954), 1275.
172. Nagarajan, V. and C. Murty. — Curr. Sci. 31 (1962), 279.
173. Johansson, K. and H. Oldenberg. — Ark. Fys. 34 (1967), 191.
174. Meal, H. — J. Am. Chem. Soc. 74 (1952), 6121.
175. Semin, G. K., T. A. Babushkina, and V. I. Robas. — Doklady AN SSSR 169 (1966), 816.
176. Dean, S. and R. Pound. — J. Chem. Phys. 20 (1952), 195.
177. Brown, R. — J. Chem. Phys. 32 (1960), 116.
178. Hirai, A. — J. Phys. Soc. Japan 15 (1960), 201.
179. Kushida, T., N. Benedek, and N. Bloembergen. — Phys. Rev. 104 (1956), 1364.
180. Duchesne, J. and A. Monfils. — Compt. rend. 238 (1954), 1801.
181. Ogawa, S. and K. Ohi. — J. Phys. Soc. Japan 15 (1960), 1064.
181a. Morino, V. and M. Toyama. — Bull. Phys. Soc. Japan 15 (1960), 288.
182. Grechishkin, V. S. and G. B. Soifer. — Fiz. Tverd. Tela 3 (1961), 2791.
183. Babushkina, T. A. and D. F. Baisa. — Fiz. Tverd. Tela 6 (1964), 2663.
184. Babushkina, T. A., D. F. Baisa, and V. I. Robas. — Zh. Strukt. Khim. 5 (1964), 136.
185. Moross, G. and H. Story. — J. Chem. Phys. 45 (1966), 3370.
186. Soutif, M. and D. Dautreppe. — J. Phys. Radium, Paris 15 (1954), 73.
187. Dautreppe, D., B. Dreyfus, and M. Soutif. — Compt. rend. 238 (1954), 2309.
188. Wang, T., C. Townes, A. Schawlow, and A. Holden. — Phys. Rev. 86 (1952), 809.
189. Kantimati, B. — Curr. Sci. 35 (1966), 359.
190. Freidlina, R. Kh., N. A. Rybakova, G. K. Semin, and E. A. Kravchenko. — Doklady AN SSSR 176 (1967), 352.
191. Ramakrishna, J. — Indian J. Pure Appl. Phys., 1 (1963), 314.
192. Pointer, R. — J. Chem. Phys. 39 (1963), 1962.
193. Grechishkin, V. S. and I. A. Kyuntsel'. — Optika Spektrosk. 15 (1963), 832.
194. Grechishkin, V. S. and I. A. Kyuntsel'. — Optika Spektrosk. 16 (1964), 161.
195. Grechishkin, V. S. and I. A. Kyuntsel'. — Trudy ENI Permsk. State Univ. 12, No. 1 (1966), 9.
196. Peterson, G. and P. Bridenbaugh. — J. Chem. Phys. 46 (1967), 2644.
197. Morino, Y., I. Miyagawa, T. Chiba, and T. Shimozawa. — J. Chem. Phys. 25 (1956), 185.
198. Duchesne, J., A. Monfils, and J. Depireux. — Compt. rend. 243 (1956), 144.
199. Morino, Y., M. Toyama, and K. Ito. — Acta crystallogr. 16 (1963), 129.
200. Kozima, K. and S. Saito. — J. Chem. Phys. 31 (1959), 560.
201. Saito, S. — J. Chem. Phys. 36 (1962), 1397.
202. Helberich, G. — Z. Naturf. 22a (1967), 761.
203. Nesmejanov, A. N., G. K. Semin, E. V. Bryuchova, T. A. Babuschkina, K. N. Anisimov, N. E. Kolobova, and Yu. V. Makarov. — Tetrahedron Lett., p. 3987, 1968.
203. Grechishkin, V. S. and G. B. Soifer. — Zh. Strukt. Khim. 4 (1963), 763.
205. Grechishkin, V. S. and G. B. Soifer. — In: "Radiospektroskopiya tverdogo tela," p. 242. Atomizdat, 1967.
206. Pavlov, B. N., I. A. Safin, G. K. Semin, E. I. Fedin, and D. Ya. Shtern. — Vest. Akad. Nauk SSSR 11 (1964), 40.
207. Grechishkin, V. S. and G. B. Soifer. — Zh. Strukt. Khim. 5 (1964), 914.
208. Myasnikova, R. M., V. I. Robas, and G. K. Semin. — Zh. Strukt. Khim. 6 (1965), 474.
209. Graybeal, J. and P. Green. 73 (1969), 2948.

210. Grechishkin, V. S. and I. A. Kyuntsel'. — Trudy ENI Perms. State Univ. 12, No. 2 (1969), 11.
211. Stone, R. and W. Flugare. — J. Molec. Spectrosc. 32 (1969), 233.
212. Voronkov, M. G., S. A. Geller, I. N. Goncharov, L. I. Mironova, and V. P. Feshin. — Izv. AN LatvSSR, chem. ser., p. 250, 1969.
213. Saha, K., R. Roy, and S. SenGupta. — Proc. Nuclear Physics and Solid State Physics Symposium, Vol. 3, 98. Bombay, 1968.
214. Smith, R. and R. West. — Tetrahedron Lett., p. 2141, 1969.
215. Voronkov, M. G., V. P. Feshin, A. P. Snyakin, V. I. Kalinin, and L. I. Zakharkin. — Chemistry of Heterocyclic Compounds, p. 565, 1970. (Russian)
216. Barton, B. — J. Chem. Phys. 51 (1969), 4670.
217. Biedenkapp, D. and A. Weiss. — J. Chem. Phys. 49 (1968), 3933.
218. Pies, W. and A. Weiss. — Z. Naturf. 24B (1969), 1268.
219. Segel, S. and R. Barnes. Catalog of Nuclear Quadrupole Interaction and Resonance Frequencies in Solids. — US Atomic Energy Commission, 1965.
220. Babushkina, T. A., A. P. Zhukov, L. S. Kobrina, G. K. Semin, and G. G. Yakobson. — Izv. Sib. Otdel. AN SSSR, ser. chem. sci., No. 12, Issue 5 (1969), 93.
221. Dinesh and P. Narasimhan. — J. Chem. Phys. 49 (1968), 2519.
222. Kantimati, B. — Indian J. Pure Appl. Phys. 7 (1969), 238.
223. Bucci, P., P. Cecchi, and A. Colligiani. — J. Chem. Phys. 50 (1969), 530.
224. Grechishkin, V. S., I. V. Izmest'ev, and G. B. Soifer. — Zh. Fiz. Khim. 43 (1969), 757.
225. Semin, G. K., T. A. Babushkina, V. M. Vlasov, and G. G. Yakobson. — Izv. Sib. Otdel. AN SSSR, ser. chem. sci., No. 12, Issue 5 (1969), 99.
226. Maksyutin, Yu. K., T. A. Babuschkina, E. N. Guryanova, and G. K. Semin. — Theoret. chim. Acta 14 (1969), 48.
227. Maksyutin, Yu. K., T. A. Babushkina, E. N. Gur'yanova, and G. K. Semin. — Zh. Strukt. Khim. 10 (1969), 1025.
228. Babushkina, T. A. and M. I. Kalinkin. — Izv. AN SSSR, chem. ser., p. 157, 1970.
229. Nesmeyanov, A. N., O. Yu. Okhlobystin, E. V. Bryukhova, V. I. Bregadze, D. N. Kravtsov, B. A. Faingor, L. S. Golovchenko, and G. K. Semin. — Izv. AN SSSR, chem. ser., p. 1928, 1969.
230. Upadysheva, A. V., T. A. Babushkina, E. V. Bryukhova, V. I. Robas, L. A. Kazitsyna, and G. K. Semin. — Izv. AN SSSR, chem. ser., p. 2068, 1969.
231. Semin, G. K., L. S. Kobrina, and G. G. Yakobson. — Izv. Sib. Otdel. AN SSSR, ser. chem. sci., No. 9, Issue 4 (1968), 84.
232. Semin, G. K., T. A. Babushkina, L. S. Kobrina, and G. G. Yakobson. — Izv. Sib. Otdel. AN SSSR, ser. chem. sci., No. 12, Issue 5 (1968), 69.
233. Semin, G. K., T. A. Babushkina, L. S. Kobrina, and G. G. Yakobson. — Ibid., p. 63.
234. Semin, G. K., T. A. Babushkina, L. S. Kobrina, and G. G. Yakobson. — Ibid., p. 73.
235. Sokolov, S. D., T. A. Babushkina, and G. K. Semin. — Izv. AN SSSR, chem. ser., p. 2065, 1969.
236. Bryukhova, E. V., F. K. Velichko, and G. K. Semin. — Izv. AN SSSR, chem. ser., p. 960, 1969.
237. Okhlobystina, L. V., T. A. Babushkina, V. M. Khutoretskii, and G. K. Semin. — Izv. AN SSSR, chem. ser., p. 1899, 1970.
238. Semin, G. K., A. A. Neimysheva, and T. A. Babushkina. — Izv. AN SSSR, chem. ser., p. 486, 1970.
239. Bryukhova, E. V., O. Yu. Okhlobystin, M. A. Kashutina, T. A. Babushkina, and G. K. Semin. — Doklady AN SSSR 183 (1968), 827.
240. Maksyutin, Yu. K., V. P. Makridin, E. N. Gur'yanova, and G. K. Semin. — Izv. AN SSSR, chem. ser., p. 1634, 1970.
241. Bryukhova, E. V., T. A. Babushkina, V. I. Svergun, and G. K. Semin, — Uchen. Zap. Mosc. Oblast Pedag. Inst. im. N. K. Krupskaya 222 (1969), 64.
242. Englin, B. A., T. A. Onishchenko, V. A. Valovoi, T. A. Babushkina, G. K. Semin, L. G. Zelinskaya, and R. Kh. Freidlina. — Izv. AN SSSR, chem. ser., p. 332, 1969.
243. Englin, B. A., B. N. Osipov, V. A. Valovoi, T. A. Babushkina, G. K. Semin, V. B. Bondarev, and R. Kh. Freidlina. — Izv. AN SSSR, chem. ser., p. 1251, 1968.

Group VII elements. ^{79}Br and ^{127}I

The data were obtained at 77°K unless stated otherwise.

Empirical formula	Structural formula	ν, MHz	s/n	References, notes
	Group I element — bromine bond			
BrD	DBr	$e^2Qq_{^{79}Br}$ = = 533 MHz $e^2Qq_{^{81}Br}$ = = 455 MHz		[1]. g
BrT	TBr	$e^2Qq_{^{79}Br}$ = = 530 MHz $e^2Qq_{^{81}Br}$ = = 443 MHz		[2]. g
BrLi	LiBr	$e^2Qq_{^{79}Br}$ = = 37,2 MHz $e^2Qq_{^{81}Br}$ = = 30.7 MHz		[3, 4]. g cf. Li
BrNa	NaBr	$e^2Qq_{^{79}Br}$ = = 58 MHz		[3, 4]. g cf. Na
BrK	KBr	$e^2Qq_{^{79}Br}$ = = 10.244 MHz $e^2Qq_{^{81}Br}$ = = 8.55 MHz		[5]. g cf. K
	Group II element—bromine bond			
Br_2Cd	$CdBr_2$	17.043	...	*[6, 7]
$C_{12}H_8Br_4CdCl_2N_4$	$[4\text{-}ClC_6H_4N_2]_2 \cdot CdBr_4$	74.26 75.90 78.01 82.42	2 2 2 2	*[189]. cf. Cl
Br_2Hg	$HgBr_2$	155,508 156.736	4 4	*[8, 170, 183]. Conversion†
$Br_2Hg + 2C_2H_6O$	$HgBr_2 \cdot 2DMSO$	153.32 161.34 169.33 169.80	8—9 7 8—10 10	*[183]. Conversion
$Br_2Hg + C_4H_{10}O_2$	$HgBr_2 \cdot DME$	140.400 141.225	10 10	*[183]
$Br_2Hg + C_3H_7NO$	$HgBr_2 \cdot DMF$	147.125 161.125 162.804	20 10 10	*[183] } Conversion

† Here and in what follows the word "conversion" means the NQR frequencies of ^{79}Br were calculated from those of ^{81}Br ($\nu_{^{79}Br} = 1.9707\ \nu_{^{81}Br}$). In this case the signal/nois ratio corresponds to the measured signal of ^{81}Br.

Empirical formula	Structural formula	ν, MHz	s/n	References, notes
$CBrCl_3Hg$	Cl_3CHgBr	146.55 156.62	1.5 1.5	*[178]. Conversion cf. C—Cl
CH_3BrHg	CH_3HgBr	121.20	15	*[9, 178]
		$e^2Qq_{79_{Br}} = 350$ MHz (g)		
		$e^2Qq_{81_{Br}} = 290$ MHz (g)		
C_2BrCl_3Hg	$Cl_2C{=}CClHgBr$	143.72 153.34	5 3	*cf. C—Cl. Conversion
$C_2H_2BrClHg$	trans-$ClHC{=}CHHgBr$	134.134	4	*[184]. cf. C—Cl
C_2H_5BrHg	C_2H_5HgBr	115.39	6	*[178]
$C_3H_{13}B_{10}BrHg$	CH₃—C——C—HgBr \O/ B₁₀H₁₀ C-methyl-C-mercurybromide- o-borane	162.37 164.52	2 2	*[178]. Conversion
$C_3H_{13}B_{10}BrHg$ + + $C_{10}H_8N_2$	CH₃—C——C—HgBr + 4,4'-bipy \O/ B₁₀H₁₀	138.72	1.5	*[178]
C_4H_9BrHg	C_4H_9HgBr	114.09	4	*[178]
$C_5H_{11}BrHg$	$C_5H_{11}HgBr$	113.87	3	*[178]
C_6BrF_5Hg	C_6F_5HgBr	147.00 142.57	8 8	*[178]. Conversion
$C_6H_4BrClHg$	$4\text{-}ClC_6H_4HgBr$	121.692	8	*[178]. cf. C—Cl
C_6H_4BrFHg	$4\text{-}FC_6H_4HgBr$	119.04	2—3	*[178]
$C_6H_4BrHgNO_2$	$4\text{-}NO_2C_6H_4HgBr$	130.72	1.5—2	*[178]. Conversion
C_6H_5BrHg	C_6H_5HgBr	119.34	2	*[178]
C_7H_7BrHg	$4\text{-}CH_3C_6H_4HgBr$	118.20	1.3	*[178]
C_7H_7BrHgO	$4\text{-}CH_3OC_6H_4HgBr$	118.86	6—7	*[178]
$C_7H_{15}BrHg$	$C_7H_{15}HgBr$	112.56	2	*[178]
$C_8H_{10}BrHgN$	$4\text{-}(CH_3)_2NC_6H_4HgBr$	129.60	5—6	*[178]
$C_8H_{15}B_{10}BrHg$	C₆H₅—C——C—HgBr \O/ B₁₀H₁₀ C-phenyl-C-mercurybromide- o-borane	170.63	4	*[178]. Conversion

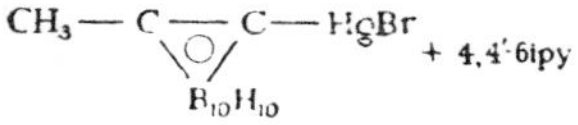

Empirical formula	Structural formula	ν, MHz	s/n	References, notes
$C_8H_{15}B_{10}BrHg$ + + $C_{10}H_8N_2$	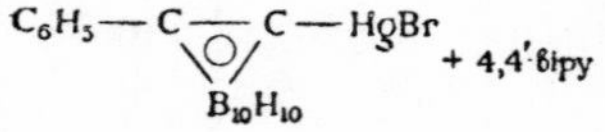	152.49 157.66	1.5 1.5	*[178]. Conversion
	Group III element — bromine bond			
BBr_3	BBr_3	175.264 175.293	. . .	[10]. $\eta = 45\%$. $e^2Qq =$ $= 340$ MHz
B_2BrH_5	B_2BrH_5	$e^2Qq_{^{79}Br} =$ $= 293$ MHz $e^2Qq_{^{81}Br} =$ $= 244$ MHz		[11]. g
$AlBr_3$	$AlBr_3(Al_2Br_6)$	97.945 113.790 115.450 At 295 °K $\eta_1 = 24.8\%$; $\eta_2 = 7.3\%$; $\eta_3 = 10.6\%$	. . .	[12, 171]. cf. Al
$AlBr_3 + C_2H_6S$	$AlBr_3 \cdot (CH_3)_2S$	95.920 99.376	30 15	*[13]
$AlBr_3 + C_4H_{10}O$	$AlBr_3 \cdot O(C_2H_5)_2$	98.192 100.512 101.016	40 35 40	*[13]
$AlBr_3 + C_5H_5N$	$AlBr_3 \cdot Py$	97.318 98.401 99.484	3 3 3	*[13]
$AlBr_3 + C_8H_{10}O$	$AlBr_3 \cdot C_2H_5OC_6H_5$	97.257 103.972 106.688	15 18 15	*[13]
$AlBr_3 + C_{12}H_{10}O$	$AlBr_3 \cdot (C_6H_5)_2O$	99.92 103.72 105.95	4 4 4	*[13]
BrGa	GaBr	$e^2Qq_{^{79}Br} =$ $= -134$ MHz		[14a]. g cf. Ga
Br_3Ga	$GaBr_3$	120.602 120.635 168.375 168.430 168.560 168.606 (Conversion)	. . .	[14]. cf. Ga
BrIn	InBr	$e^2Qq_{^{79}Br} =$ $= -138$ MHz		[14a]. g cf. In

Empirical formula	Structural formula	ν, MHz	s/n	References, notes
Br_3In	$InBr_3$	104.112 128.110 128.608	. . .	[14, 15]. cf. In
BrTl	TlBr	$e^2Qq_{^{79}Br} =$ $= -130$ MHz		[14a] g
Br_3Tl	$TlBr_3$	125.9 125.7	. . .	[73a]. $T = 297$ °K
Br_3Nd	$NdBr_3$	53.373 28.185	1 2	[16]. Room temp.
		$e^2Qq^{81}{}_{Br} = 89.08$ MHz; $\eta = 68\%$ $e^2Qq_{^{81}Br} = 45.35$ MHz; $\eta = 48.7\%$		
$Br_4O_2Rb_2U + 2H_2O$	$Rb_2(UO_2Br_4) \cdot 2H_2O$	52.30 45.75	. . .	[185]
	Group IV element — bromine bond			
$CBrF_3$	CF_3Br	301.984	. . .	[17—19]. Conversion
		$e^2Qq_{^{79}Br} = 619$ MHz (g) $e^2Qq_{^{81}Br} = 517$ MHz (g)		
CBrN	BrCN	354.637	. . .	[20—26]. Conversion cf. N
		$e^2Qq_{^{79}Br} = 686.1$ MHz (g) $e^2Qq_{^{81}Br} = 572.27$ MHz (g)		
$CBrN_3O_6$	$BrC(NO_2)_3$	359.394	4	*[27]
CBr_2F_2	CF_2Br_2	301.219 306.076	. . .	[19]. Conversion
$CBr_2N_2O_4$	$Br_2C(NO_2)_2$	344.913	7	*[27]
CBr_3F	$CFBr_3$	307.166 314.084	. . .	[19]. Conversion
CBr_4	CBr_4	318.01 318.33 319.14 319.33 319.41 319.46 319.96 320.83 320.91 321.14	10 10 10 10 10 10 10 20 10 10	*[20, 27—31]

Empirical formula	Structural formula	ν, MHz	s/n	References, notes
CBr_4 (contd.)		321.36 322.38 322.59 322.81	10 10 20 10	
$CBr_4 + C_6H_6$	$CBr_4 \cdot C_6H_6$	319.11 321.82	...	[32]. Conversion
$CBr_4 + C_7H_9N$	CBr_4 + 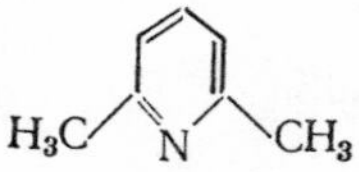	319.53 322.47	...	The same
$CBr_4 + C_8H_{10}$	$CBr_4 \cdot 1,2\text{-}(CH_3)_2C_6H_4$	321.24 321.85 322.52	...	»
	$CBr_4 \cdot 1,4\text{-}(CH_3)_2C_6H_4$	317.00 322.09	...	[30, 32]
$CBr_4 + C_8H_{11}N$	$CBr_4 \cdot$ 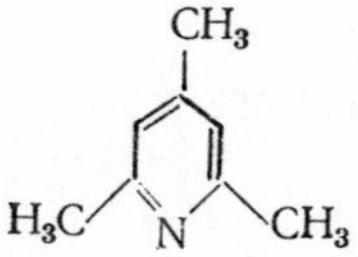	316.78 317.54 318.74	...	[32]. Conversion
$CBr_4 + C_9H_{12}$	$CBr_4 \cdot 1,3,5\text{-}(CH_3)_3C_6H_3$	319.35 320.73 321.51	...	The same
$CBr_4 + C_{10}H_{14}$	$CBr_4 \cdot 1,2,4,5\text{-}(CH_3)_4C_6H_2$	320.40 322.44	...	»
$CHBr_3$	$CHBr_3$	299.906 300.072 301.633	...	[19, 33]. Conversion
CH_2BrF	CH_2BrF	$e^2Qq_{^{79}Br}$ = 547.9 MHz $e^2Qq_{^{81}Br}$ = 459.3 MHz		[34]. g
CH_2Br_2	CH_2Br_2			*[19, 28, 29, 35]
	α-phase	282.072	3	
	β-phase	280.899 281.099	6 20	
CH_3Br	CH_3Br	264.508	...	[19, 28, 36—42]
		$e^2Qq_{^{79}Br}$ = 577.15 MHz (g); $e^2Qq_{^{81}Br}$ = 482.16 MHz (g)		

Empirical formula	Structural formula	ν, MHz	s/n	References, notes
CD_3Br	CD_3Br	264.026 $e^2Qq_{^{79}Br} = 574.6$ MHz (g); $e^2Qq_{^{81}Br} = 479.8$ MHz (g)	...	[37, 43]
C_2Br_4	$Br_2C{=}CBr_2$	303.46 307.05 309.02 311.24	...	*[191]. Conversion
C_2Br_6	Br_3CCBr_3	320.629 322.980 324.866 324.914	5 10 5 10	*[44]
$C_2HBr_3O_2$	Br_3CCOOH	319.870 322.939 323.813	4 4 6	*
C_2HBr_5	Br_2HCCBr_3	317.249 316.197 312.917 305.002 303.877	7 5 6 7 7	*[29, 44]
$C_2H_2Br_2O$	$BrCH_2COBr$	292.50 234.87		[73a]
$C_2H_2Br_4$	$Br_2CHCHBr_2$	300.5	...	*[73a]. $\Delta\nu \sim 3.5$ MHz
C_2H_3Br C_2H_2DBr	$H_2C{=}CHBr$ $H_2C{=}CDBr$	$e^2Qq_{^{79}Br} = 550.74$ MHz; $\eta = 7.5\%$		[11, 45, 46]. g
C_2HD_2Br	$D_2C{=}CHBr$	$e^2Qq_{^{79}Br} = 558.2$ MHz; $\eta = 8.8\%$		[45]. g
C_2H_3BrO	CH_3COBr	213.182 214.922 215.646 217.094	7 7 7 7	[47, 48]. Conversion $e^2Qq_{^{79}Br} = 464$ MHz (g) $e^2Qq_{^{81}Br} = 388$ MHz (g) $\eta = 21{,}1\%$ (g)
$C_2H_3BrO_2$	$BrCH_2COOH$	284.760 287.014	15 15	*[29, 47, 49—51]
$C_2H_4Br_2$	$BrCH_2CH_2Br$	259.502 262.080	9 9	*[19, 28, 44, 49, 52]

Empirical formula	Structural formula	ν, MHz	s/n	References, notes
$C_2H_4Br_2$	CH_3CHBr_2	232.0	...	[53]. Conversion $T = 90\ °K$
C_2H_5Br	CH_3CH_2Br	248.745	...	[19, 28, 33, 54, 55].
		$e^2Qq_{^{79}Br} = 541$ MHz (g) $e^2Qq_{^{81}Br} = 450$ MHz (g); $\eta = 1.3\%$ (g)		
$C_2H_6BrN + HBr$	$BrCH_2CH_2NH_2 \cdot HBr$	263.874	10	[56]
$C_3H_2Br_2N_4O_8$	$BrC(NO_2)_2CH_2CBr(NO_2)_2$	335.552	6	*[27, 29, 57]
C_3H_3Br	$BrC{\equiv}CCH_3$	$e^2Qq_{^{79}Br} =$ $= 647$ MHz $e^2Qq_{^{81}Br} =$ $= 539$ MHz		[18]. g
	$BrCH_2C{\equiv}CH$	$e^2Qq_{^{79}Br} =$ $= 587$ MHz $\eta_{^{79}Br} = 2.6\%$ $e^2Qq_{^{81}Br} =$ $= 484$ MHz $\eta_{^{81}Br} = 1.7\%$		[58, 59]. g
C_3H_4BrN	$BrCH_2CH_2CN$	261.822	20	*
$C_3H_5Br_3$	$BrCH_2CHBrCH_2Br$	265.250 268.634 269.653	5—6 7—8 7—8	*[57]
$C_3H_5BrO_2$	$BrCH_2CH_2COOH$	260.010	2	*[29, 56]
	$H_3CCHBrCOOH$	276.74	1.5	*[29]
$C_3H_5BrN_2O_4$	$BrC(NO_2)_2CH_2CH_3$	324.504	6	*[27, 29]
C_3H_6BrNO	$BrCH_2CH_2CONH_2$	255.874	8	*
C_3H_7Br	$CH_3CH_2CH_2Br$	251.72	...	[19, 33]. Conversion
	$(CH_3)_2CHBr$	$e^2Qq_{^{79}Br} =$ $= 529.8$ MHz (g) $\eta_{^{79}Br} = 0$ $e^2Qq_{^{81}Br} =$ $= 514.2$ MHz (g) $\eta_{^{81}Br} = 3.03\%$		[60]
$C_4H_7BrO_2$	$BrCH_2CH_2COOCH_3$	259.629	3	*[56]
	$CH_3COOCH_2CH_2Br$	261.785	~3	

Empirical formula	Structural formula	ν, MHz	s/n	References, notes
$C_4H_8Br_2$	$Br(CH_2)_4Br$	251.117	~1	[56]. Conversion
C_4H_9Br	C_4H_9Br	249.96	...	[33, 56]
	$(CH_3)_3CBr$	233.5	...	[61, 62]
		$e^2Qq_{^{79}Br} = 511.6$ MHz (g)		
		$e^2Qq_{^{81}Br} = 427.4$ MHz (g)		
$C_4H_{14}B_{10}Br_2$	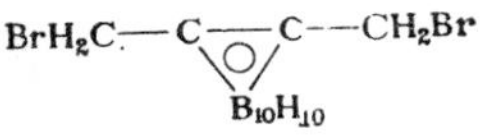 C,C-di(bromomethyl)-o-borane	286.344	10	*[63]
$C_5H_3Br_2N$	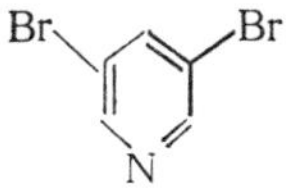	278.005	...	[64, 172]
C_5H_4BrN	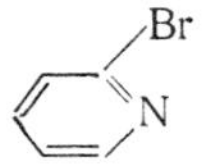	265.213	...	[64, 172]
$C_5H_6Br_2$	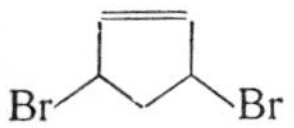 (trans)	248.538 251.920	...	[65]
$C_5H_8Br_2Cl_2$	$BrH_2CCH_2CCl_2CH_2CH_2Br$	262.662	10	*cf. C—Cl
$C_5H_8Br_4$	$BrCH_2CH_2CBr_2CH_2CH_2Br$	264.336 275.834	8 3	*
C_5H_9Br	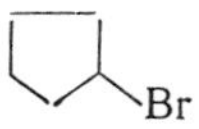	244.15	...	[65]
$C_5H_{10}Br_2$	$BrCH_2(CH_2)_3CH_2Br$	254.398	~1	[56]. Conversion
$C_6Br_4Cl_2$	1,4-$Cl_2C_6Br_4$	307.06	10	*Conversion $\Delta\nu \approx 2.5$ MHz cf. C—Cl
$C_6Br_4F_2$	1,3-$F_2C_6Br_4$	303.393 308.355 309.933	7 15 7	*[29]
	1,4-$F_2C_6Br_4$	303.959 309.894	4 4	*[29]
$C_6Br_4O_2$	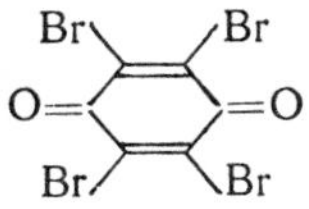	298.400 298.974	10 10	*[29, 66]

Empirical formula	Structural formula	ν, MHz	s/n	References, notes
C_6Br_5F	C_6Br_5F	303.873 303.918 305.706 305.754 306.135 306.440 307.392 307.422 309.353 309.912	3 3 3 3 3 3 3 3 3 3	*[66—68]
C_6Br_6	C_6Br_6	305.200 305.298 305.880	2 2 3	*[29, 67, 69, 70, 74]
$C_6HBr_4NO_2$	2,3,5,6-$Br_4C_6HNO_2$	299.050 301.216 306.234 307.513	3 3 3 3	*[29]
C_6HBr_5	C_6HBr_5	294.303 299.252 302.429	3 3.5 4	*[29, 67]
C_6HBr_5O	C_6Br_5OH	293.860 298.410 303.808 305.760	3 2 6 2	*[29, 56, 67]
$C_6H_2Br_2ClNO$	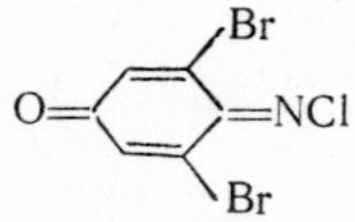	294.695 295.368	...	[64]. Conversion
$C_6H_2Br_2NNaO_3$	2,6-Br_2-4-$NO_2C_6H_2ONa$	270.32 270.71	4 4	[15]. $T = 298{,}5$ °K
$C_6H_2Br_4$	1,2,3,5-$Br_4C_6H_2$	282.75 288.79 298.56	2 4 2	*[69].
	1,2,4,5-$Br_4C_6H_2$	289.675 290.572	12 12	*[69, 71]
$C_6H_3BrN_2O_4$	2,4-$(NO_2)_2C_6H_3Br$	301.270	6	[71]
$C_6H_3Br_2NO_2$	2,5-$Br_2C_6H_3NO_2$	282.038 300.788	4 4	*[28, 29, 72]
		At 296 °K $\eta_1 = 8\%$; $\eta_2 = 10\%$		
	3,4-$Br_2C_6H_8NO_2$	288.809 290.027 291.817	4.5 2.5 2.5	*[29]

Empirical formula	Structural formula	ν, MHz	s/n	References, notes
$C_6H_3Br_2NO_3$	2,6-Br_2-4-$NO_2C_6H_2OH$	294.80 294.24 281.16 280.47	4 4 4 4	[15]. $T = 297$ °K
$C_6H_3Br_3$	1,2,4-$Br_3C_6H_3$	278.12 282.90 286.93		*[71]
	1,3,5-$Br_3C_6H_3$	279.214 280.210 280.834	10—15 10—15 10—15	*[28, 29, 71]
$C_6H_3Br_3O$	2,4,6-$Br_3C_6H_2OH$	279.818 281.471 288.905	2 2 2.5	*[15, 29]
$C_6H_4BBrF_4N_2$	[4-$BrC_6H_4N_2$]BF_4	282.427	1.5	*[29]
$C_6H_4BrClO_2S$	4-$BrC_6H_4SO_2Cl$	283.284	25	[28, 50, 51, 56]. Conversion cf. Cl
C_6H_4BrF	3-BrC_6H_4F	272.35	5	*[186]
	4-BrC_6H_4F	271.077 275.234 275.741	2—3 2—3 2—3	*[56, 73]
$C_6H_4BrNO_2$	2-$BrC_6H_4NO_2$	296.486	1.5—2	*[28, 29, 73]
	3-$BrC_6H_4NO_2$	279.673	2—3	The same
	4-$BrC_6H_4NO_2$	278.582	1.5	[15,28,29,73]
$C_6H_4Br_2$	1,2-$Br_2C_6H_4$	282.455 282.574	. . .	*[69, 73, 74, 186]
	1,3-$Br_2C_6H_4$	276.454 277.220 277.753	3.5 2.5 2.5	*[73, 186]
	1,4-$Br_2C_6H_4$	271.125	25—30	*[20, 29, 66, 71, 73, 75]
		At	299 °K	$\eta = 5\%$
$C_6H_4Br_2N_2O_2$	2,6-Br_2-4-NO_2-$C_6H_4NH_2$	282.458	10	*[29, 66]
$C_6H_4Br_2O$	2,4-$Br_2C_6H_3OH$	274.122 275.688 275.722 277.302 278.212 281.025	7 7 7 7 7 5.5	*[29, 73a]
	2,6-$Br_2C_6H_3OH$	273.090	. . .	[73a]. Conversion

Empirical formula	Structural formula	ν, MHz	s/n	References, notes
$C_6H_4Br_3N$	2,4,6-$Br_3C_6H_2NH_2$	275.448 276.233 279.598	14 14 14	[15, 50, 51, 71]. Conversion
C_6H_5Br	C_6H_5Br	268.856	5—7	[33,71,76,77]
		$e^2Qq_{^{79}Br}$ = 567 MHz (g); $\eta_{^{79}Br}$ = 4.9% (g)		
		$e^2Qq_{^{81}Br}$ = 480 MHz (g); $\eta_{^{81}Br}$ = 2,9% (g)		
$C_6H_5Br + Br_2$	$C_6H_5Br \cdot Br_2$	258.11	...	[78, 79]. cf. Br_2
C_6H_5BrO	2-BrC_6H_4OH	275.884	15	[47]
	3-BrC_6H_4OH	271.420 272.745	5—6 <2	[47]
	4-BrC_6H_4OH Phase I	268.378 269.830	12 12	[15, 28, 29, 66, 71, 73, 76, 80, 81, 173, 174, 192]. p.t. at 160 °K
	Phase II	264.733	2—4	T = 297 °K. η = 5,3%
	Phase III	271.500 274.795		
$C_6H_5Br_2N$	2,4-$Br_2C_6H_3NH_2$	270.735 272.597	4 4	*[29, 73a, 175].
		At 295 °K η_1 = 4.7%; η_2 = 5.3%		
	2,5-$Br_2C_6H_3NH_2$	269.204 264.894	...	[73a]. Conversion
	2,6-$Br_2C_6H_3NH_2$	270.606 272.012	10 10	*[29, 73a]
	3,5-$Br_2C_6H_3NH_2$	273.658 275.692	3—4 3—4	*[29]
$C_6H_5Br_2NO$	2,6-Br_2-4-$NH_2C_6H_2OH$	274.79 277.80	20 15	[15]. T = 297,7 °K
C_6H_6BrN	2-$BrC_6H_4NH_2$	263.667	3	[50, 71]. Conversion
	3-$BrC_6H_4NH_2$	266.357 266.640	5 5	*[29, 50, 71, 73]

Empirical formula	Structural formula	ν, MHz	s/n	References, notes
$_6H_6BrN$	$4\text{-}BrC_6H_4NH_2$	265.525	15	*[15, 28, 29, 50, 51, 66, 71, 73, 76, 80, 82, 173]. cf. N
		At 299 °K $\eta = 8\%$		
$_6H_6BrN + HCl$	$4\text{-}BrC_6H_4NH_2 \cdot HCl$	277.779	20	[50]. Conversion
$_6H_6Br_6$	$1,2,3,4,5,6\text{-}Br_6C_6H_6$	284.485 286.678 288.975	9 5 5	[56]. Conversion
$_6H_7BrN_2 + HCl$	$4\text{-}BrC_6H_4NHNH_2 \cdot HCl$	280.131	2	[56]
$_6H_{10}Br_2$	 (cis)	245.425 246.296	...	[65, 83]
	 (trans)	248.2	...	[65, 83]. Doublet with splitting 110 kHz
$_6H_{11}Br$	$H_2C(CH_2)_4CHBr$	242.108	...	[73a]
$_6H_{13}Br$	$CH_3(CH_2)_4CH_2Br$	249.261	2	*[56]
$_7H_3Br_3O_2$	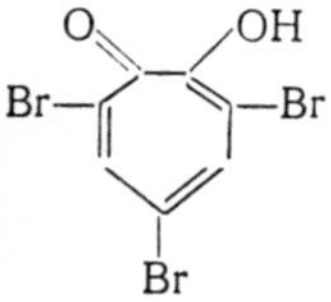	289.0	...	[84]. Conversion
$_7H_3Br_5O$	$Br_5C_6OCH_3$	300.49 302.54 302.90 304.75	2 1 1 1	*[67]
$_7H_4BrClO$	$3\text{-}BrC_6H_4COCl$	277.36	...	[73a]
$_7H_4BrNO_4$	$2\text{-}Br\text{-}3NO_2C_6H_3COOH$	302.54	1	[15]. T = 296,7 °K
$_7H_4Br_2O_2$	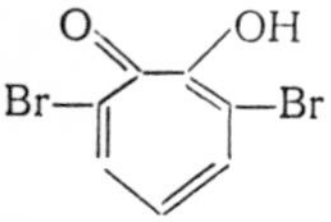	283.96 284.74	10 10	[84]. Conversion
	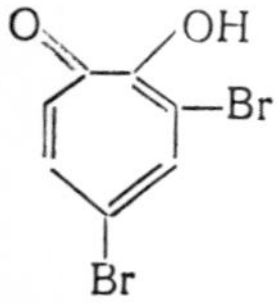	279.13 288.33	10 10	The same

Empirical formula	Structural formula	ν, MHz	s/n	References, notes
C_7H_5BrO	C_6H_5COBr	229.685	...	[47]. Conversion
$C_7H_5BrO_2$	2-BrC_6H_4COOH	285.286	2	*[29, 50, 51, 73]
	3-BrC_6H_4COOH	272.805	1.5—2	*[15]
	4-BrC_6H_4COOH	271.093	3—5	*[15, 29, 73]
	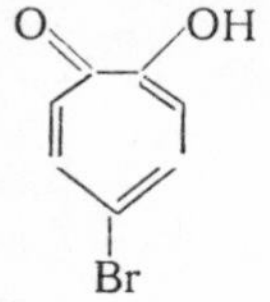	270.72	2	[84]. Conversion
	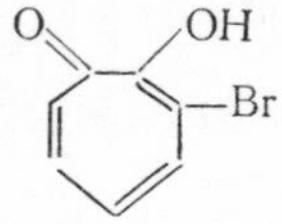	277.79	...	The same
	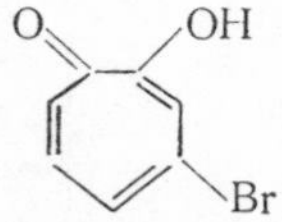	271.69	2	»
$C_7H_5BrO_3$	2-(OH)-5-BrC_6H_3COOH	271.820	6—7	[71]
$C_7H_5Br_3O$	2,4,6-$Br_3C_6H_2OCH_3$	280.632 283.888 289.194	10 10 10	[15, 71]
$C_7H_6Br_2$	2,5-$Br_2C_6H_3CH_3$	271.173 273.261	5 6	[56]
C_7H_7BrO	2-$BrC_6H_4OCH_3$	279.224	6	[50, 71]. Conversion
	4-$BrC_6H_4OCH_3$	271.371	6	The same
C_7H_8BrN	2-CH_3-4-$BrC_6H_3NH_2$	263.616	20—25	[71]
	2-Br-4-$CH_3C_6H_3NH_2$	263.222 265.974	5 5	[47]
$C_7H_{15}Br$	$CH_3(CH_2)_5CH_2Br$	249.40	~1	[56]. Conversion
C_8H_6BrClO	4-$ClC_6H_4COCH_2Br$	279.10		[50]. Conversion
$C_8H_6Br_2O$	4-$BrC_6H_4COCH_2Br$	272.428 278.002	1.5 1.3	*[29]
C_8H_7Br	$C_6H_5CH{=}CHBr$	267.364 268.043	5 5	[56]

Empirical formula	Structural formula	ν, MHz	s/n	References, notes
$_8H_7BrO$	$C_6H_5COCH_2Br$	276.118	10	[51]. Conversion
	4-$BrC_6H_4COCH_3$	276.135	4	[50, 51, 71]
$_8H_7BrO_2$	2-$BrC_6H_4COOCH_3$	286.230	4	*[73]
	3-$BrC_6H_4COOCH_3$	271.772	5—8	[71]
	4-$BrC_6H_4COOCH_3$	275.960	12—15	[71]
$_8H_8BrNO$	4-$BrC_6H_4NHCOCH_3$	271.508	8—10	[50, 71, 85]. Conversion
		At 303 °K $\eta = 6\%$		
$_8H_8Br_2$	1,4-$(BrCH_2)_2C_6H_4$	257.697	10—12	*[29, 66]
$_8H_9Br$	4-$CH_3C_6H_4CH_2Br$	254.290	8—10	[71]
	2,5-$(CH_3)_2C_6H_3Br$	263.332	...	[47]
$_8H_9BrO$	4-$BrC_6H_4OC_2H_5$	266.818	5—8	[47]
	$C_6H_5OCH_2CH_2Br$	262,548	20	[56]
$_8H_{10}BrN$	4-$BrC_6H_4N(CH_3)_2$	254.20 255.10 256.03	1.5 2 2.5	[15]. $T = 299$ °K
$_8H_{15}B_{10}Br$	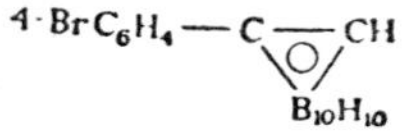 C-(4-Bromophenyl)-o-borane	273.945 274.404	4 4	*[63]
$_8H_{17}Br$	$CH_3(CH_2)_6CH_2Br$	249.285	6	[56]
$_9H_5Br_2NO$	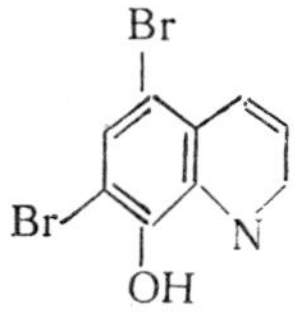	279.960 282.676	...	[86]
$_9H_6BrN$	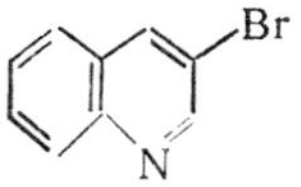	275.187	...	[64]
$_9H_{10}BrNO$	2-Br-4-$CH_3C_6H_3NHCOCH_3$	275.230	6	[56]
	4-Br-2-$CH_3C_6H_3NHCOCH_3$	269.892	12	[56]
$_9H_{11}Br$	2,4,6-$(CH_3)_3C_6H_2Br$	261,969	4	*[29]
$_{10}H_6Br_2O$	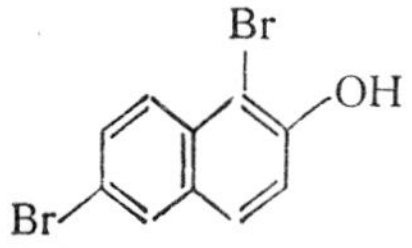	265.15 267.87	2 4	[15]. $T = 299$ °K

Empirical formula	Structural formula	ν, MHz	s/n	References, notes
$C_{10}H_7Br$	1-$BrC_{10}H_7$	267.895	6	[47]
$C_{10}H_{14}BrN$	4-$BrC_6H_4N(C_2H_5)_2$	263.444 262,906	8—10 8—10	[47]
$C_{10}H_{15}BrO$	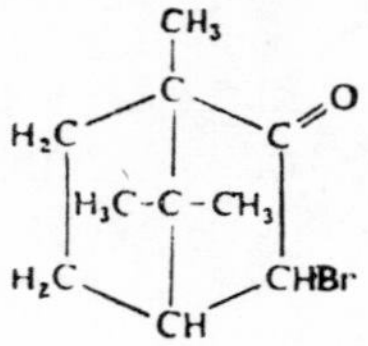	270,564 At room temp. $\eta = 10\%$	10	[56, 87]
$C_{10}H_{20}Br_2$	$BrCH_2(CH_2)_8CH_2Br$	250,737	7—8	*[29]
$C_{12}H_8Br_2$	4-$BrC_6H_4C_6H_4Br$-4′	267,608 269,610 271,290 271,978	2 2 2 2	*[28, 29, 71]
$C_{12}H_8Br_2O$	4-$BrC_6H_4OC_6H_4Br$-4′	278,461	20	[56]
$C_{12}H_9Br$	4-$BrC_6H_4C_6H_5$	268.70	1	[15]. $T=298,7$ °K
$C_{12}H_9BrO$	4-$BrC_6H_4OC_6H_5$	272,500	20	[47]
$C_{13}H_9Br_3O_2S$	2,4,6-$Br_3C_6H_2SO_2C_6H_4CH_3$-4′	286,631 292,600 300,166	7 7 7	[15, 56]. Conversion
$C_{14}H_8Br_2$	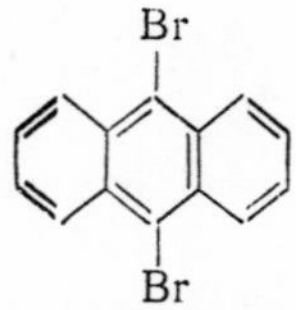	273.116 275.188	6 10	*[15, 29, 71]
$C_{15}H_{11}BrO$	$C_6H_5CH{=}CBrCOC_6H_5$	276,30	6	[15]. $T = 297,8$ °K
$C_{19}H_{15}Br$	$(C_6H_5)_3CBr$	266.351 267.266	1.3 1.5	*
$BrCl_3Si$	Cl_3SiBr	175,03	...	[90]. Conversion
BrF_3Si	F_3SiBr	$e^2Qq_{^{79}Br} = 440$ MHz $e^2Qq_{^{81}Br} = 370$ MHz		[88]. *g*

Empirical formula	Structural formula	ν, MHz	s/n	References, notes
BrH_3Si	H_3SiBr	$e^2Qq_{^{79}Br}$ = 336 MHz $e^2Qq_{^{81}Br}$ = 278 MHz		[89]. g
Br_4Si	$SiBr_4$	176.585 176.663	...	[20, 33]. Conversion
CH_3Br_3Si	CH_3SiBr_3	162.66	...	[90]. Conversion
$C_6H_5Br_3Si$	$C_6H_5SiBr_3$	163.58 163.02 164.13 164.89	1 1 2 2	The same
$C_6H_{15}BrSi$	$(C_2H_5)_3SiBr$	131.85	30	*[187]
$C_8H_{11}BrSi$	$(CH_3)_2(C_6H_5)SiBr$	137.15	8—10	*[187]
$C_{18}H_{15}BrSi$	$(C_6H_5)_3SiBr$	144.00 145.25	3 3	*[187]
BrH_3Ge	H_3GeBr	$e^2Qq_{^{79}Br}$ = 380 MHz $e^2Qq_{^{81}Br}$ = 321 MHz		[91]. g
Br_4Ge	$GeBr_4$	210.192 210.629	...	[20]. Conversion
CH_3Br_3Ge	CH_3GeBr_3	188.26	10	*[187]. Conversion
$C_2H_6Br_2Ge$	$(CH_3)_2GeBr_2$	168.79 167.35 165.12 164.66	2 3 4 3	The same
$C_6H_{15}BrGe$	$(C_2H_5)_3GeBr$			*[187]
	Phase I	140.82	20	
	Phase II	139.58 141.50	5—7 5—7	} Conversion
$C_{18}H_{15}BrGe$	$(C_6H_5)_3GeBr$	158.32 160.35	3 3	*[187]. Conversion
$CsBr_3Ge$	$CsGeBr_3$	97.84		[185]
Br_2Sn	$SnBr_2$	133.94 151.212 166.439	2 6 3	*[92] } Conversion
Br_3CsSn	$CsSnBr_3$	132.35	10	*

Empirical formula	Structural formula	ν, MHz	s/n	References, notes
Br_4Sn	$SnBr_4$	193.964 197.778 198.002 198.283	15 20 20 20	*[20, 33, 52]. Conversion
$Br_6H_8N_2Sn$	$[NH_4]_2SnBr_6$	126.53	...	[93]. T=296 °K
Br_6K_2Sn	K_2SnBr_6	129.17 130.53 132.00	...	[93]
Br_6Rb_2Sn	Rb_2SnBr_6	128,91	...	[94]
$C_4H_{10}Br_2Sn$	$Br_2Sn(C_2H_5)_2$	128.35 134 88	6 9	*[92]
$C_6H_{15}BrSn$	$BrSn(C_2H_5)_3$	118,77	6	*[92]
$C_{18}H_{15}BrSn$	$BrSn(C_6H_5)_3$	139.77 141,86	2 2	*[92]
Br_4Ti	$TiBr_4$ Phase I	46,943 47,060 47,454 47,999	50 50 40—45 40	*[95, 96, 190]. Conversion
	Phase II	46.309 47.127		
$C_2H_5Br_3OTi$	$Br_3TiOC_2H_5$	68.928 74,844 75.400 84,672 86.819	10 5 6 5 4	*[190]
$C_4H_{10}Br_2O_2Ti$	$Br_2Ti(OC_2H_5)_2$	73.38 85.03	4—5 6	*[190]
$C_5H_5Br_3Ti$	$Br_3TiC_5H_5$	66.011 66.231 66,704	20 20 20	*[190]
$C_6H_{15}BrO_3Ti$	$BrTi(OC_2H_5)_3$	71.32 72,52	5 5	*[190]. Conversion
$C_7H_{10}Br_2OTi$	$Br_2Ti(C_5H_5)(OC_2H_5)$	71.50 73,79 74,11 74,26 75,20 75.51 76,10 76.90	...	The same
$C_9H_{15}BrO_2Ti$	$BrTi(C_5H_5)_2(OC_2H_5)_2$	78.24 78,28	6 6	

Empirical formula	Structural formula	ν, MHz	s/n	References, notes
$C_{10}H_{10}Br_2Ti$	$Br_2Ti(C_5H_5)_2$	96.33 97.58 97.81 100.00	6 6 6 6	*[190]
Br_4Zr	$ZrBr_4$	60.09	...	[73a]
	Group V element — bromine bond			
C_2H_4BrNO	$CH_3CONHBr$	344.1	...	[182]
$C_4H_4BrNO_2$	$H_2CCONBrCOCH_2$ (ring)	358.73	...	[182]
C_4H_6BrNO	$H_2CCONBrCH_2CH_2$ (ring)	343.3	...	[182]
$C_6H_{10}BrNO$	$H_2C(CH_2)_4NBrCO$ (ring)	341.9	...	[182]
$C_8H_4BrNO_2$	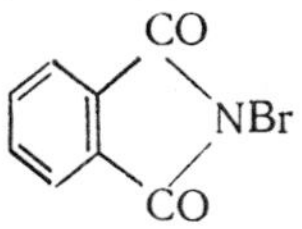	370.8	...	[182]
Br_3P	PBr_3	218.635 220.575	...	[33, 97]
Br_5P	PBr_5	238.427 237.640 236.771 236.680	...	[73a]
$AsBr_3$	$AsBr_3$	205.278 206.773 207.270	10 10 10	[98, 99]. Conversion cf. As
$BrSSb$	$SbSBr$	74.67	...	*[176, 188]. cf. Sb
Br_3Sb	$SbBr_3$ Phase I	 164.492 164.768 171.576	...	[20, 97, 100—102]. cf. Sb
	Phase II	159.81 169.896 170.733	... ...	Conversion
$Br_3Sb + {}^1/_2C_6H_6$	$2SbBr_3 \cdot C_6H_6$	160.165 170.068 170.872	...	[101—105]. cf. Sb

Empirical formula	Structural formula	ν, MHz	s/n	References, notes
$Br_3Sb + C_7H_8O$	$SbBr_3 \cdot C_6H_5OCH_3$	161.451 168.834 170.748	3 8 8	[101, 103, 104]. cf. Sb
$Br_3Sb + C_8H_{10}$	$SbBr_3 \cdot 1,2\text{-}C_6H_4(CH_3)_2$	161.338 162.358 164.268	3 3 15	The same
	$SbBr_3 \cdot C_6H_5C_2H_5$	162.894 166.520 168.286	4 4 3.5	[101—104]. cf. Sb
$Br_3Sb + {}^1/_2C_8H_{10}$	$2SbBr_3 \cdot 1,4\text{-}C_6H_4(CH_3)_2$	153.806 158.715 159.135 163.358 174.39 178.14	3 2.5 3 3.5 4 4	[101, 103, 104]. cf. Sb } Conversion
$Br_3Sb + C_8H_{10}O$	$SbBr_3 \cdot C_6H_5OC_2H_5$	163.960 166.145 169.483	4 3 5	[101, 103, 104]. cf. Sb
$Br_3Sb + {}^1/_2C_{10}H_8$	$2SbBr_3 \cdot C_{10}H_8$	175.77 167.237 164.512	1.5 2 2	The same
$Br_3Sb + {}^1/_2C_{12}H_{10}$	$2SbBr_3 \cdot C_6H_5C_6H_5$	162.305 166.683	. . .	[101, 103, 104]. $T = 292\ °K$. cf. Sb
$Br_3Sb + {}^1/_2C_{13}H_{12}$	$2SbBr_3 \cdot (C_6H_5)_2CH_2$	163.709 170.044 171.472	20 6 4	[101]. cf. Sb
$Br_3Sb + C_{14}H_{10}$	$SbBr_3 \cdot$ 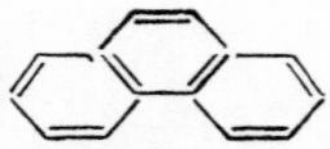	172.69 171.49 169.85	. . .	[102]. Conversion
$C_3H_9Br_2Sb$	$(CH_3)_3SbBr_2$	$e^2Qq =$ $= 192.254$ MHz		[106]. cf. Sb
$C_{12}H_{27}Br_2Sb$	$(C_4H_9)_3SbBr_2$	113.10	2	* Conversion cf. Sb
	$(iso\text{-}C_4H_9)_3SbBr_2$	110.80	3	* cf. Sb
Br_3NbO	$NbOBr_3$	72.930 73.428 73.905 94.245	4—5 10—12 3—4 10—12	*[181]

Empirical formula	Structural formula	ν, MHz	s/n	References, notes
Br_5Nb	$NbBr_5$	59.17 59.76 106.99 107.54 108.04	10 5 5 5 5	*[181]. Conversion
Br_5Ta	$TaBr_5$	64.98 65.13 65.23 65.87 66.21 109.30 109.44 109.62	10 10 20 10 10 5 5 5	*[181]
	Group VI element-bromine bond			
$AgBrO_3$	$AgBrO_3$	169.51	...	[49, 107]. $T = 195$ °K
$BaBr_2O_6 + H_2O$	$Ba(BrO_3)_2 \cdot H_2O$	176.223	20	*[49, 66, 107]
$BrCsO_3$	$CsBrO_3$	174.372	...	[20]. Conversion
$BrHgO_3$	$HgBrO_3$	168.91	...	[107]. $T = 285$ °K
$BrKO_3$	$KBrO_3$	178.800	...	*[20, 49, 107, 177]. Conversion
$BrLiO_3$	$LiBrO_3$	180.300 180.240	...	*[49, 107].
$BrNaO_3$	$NaBrO_3$	181.843	15—20	*[20, 49, 66, 107, 108, 177]. cf. Na
$Br_2CaO_6 + H_2O$	$Ca(BrO_3)_2 \cdot H_2O$	176.85	...	[49, 107]. $T = 293$ °K
$Br_2CdO_6 + H_2O$	$Cd(BrO_3)_2 \cdot H_2O$	176.25	...	[107]. $T = 195$ °K
$Br_2CoO_6 + H_2O$	$Co(BrO_3)_2 \cdot H_2O$	181.07	...	[107, 109]
$Br_2CuO_6 + 6H_2O$	$Cu(BrO_3)_2 \cdot 6H_2O$	181.82	...	[107, 109]. $T = 20$ °K
Br_2HgO_6	$Hg(BrO_3)_2$	172.62	...	[107]. $T = 285$ °K

Empirical formula	Structural formula	ν, MHz	s/n	References, notes
Br_2MgO_6	$Mg(BrO_3)_2$	176.58	...	[52]. $T = 20\,°K$
$Br_2MgO_6 + H_2O$	$Mg(BrO_3)_2 \cdot H_2O$	178.70	...	[107]. $T = 195\,°K$
$Br_2NiO_6 + 6H_2O$	$Ni(BrO_3)_2 \cdot 6H_2O$	181.23	...	[107, 109]
$Br_2O_6Pb + H_2O$	$Pb(BrO_3)_2 \cdot H_2O$	174.30	...	[107]. $T = 285\,°K$
$Br_2O_6Sr + H_2O$	$Sr(BrO_3)_2 \cdot H_2O$	173.76	...	[49, 107]. $T = 293\,°K$
$Br_2O_6Zn + 6H_2O$	$Zn(BrO_3)_2 \cdot 6H_2O$	181.34	...	[107, 109].
BrF_5S	SF_5Br	$e^2Qq_{^{79}Br} =$ $= 800$ MHz $e^2Qq_{^{81}Br} =$ $= 705$ MHz		[110]. g
Br_6K_2Se	K_2SeBr_6	173.599 173.285 172.389	...	[111]
Br_6Cs_2Se	Cs_2SeBr_6	177.44	...	[111]
$Br_6H_8N_2Se$	$[NH_4]_2SeBr_6$	172.623	...	[111]. $T = 202\,°K$
Br_6Cs_2Te	Cs_2TeBr_6	135.962	...	[112]
$Br_6H_8N_2Te$	$[NH_4]_2TeBr_6$	129.946 134.258 135.062	...	[112]
Br_6K_2Te	K_2TeBr_6	134.496 135.670 135.946	...	[112]
$C_8H_{24}Br_6N_2Te$	$[N(CH_3)_4]_2TeBr_6$	142.58	...	[112]
Br_3Cr	$CrBr_3$	102.62	...	[113]. $T = 297\,°K$

Empirical formula	Structural formula	ν, MHz	s/n	References, notes
	Group VII element — bromine bond			
BrCl	BrCl	$e^2Qq_{^{79}Br} = 876.8$ MHz $e^2Qq_{^{81}Br} = 732.9$ MHz		[114]. g cf. Cl
BrF	BrF	$e^2Qq_{^{79}Br} = 1089.0$ MHz $e^2Qq_{^{81}Br} = 909.2$ MHz		[115]. g
BrF_3	BrF_3	$e^2Qq = -1109.35$ MHz $\eta = 9.5\%$		[116]. g
BrI	IBr	$e^2Qq_{^{79}Br} = 722$ MHz $e^2Qq_{^{81}Br} = 603$ MHz		[117]. g
Br_2	Br_2	382.520	30—40	*[29, 30, 78, 79, 118—120]. Conversion
		At 253 °K $\eta = 20\%$		
$Br_2 + C_3H_6O$	$Br_2 \cdot CH_3COCH_3$	381.080	...	[78, 79]. Conversion
$Br_2 + C_4H_8O_2$	Br·diox	376.81	...	[79]
$Br_2 + C_4H_{10}O$	$Br_2 \cdot C_2H_5OC_2H_5$	383.669	...	[78, 79].
$Br_2 + C_6H_5Br$	$Br_2 \cdot C_6H_5Br$	380.442	...	[78, 79]. cf. C—Br
$Br_2 + C_6H_5Cl$	$Br_2 \cdot C_5H_6Cl$	382.453	...	[78, 79]. cf. C—Cl
$Br_2 + C_6H_5F$	$Br_2 \cdot C_6H_5F$	382.436	...	[78, 79]
$Br_2 + C_6H_6$	$Br_2 \cdot C_6H_6$	382.612	...	[30, 78, 79].
Br_3Cs	$CsBr_3$	251.7	...	[121]. $T = 273$ °K
		407.0	...	[121]. $T = 283$ °K

Empirical formula	Structural formula	ν, MHz	s/n	References, notes
Br_3H_4N	$[NH_4]Br_3$	408.2	...	[121]. $T = 283$ °K
Br_7P	PBr_7	286,6 403,0	...	The same
$C_4H_{12}Br_2IN$	$[N(CH_3)_4]IBr_2$	162,2	10	[122]
Br_6Cs_2Re	$Cs_2[ReBr_6]$	118.41	...	[123]
$Br_6H_8N_2Re$	$[NH_4]_2[ReBr_6]$	113,95	...	[123]
Br_6K_2Re	$K_2[ReBr_6]$	115.09 115,86 116,06	...	[125, 126, 128]
Br_6Rb_2Re	$Rb_2[ReBr_6]$	115,73	...	[123]

Group VIII element — bromine bond

Br_4K_2Pd	$K_2[PdBr_4]$	129,77	2	[127]. $T = 200$ °K
Br_6K_2Pd	$K_2[PdBr_6]$	205.34	8	[127]
$C_6H_{12}Br_2Pd_2$	$(CH_3{=}CH{=}CH_2)_2Pd_2Br_2$	73,776	5	*
Br_4K_2Pt	$K_2[PtBr_4]$	139,84	2—3	[127]
Br_6Cs_2Pt	$Cs_2[PtBr_6]$	208,32	...	[124]
$Br_6H_2Pt + 9H_2O$	$H_2[PtBr_6]\cdot 9H_2O$	201.52	...	[124]. $T = 292.8$ °K
$Br_6H_8N_2Pt$	$[NH_4]_2[PtBr_6]$	203.983	...	[124]
Br_6K_2Pt	$K_2[PtBr_6]$	202.25 204.11	3 15	[129]
$Br_6Na_2Pt + 6H_2O$	$Na_2[PtBr_6]\cdot 6H_2O$	207,44	...	[124]
Br_6PtRb_2	$Rb_2[PtBr_6]$	205.63	...	[124]

Empirical formula	Structural formula	T, °K	$\nu_{1/2-3/2}$, MHz	s/n	$\nu_{3/2-5/2}$, MHz	s/n	e^2Qq, MHz	η, %	References, notes
			Group I element — iodine bond						
DI	DI	—	—	—	—	—	—1823	...	[130, 131]. g
IT	TI	—	—	—	—	—	—1822.6	...	[2]. g
ILi	LiI	—	—	—	—	—	—198.2	...	[3, 4]. g cf. Li
INa	NaI	—	—	—	—	—	—259.9	...	[3, 4]. g cf. Na
IK	KI	—	—	—	—	—	—60	...	[3, 4]. g
			Group II element — iodine bond						
I_2Zn	ZnI_2	90	182.18 183.75 188.88	...	... ...	...	... ...	... ...	[132]
$C_{12}H_8CdCl_2I_4N_4$	$[4\text{-}ClC_6H_4N_2]_2CdI_4$	77	85.78 94.48 98.15	6 3 3	... ...	...	... ...	... ...	*[189]. cf. C—Cl
CdI_2	CdI_2	77	14.752	10	29.474	20	98.27	2.8	*[6, 7]
CH_3HgI	CH_3HgI	77 —	127.42 —	10 —	254.85 —	7 —	849.5 —1674 (g)	0 ...	*[133]
CH_3HgI + + $^1/_2CH_3ClHg$	$2CH_3HgI \cdot CH_3HgCl$	77	127.33	10	...	...	...	...	*[178]
$C_4H_{12}HgI_3N$	$[(CH_3)_4N]HgI_3$	77	102.04 103.72 122.80 137.54 139.04 152.94	1.5—2 1.5—2 10 10 10 10	159.03 174.96 243.60 270.00 276.30 305.78	3 3 10 10 10 10	553.0 599.8 813.0 916.7 921.9 1019.3	48.6 38.9 8.0 1.5 7.1 1.6	*
C_6H_4ClHgI	$4\text{-}ClC_6H_4HgI$	77	122.59	5	245.44	5	817.7	0	*[178]. cf. C—Cl

Empirical formula	Structural formula	T, °K	$\nu_{1/2-3/2}$, MHz	s/n	$\nu_{3/2-5/2}$, MHz	s/n	e^2Qq, MHz	η, %	References, notes
C_6H_5HgI	C_6H_5HgI	77	125,60	2—3	...	...	...	...	*[178]
C_7H_7HgI	$4\text{-}CH_3C_6H_4HgI$	77	124,50	2	...	...	...	...	*[178]
C_7H_7HgIO	$4\text{-}CH_3OC_6H_4HgI$	77	119,52	15	238.89	10	796.4	2.2	*[178]
$C_8H_{10}HgIN$	$4\text{-}(CH_3)_2NC_6H_4HgI$	77	119.65	5	239.3	2	797.5	~0	*[178]
HgI_2	HgI_2	77 249	151.625 151,25 151.12	50	237.60 230,03 229.47	20	825.0 800.7 799.0	48.1 52.5 52.7	*[132, 170]
$HgI_3K + 2H_2O$	$K[HgI_3]\cdot 2H_2O$	77	108,39 125,27 134,45	20 30 15	154.49 248.35 267.94	20 20 20	545.7 829.0 893,6	59.1 8.2 5.2	*
HgI_4K_2	$K_2[HgI_4]$	77	108.20 125,33 134.50	1,5—2 6 3	... ...	...	... ...	... ...	*[8]
			Group III element — iodine bond						
BI_3	BI_3	78	214.0	...	342,5	...	1242.4	45,5	[134, 135]. All the lines are doublets with intensity ratio 1:4
B_5H_8I	B_5H_8I	78	196,2	...	...	...	...	...	[135]
$B_{10}H_{13}I$	$B_{10}H_{13}I$	78	190,8	...	...	...	...	...	[135]
AlI_3	AlI_3	77	112,314 131.371 131.844	...	215,614 263,228 263,920	...	723.38 876.52 879.33	18.1 0 0	*[136]
GaI_3	GaI_3	77	135,719 176,496 177,438	...	253,347 352.948 354.519	...	853.92 1176.53 1186.91	23,7 0.9 2.8	[136]. cf. Ga

GaI	GaI	—	—	—	—	—	—549 (g)	...	[14a]. cf. Ga
I_3In	InI_3	77	...	...	...	...	...	...	[15, 136]. cf. In
			176.763		353,466		1178.0	3.4	
			177.146		354.580		1181.6	0	
		297	122,728	...	229.190	...	772,25	23.7	
			173.177		346.289		1154,33	1,1	
			173.633		347.234		1157.46	0	
I_3U	UI_3	6	56.197	...	109,086	...	365.35	15.4	[16]
			39,615		50,960		183.49	70,6	
ITl	TlI	—	—	—	—	—	—537 (g)	...	[14a]
Group IV element — iodine bond									
CF_3I	CF_3I	83	310.375	...	...	...	~2069	...	[18, 97, 137—139]
		—	—	—	—	—	—2143.8 (g)	...	
CHI_3 + 24S	$CHI_3 \cdot 3S_8$	77	307.106	...	613,949	...	2046.63	1.82	[97]
CH_2ClI	CH_2ClI	77	281,750	...	...	...	...	...	*[35]. cf. C—Cl
CH_2I_2	CH_2I_2	77	286.883	65	568.362	...	1897.37	2.72	*[19, 28, 35, 97]
			289.149	1.2	...		...	...	
CH_3I	CH_3I	77	265.102	...	529.670	...	1765.85	2.93	[19, 28, 37—39, 42, 76, 97, 139—141]
		—	—	—	—	—	—1934 (g)	...	
CD_3I	CD_3I	77	264.591	...	...	...	...	...	[37—39, 42, 43]
		—	—	—	—	—	—1929 (g)	...	
ClN	ICN	293	382.4	...	...	...	...	...	[23, 141]. cf. N
		—	—	—	—	—	—2420 (g)	...	
CI_4	CI_4	77	319,549	...	...	...	...	...	*[97]
$C_2F_4I_2$	$I(CF_2)_2I$	77	315.625	20	...	...	...	...	*[142]
$C_2F_4I_2 + C_4H_{11}N$	$I(CF_2)_2I \cdot NH(C_2H_5)_2$	77	308.834	10	...	...	...	...	*[142]

Empirical formula	Structural formula	T, °K	$\nu_{1/2-3/2}$, MHz	s/n	$\nu_{3/2-5/2}$, MHz	s/n	e^2Qq, MHz	η, %	References, notes
$C_2H_2Cl_2FI$	CCl_2ICH_2F	77	291.600	15	...	...	...	...	*cf. C—Cl
$C_2H_2I_2$	trans-IHC═CHI	83	277.1	...	553.82	...	1846.3	2.3	[141]
C_2H_3I	CH_2═CHI	—	—	—	—	—	—1847	4.2	[143, 179]. g
C_2H_3IO	CH_3COI	—	—	—	—	—	—1570	17.3	[144]. g
C_2D_3IO	CD_3COI	—	—	—	—	—	—1568	17.3	[144]. g
$C_2H_3IO_2$	ICH_2COOH	77	298.642	10—12	...	...	...	...	*[29, 140]
			299.483	10—12	...	...	...	...	
		83	297.86	...	592.79	...	1978.0	6.1	
			298.87		596.12		1988.0	4.6	
$C_2H_4I_2$	$I(CH_2)_2I$	77	260.702	...	...	...	...	...	*[44, 142, 145]
C_2H_5I	C_2H_5I	77	247.374	...	...	...	1649	0	[19, 28, 140, 146, 147]
		83	247.115		494.058		1647.0	1.6	
		—	—	—	—	—	—1771 (g)	20.5 (g)	
C_3H_3I	IC≡CCH_3	—	—	—	—	—	—2230	...	[18]. g
$C_3H_5IO_2$	ICH_2CH_2COOH	90	260.9	8	...	...	...	...	[76]
C_3H_7I	C_3H_7I	77	250.962	...	...	...	1673.1	...	[19, 140]
		83	250.810	...	499.810	...	1666.9	5.2	
	$(CH_3)_2CHI$	77	235.921	...	...	...	1572.8	...	[19]
$C_4F_8I_2$	$I(CF_2)_4I$	77	313.701	20	...	...	...	...	*[142]
$C_4H_4F_4I_2$	$I(CH_2)_2(CF_2)_2I$	77	276.340	10	...	...	...	...	*[142]
			317.714	10	...	...	...	...	
$C_4H_8I_2$	$I(CH_2)_4I$	77	253.622	10	...	...	...	...	*[142]
C_4H_9I	C_4H_9I	83	249.062	...	497.539	...	1658.8	3.0	[140]
$C_6F_{12}I_2$	$I(CF_2)_6I$	77	313.749	5	...	...	...	...	*[142]
$C_6H_3I_2NO_3$	2,6-I_2-4-$NO_2C_6H_2OH$	77	297.698	2—3	...	...	...	...	*[29, 148]

			308.762	2—3	...	...	...	...	
		300	293.733	...	587.05	...	1957.03	2.4	
			305.086	...	598.50		2001.01	12.3	
$C_6H_3I_3$	1,3,5-$I_3C_6H_3$	77	286.111	...	...	...	...	...	*[29, 186]
			285.734		...		...	...	
			280.478		...		...	...	
C_6H_4FI	2-IC_6H_4F	77	290.50	20	...	...	...	...	*[186]
	3-IC_6H_4F	77	281.19	3	...	...	...	...	*[186]
	4-IC_6H_4F	77	280.872	2	552.8	4	1847.3	11.2	*[29, 73]
$C_6H_4INO_2$	2-$IC_6H_4NO_2$	299.3	315.19	6	613.88	3	2055.0	14.47	*[15]
			315.97	6	615.26	3	2059.9	14.54	
	3-$IC_6H_4NO_2$	77	291.160	...	...	...	...	...	*[15, 51, 76]
		300	285.63	20	570.29	15	1901.5	5.05	
			286.66	20	571.43	15	1905.8	3.65	
	4-$IC_6H_4NO_2$	77	288.410	1.5	...	...	...	...	*[15, 29, 73]
		300.7	284.36	2	565.82	2	1887.7	6.30	
$C_6H_4I_2$	1,2-$I_2C_6H_4$	77	284.036	...	564.54	...	1888.3	8.8	*[29, 69, 74, 149]
			285.531		565.20				
			286.285		566.08				
			286.923		566.33				
	1,3-$I_2C_6H_4$	77	281.579	10	562.1	5	1874.2	3.8	*[28, 29]
	1,4-$I_2C_6H_4$	77	280.147	...	559.104	...	1864.3	4.1	*[28, 29, 15, 73, 74, 149]
$C_6H_4I_2N_2O_2$	2,6-I_2-4-$NO_2C_6H_2NH_2$	77	289.681	3	...	...	...	...	*[29]
		300	286.246	1.5	571.69	8	1906.02	3.22	
$C_6H_4I_3N$	2,4,6-$I_3C_6H_2NH_2$	77	278.254	2—3	...	...	...	...	*[29]
			282.719	2—3	...	...	...	...	
			283.820	2—3	...	...	...	...	
C_6H_5I	C_6H_5I	77	274.465	8	545.65	10	1820.4	6.9	*[28, 29, 66, 73, 150]
			274.978	8	546.60	10	1823.7	6.9	
$C_6H_5IN_2O_2$	2-I-4-$NO_2C_6H_3NH_2$	77	283.674	3	...	...	...	...	*[29]

Empirical formula	Structural formula	T, °K	$\nu_{1/2-3/2}$, MHz	s/n	$\nu_{3/2-5/2}$, MHz	s/n	e^2Qq, MHz	η, %	References, notes
C_6H_5IO	2-IC_6H_4OH	77	282.20	10	560.6	5	1870.7	7.2	*[15]
	3-IC_6H_4OH	77	277.80	10	553.8	10	1846.9	5.0	*[15, 28]
	4-IC_6H_4OH	77	273.708	10	546.8	10	1823.1	2.9	*[28, 29, 73, 192]
			274.777	10	549.0	10	1830.2	2.8	
$C_6H_5IO_2$	$C_6H_5IO_2$	294.7	162.23	2	311.70	3	1045.9	17.9	[15]
C_6H_6IN	2-$IC_6H_4NH_2$	77	267.320	2.5	534.0	2	1780.4	3.0	*[28, 29, 73]
	3-$IC_6H_4NH_2$	77	270.65	20	...	...	...	...	*[73]
			270.39	20	...	...	...	...	
	4-$IC_6H_4NH_2$	77	264.280	15	527.3	10	1758.3	4.3	*[15, 28, 29, 66, 73] cf. N
$C_6H_6IN + HCl$	4-$IC_6H_4NH_2 \cdot HCl$	296	288.28	35	572.15	12	1909.5	7.71	[15]
$C_6H_8F_4I_2$	$CF_2CH_2CH_2I$ | $CF_2CH_2CH_2I$	77	267.712	10	...	...	...	...	*[142]
$C_6H_{10}I_2$	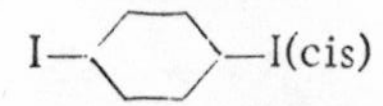	77	244.848	...	...	...	...	...	[65, 83]
			248.458		...		...	...	
	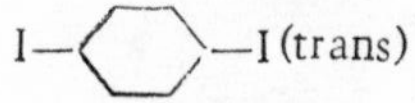	77	246.890	...	...	...	...	...	[65, 83]
$C_6H_{11}I$	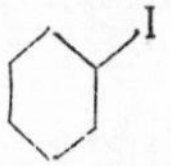	77	240.043	...	...	...	...	...	[65]
C_6I_6	C_6I_6	77	309.068	10	610.32	3	2038.4	10.0	*[29, 70]
			311.678	10	610.50	3	2041.7	12.8	
			312.642	10	612.72	3	2049.0	12.6	
$C_7H_3F_3I_2$	2,5-$I_2C_6H_3CF_3$	83	287.2	...	...	...	...	...	[28]
C_7H_4IN	2-IC_6H_4CN	77	285.250	1.2	...	...	...	...	*[29, 73]
		300	282.210	2	559.40	10	1867.29	8.32	

$C_7H_4INaO_2$	2-IC_6H_4COONa	296.3	294.81	18	...	...	...	...	[15]
	3-IC_6H_4COONa	298	272.73	18	545.17	2	1817.4	2.03	[15]
			273.76	12	547.42	4	1824.8	1.27	
	4-IC_6H_4COONa	295.7	277.59	...	...	...	...	...	[15]
			280.82		...		...	...	
			282.25		...		...	...	
$C_7H_4I_2O_3$	2-OH-3,5-$I_2C_6H_2COOH$	296.6	284.19	2	...	...	...	...	[15]
			295.14	2	...		...	...	
$C_7H_5IO_2$	2-IC_6H_4COOH	77	297.791	10	579.1	10	1939.38	14.7	*[15, 51]
	3-IC_6H_4COOH	77	278.570	10	...	...	...	...	[15, 28, 51]
		301	275.67	3	548.56	2	1830.1	6.27	
	4-IC_6H_4COOH	77	275.824	50	549.8	10	1833.6	5.1	*[15]
C_7H_7IO	2-$IC_6H_4OCH_3$	77	287.216	...	...	...	...	...	[73a]
	3-$IC_6H_4OCH_3$	77	274.593	...	...	...	...	...	[73a]
C_8H_7IO	4-$IC_6H_4COCH_3$	77	288.84	1.5	...	...	...	...	*
$C_8H_7IO_2$	3-$IC_6H_4COOCH_3$	301	276.24	80	550.16	80	1835.0	5.71	[15]
	4-$IC_6H_4COOCH_3$	301	284.03	20	564.21	12	1882.7	7.26	[15]
C_8H_8INO	4-$IC_6H_4NHCOCH_3$	301.3	273.74	5	546.06	4	1820.9	4.49	[15, 28]
C_8H_9I	2,5-$(CH_3)_2C_6H_3I$	77	270.770	...	...	...	...	...	[73a]
	3,5-$(CH_3)_2C_6H_3I$	77	273.353	...	...	...	...	...	[73a]
$C_8I_4O_3$	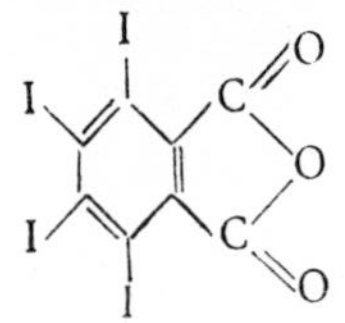	77	316.578	7	603.9	3	2027.9	19.1	*[15]
			319.347	7	605.5	3	2035.0	20.3	
			327.829	5	613.1	3	2064.9	23.0	
			332.358	5	650.7	3	2175.7	12.4	
$C_9H_7INNaO_3$	2-$IC_6H_4CONHCH_2COONa$	77	291.000	1.2	...	...	...	...	*

Empirical formula	Structural formula	T, °K	$\nu_{1/2-3/2}$, MHz	s/n	$\nu_{3/2-5/2}$, MHz	s/n	e^2Qq, MHz	η, %	References, notes
$C_9H_8INO_3$	2-$IC_6H_4CONHCH_2COOH$	77	291.228	1,2	...	...	...	...	*
			290.465	5	...		...	...	
$C_{12}H_9I$	4-$IC_6H_4C_6H_5$	300.4	275.40	1	...	...	...	...	[15]
C_3H_9ISi	$(CH_3)_3SiI$	77	143.22	30	286.30	10	954.41	2	*[187]
$C_6H_{15}ISi$	$(C_2H_5)_3SiI$	77	142.25	100	284.50	10	948.33	0	*[187]
$C_8H_{11}ISi$	$(CH_3)_2(C_6H_5)SiI$	77	145.97	30	284,45	10	971.34	4.2	*[187]
F_3ISi	SiF_3I	—	—	—	—	—	—1450 (g)	...	[151]
I_4Si	SiI_4	77	198.736	...	397.430	...	1324.79	0.90	[97]
			199.999		399.993		1333.31	0.33	
$C_2H_6GeI_2$	$(CH_3)_2GeI_2$	77	178.65	50	...	...	...	...	*[187]
GeI_4	GeI_4	77	222.672	9	445.294	...	1484.34	0.93	*[20, 97]
			225.088	3	450.173		1500.58	0.22	
$C_4H_{10}I_2Sn$	$(C_2H_5)_2SnI_2$	77	149.40	10	293.30	10	980.45	12.1	*[92]
$C_{18}H_{15}ISn$	$(C_6H_5)_3SnI$	77	153.75	2	...	...	...	...	*[92]
I_4Sn	SnI_4	77	207.683	15	415.320	10	1384.42	0.92	*[20, 52, 92, 97, 100, 141, 180]
			209.133	5	418.257	3	1394.19	0.40	
		77	138.678†	...	208.025††	...	970.79	0.9	[152, 153]
			139.660†		209.492††		977.64	0.5	
$I_4Sn + 16S$	$SnI_4 \cdot 2S_8$	77	207.458	...	414.907	...	1383.03	0.2	[100]
			210.746		420.675		1402.68	3.9	
$I_4Sn + 32S$	$SnI_4 \cdot 4S_8$	77	204.64	...	409.35	...	1364.3	0.0	[100]
			205.25		410.50		1368.3	0.0	
			210.99		421.40		1405.0	3.0	
			219.04		...		...	...	
I_6Pb_2Sn	Pb_2SnI_6	77	132.57	...	...	...	...	...	[94]
I_4Ti	TiI_4	77	44.856	20	89.780	30	299.2	~0	*
			43,650	7	87.298	10	291.0	~0	

Group V element — iodine bond

AsI_3	AsI_3	77 383	207.023 206.20 208.45 211.74	... ...	395.979 403.16 406.96 412.46	... ...	1330.23 1348.53 1361.48 1380.50	18.91 13.1 13.5 14.1	[98, 100, 140]. p.t. at 365°K cf. As
$AsI_3 + 24S$	$AsI_3 \cdot 3S_8$	77	227.582	...	455.114	...	1517.07	0.89	[97, 100]. cf. As
ISSb	SbSI	77	137.38	10	186.2	10	664.2	64.7	*[188]. cf. Sb
$C_3H_9I_2Sb$	$(CH_3)_3SbI_2$	...	...	...	...	...	845.40	...	[106]. cf. Sb
I_3Sb	SbI_3	77	174.356	...	254.637	...	895.83	56.5	[20, 99, 100, 154]. cf. Sb
$I_3Sb + 24S$	$SbI_3 \cdot 3S_8$	77	184.155	...	367.830	...	1226.35	3.2	[97, 100]. cf. Sb
BiIS	BiSI	77	51.80	10	62.05	15	224.08	78.8	*[188]. cf. Bi
BiI_3	BiI_3	77	111.320	3	201.380	10	682.18	29.0	[154, 155]

Group VI element — iodine bond

$CaI_2O_6 + 6H_2O$	$Ca(IO_3)_2 \cdot 6H_2O$	77 297	159.49 150.75 150.97 ...	10 4 4 ...	... 297.470 298.709 307.612	... 5 8 2	... ...	... ...	*[15]
$CsIO_3$	$CsIO_3$	77	146.91	20	293.75	5	979.2	1.4	*[156]
$CuI_2O_6 + 2H_2O$	$Cu(IO_3)_2 \cdot 2H_2O$	77	154.5	...	...	...	...	...	[157]
F_5IO	IOF_5	—	—	—	—	—	552.9	—	[158]. g

† Data for ^{129}I; transition $3/2 - 5/2$.
†† Data for ^{129}I; transition $5/2 - 7/2$.

Empirical formula	Structural formula	T, °K	$\nu_{1/2-3/2}$, MHz	s/n	$\nu_{3/2-5/2}$, MHz	s/n	e^2Qq, MHz	η, %	References, notes
HIO_3	HIO_3	77	203.12	50	330,84	30	1141.0	43,37	*[15, 159—161]. At 297.1°K transition ($\pm^1/_2-\pm^5/_2$=528.746)
H_4INO_3	NH_4IO_3	77	150,48	50	300,75	20	1002,6	2,2	*[156]
IKO_3	KIO_3	77	149,200 149,275 149,950 150.075	50 50 20 20	298.200 298,350 299,475 299,925	4 4 3 3	994.1 994.6 998,5 999,9	2,2 2.2 3,3 2.4	*[15, 156, 162]. p.t. at 348 and 493 °K
$ILiO_3$	$LiIO_3$	77	153,488	40	306.615	20	1022.4	2.8	*[156]. cf. Li
$INaO_3$	$NaIO_3$	77	151,44	10	300.48	3	1002,8	7.9	*[156]
IO_3Rb	$RbIO_3$	77	148,75	50	297.43	10	991.5	1.4	*[156, 157]
$I_2NiO_6 + 2H_2O$	$Ni(IO_3)_2 \cdot 2H_2O$	4,2	148.97	...	297.05	...	990.64	4.8	[109, 163]
I_2O_5	I_2O_5	77	177.96 203,08 210,3	10 10 10	308.10 316,32 318,49	10 10 10	1052.2 1099.6 1112.7	35,3 48.8 52.1	*[49, 132]
$C_2H_6I_2Te$	$(CH_3)_2TeI_2$	77	167,50	...	334.20	...	1114.43	4.3	[164]
Cs_2I_6Te	$Cs_2[TeI_6]$	77	154.140	...	308,28	...	1027.60	0.0	[112]
$H_8I_6N_2Te$	$[NH_4]_2[TeI_6]$	77	148.173 152.165 153,831	...	295.123 302.335 306,192	...	984,39 1008,82 1021,41	5.7 7,1 6.1	[112]
I_6K_2Te	$K_2[TeI_6]$	77	151.878 154.095 154,250	...	303,230 306.603 307.353	...	1011.04 1022.85 1024.11	3.7 6,3 5,4	[112]
I_6Rb_2Te	$Rb_2[TeI_6]$ Phase IV	77	151,995	...	303.688	...	1012.45	2.8	[124]

	Phase III		153.240	...	305.636	...	1019.24	4.6	
			153.903		306.619		1022.68	5.5	
	Phase II	292	148.789	...	297.57	...	991.91	0.4	
	Phase I		150.934		301.50		1005.2	3	
			Group VII element — iodine bond						
ClI	ClI	77	457.40	...	913.75	...	3046.4	2.98	[132, 141, 165—168]. cf. Cl
			460.24	...	...	...	...	...	
			—	—	—	—	—2930 (g)		
Cl_2CsI	$CsICl_2$	298	464.949	440	...	...	...	...	[166]. cf. Cl
Cl_2IK	$KICl_2$	298	462.060	320	...	...	...	...	The same
			469.160	290	...	...	...	...	
$Cl_2IK + H_2O$	$KICl_2 \cdot H_2O$	298	471.051	60	...	...	...	...	»
Cl_2IRb	$RbICl_2$	298	463.988	250	...	...	...	...	»
Cl_4CsI	$CsICl_4$	77	469.9	10	938.1	~3	3127.8	3.7	*[166]. cf. Cl
		298	453.123	50	...	...	...	...	
Cl_4IK	$KICl_4$	298	477.506	15	...	...	...	...	[166]. cf. Cl
$Cl_4IK + H_2O$	$KICl_4 \cdot H_2O$	298	458.863	100	...	...	...	...	The same
$Cl_4INa + 2H_2O$	$NaICl_4 \cdot 2H_2O$	298	461.702	360	...	...	...	...	»
Cl_4IRb	$RbICl_4$	77	468.4	3	933.2	3	3112.6	5.4	*[166]. cf. Cl
			476.3	3	950.7	3	3170.0	3.9	
		298	458.988	17	...	...	...	...	
			470.617	20	...	...	...	...	
Cl_6I_2	I_2Cl_6	77	458.19	...	909.37	...	3034.9	7.72	[166]. cf. Cl
BrI	BrI	—	—	—	—	—	2731 (g)	...	[117]. cf. Br
Br_2H_4IN	$[NH_4]IBr_2$	77	429.6	200	...	...	...	...	[122]. cf. Br

Empirical formula	Structural formula	T, °K	$\nu_{1/2-3/2}$, MHz	s/n	$\nu_{3/2-5/2}$, MHz	s/n	e^2Qq, MHz	η, %	References, notes
$C_4H_{12}Br_2IN$	$[(CH_3)_4N]IBr_2$	77	421.5	100	...	...	...	...	[122]. cf. Br
			420.5	100	...		...	...	
			419.1	100	...		...	...	
$C_4H_{12}I_3N$	$[(CH_3)_4N]I_3$	77	362.0	200	...	...	2410	...	[122, 132]
			177.6		354.2	80	1181	5	
			176.7		352.4	40	1175	5	
			172.5		343.8	80	1147	5	
$C_4H_{12}I_5N$	$[(CH_3)_4N]I_5$	90	361.5	...	...	...	...	...	[132]
$C_8H_{20}I_3N$	$[(C_2H_5)_4N]I_3$	77	356.6	200	...	...	2374	...	[122]
			175.2		350.2	150	1167	2	
			167.5		334.7	150	1116	2	
$C_{16}H_{36}I_3N$	$[(C_4H_9)_4N]I_3$	77	361.1	180	...	...	2404	...	[122]
			359.6	180	...	...	2394	...	
			167.6		334.7	10	1116	3	
CsI_3	CsI_3	77	122.9	...	245.7	...	819.0	2	[122, 169]
			215.89		430.86		1436.6	4.1	
			371.63		...		2477.5	...	
Cs_2I_8	Cs_2I_8	77	373.0	200	...	...	2484	...	[122]
		298	370.0	100	739.4		2465	3	
H_4I_3N	$[NH_4]I_3$	77	70.6	...	138.1	...	466.8	13	[122, 132, 165, 169]
			265.1		515.1		1725.0	15.1	
			368.8		...		2458.7	...	
I_2	I_2	77	333.96	20—30	643.32	10	2157.18	17.3	*[97, 141]

I_3Rb	RbI_3	77	116.5 218.3 369.9	...	232.1 434.7 ...	...	774.2 1449.8 2465.7	5.6 5.6 ...	[169]
Cs_2I_6Re	$Cs_2[ReI_6]$	77	124.49	...	249.01	...	829.4	0	[123]
$H_8I_6N_2Re$	$[NH_4]_2[ReI_6]$	77	119.73 122.10 122.62	...	239.42 241.03 242.03	...	796.9 804.0 811.0	1.2 10.1 10.1	[123]
I_6K_2Re	$K_2[ReI_6]$	77	122.92 123.73 124.58	...	245.52 246.55 247.45	...	818.50 822.32 825.73	3.2 5.3 7.3	[125, 126, 128]
I_6Rb_2Re	$Rb_2[ReI_6]$	77	122.4 122.8	...	244.8 ...	...	812.4 ...	0 ...	[123]

Group VIII element — iodine bond

$C_7H_5FeIO_2$	$C_5H_5FeI(CO)_2$	77	159.30 160.85	15 15	... ...	... ...	... ...	... ...	*
$H_8I_6N_2Pt$	$[NH_4]_2[PtI_6]$	77	205.48 205.69	...	... ...	...	... ...	... ...	[124]
I_6K_2Pt	$K_2[PtI_6]$	77	202.60 203.70 204.68	3 20 20	405.02 407.18 408.28	4 25 25	1350.20 1357.38 1361.50	2.0 2.0 4.5	[129]
I_6PtRb_2	$Rb_2[PtI_6]$	77	205.33	...	410.66	...	1368.87	0.0	[124]

BIBLIOGRAPHY

1. Gordy, W. and C. Burrus. — Phys. Rev. 93 (1954), 419.
2. Burrus, C. — Phys. Rev. 97 (1955), 1661.
3. Honig, A., M. Mandel, M. Stitch, and C. Townes. — Phys. Rev. 93 (1954), 953.
4. Honig, A., M. Mandel, M. Stitch, and C. Townes. — Phys. Rev. 96 (1954), 629.
5. Lee, C., N. Fabrikand, R. Carlson, and I. Rabi. — Phys. Rev. 91 (1953), 1395.
6. Segel, S., R. Barnes, and W. Jones. — Bull. Am. Phys. Soc., ser. 2, 5 (1960), 412.
7. Barnes, R., S. Segel, and W. Jones. — J. Appl. Phys., Vol. 33, Suppl. 1, p. 296, 1962.
8. Nakamura, D., Y. Uehara, Y. Kurita, and M. Kubo. — J. Chem. Phys. 31 (1959), 1433.
9. Gordy, W. and J. Sheridan. — J. Chem. Phys. 22 (1954), 92.
10. Chiba, T. — J. Phys. Soc. Japan 13 (1958), 860.
11. Cornwell, C. — J. Chem. Phys. 18 (1950), 1118L.
12. Barnes, R. and S. Segel. — J. Chem. Phys. 25 (1956), 180.
13. Maksyutin, Yu. K., E. V. Bryukhova, E. N. Gur'yanova, and G. K. Semin. — Izv. AN SSSR, chem. ser., p. 2658, 1968.
14. Barnes, R., S. Segel, P. Bray, and P. Casabella. — J. Chem. Phys. 26 (1957), 1345.
14a. Barrett, A. and M. Mandel. — Phys. Rev. 109 (1958), 1572.
15. Ludwig, G. — J. Chem. Phys. 25 (1956), 159.
16. Parks, S. and W. Moulton. — Phys. Rev. Lett. 26A (1967), 63.
17. Sharbough, A., B. Pritchard, and T. Madison. — Phys. Rev. 77 (1950), 302.
18. Sheridan, J. and W. Gordy. — J. Chem. Phys. 20 (1952), 735.
19. Zeldes, H. and R. Livingston. — J. Chem. Phys. 21 (1953), 1418.
20. Schawlow, A. — J. Chem. Phys. 22 (1954), 1211.
21. Smith, A., H. Ring, W. Smith, and W. Gordy. — Phys. Rev. 74 (1948), 370.
22. Smith, A., H. Ring, W. Smith, and W. Gordy. — Ibid., p. 123.
23. Townes, C., A. Holden, and F. Meritt. — Phys. Rev. 74 (1948), 1113.
24. Tetenbaum, S. — Phys. Rev. 82 (1951), 323.
25. Tetenbaum, S. — Phys. Rev. 86 (1952), 440.
26. Oka, T. and H. Hirakawa. — J. Phys. Soc. Japan 12 (1957), 820.
27. Semin, G. K. and A. A. Fainzil'berg. — Zh. Strukt. Khim. 6 (1965), 213.
28. Hatton, J. and B. Rollin. — Trans. Faraday Soc. 50 (1954), 358.
29. Babushkina, T. A., V. I. Robas, and G. K. Semin. — In: "Radiospektroskopiya tverdogo tela," p. 221. Atomizdat, 1967.
30. Hooper, H. — J. Chem. Phys. 41 (1964), 599.
31. Alderdice, D. and T. Iredale. — Trans. Faraday Soc. 62 (1966), 1370.
32. Gilson, D. and C. O'Konski. — J. Chem. Phys. 48 (1968), 2767.
33. Kojima, S., K. Tsukada, S. Ogawa, and A. Shimauchi. — J. Chem. Phys. 21 (1953), 1415.
34. Curnuk, P. and J. Sheridan. — Nature 202 (1964), 591.
35. Babushkina, T. A. and G. K. Semin. — Zh. Strukt. Khim. 7 (1966), 114.
36. Dehmelt, H. and H. Krüger. — Z. Phys. 130 (1951), 401.
37. Duchesne, J., A. Monfils, and J. Garson. — Physica 22 (1956), 816.
38. Gordy, W., J. Simmons, and A. Smith. — Phys. Rev. 74 (1948), 243.
39. Gordy, W., J. Simmons, and A. Smith. — Ibid. p.1246
40. Mulliken, R., C. Rieke, D. Orloff, and H. Orloff. — J. Chem. Phys. 17 (1949), 510.
41. Kraitchman, J. and B. Dailey. — J. Chem. Phys. 22 (1954), 1477.
42. Tokuhiro, T. — J. Chem. Phys. 47 (1967), 2353.
43. Simmons, J. and J. Goldstein. — J. Chem. Phys. 20 (1952), 122.
44. Semin, G. K. — Zh. Strukt. Khim. 6 (1965), 916.
45. Goedertier, R. — J. Phys. 24 (1963), 633.
46. Benz, H., A. Bouder, and H. Günthard. — J. Molec. Spectrosc. 21 (1966), 165.
47. Bray, P. — J. Chem. Phys. 22 (1954), 1787.
48. Krisher, L. — J. Chem. Phys. 33 (1960), 1237.
49. Shimomura, K., T. Kushida, and N. Inoue. — J. Chem. Phys. 22 (1954), 350.
50. Bray, P. — J. Chem. Phys. 22 (1954), 950.
51. Barnes, R., O. Miller, and F. Wooten. — J. Chem. Phys. 22 (1954), 946.
52. Shimomura, K. — J. Sci. Hiroshima Univ. A17 (1954), 383.
53. Dodgen, H. and J. Ragle. — J. Chem. Phys. 25 (1956), 376.

54. Solimene, N. — J. Chem. Phys. 26 (1957), 1593.
55. Flangan, R. and L. Pierce. — J. Chem. Phys. 38 (1963), 2963.
56. Hooper, H. and P. Bray. — J. Chem. Phys. 33 (1960), 334.
57. Semin, G. K. — Doklady AN SSSR 158 (1964), 1169.
58. Kikuchi, Y., E. Hirota, and Y. Morino. — J. Chem. Phys. 31 (1959), 1129.
59. Kikuchi, Y., E. Hirota, and Y. Morino. — Bull. Chem. Soc. Japan 34 (1961), 348.
60. Schwendeman, R. and F. Tobiason. — J. Chem. Phys. 43 (1965), 201.
61. Zeil, W., M. Winnerwisser, and W. Hüttner. — Z. Naturf. 16a (1961), 1248.
62. Zeil, W., H. Heel, H. Pförtner, and A. Schmitt. — Z. Elektrochem. 63 (1959), 1009.
63. Semin, G. K., V. I. Robas, V. I. Stanko, and V. A. Brattsev. — Zh. Strukt. Khim. 6 (1965), 305.
64. Segel, S., R. Barnes, and P. Bray. — J. Chem. Phys. 25 (1956), 1286.
65. Tokuhiro, T. — Bull. Tokyo Inst. Technol. 57 (1964), 31; J. Chem. Phys. 41 (1964), 438.
66. Semin, G. K. — Zh. Strukt. Khim. 2 (1961), 370.
67. Semin, G. K., V. I. Robas, L. S. Kobrina, and G. G. Yakobson. — Zh. Strukt. Khim. 5 (1964), 915.
68. Khotsyanova, T. L., V. I. Robas, and G. K. Semin. — Zh. Strukt. Khim. 5 (1964), 644.
69. Casabella, P., P. Bray, S. Segel, and R. Barnes. — J. Chem. Phys. 25 (1956), 1280.
70. Babushkina, T. A., T. L. Khotsyanova, and G. K. Semin. — Zh. Strukt. Khim. 6 (1965), 307.
71. Bray, P. and R. Barnes. — J. Chem. Phys. 22 (1954), 2023.
72. Rama Rao, K. and C. Murty. — J. Phys. Soc. Japan 21 (1966), 1627.
73. Semin, G. K. — Zh. Strukt. Khim. 3 (1962), 292.
73a. Segel, S. and R. Barnes. — Catalog of Nuclear Quadrupole Interaction and Resonance Frequencies in Solids. — US Atomic Energy Commission, 1965.
74. Kojima, S. — J. Jap. Chem. (Kagaku-no-ryoiki) 13 (1962), 619.
75. Shimomura, K. — J. Phys. Soc. Japan 14 (1959), 235.
76. Alderdice, D., R. Brown, and T. Iredale. — J. Chem. Phys. 30 (1959), 344.
77. Rosental, E. and B. Dailey. — J. Chem. Phys. 43 (1965), 2093.
78. Read, M., R. Cahay, P. Cornil, and J. Duchesne. — Compt. rend. 257 (1963), 1778.
79. Cornil, P., J. Duchesne, M. Read, and R. Cahay. — Bull. Acad. r. Belg. Cl. Sci., ser. 5, 50 (1964), 235.
80. Rama Rao, K. and C. Murty. — Proc. Phys. Soc. 89 (1966), 711.
81. Rama Rao, K. and C. Murty. — Curr. Sci. 34 (1965), 660.
81a. Bucci, P., P. Cecchi, A. Colligiani, and R. Meschia. — Ectr. Boll. sci. Bologna 23 (1965), 307.
82. Rama Rao, K. and C. Murty. — Curr. Sci. 35 (1966), 252.
83. Tokuhiro, T. — Bull. Chem. Soc. Japan 35 (1962), 1293.
84. Ikeda, R., D. Nakamura, K. Ito, H. Sugiyama, and M. Kubo. — Tetrahedron Lett. 20 (1964), 909.
85. Rama Rao, K. and C. Murty. — Chem. Phys. Letters 1 (1967), 323.
86. Bray, P., S. Moskowitz, H. Hooper, R. Barnes, and S. Segel. — J. Chem. Phys. 28 (1958), 99.
87. Rama Rao, K. and C. Murty. — Phys. Rev. Lett. 23 (1966), 404.
88. Sheridan, J. and W. Gordy. — J. Chem. Phys. 19 (1951), 965.
89. Sharbough, A., J. Bragg, T. Madison, and V. Thomas. — Phys. Rev. 76 (1949), 1419.
90. Biryukov, I. P., E. Ya. Lukevets, M. G. Voronkov, and I. A. Safin. — Izv. AN LatvSSR, p. 754, 1967.
91. Sharbough, A., B. Pritchard, V. Thomas, J. Mays, and B. Dailey. — Phys. Rev. 79 (1950), 189.
92. Bryuchova, E. V., G. K. Semin, V. I. Goldanskij, and V. V. Chrapov. — Chem. Communs., p. 491, 1968.
93. Nakamura, D., K. Ito, and M. Kubo. — Inorg. Chem. 1 (1962), 592.
94. Nakamura, D. — Bull. Chem. Soc. Japan 36 (1963), 1662.
95. Barnes, R. and R. Engardt. — J. Chem. Phys. 28 (1958), 248.
96. Hamlen, R. and W. Koski. — J. Chem. Phys. 25 (1956), 360.
97. Robinson, H., H. Dehmelt, and W. Gordy. — J. Chem. Phys. 22 (1954), 511.
98. Kojima, S., K. Tsukada, S. Ogawa, A. Shimauchi, and Y. Abe. — J. Phys. Soc. Japan 9 (1954), 805.
99. Barnes, R. and P. Bray. — J. Chem. Phys. 23 (1955), 407.
100. Ogawa, S. — J. Phys. Soc. Japan 13 (1958), 618.
101. Grechishkin, V. S. and I. A. Kyuntsel'. — Trudy ENI Permsk. State Univ. 12, Issue 1 (1966), 9.

102. Negita, H., T. Okuda, and M. Kashima. — J. Chem. Phys. 45 (1966), 1076.
103. Grechishkin, V. S. and I. A. Kyuntsel'. — Optika Spektrosk. 15 (1963), 832.
104. Grechishkin, V. S. and I. A. Kyuntsel'. — Optika Spektrosk. 16 (1964), 161.
105. Grechishkin, V. S. and I. A. Kyuntsel'. — Zh. Strukt. Khim. 5 (1964), 53.
106. Parker, D. — Diss. Abstr., Sect. B, 20 (1953), 2044.
107. Shimomura, K., T. Kushida, N. Inoue, and I. Imaeda. — J. Chem. Phys. 22 (1954), 1944.
108. Tipsword, R. and W. Moulton. — J. Chem. Phys. 39 (1963), 2730.
109. Burgiel, J., V. Jaccarino, and A. Schawlow. — Phys. Rev. 122 (1961), 429.
110. Neuvar, E. and A. Jache. — J. Chem. Phys. 39 (1963), 596.
111. Nakamura, D., K. Ito, and M. Kubo. — Intern. Sympos. on Molecular Structure Spectroscopy. Tokyo, 1962; Inorg. Chem. 2 (1963), 61.
112. Nakamura, D., K. Ito, and M. Kubo. — J. Am. Chem. Soc. 84 (1962), 163.
113. Barnes, R. and S. Segel. — Phys. Rev. Lett. 3 (1959), 462.
114. Smith, D., M. Tidwell, and D. Williams. — Phys. Rev. 79 (1950), 1007.
115. Smith, D., M. Tidwell, and D. Williams. — Ibid., p. 420.
116. Magnuson, D. — J. Chem. Phys. 21 (1953), 609.
117. Jaseja, T. — J. Molec. Spectrowc. 5 (1960), 445.
118. Kojima, S., K. Tsukada, and Y. Hinaga. — J. Phys. Soc. Japan 9 (1954), 795.
119. Kojima, S., Y. Abe, M. Minematsu, K. Tsukada, and A. Shimauchi. — J. Phys. Soc. Japan 12 (1957), 1225.
120. Dehmelt, H. — Z. Phys. 130 (1951), 480.
121. Breneman, G. and R. Willett. — J. Phys. Chem. 71 (1967), 3684.
122. Bowmaker, G. and S. Hacobian. — Aust. J. Chem. 21 (1968), 551.
123. Ikeda, R., A. Sasane, D. Nakamura, and M. Kubo. — J. Phys. Chem. 70 (1966), 2926.
124. Nakamura, D., K. Ito, and M. Kubo. — J. Phys. Chem. 68 (1964), 2986.
125. Ikeda, R., D. Nakamura, and M. Kubo. — J. Phys. Chem. 69 (1965), 2101.
126. Ikeda, R., D. Nakamura, and M. Kubo. — Bull. Chem. Soc. Japan 36 (1963), 1056.
127. Nakamura, D., Y. Kurita, K. Ito, and M. Kubo. — J. Am. Chem. Soc. 83 (1961), 4526.
128. Nakamura, D. and M. Kubo. — Proc. VIII International Conference on Coordinate Chemistry, p. 93. Vienna, 1964.
129. Nakamura, D., Y. Kurita, K. Ito, and M. Kubo. — J. Am. Chem. Soc. 82 (1960), 5783.
130. Klein, J. and A. Nethercot. — Phys. Rev. 91 (1953), 1018.
131. Burkhard, D. — J. Chem. Phys. 21 (1953), 1541.
132. Kojima, S., K. Tsukada, S. Ogawa, and A. Shimauchi. — J. Chem. Phys. 23 (1955), 1963.
133. Feige, C. and H. Hartmann. — Z. Naturf. 22a (1967), 1286.
134. Laurita, W. and W. Koski. — J. Am. Chem. Soc. 81 (1959), 3179.
135. Laurita, W. and W. Koski. — Mém. Acad. r. Belg. Cl. Sci. 33 (1961), 321.
136. Barnes, R. and S. Segel. — J. Chem. Phys. 25 (1956), 578.
137. Sheridan, J. and W. Gordy. — Phys. Rev. 77 (1950), 292.
138. Sterzer, F. — J. Chem. Phys. 22 (1954), 2049.
139. Sterzer, F. and P. Beers. — Phys. Rev. 100 (1955), 1174.
140. Kojima, S., K. Tsukada, and A. Shimauchi. — J. Chem. Phys. 21 (1953), 2237.
141. Dehmelt, H. — Z. Phys. 130 (1951), 356.
142. Babushkina, T. A., G. K. Semin, S. P. Khrlakyan, V. V. Shokina, and I. L. Knunyants. — Teor. Eksperim. Khim. 4 (1968), 275.
143. Morgan, H. and J. Goldstein. — J. Chem. Phys. 22 (1954), 1427.
144. Moloney, M. and L. Krisher. — J. Chem. Phys. 45 (1966), 3277.
145. Dodgen, H. and R. Anderson. — J. Chem. Phys. 31 (1959), 851.
146. Kasuya, T. and T. Oka. — J. Phys. Soc. Japan 14 (1959), 980.
147. Kasuya, T. — J. Phys. Soc. Japan 15 (1960), 296.
148. Semin, G. K. and V. I. Robas. — In: "Radiospektroskopiya tverdogo tela," p. 229. Atomizdat, 1967.
149. Bray, P., R. Barnes, and R. Bersohn. — J. Chem. Phys. 25 (1956), 813.
150. Babushkina, T. A. and H. Zeldes. — Phys. Rev. 90 (1953), 609.
151. Sams, L. and A. Jache. — J. Chem. Phys. 47 (1967), 1314.
152. Dehmelt, H. — Naturwiss. 37 (1950), 398.
153. Livingston, R. and H. Zeldes. — Phys. Rev. 90 (1953), 609.
154. Barnes, R. and P. Bray. — J. Chem. Phys. 23 (1955), 1177.
155. Barnes, R. and P. Bray. — Ibid., p. 1182.

156. Herlach,F., H. Granicher, D. Itschner, and P. Kesselring. — V International Congress and Symposium, International Union of Crystallography, p. 98. Cambridge, 1960.
157. Herlach,F. — Helv. phys. Acta 34 (1961), 305.
158. Pierce,S. and C. Cornwell. — J. Chem. Phys. 47 (1967), 1731.
159. Livingston,R. and H. Zeldes. — J. Chem. Phys. 26 (1957), 351.
160. Shimomura, K. and N. Inoue. — J. Phys. Soc. Japan 14 (1959), 86.
161. Ramakrishna, J. — Proc. Natn. Inst. Sci. India A29 (1963), 643.
162. Herlach,F., H. Granicher, and D. Itschner. — Arch. sci., Vol. 12, fasc. spec., p. 182, 1959.
163. Burgiel,J., V. Jaccarino, and A. Schawlow. — Bull. Am. Phys. Soc., ser. 2. 4 (1959), 424.
164. Zeil,W. — Angew. Chem. 76 (1964), 654.
165. Kojima,S., A. Shimauchi, S. Hagiwara, and Y. Abe. — J. Phys. Soc. Japan 10 (1955), 930.
166. Yamasaki,R. and C. Cornwell. — J. Chem. Phys. 30 (1959), 1265.
167. Townes,C., F. Merritt, and B. Wright. — Phys. Rev. 73 (1948), 1334.
168. Townes,C., F. Merritt, and B. Wright. — Ibid., p. 1249.
169. Sasane,A., D. Nakamura, and M. Kubo. — J. Phys. Chem. 71 (1967), 3249.
170. Brill,T. — J. Inorg. Nucl. Chem. 32 (1970), 1869.
171. Okuda,T., H. Terao, O. Ege, and H. Negita. — J. Chem. Phys. 52 (1970), 5489.
172. Bowmaker,G. and S. Hacobian. — Aust. J. Chem. 22 (1969), 2047.
173. Rama Rao,K. and C. Murty. — J. Phys. Soc. Japan 25 (1968), 1423.
174. Bucci,P., P. Cecchi, and A. Colligiani. — J. Chem. Phys. 50 (1969), 530.
175. Ambrosetti,R., P. Bucci, P. Cecchi, and A. Colligiani. — J. Chem. Phys. 51 (1969), 852.
176. Popov,S.N., N.N. Krainik, and I.E. Myl'nikova. — Izv. AN SSSR, phys. ser., 33 (1969), 271.
177. Tipsword,R., J. Allender, E.A. Stahl, and C.D. Williams. — J. Chem. Phys. 49 (1968), 2464.
178. Nesmeyanov,A.N., O.Yu. Okhlobystin, E.V. Bryukhova, V.I. Bregadze, D.N. Kravtsov, B.A. Faingor, L.S. Golovchenko, and G.K. Semin. — Izv. AN SSSR, chem. ser., p. 1928, 1969.
179. Moloney,M. — J. Chem. Phys. 50 (1969), 1981.
180. Ward,R., C. Williams, and R. Tipsword. — J. Chem. Phys. 51 (1969), 823.
181. Buslaev,Yu.A., E.A. Kravchenko, S.M. Sinitsyna, and G.K. Semin. — Izv. AN SSSR, chem. ser., p. 2816, 1969.
182. Kashiwagi,H. — Tetrahedron Lett. 21 (1965), 1095.
183. Maksyutin,Yu.K., E.N. Gur'yanova, and G.K. Semin. — Uspekhi Khimii 39 (1970), 727.
184. Bryukhova,E.B., T.A. Babushkina, M.A. Kashutina, O.Yu. Okhlobystin, and G.K. Semin. — Doklady AN SSSR 183 (1968), 827.
185. Biryukov,I.P., M.G. Voronkov, and I.A. Safin. Frequency Tables of the Nuclear Quadrupole Resonance. — "Khimiya," 1968. (Russian)
186. Semin,G.K., L.S. Kobrina, and G.G. Yakobson. — Izv. Sib. Otdel. AN SSSR, ser. chim. sci., No. 9, Issue 4 (1968), 84.
187. Bryukhova,E.V., T.A. Babushkina, V.I. Svergun, and G.K. Semin. — Uchen. Zap. Mosc. Oblast. Pedag. Inst. im. N.K. Krupskaya 222 (1969), 64.
188. Belyaev,L.M., V.A. Lyakhovitskaya, E.V. Bryukhova, T.A. Babushkina, A.P. Zhukov, A.S. Naumov, D.A. Tambovtsev, and G.K. Semin. — Reports of the 6th All-Union Conf. on Ferroelectricity, p. 36. Riga, 1968. (Russian)
189. Upadysheva,A.V., T.A. Babushkina, E.V. Bryukhova, V.I. Robas, L.A. Kazitsyna, and G.K. Semin. — Izv. AN SSSR, chem. ser., p. 2068, 1968.
190. Semin,G.K., O.V. Nogina, V.A. Dubovitskii, T.A. Babushkina, E.V. Bryukhova, and A.N. Nesmeyanov. — Doklady AN SSSR 194 (1970), 101.
191. Semin,G.K. and A.A. Boguslavskii. — Uchen. Zap. Mosc. Oblast. Pedag. Inst. im. N.K. Krupskaya 222 (1969), 64.
192. Babushkin,A.A., T.A. Babushkina, E.A. Orlova, I.B. Sperantova, and G.K. Semin. — Zh. Fiz. Khim. 43 (1969), 1999.

Group VII and VIII metals

Empirical formula	Structural formula	T, °K	ν, MHz		e^2Qq, MHz	η, %	References, notes
			1/2—3/2	3/2—5/2			
		^{55}Mn					
$C_8H_4ClMnO_5S$	$C_5H_4SO_2ClMn(CO)_3$	77	...	18.486	...	...	*[1]. cf. Cl
$C_8H_4IMnO_3$	$C_5H_4IMn(CO)_3$	77	10.07	20.094	67.00	4.3	*[1]. cf. I (sup.)
$C_8H_4MnNaO_6S$	$C_5H_4SO_3NaMn(CO)_3$	77 Room	9.88 ...	19.740 19.523	65.81 ...	2.9 ...	*[1]
$C_8H_5MnO_3$	$C_5H_5Mn(CO)_3$	77 293	9.798 9.600	19.572 19.280	65.2 64.18	1.9 0	*[1, 2, 32]
$C_8H_6MnNO_5S$	$C_5H_4SO_2NH_2Mn(CO)_3$	77	9.58	19.050	62.56	6.9	*[1]
$C_9H_4ClMnO_4$	$C_5H_4COClMn(CO)_3$	77	...	17.960	...	...	*[1]. cf. Cl
$C_9H_5MnO_5$	$C_5H_4COOHMn(CO)_3$	77	9.34	18.682	62.27	0	*[1]
$C_9H_6ClMnO_3$	$C_5H_4CH_2ClMn(CO)_3$	77	9.77	19.476	64.95	5.1	*[1]. cf. Cl
$C_9H_7MnO_3$	$C_5H_4CH_3Mn(CO)_3$	77	9.75	19.452	64.86	4.4	*[1]
$C_{10}H_6BrMnO_4$	$C_5H_4COCH_2BrMn(CO)_3$	77	...	18.150	...	...	*[1]
$C_{10}H_6ClMnO_4$	$C_5H_4COCH_2ClMn(CO)_3$	77	...	18.180	...	...	*[1]. cf. Cl
$C_{10}H_7ClMnO_3$	$C_5H_3(CH_2Cl)_2Mn(CO)_3$	77	...	18.820	...	...	*[1]. cf. Cl
$C_{10}H_7MnO_5$	$C_5H_4COOCH_3Mn(CO)_3$	77	9.21	18.348	61.20	5.6	*[1]

$C_{10}H_9MnO_3$	$C_5H_4C_2H_5Mn(CO)_3$	77	...	19.380	...	...	*[1]
			...	19,760	...	...	
$C_{10}H_9MnO_6S$	$C_5H_4SO_3C_2H_5Mn(CO)_3$	77	...	19,30	...	...	*[1]
$C_{10}Mn_2O_{10}$	$Mn_2(CO)_{10}$	77	—	—	4,38	21,1	[2,3,33]. NMR cr.
		Room	—	—	3,4	36,4	
$C_{11}H_4F_3MnO_4$	$C_6H_4COCF_3Mn(CO)_3$	77	9.09	17,736	59.35	13.9	*[1]
$C_{11}H_7MnO_4$	$C_6H_4COCH_3Mn(CO)_3$	77	9.29	18.472	61.63	6.8	*[1]
$C_{15}H_9MnO_4$	$C_5H_4COC_6H_5Mn(CO)_3$	77	9.50	18.672	62.41	11.8	*[1]
$C_{25}H_{20}AsMnO_2$	$C_5H_5Mn(CO)_2As(C_6H_5)_3$						* cf. As
	Phase I	77	...	19.464	...	...	
	Phase II	77	...	19.656	...	...	
			...	19.904	...	...	
$FMnO_3$	MnO_3F	—	—	—	16.8	...	[4—6]. g
Mn	α-Mn	Room	...	2.456	8.48	41	[7]
			...	2,965	10.08	28	
	β-Mn	295	—	—	1.71	...	[8]. NMR cr.
	Mn (in corundum)	Room	—	—	0.138	...	[9]. NMR cr.
Mn_2Y	YMn_2		—	—	15,3	...	[10]. NMR cr.
Tc							
Tc	Tc (met.)	303	—	—	5,7	...	[11]. NMR cr.

Empirical formula	Structural formula	Isotope	T, °K	ν, MHz $^1/_2$—$^3/_2$	ν, MHz $^3/_2$—$^5/_2$	e^2Qq, MHz	η, %	References, notes
				Re				
$C_8H_4IO_3Re$	$IC_5H_4Re(CO)_3$	185	77	98.38	...	635.9	17.1	*
		187		93.12	179.52	601.9	17.1	
$C_8H_5O_3Re$	$C_5H_5Re(CO)_3$	185	77	93.72	185.64	619.72	8.8	*[1, 12]
		187		88.70	175.74	586.64	8.7	
$C_9H_4ClO_4Re$	$C_5H_4COClRe(CO)_3$	185	77	87.42	147.37	505.24	39.1	[1, 12]. cf. Cl
		187		82.75	139.50	478.25	39.1	
$C_9H_5O_5Re$	$C_5H_4COOHRe(CO)_3$	185	77	88.60	165.66	558.23	23.4	*[1, 12]
		187		83.86	156.84	528.44	23.4	
$C_9H_7O_3Re$	$CH_3C_5H_4Re(CO)_3$	185	77	99.81	192.43	645.17	17.1	*[1, 12]
		187		94.452	182.10	610.53	17.1	
$C_{10}H_7O_4Re$	$C_5H_4COCH_3Re(CO)_3$	185	77	95.27	170.76	579.42	30.5	*[1, 12]
		187		90.18	161.64	548.47	30 5	
$C_{10}O_{10}Re_2$	$Re_2(CO)_{10}$	185	77	35.736	39.725	147.11	87.35	*[1, 12, 13, 30, 31].
			297	28.836	39.650	140.17	66.1	$\nu_{1/2-5/2}$ = 75, 45 MHz;
		187	77	33.828	37.597	139.22	87.35	$\nu_{1/2-5/2}$ = 71, 51 MHz
			297	27.191	37.730	132.7	66.1	
$C_{12}H_4ClO_6Re$	$4\text{-}ClC_6H_4CORe(CO)_5$	185	77	107.75	...	...	...	*[1, 12]. cf. Cl
		187		102.00	...	...	...	
$C_{15}H_9O_4Re$	$C_5H_4COC_6H_5Re(CO)_3$	185	77	86.49	142.65	490.91	41.9	*[1, 12]
		187		81.84	135.02	464.66	41.8	
$C_{15}H_{14}NO_6ReS$	$[C_5H_4SO_3Re(CO)_3]\cdot$ $\cdot[4\text{-}CH_3C_6H_4NH_3]$	185	77	100.04	...	658.2	9.9	*[1, 12]
				104.63	204.16	681.1	13.4	
		187		94.71	187.07	624.8	9.9	
				99.03	193.20	646.5	13.4	
$C_{26}H_{19}O_3ReSn$	$(C_6H_5)_3SnC_5H_4Re(CO)_3$	185	77	87.52	170.13	569.7	15.0	*
		187		82.83	161.00	539.1	15.0	

ClO_3Re	$ReClO_3$	185	—	—	—	270	...	[5, 6, 14]. g cf. Cl
		187	—	—	—	253	...	
FO_3Re	$ReFO_3$	185	—	—	—	—53	...	[6, 15]. g
KO_4Re	$KReO_4$	185	77	29.386	58.746	195.83	1.9	[28]
			296	28.312	56.600	185.52	1.8	
		187	77	27.839	55.651	188.68	1.9	
			296	26,825	53,626	178.77	1,8	
Re	Re (met.)	185	0.5	—	—	301.3	...	[16, 17, 29]. NMR cr.
			293	—	—	274	...	

Empirical formula	Structural formula	T, °K	ν, MHz			e^2Qq, MHz	η, %	References, notes
			$^1/_2$—$^3/_2$	$^3/_2$—$^5/_2$	$^5/_2$—$^7/_2$			
			^{59}Co					
$CH_{12}CoN_5O_6$ + + $^1/_2H_2O$	$[Co(NH_3)_4CO_3]NO_3 \cdot {}^1/_2H_2O$	Room	—	—	—	29.3	—	[18]. NMR 1. cf. N
$CH_{15}Cl_2CoN_6O_8$ + + $^1/_2H_2O$	$[Co(NH_3)_5CN](ClO_4)_2 \cdot {}^1/_2H_2O$	»	—	—	—	48,8	—	The same
$C_4Cl_3CoGeO_4$	$Cl_3Ge[Co(CO)_4]$	77	...	23.180	34.800	162.43	4 4	*[34—36]
			...	23.448	35.244	164.54	6.8	
		298	...	22.770	34.150	159.37	0	
			...	23.090	34.620	161.56	0	
$C_4H_{10}Br_3CoN_2$	trans-$[CoBr_2(en)_2]Br$	Room	...	8.392	12.860	60.36	23.5	[19]
$C_4H_{10}Cl_2CoN_3O_3$	trans-$[CoCl_2(en)_2]NO_3$	»	4.777	8.877	13 426	62.78	13.2	[19]
$C_4H_{10}Cl_3CoN_2$	cis-$[CoCl_2(en)_2]Cl$	»	—	—	—	61.4	—	[18]. NMR 1.
	trans-$[CoCl_2(en)_2]Cl$	»	5.367	8.368	12.893	60.63	27.2	[19]. cf. N
			—	—	—	75.8	—	[18]. NMR 1.

Empirical formula	Structural formula	T, °K	ν, MHz $1/2$—$3/2$	ν, MHz $3/2$—$5/2$	ν, MHz $5/2$—$7/2$	eQq, MHz	η, %	References, notes
$C_4H_{10}Cl_3CoN_2$ + HCl + xH_2O	trans-$[CoCl_2(en)_2]Cl \cdot HCl \cdot xH_2O$	Room	5.979	10.001	15.297	71.73	22.2	[20, 37]. cf. Cl
$C_4H_{10}Cl_3CoN_2$ + $3H_2O$	$[Co(en)_2Cl_2]Cl \cdot 3H_2O$	»	—	—	—	—2.63	—	[18]. NMR 1.
$C_4H_{10}CoN_5O_6$	trans-$[Co(en)_2(NO_2)_2]NO_2$	»	—	—	—	21.8	—	[18]. NMR 1.
$C_6CoK_3N_6$	$K_3[Co(CN)_6]$	4.2	—	—	—	7.3	75	[21]. NMR cr.
$C_8Cl_2Co_2GeO_8$	$Cl_2Ge[Co(CO)_4]_2$	77	...	22.540	33.852	158.02	5.3	*[34, 35]. cf. Cl
			...	20.704	31.368	146.73	15.4	
		298	...	22.20	33.34	155.6	4	
			...	20.40	30.99	145.0	17	
$C_8Cl_2Co_2O_8Sn$	$Cl_2Sn[Co(CO)_4]_2$	77	...	22.040	33.090	154.44	4.5	*[34, 35, 38]
			...	21.700	32.652	152.48	8.4	
			...	20.912	31.640	147.95	1.4	
			...	19.880	30.096	140.75	1.5	
		298	10.81	21.64	32.49	151.6	0	
			...	21.25	31.94	149.0	6	
			...	20.48	30.94	146.6	13	
			10.46	19.58	29.58	138.2	12	
$C_8Co_2O_8$	$Co_2(CO)_8$	77	...	12.34	19.14	90.23	31.5	*[2, 34, 39]
			10.22	11.82	18.67	89.23	47.7	
$C_{10}H_5Cl_2CoO_4Sn$	$(C_6H_5)Cl_2Sn[Co(CO)_4]$	77	...	20.728	31.110	145.25	3.7	*[34]
			...	20.400	30.714	143.44	9.1	
			...	20.440	30.756	143 63	8.4	
			...	20.200	30.420	142.04	9.4	
$C_{10}H_{10}Br_3Co$	$[(C_5H_5)_2Co]Br_3$	77	12.031	24.064	36.090	168.43	0	*[34]
$C_{10}H_{10}ClCoO_4$	$[(C_5H_5)_2Co] \cdot ClO_4$	78	...	24.08	36 12	168.56	0	[2, 22]
			12.25	24.50	36.74	171.50	0	

$C_{12}ClCo_3O_{12}Sn$	$ClSn[Co(CO)_4]_3$	77	...	19.880	29.900	139.62	7.8	*[34, 35]
			...	19.158	28.750	134.18	3.2	
			...	19.128	28.720	134.05	4,7	
		298	...	19.48	29.34	135.9	9	
			...	18.89	28.40	132.6	6	
$C_{16}H_{10}ClCoO_4Sn$	$(C_6H_5)_2ClSn[Co(CO_4)]$	77	...	17.064	25.635	119.66	5,8	*[34]
$C_{22}H_{15}CoGeO_4$	$(C_6H_5)_3Ge[Co(CO)_4]$	77	...	15.744	23.640	110.34	4.7	*[34]
$C_{39}H_{30}CoGeO_3P$	$(C_6H_5)_3Ge \cdot [Co(CO)_3P(C_6H_5)_3]$	77	...	...	24.220	...	...	*[34]
$C_{39}H_{30}CoO_3PSn$	$(C_6H_5)_3Sn \cdot [Co(CO)_3P(C_6H_5)_3]$	77	...	16.28	24.52	114.53	9.6	*[34]
$CeCo_2$	$CeCo_2$	Room	—	—	—	17.16	...	[23]. NMR cr.
$ClCoH_{12}N_6O_4$	cis-$[Co(NH_3)_4(NO_2)_2]Cl$	»	—	—	—	18.2	—	[18]. NMR l.
	trans-$[Co(NH_3)_4(NO_2)_2]Cl$	»	—	—	—	18.2	—	[18]. NMR l. cf. N
$Cl_2CoH_{15}N_6O_2$	$[Co(NH_3)_5NO_2]Cl_2$	»	—	—	—	14,9	—	The same
$Cl_3CoH_{12}N_4$	trans-$[CoCl_2(NH_3)_4]Cl$	»	...	8.370	12.665	59.23	13.6	[19]
$Cl_3CoH_{15}N_5$	$[Co(NH_3)_5Cl]Cl_2$	»	—	—	—	35.2	—	[18]. NMR l.
$Cl_3CoH_{16}N_4O_{14}$	$[Co(NH_3)_4(H_2O)_2](ClO_4)_3$	»	—	—	—	61.5	—	[18]. NMR l. cf. N
$Cl_3CoH_{17}N_5O_{13}$	$[Co(NH_3)_5H_2O](ClO_4)_3$	121	—	—	—	56.7	...	The same
		182	—	—	—	55.7	...	
		232	—	—	—	55.3	...	
		266	—	—	—	55.1	...	
		298	—	—	—	54.8	...	
$Cl_3CoH_{18}N_6$	$[Co(NH_3)_6]Cl_3$	4.2	—	—	—	1.5	0	[21]. NMR cr. cf. N
Co	Co (met.)	4.2	—	—	—	0.182	...	[17, 24—27].
		293	—	—	—	0.112	...	NMR cr.
$CoH_9N_6O_6$	trans-$[Co(NH_3)_3](NO_2)_3$	Room	—	—	—	22.7	—	[18]. NMR l.
$CoH_{10}N_7O_8$	$(NH_4)[Co(NO_2)_4(NH_3)_2]$	»	—	—	—	29.8	—	[18]. NMR l. cf. N

Empirical formula	Structural formula	T, °K	ν, MHz			eQq, MHz	η, %	References, notes
			$^1/_2$—$^3/_2$	$^3/_2$—$^5/_2$	$^5/_2$—$^7/_2$			
$CoH_{15}N_5O_3S_3$ + + $2H_2O$	$[Co(NH_3)_5SO_3]_2SO_3 \cdot 2H_2O$	Room	—	—	—	23.7	—	[18]. NMR 1.
$CoH_{16}N_7O_7$	$[Co(NH_3)_5OH](NO_3)_2$	»	—	—	—	38.8	—	[18] NMR 1.
Co_2Er	Co_2Er	»	—	—	—	8.66	...	[23]. NMR cr.
Co_2Lu	Co_2Lu	»	—	—	—	6.08	...	The same
Co_2Nd	Co_2Nd	»	—	—	—	6.88	...	»
Co_2Sc	Co_2Sc	»	—	—	—	9.59	...	»
Co_2Tm	Co_2Tm	»	—	—	—	5.82	...	»
Co_2U	Co_2U	»	—	—	—	14.00	...	»
Co_2Y	Co_2Y	»	—	—	—	2.18	...	»
Co_2Zr	Co_2Zr	»	—	—	—	14.04	...	»

BIBLIOGRAPHY

1. Nesmeyanov, A. N., G. K. Semin, E. V. Bryukhova, T. A. Babushkina, K. N. Anisimov, N. E. Kolobova, and Yu. V. Makarov. — Izv. AN SSSR, chem. ser., p. 1953, 1968.
2. Voitlander, J., H. Klocke, R. Longino, and H. Thieme. — Naturwissenschaften 49 (1962), 491.
3. Calderazzo, F., E. Lucken, and D. Williams. — J. Chem. Soc. (A), p. 154, 1967.
4. Javan, A. and A. Grosse. — Phys. Rev. 87 (1952), 227.
5. Javan, A. and A. Engelbrecht. — Phys. Rev. 96 (1954), 649.
6. Lotspeich, J. — J. Chem. Phys. 31 (1959), 643.
7. Anderson, L. — Phys. Rev. Lett. 26a (1968), 279.
8. Drain, L. — Proc. Phys. Soc. 88 (1966), 111.
9. Lawrence, N. and N. Lambe. — Phys. Rev. 132 (1963), 1029.
10. Segel, S. and R. Barnes. — Bull. Am. Phys. Soc., ser. 2, 8 (1963), 260.
11. Jones, W. and F. Milfort. — Phys. Rev. 125 (1962), 1259.
12. Semin, G. K. and E. V. Bryukhova. Nuclear Physics, p. 1346, 1968. (Russian)
13. Segel, S. and R. Barnes. — Phys. Rev. 107 (1957), 638.
14. Javan, A., G. Silvey, C. Townes, and H. Grosse. — Phys. Rev. 91 (1953), 222.
15. Lotspeich, J. and A. Javan. — Phys. Rev. 94 (1954), 789.
16. Keesonm, P. — Phys. Rev. Lett. 2 (1959), 260.
17. Das, T. — Phys. Rev. 123 (1961), 2070.
18. Hartmann, H. and H. Sillescu. — Theoret. chim. Acta 2 (1964), 371.
19. Watanabe, T. and Y. Yamagata. — J. Chem. Phys. 46 (1967), 407.
20. Hartmann, H., M. Fleissner, and H. Sillescu. — Naturwissenschaften 50 (1963), 591.
21. Sugawara, T. — J. Phys. Soc. Japan 14 (1959), 858.
22. Voitlander, J. and R. Longino. — Naturwissenschaften 46 (1959), 664.
23. Barnes, R. and R. Lecandre. — J. Phys. Soc. Japan 22 (1967), 930.
24. Riedi, P. and R. Scurlock. — Phys. Rev. 24a (1967), 42.
25. Koi, J. and A. Tsujima. — J. Phys. Soc. Japan 15 (1960), 2100.
26. Street, R. — Phys. Rev. 121 (1961), 84.
27. Street, R. — J. Appl. Phys. 32 (1961), 1225.
28. Rogers, M. T. and K. Rama Rao. — J. Chem. Phys. 49 (19680, 1229.
29. Rockwood, S., E. Gregory, and D. Goodstein. — Phys. Rev. Lett. 30a (1969), 225.
30. Harris, C. — Inorg. Chem. 7 (1968), 1691.
31. Segel, S. and L. Anderson. — J. Chem. Phys. 49 (1968), 1407.
32. Fayer, M. and C. Harris. — Inorg. Chem. 8 (1969), 2792.
33. Segel, S. — J. Chem. Phys. 51 (1969), 848.
34. Nesmeyanov, A. N., G. K. Semin, E. V. Bryukhova, K. N. Anisimov, N. E. Kolobova, and V. N. Khandozhko. — Izv. AN SSSR, chem. ser., p. 1936, 1969.
35. Spencer, D., J. Kirsch, and T. Brown. — Inorg. Chem. 9 (1970), 235.
36. Brown, T., P. Edwards, C. Harris, and J. Kirsch. — Inorg. Chem. 8 (1969), 763.
37. Spiess, H., H. Haas, and H. Hartmann. — J. Chem. Phys. 50 (1969), 3057.
38. Graybeal, J., S. Ing, and M. Hsu. — Inorg. Chem. 9 (1970), 678.
39. Moobery, E., H. Spiess, B. Garrett, and R. Sheline. — J. Chem. Phys. 51 (1969), 1970.

SUPPLEMENT

Empirical formula	Structural formula	T, °K	e^2Qq, MHz	η, %	References, notes
		^2D			
AlD_4Li	$LiAlD_4$	Room	0.072	0	[1]. NMR cr.
BrD	DBr	96	0.153	...	[2]. NMR cr.
$CDCl_3$	$CDCl_3$	77	0.168	...	[3]. cf. Cl
CD_2Cl_2	CD_2Cl_2	77	0.169	...	[3]. cf. Cl
CD_3Br	CD_3Br	Room	0.171	—	[4]. Nematic l.
$CHDO_2$	DCOOH	—	0.249	...	[5]. g
	HCOOD	—	0.391	...	[5]. g
C_2D_2O	$CD_2{=}C{=}O$	—	0.240	...	[6]. g
$C_2D_3Cl_3$	CD_3CCl_3	77	0.174	...	[3]. cf. Cl
C_2D_3N	CD_3CN	Room	0.165	—	[4, 4a]. Nematic l.
$C_2D_4Cl_2$	CD_2ClCD_2Cl	77	0.172	...	[3]. cf. Cl
$C_3D_6Cl_2$	$CD_3CCl_2CD_3$	77	0.175	...	[3]. cf. Cl
$C_3H_3DO_4$	$DHC(COOH)_2$	Room	0.182 0.179 0.168 0.165	10.9 12.5 3.8 1.0	[7]. NMR cr.
$C_3D_4O_4$	$CD_2(COOD)_2$	»	0.169	—	[8]. ENDOR
$C_3H_9D_3BN$	$D_3B \cdot N(CH_3)_3$	»	0.105	...	[9]. NMR cr.
$C_6D_4Br_2$	$1,4\text{-}Br_2C_6D_4$	»	0.179	...	[4]. NMR cr.
C_6D_5Br	C_6D_5Br	»	0.179	...	[4]. NMR cr.
$C_6H_4DNO_2$	$4\text{-}DC_6H_4NO_2$	»	0.180	—	[10]. NMR l.
$C_{10}D_{10}Fe$	$(\pi\text{-}C_5D_5)_2Fe$	»	0.198	...	[11]. NMR cr.

$C_{14}D_{10}$	$C_6D_5C{\equiv}CC_6D_5$	Room	0.215	...	[11]. NMR cr.
ClD	DCl	77	0.149	...	[2,3]. cf. Cl
		96	0.190	...	
DH_2N	NH_2D	—	0.291	...	[5, 5a]. g
DI	DI	96	0.113	...	[2]. NMR cr.
$D_2O + CaO_4S$	$CaSO_4 \cdot D_2O$	Room	0.118	87.7	[12]. NMR cr.
$D_2O + {}^1/_4C_2D_2CuO_4$	$Cu(DCOO)_2 \cdot 4D_2O$	153	0.216	7.4	[13, 14]. NMR cr.
			0.217	7.9	
			0.214	13.6	
			0.221	9.1	
			0.214	10.4	
			0.214	9.4	
			0.221	11.2	
			0.215	13.4	
			0.161	3.4	
			0.160	2.5	
		244	0.158	5.7	
			0.157	5.0	
		253	0.157	5.3	
		291	0.155	8.5	
$D_2O + {}^1/_2Cl_2Cu$	$CuCl_2 \cdot 3D_2O$	Room	0.120	80	[7]. NMR cr.
$D_2O + {}^1/_6Cl_6NiSn$	$NiSnCl_6 \cdot 6D_2O$	220	0.129	88	[15]. NMR cr.
		298	0.121	90.6	
		328	0.118	90.6	
$D_2O + {}^1/_2D_4NNaO_4S$	$NaND_4SO_4 \cdot 2D_2O$	97	0.176	0	[16]. NMR cr. cf. Na
			0.058	0	
		298	0.125	79.2	
$D_2O + {}^1/_6F_6NiSi$	$NiSiF_6 \cdot 6D_2O$	157	0.238	10	[15]. NMR cr.
			0.236	11	
		296	0.121	95	
		343	0.115	96.3	
		423	0.045	0	

Empirical formula	Structural formula	T, °K	e^2Qq, MHz	η, %	References, notes
D_2S	D_2S	77	0.147	...	[17]. NMR cr.
		110	0.094	32	
		125	0.081	28	
D_4Ge	GeD_4	49	0.082	—	[18]. NMR 1.
		^{7}Li			
$AlLiO_4Si$	$LiAlSiO_4$	Room	0.132	37	[19]. NMR cr.
$D_3LiO_6Se_2$	$LiD_3(SeO_3)_2$	293	0.034	59.3	[20]. NMR cr. cf.. D
Li_3Sb	Li_3Sb	Room	0.075	0	[21]. NMR cr.
			0.018	...	
		^{23}Na			
$AlNaO_2$	$NaAlO_2$	Room	3.045	...	[22]. NMR cr.
$AsNa_3$	Na_3As	»	1.70	0	[21]. NMR cr.
			0.53	...	
$ClNaO_4$	$NaClO_4$	293	0.836	<10	[23]. NMR cr.
$CrNaO_2$	$NaCrO_2$	Room	3.295	...	[22]. NMR cr.
$CrNaS_2$	$NaCrS_2$	»	1.333	...	The same
$CrNaSe_2$	$NaCrSe_2$	»	1.034	...	»
$HNa_2O_3P + 5H_2O$	$Na_2HPO_3 \cdot 5H_2O$	»	1.750	98.8	[24]. NMR cr.
			1.200	44.5	

$H_3NaO_6Se_2$	$NaH_3(SeO_3)_2$	91	1,321	19.7	[25]. NMR cr.
			1,342	67.9	
		111	1.299	33.1	
			1,280	26.0	
			1,245	22.5	
			1,280	27.4	
		293	1.198	8.2	
$D_3NaO_6Se_2$	$NaD_3(SeO_3)_2$	243	1,248	6	[25]. NMR cr. cf.. D
			1,285	63	
		293	1,215	14	
		256	1.24	61	[26]
			1,22	13	
		298	1.19	9	
$H_4NNaO_4S + 2H_2O$	$NaNH_4SO_4 \cdot 2H_2O$	298	0,697	61,8	[16]. NMR cr.
$INaO_4$	$NaIO_4$	Room	0,049	1.2	[27]. NMR cr.
$InNaO_2$	$NaInO_2$	»	1,925	...	[22]. NMR cr.
$InNaS_2$	$NaInS_2$	»	0.687	...	The same
$InNaSe_2$	$NaInSe_2$	»	0.583	...	»
NaO_2Se	$NaSeO_2$	»	2.362	...	»
NaO_2Ti	$NaTiO_2$	»	3,201	...	[22]. NMR cr.
NaO_2Tl	$NaTlO_2$	»	0.538	...	The same
Na_2O_4S	Na_2SO_4	»	2.590	60,5	[28]. NMR cr.
Na_3Sb	Na_3Sb	»	1,49	0	[21]. NMR cr.
			0.44	...	
$Na_4O_7P_2 + 10H_2O$	$Na_4P_2O_7 \cdot 10H_2O$	»	1,907	14.3	[24]. NMR cr.
			0.467	87.8	

Empirical formula	Structural formula	T, °K	e^2Qq, MHz	η, %	References, notes
		^{87}Rb			
AsH_2O_4Rb	RbH_2AsO_4	108 370	5.76 6.90	85 0	[29]. NMR cr.
$Cl_4CuRb_2 + 2H_2O$	$Rb_2CuCl_4 \cdot 2H_2O$	Room	1.86	0	[30]. NMR cr.
HORb	RbOH	—	35	...	[31]. g
H_2O_4PRb	RbH_2PO_4	150 300	9.6 9.98	... ...	[32]. NMR cr.
		^{63}Cu			
$C_{12}H_{18}CuO_4$	$Cu(acac)_2$ in $Pd(acac)_2$	298 300	96 84	... ...	[33]. EPR
Cu_2HgI_4	$Cu_2[HgI_4]$	Room	0.82	...	[34]. NMR cr.
		^{197}Au			
AuI	AuI	77 300	512.554 516.420 516.620 508.296	... ...	[35]
		^{137}Ba			
$BaCl_2 \cdot 2H_2O$	$BaCl_2 \cdot 2H_2O$	300	ν=16.303 MHz		[36]
		^{11}B			
BBr_3	BBr_3	77	2.46	...	[37]. NMR cr.
BCl_3	BCl_3	77	2.54	...	The same
BI_3	BI_3	77	2.40	...	»

$C_2HD_6BO_2$	$HB(OCD_3)_2$	Room	3.0	—	[38, 38a]. NMR l.
$C_4H_{12}B_2Br_4N_2$	$[Br_2BN(CH_3)_2]_2$	»	0.406	100	[39]. NMR cr.
			or		
			0,812	0	The same
$C_4H_{12}B_2Cl_4N_2$	$[Cl_2BN(CH_3)_2]_2$	»	0.314	100	»
			or		
			0.628	0	»
$C_4H_{12}B_2F_4N_2$	$[F_2BN(CH_3)_2]_2$	»	0.140	51	»
$C_4H_{16}B_2N_2$	$[H_2BN(CH_3)_2]_2$	»	1.480	29	»
$C_8H_{24}B_2N_2$	$[(CH_3)_2BN(CH_3)_2]_2$	»	3.5	0	»
		^{27}Al			
$AlGdO_3$	$GdAlO_3$	293	0.544	87.6	[40]. NMR cr.
$Al_2Ca_3O_{12}Si_3$	$Ca_3Al_2(SiO_4)_3$	293	3.609	0	[41]. NMR cr.
$Al_2Dy_3O_{12}Si_3$	$Dy_3Al_2(SiO_4)_3$	293	5.873 0,447	... ...	[42]. NMR cr.
$Al_2O_{12}Si_3Tb_3$	$Tb_3Al_2(SiO_4)_3$	293	5.731 0.320	... ...	[42]. NMR cr.
$Al_2O_{12}Si_3Tm_3$	$Tm_3Al_2(SiO_4)_3$	Room	6,155 0.892	... ...	[43]. NMR cr.
$Al_2O_{12}Si_3Y_3$	$Y_3Al(SiO_4)_3$	293	6,020 0.630	... ...	[44]. NMR cr.
$Al_2O_{12}Si_3Yb_3$	$Yb_3Al_2(SiO_4)_3$	293	6.300 1.053	... ...	[42]. NMR cr.
$Al_3Ca_2O_{12}Si_3$	$Ca_2Al_3(SiO_4)_3$	Room	8.05 18.50	... ...	[45]. NMR cr.
$CH_6AlNO_8S_2 + 12H_2O$	$[CH_3NH_3]Al(SO_4)_2 \cdot 12H_2O$	138 293	0.640 0,315	0 ...	[46, 47]. NMR cr. cf. N
C_3H_9Al	$(CH_3)_3Al$	196	23.546	78.4	[47a]

Empirical formula	Structural formula	T, °K	Transition	ν, MHz	e^2Qq, MHz	η, %	References, notes
	^{69}Ga						
C_3H_9Ga	$Ga(CH_3)_3$						*
	Phase I	77	$^1/_2$—$^3/_2$	80.992	...	...	
	Phase II			80.352	...	...	
				81,008	...	...	
				81.168	...	...	
				81.472	...	...	
$C_6H_{15}Ga$	$Ga(C_2H_5)_3$	77	$^1/_2$—$^3/_2$	81.57	...	...	*[48]
				82.18	...	...	
$C_6H_{15}Ga + C_4H_{10}O$	$Ga(C_2H_5)_3 \cdot O(C_2H_5)_2$	77	$^1/_2$—$^3/_2$	62.96	...	...	*[48]
$C_{18}H_{15}Ga$	$Ga(C_6H_5)_3$	77	$^1/_2$—$^3/_2$	79.200	...	...	*[48]
		300	$^1/_2$—$^3/_2$	76,899	...	...	
	^{115}In						
BrOIn	InOBr	77	$^1/_2$—$^3/_2$	16,27	137.0	70.2	*[49]. cf. Br
			$^3/_2$—$^5/_2$	11,57			
			$^5/_2$—$^7/_2$	15.45			
			$^7/_2$—$^9/_2$	21.84			
$C_7H_8InNO_2$	$(CH_3)_2InCH_2NO_2$	77	$^3/_2$—$^5/_2$	73.07	942.5	33.1	*
			$^5/_2$—$^7/_2$	115.28			
C_3H_9In	$In(CH_3)_3$	77	$^1/_2$—$^3/_2$	53.90	1115.12	14	*[48]
			$^3/_2$—$^5/_2$	90.90			
			$^5/_2$—$^7/_2$	138.91			
		296	$^1/_2$—$^3/_2$	50.94	1103.3	11.3	
			$^3/_2$—$^5/_2$	90.39			
			$^5/_2$—$^7/_2$	137.35			
$C_3H_9In + C_4H_{10}O$	$In(CH_3)_3 + O(C_2H_5)_2$	77	$^1/_2$—$^3/_2$	43.001	1020.87	3.6	*[48]
			$^3/_2$—$^5/_2$	84.938			
			$^5/_2$—$^7/_2$	127.017			

$C_4H_{10}BrIn$	$(C_2H_5)_2InBr$	77	$^1/_2$—$^3/_2$ $^3/_2$—$^5/_2$ $^5/_2$—$^7/_2$ $^7/_2$—$^9/_2$	54.30 94.73 144,25 192,64	1157,2	12.2	* cf. Br
$C_4H_{10}ClIn$	$(C_2H_5)_2InCl$	77	$^1/_2$—$^3/_2$ $^3/_2$—$^5/_2$ $^5/_2$—$^7/_2$	57.90 92.45 142.13	1142.4	16.4	*
$C_6H_{15}In$	$In(C_2H_5)_3$	77	$^1/_2$—$^3/_2$ $^3/_2$—$^5/_2$ $^5/_2$—$^7/_2$ $^7/_2$—$^9/_2$	51.44 52,12 93,06 95.53 141.40 143.16 188,50 193.16	1135.2 1159.6	10.7 9.6	*[48]
$C_7H_{16}InNO_2$	$(C_2H_5)_2InC(CH_3)_2NO_2$	77	$^1/_2$—$^3/_2$ $^3/_2$—$^5/_2$ $^5/_2$—$^7/_2$	72.36 73.19 74,25 74.87 117.00 118.04	961.9 970,2	36,6 36.8	*
FIn	InF	—	—	—	—723.74	. . .	[51]. g
InN_3O_9	$In(NO_3)_3$	77	$^1/_2$—$^3/_2$ $^3/_2$—$^5/_2$ $^5/_2$—$^7/_2$ $^7/_2$—$^9/_2$	7.4 . . . 11.25 15.50	95	51.5	*[49]
$InO_4P + H_3O_4P +$ $+ 4H_2O$	$InPO_4 \cdot H_3PO_4 \cdot 4H_2O$	77	$^1/_2$—$^3/_2$ $^3/_2$—$^5/_2$ $^5/_2$—$^7/_2$ $^7/_2$—$^9/_2$	10.34 8.72 13.356 18.354	112.34	49.3	*
InSe	InSe	77	$^1/_2$—$^3/_2$ $^3/_2$—$^5/_2$ $^5/_2$—$^7/_2$ $^7/_2$—$^9/_2$	10.50 21.00 31.50 42.00	252.0	0	*

Empirical formula	Structural formula	T, °K	Transition	ν, MHz	e^2Qq, MHz	η, %	References, notes
In_2O_3	In_2O_3	77	$^1/_2$—$^3/_2$	18.1	129	85	*[50]
				7,65	182	0,8	
			$^3/_2$—$^5/_2$	12,5			
				15,2			
			$^5/_2$—$^7/_2$	13,5			
				23.9			
			$^7/_2$—$^9/_2$	19,3			
				30.5			
In_2Se_3	α-In_2Se_3	77	$^3/_2$—$^5/_2$	11,77	142	3,5	*[50]
				9.53	113	4.0	
			$^5/_2$—$^7/_2$	17,72			
				13,91			
			$^7/_2$—$^9/_2$	23,62			
				18 60			
	β-In_2Se_3	77	$^1/_2$—$^3/_2$	16,92	138,70	72,7	*
				...	205,47	42,9	
			$^3/_2$—$^5/_2$	11.93			
				15.74			
			$^5/_2$—$^7/_2$	15,34			
				24,81			
			$^7/_2$—$^9/_2$	22,20			
				33,83			
		300	$^1/_2$—$^3/_2$	17,10	138,0	73.6	*
				17,20	206.3	43.3	
			$^3/_2$—$^5/_2$	11,88			
				15,9			
			$^5/_2$—$^7/_2$	15,39			
				24,82			
			$^7/_2$—$^9/_2$	21,96			
				33,86			
In_2Te_3	In_2Te_3	77	$^1/_2$—$^3/_2$	23.77	177.1	83,5	*
				29,98	251.3	67,9	
			$^3/_2$—$^5/_2$	16.27			

			$5/2—7/2$	20.71 19.20			
			$7/2—9/2$	28.18 27.82 39.97			
In_2Te_5	In_2Te_5	77	$1/2—3/2$	14.2 18.9	182 184	39.0 58.0	*
			$3/2—5/2$	14.02 14.6			
			$5/2—7/2$	22.08 21.45			
			$7/2—9/2$	30,00 29.82			
			^{45}Sc				
Mn_2Sc	$ScMn_2$	4.2	—	—	0.777	0	[52]. NMR cr. cf. Mn
			Gd				
$Gd_3Fe_2O_{12}Si_3$	$Gd_3Fe_2(SiO_4)$	4.2	—	—	24	...	[53]. NMR cr.

Empirical formula	Structural formula	T, °K	ν_+, MHz	ν_-, MHz	e^2Qq, MHz	η, %	References, notes
			^{14}N				
$BaN_2O_4 + H_2O$	$Ba(NO_2)_2 \cdot H_2O$	77	4,671 4,639	3.686 3.672	5.571 5.541	35.4 34.9	[55, 56]
$Br_2H_6N_2$	$(H_3N^+N^+H_3)2Br^-$	77	3.0449	2,9336	3.9857	5.585	[57]
CH_2N_2	NH_2CN NH_2 CN	$e^2Qq_{aa} = 3.05$ MHz; $e^2Qq_{bb} = 1.85$ MHz $e^2Qq_{aa} = 3.30$ MHz; $e^2Qq_{bb} = 2.86$ MHz					[58]. g
CH_3F_2N	F_2NCH_3	—	—	—	—7.05	...	[59]. g

Empirical formula	Structural formula	T, °K	ν_+, MHz	ν_-, MHz	e^2Qq, MHz	η, %	References, notes
CH_3N	$H_2C{=}NH$	$e^2Qq_{aa} = 0.685$ MHz; $e^2Qq_{bb} = 3.004$ MHz; $e^2Qq_{cc} = -3.689$ MHz					[60]. g
CH_3NO	CH_3NO	$e^2Qq_{aa} = 0.5$ MHz; $e^2Qq_{bb} = -6.016$ MHz; $e^2Qq_{cc} = +5,518$ MHz					[61]. g
CD_3NO	CD_3NO	$e^2Qq_{aa} = 0,6$ MHz; $e^2Qq_{bb} = -6.007$ MHz; $e^2Qq_{cc} = 5.411$ MHz					[61]. g
$CH_6AlN_3O_4S + 6H_2O$	$[C(NH_2)_3{}^+]AlSO_4 \cdot 6H_2O$	77	3,0344	2.3287	3,5754	39.5	[62]. cf. Al
$CH_6AlN_3O_4Se + 6H_2O$	$[C(NH_2)_3{}^+]AlSeO_4 \cdot 6H_2O$	77	3,0304	...	...	...	[62]. cf. Al
CH_6ClNO	$(CH_3H_2N^+OH)Cl^-$	77	3.2381	2.6039	3.8947	32.567	[57]
$CH_6CrN_3O_4S + 6H_2O$	$[C(NH_2)_3{}^+]CrSO_4 \cdot 6H_2O$	77	3.0242	2.3148	3.5593	39.9	[62]
$CH_6GaN_3O_4S + 6H_2O$	$[C(NH_2)_3^+]GaSO_4 \cdot 6H_2O$	77	3.0231	2.3135	3.5756	39.9	[62]. cf. Ga
C_2H_3NO	H_3CCNO	—	—	—	0.495	...	[64]. g
C_2H_4ClN	CH_2CH_2NCl (ring)	77	4.1939	3.4598	5,1	29,5	[65]. cf. Cl
C_2H_5N	CH_2CH_2NH (ring)	77	3.1612 3,1599 3,1563	2.2182 2.207	3,5811	52.86	[65, 66]
C_2H_5NO	cis-CH_3CHNOH	$e^2Qq_{aa} = -2.8$MHz; $e^2Qq_{bb} - e^2Qq_{cc} = 0.6$MHz					[67]. g
$C_2H_6NO_2$	trans-CH_3CHNOH	$e^2Qq_{aa} = 4.07$MHz; $e^2Qq_{bb} - e^2Qq_{cc} = -5.0$MHz					[67]. g
	$H_3N^+CH_2COO^-$	Room	—	—	1.20	50	[68]. NMR cr.
$C_2H_6N_2O_2$	$(CH_3)_2NNO_2$	»	—	—	0.4	—	[69]. NMR l.

C_2H_7N	$NH(CH_3)_2$	—	—	—	—5.05	∼20	[70]. g
		77	3.8308 3.8805	3.0940 3.1118	4.6390	32.5	[71]
	$C_2H_5NH_2$	77	3.3593	2.7296	4.0592	30.9	[71]
$C_2H_7NO_3$	$(H_3N^+OH)OCOCH_3$	77	3.4423	3.0996	4.3613	15.715	[57]
C_2H_8ClNO	$(H_3N^+OCH_2CH_3)Cl^-$	77	3.1604	2.9608	4.0808	9.782	[57]
$C_2H_{10}Cl_2N_2$	$(CH_3H_2N^+N^+H_2CH_3)2Cl^-$	77	3.2265	2.3795	3.7372	45.328	[57]
C_2HgN_2	$Hg(CN)_2$	77 194 294.5	2.9945 2.9688 2.9430	2.9325 2.9056 2.8794	3.9513 3.9163 3.8816	3.14 3.23 3.28	[72]
C_3H_5N	C_2H_5NC	Room	—	—	0.30	—	[73]. NMR 1.
$C_3H_6N_2$	$(CH_3)_2CN_2$	$e^2Qq_{aa} = -0{,}94$ MHz; $e^2Qq_{bb} = 3.20$ MHz; $e^2Qq_{cc} = -2.3$ MHz					[74]. g
$C_3D_6N_2$	$(CD_3)_2CN_2$	$e^2Qq_{aa} = -1.2$ MHz; $e^2Qq_{bb} = 3.4$ MHz; $e^2Qq_{cc} = -2.2$ MHz					[74]. g
C_3H_7N	$CH_2CH_2NCH_3$ (ring)	77	3.5523 3.5462	... ...	... ...	... ...	[66]
	$CH_2CH_2CH_2NH$ (ring)	77	3.6207 3.6150	2.9439 2.9320	4.53	30	[65, 66]
			3.5928 3.5687	... ...	... ...	... ...	
		112	3.5502 3.5168	... ...	... ...	... ...	
		141	3.4985	...	...	...	
		183	3.5212 3.4466 3.4361	2.9192 2.9045 2.8975	4.250	26.4	
$C_3H_7NO_2$	$(CH_3)_2CHNO_2$	Room	—	—	1.55	—	[69]. NMR 1.
$C_4CdK_2N_4$	$K_2Cd(CN)_4$	77 193 294	3.1491 3.0643 2.9813	— — —	4.1988 4.0857 3.9751	0 0 0	[72]
$C_4CuK_3N_4$	$K_3Cu(CN)_4$	77	3.0021	2.9438	3.9639	2.94	[72]

Empirical formula	Structural formula	T, °K	ν_+, MHz	ν_-, MHz	e^2Qq, MHz	η, %	References, notes
$C_4H_6N_2S$	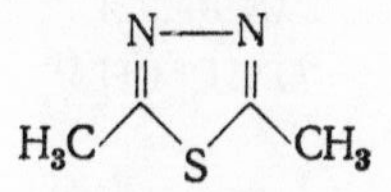	295	3.5018	2.7090	4.140	38.3	[75]
C_4H_9N	$CH_2CH_2CH_2CH_2NH$ (ring)	77	3.6380	2.9016	4.3597	33.8	[65, 66]
		141	3.6164	2.9016	4.3453	32.9	
		161	3.6075	2.9016	4.3393	32.5	
		195	3.5883	2.9016	4.3266	31.7	
	$(CH_3)_2CCH_2NH$ (ring)	77	3.1585	2.1393	3.53	58	[65]
C_4H_9NO	$CH_2CH_2OCH_2CH_2NH$ (ring)	77	3.6630	3.0603	4.4822	26.8	[65, 66]
		141	3.6501	3.0585	4.4724	26.4	
		195	3.6357	3.0570	4.4618	25.9	
$C_4H_{10}N_2O_2$	$(C_2H_5)_3NNO_2$	Room	—	—	0.4	—	[68]. NMR 1.
$C_4H_{11}N$	$(C_2H_5)_2NH$	77	3.644	3.103	4.618	24.5	[71]
$C_4HgK_2N_4$	$K_2Hg(CN)_4$	77	3.0356	...	4047.5	0	[72]
			3.0694	3.0414	4073.9	1.37	
		193	2.9676	...	...	...	
		294	2.8993		...	...	
$C_4K_2N_4Pt + 3H_2O$	$K_2Pt(CN)_4 \cdot 3H_2O$	77	2.6412	2.5947 }	3.467	3.2	[72]
			2.6142	2.5497 }			
$C_4K_2N_4Zn$	$K_2Zn(CN)_4$	77	3.1042	—	4.1390	0	[72]
		193	3.0485	—	4.0647	0	
		294	2.9925	—	3.9900	0	
C_5H_4FN	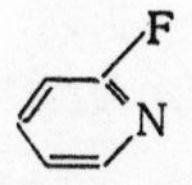	77	3.39993	2.90353	4.20231	23.63	[76]

$C_5H_6N_2$	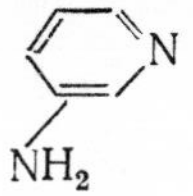	77	3.8104	2.9344	4.4965	38.96	[77, 78]
			3.1382	2.4269	3.7100	38.35	
		293	3.723	2.8891	4.4081	37.83	
			3.132	2.4487	3.7205	36.73	
	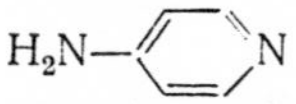	77	2.9679	2.2917	3.5064	38.57	[77, 78]
			2.9155	2.7555	3.7807	8.46	
		293	2.913	2.7358	3.7659	9.41	
			2.948	2.2957	3.4948	37.32	
$C_5H_{11}N$	$CH_2CH_2CH_2CH_2NCH_3$ (ring)	77	3.7722	3.7488	5.0140	9.3	[66]
$C_5H_{11}NO$	$CH_2CH_2OCH_2CH_2NCH_3$ (ring)	77	3.8169	3.7648	5.0545	2.06	[66]
$C_5H_{12}N_2$	$CH_2CH_2NHCH_2CH_2NCH_3$ (ring)	77	3.6824 3.6650	3.0256 2.9842	4.4524	30.0	[66]
			3.8242 3.8173	3.7758 3.7737	5.0637	18.2	
C_6H_6IN	4-$IC_6H_4NH_2$	292	3.1360	2.5120	3.765	33.15	[79]
C_6H_7N	$C_6H_5NH_2$	77	3.2432 3.1837	2.7207 2.6498	3.932	26.9	[71, 77, 79]
		143	3.2345 3.1762	2.7160 2.6400	3.922	26.9	
		195	3.2112 3.1605	2.7100 2.6180	3.900	26.8	
		228	3.2017 3.1460	2.7043 2.593	3.882	27.06	
		250	... 3.1340	2.6990 2.5780	3.848	25.8	
$C_6H_6NO_2S$	$C_6H_5SO_2NH_2$	Room	3.172	—	4.229	0	[80]

Empirical formula	Structural formula	T, °K	ν_+, MHz	ν_-, MHz	e^2Qq, MHz	η, %	References, notes
$C_6H_8N_2$	1,2-$(NH_2)_2C_6H_4$	77	3.1506 3.1316	2.5360 2.4973	3.772	33.1	[79]
		143	3.1428 3.1215	2.5338 2.4923	3.764	32.9	
		195	3.1340 3.1091	2.5255 2.4842	3.751	32.9	
		292	3.1075 3.0740	2.5000 2.4550	3.712	33.0	
	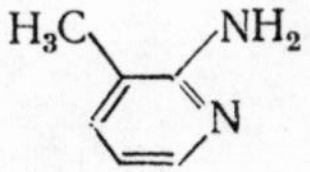	77	3.015 3.007	2.892 2.4365	3.938 3.629	6.2 20.9	[78]
		293	2.852 2.8726	2.7442 ...	3.7308	5.8	
	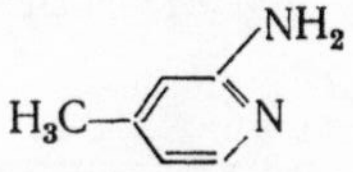	77	2.8369 2.9338	2.7223 ...	3.706	6.2	[78]
	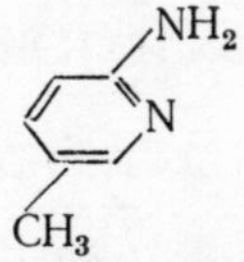	77	2.8674 3.0436	2.793 2.419	3.774 3.6417	3.95 22.9	[78]
	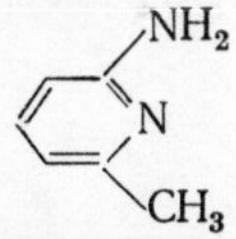	77	2.838 2.997	2.740 ...	3.719 ...	3.4 ...	[78]
$C_6H_{12}N_2$	$N(C_2H_4)_3N$	78.8 294.9	3.6930 3.6562	... ...	... ...	... ...	[81]
$C_6H_{13}N$	$CH_2CH_2CH(CH_3)CH_2CH_2NH$ (ring)	77	3.6490	2.9710	4.4133	30.7	[66]

	$CH_2CH_2CH_2CH_2CH_2NCH_3$ (ring closed by bracket)	77	3.7840	3.6945	4.9857	3.59	[66]
$C_6H_{18}N_3OP$	$[(CH_3)_2N]_3PO$	77	4.0010 3.9400 3.8250	3.4255 ... 3.3620	4.951 ... 4.791	23.3 ... 19.3	[75]
C_6N_4	$(CN)_2C{=}C(CN)_2$	77	3.2131 3.2102	3.1483 } 3.0905 }	4.331	4.4	[82]
C_7H_7N	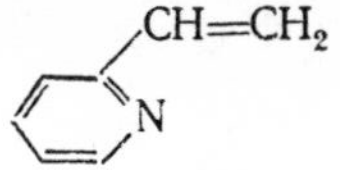	77	3.74569	3.04544	4.52742	30.93	[76]
C_7H_7NO	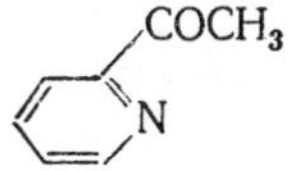	77	3.59614	3.03796	4.68940	40.87	[76]
	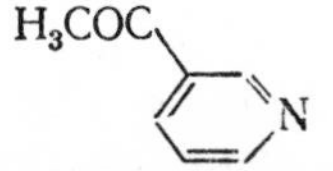	77	3.90005	3.03791	4.62531	37.28	[76]
	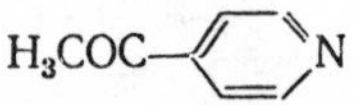	77	4.08876	3.05381	4.76171	43.47	[76]
$C_7H_7NO_2$	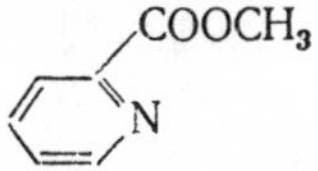	77	4.05497 4.04027	3.12708 } 3.12598 }	4.78276	38.52	[76]
	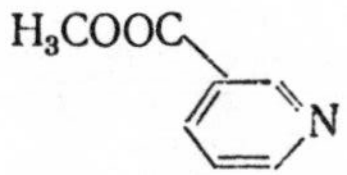	77	3.94643 3.91679	3.05213 } 3.04489 }	4.65341	37.95	[76]
	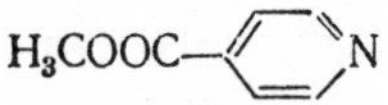	77	4.09819	3.06954	4.77849	43.05	[76]
C_7H_8IN	$2\text{-}I\text{-}4\text{-}CH_3C_6H_3NH_2$	296	3.0790	...	...	...	[75]
C_7H_9N	$4\text{-}CH_3C_6H_4NH_2$	293	3.2270 3.1475	2.9270 ...	4.1027 ...	14.2 ...	[75]

Empirical formula	Structural formula	T, °K	ν_+, MHz	ν_-, MHz	e^2Qq, MHz	η, %	References, notes
C_7H_9NO	$4\text{-}CH_3OC_6H_4NH_2$	293	3.2870	2.7220	4.006	28.2	[75]
$C_7H_{10}N_2$	$2,6\text{-}(NH_2)_2C_6H_3CH_3$	294	3.1035 3,0700	2.5095 2.4900	3,742 3.707	31.7 31.3	[75]
	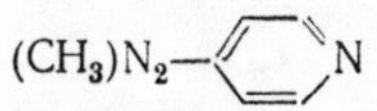	77 293	3.1668 3.6297 3.1482 3.5283	2.8661 3,5689 2.837 3.5169	4.0219 4,7990 3.9901 4.6968	14,95 2.53 15,60 0.5	[78]
$C_8H_6N_2S$	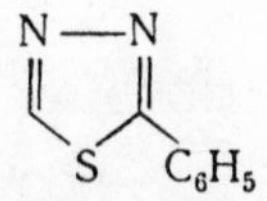	296	2.7000 2.6750	2.4560 2.4415	3.437 3.411	14.2 13.7	[75]
C_8H_7NO	$4\text{-}CH_3OC_6H_4CN$	77	3,01590	3.01464	4.02034	6	[76]
$C_9H_9NO_2$	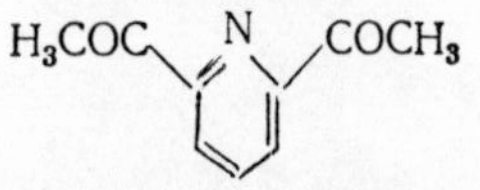	77	4.13331	3.08872	4.81469	43.39	[76]
$C_8H_{12}N_2$	$NCC(CH_3)_2C(CH_3)_2CN$	77	2,9761	2.9038	3.9199	3.69	[63]
$C_8H_{15}N$	$C_7H_{15}CN$	77	2.8972	2.8069	3,8027	4.75	[63]
$C_{12}H_4N_4$	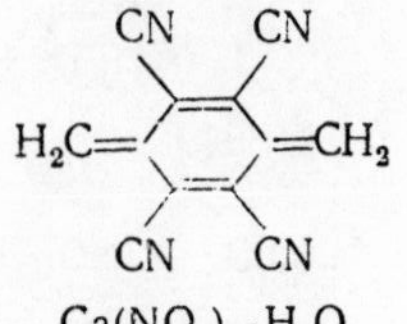	77	3.1516 3.1429	2.8887 2.8795	4.021	13.1	[82]
$CaN_2O_4 + H_2O$	$Ca(NO_2)_2 \cdot H_2O$	77	4.594 4.698 4.707 4.924	3.668 3,720 3,756 3.865	5.655	34.6	[55]
ClH_4NO	$(H_3N^+OH)Cl^-$	77	3.0531	2.9667	4.0132	4.306	[57]

$Cl_2D_6N_2$	$(D_3N^+N^+D_3)2Cl^-$	77	2,9302	—	3,9069	0	[57]
F_2HN	F_2NH	—	—	—	—8,9	44	[83]. g
F_3NO	NF_3O	—	—	—	—1.52	. . .	[84]. g
$H_6F_2N_2$	$(H_3N^+N^+H_3)2F^-$	77	3.0908	—	4,1211	0	[57]
$H_6N_2O_4S$	$(H_3N^+N^+H_3)SO_4^{2-}$	77	3.0340 3,0819	2.7992 2.9668	3.8890 4.0324	12,075 5.709	[57]
$H_8N_2O_6S$	$(H_3N^+OH)_2SO_4^{2-}$	77	3,2264 3.2577	3,0617 3,1624	4,1921 4,2801	7,858 4.453	[57]
$LiNO_2 + H_2O$	$LiNO_2 \cdot H_2O$	77	4.642	3.648	5.5267	35.97	[55]
$N_2O_4Sr + H_2O$	$Sr(NO_2)_2 \cdot H_2O$	77	4,784	3,757	5,6940	36.07	[55]

Empirical formula	Structural formula	T, °K	ν, MHz	s/n	References, notes
	^{75}As				
AsSTl	TlAsS	77	68.93	10	*
As_2Pt	$PtAs_2$	4,2	26.48	10	[54]
As_2Te_3	As_2Te_3	77	43.76	10	*[85]
$As_4I_2Te_5$	$As_4Te_5I_2$	77	46,28	8	*
CH_3AsCl_2	Cl_2AsCH_3	77	170,807	30	*[86]. cf. Cl
$CH_{11}AsB_{10}$	 o-arsenoborane	77	97.3	3	*$\Delta\nu \approx 2$ MHz
	$AsB_{10}H_{10}CH$ m-arsenoborane	77	58,2	3	The same

Empirical formula	Structural formula	T, °K	ν, MHz	s/n	References, notes
$CH_{11}AsB_{10}$ (contd.)	$AsB_{10}H_{10}CH$ p-arsenoborane	77	61.36	3	*
$C_2H_5AsCl_2$	$Cl_2AsC_2H_5$	77	172.28	10	*[86]. cf. Cl
$C_3H_7AsCl_2$	$Cl_2AsC_3H_7$	77	172.64	3	The same
$C_6F_{15}As$	$(C_2F_5)_3As$	77	108.92	15	*
$C_6H_3F_{12}As$	$(F_3CCFH)_3As$	77	103.28	5	*
$C_6H_{15}As$	$As(C_2H_5)_3$	77	89.613	50	*[86]
$C_{12}H_{10}AsCl$	$ClAs(C_6H_5)_2$	77	161.94	10	*[86]. cf. Cl
$C_{14}H_{13}As$	$(C_6H_5)_2AsCH{=}CH_2$	77	98.411	5	*[86]
$C_{19}H_{15}AsO_2$	$3\text{-}COOHC_6H_4As(C_6H_5)_2$	77	100.04	20	*[86]
	$4\text{-}COOHC_6H_4As(C_6H_5)_2$	77	99.18	25	*[86]
$C_{21}H_{21}As$	$(2\text{-}CH_3C_6H_4)_3As$	77	101.100 102.264	8 8	*[86]
	$(4\text{-}CH_3C_6H_4)_3As$	77	99.300	10	*[86]
$C_{25}H_{20}AsMnO_2$	$C_5H_5Mn(CO)_2 \cdot As(C_6H_5)_3$	77	75.12	10	*cf. Mn

Empirical formula	Structural formula	Isotope	T, °K	ν, MHz $^{1}/_{2}$—$^{3}/_{2}$	ν, MHz $^{3}/_{2}$—$^{5}/_{2}$	ν, MHz $^{5}/_{2}$—$^{7}/_{2}$	e^2Qq, MHz	η, %	References, notes
Sb									
AgS_2Sb	$AgSbS_2$	121	77	54.00	94.44	—	321.7	34.1	[87]
		123	77	39.57	55.80	86.82	410.1	34.1	
		121	300	52.80	91.83	—	313.2	34.8	
		123	300	...	54.20	84.47	399.2	35.1	
C_3H_9Sb	$(CH_3)_3Sb$	121	77	74.28	148.35	—	494.6	3.3	*
		123	77	45.20	90.03	135.09	630.5	3.2	
$C_6H_6Br_2Cl_3Sb$	(cis-ClCH=CH)$_3$SbBr$_2$	121	77	74.880	147.936	—	494.0	9.8	*[152]. cf.
				73.437	146.616	—	488.8	3.7	Br and Cl—C
		123	77	46.548	89.49	...	629.7	9.8	
				44.793	88.92	...	623.0	3.7	
$C_6H_6Cl_3Sb$	(trans-CH=CHCl)$_3$Sb	121	77	79.97	159.90	—	533.0	1.3	
		123	77	48.564	97.052	145,58	679.4	1.3	*cf. Cl—C
$C_6H_6Cl_5Sb$	(cis-ClCH=CH)$_3$SbCl$_2$	121	77	80.30	160.176	—	534.6	3.8	*[152]. cf.
		123	77	48.93	97.272	146,1	681,5	3.8	Cl—C,Cl—Sb
	(trans-CH=CHCl)$_2$·Sb·(cis-CH=CHCl)Cl$_2$	121	77	86.53	...	—	570.2	7.0	*cf.
				85.98	...	—	573.8	7.0	Cl—C,Cl—Sb
		123	77	52.83	103.56	...	726.8	7.0	
				53.17	104.20	...	731.5	7.0	
$C_6H_{33}B_{30}Sb$	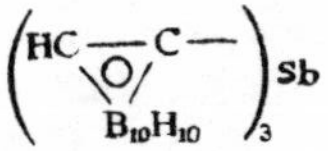 antimony tri-(o-borane)	123	77	...	83.19	...	...	...	*
$C_{12}H_{10}ClSb$	$(C_6H_5)_2SbCl$	121	77	180.63	209.26	—	768.8	82.5	*
		123	77	162.72	127,95	194,75	979.9	82.5	
$C_{12}H_{27}Br_2Sb$	(iso-C$_4$H$_9$)$_3$SbBr$_2$	121	77	100,28	...	—	666.64	5.2	*[152]. cf.
		123	77	60.84	120.36	...	843.8	5 2	Br

Empirical formula	Structural formula	Isotope	T, °K	ν, MHz $1/2-3/2$	ν, MHz $3/2-5/2$	ν, MHz $5/2-7/2$	e^2Qq, MHz	η, %	References, notes
$C_{15}H_{45}B_{30}Sb$	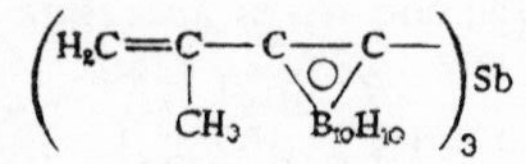	121	77	73.52	...	—			*
				74.58	...	—			
$C_{18}H_{12}Cl_3Sb$	$(4\text{-}ClC_6H_4)_3Sb$	121	77	...	157.69	—	517.3	3,8	*
		123	77	47.35	94.13	...	659.5	3.8	
$C_{19}H_{15}MnO_2Sb$	$C_5H_5Mn(CO)_2 \cdot Sb(C_6H_5)_2$	121	77	62.10	123.16	—	411.08	8.2	* cf. Mn
		123	77	...	74.62	112.19	524.02	8.2	
$C_{21}H_{21}O_3Sb$	$(4\text{-}CH_3OC_6H_4)_3Sb$	121	77	79.52	159.03	—	530.1	0.4	*
		123	77	48.27	96.53	144.80	675,7	0.4	
$C_{22}H_{24}Cl_2NO_2Sb$	$(2\text{-}C_2H_5OC_6H_4)_2Sb \cdot$	121	77	82.59	...	—	552.11	7.7	* cf. Cl
	$\cdot (2\text{-}NH_2C_6H_4)Cl_2$	123	77	51.35	100.20	...	703.8	7.7	
$C_{26}H_{20}Sb_2$	$[(C_6H_5)_2SbC\equiv]_2$	121	77	106.00	163.53	—	569,32	50	*
		123	77	85.76	95.87	...	725.75	50	
CuS_2Sb	$CuSbS_2$	121	77	56.922	113,802	—	379.3	1.6	[88]
			300	55.36	110.69	—	368.9	1,4	
		123	77	34.582	69.072	103,630	482.14	1.8	
S_2SbTl	$TlSbS_2$	121	77	75,288	102.306	—	364.6	64.5	*
				75,764	112.09	—	392.9	55.3	
		123	77	64,790	60.400	95.248	464.8	64.5	
				62.500	65,505	103,734	499.1	55.3	

Empirical formula	Structural formula	T, °K	ν, MHz				e^2Qq, MHz	η, %	References, notes
			1/2—3/2	3/2—5/2	5/2—7/2	7/2—9/2			
			^{209}Bi						
BiBrO	BiOBr	77	...	10.3	15.5	20.7	101.0	0	*
BiClO	BiOCl	77	6.59	13.18	19.77	26.36	131.8	0	*
$BiCl_6Cs_2K$	Cs_2KBiCl_6	77	...	9.082	13.690	18,240	109.6	6.6	*cf. Cl
BiIO	BiOI	77	...	...	12.1	16.1	80.0	0	*
$Bi_2Mo_3O_{12}$	$Bi_2Mo_3O_{12}$	77 300	20.68 28.74 33.7	35.9 26.5 25.7	54.7 41.4 37.3	73.06 56.45 52.0	439.0 344.0 321.3	12.5 42.9 59.2	*
Bi_2WO_6	Bi_2WO_6	77	35.03 37,02	34.996 34.62	55.107 54.19	74.75 73.82	453.84 449.4	38.1 42.0	*
$C_{18}H_{12}BiBr_3$	$(4\text{-}BrC_6H_4)_3Bi$	77	...	55.804	84.005	111.993	672.54	6.3	[89]
$C_{18}H_{12}BiCl_3$	$(4\text{-}ClC_6H_4)_3Bi$	77	...	57.083	85,777	...	686.50	4.7	[89]
$C_{21}H_{21}Bi$	$(2\text{-}CH_3C_6H_4)_3Bi$	77	...	56.772	85.286	...	682.79	4.0	[89]
	$(3\text{-}CH_3C_6H_4)_3Bi$	77	...	55.760	83.835	111.836	671.13	4.9	[89]
	$(4\text{-}CH_3C_6H_4)_3Bi$	77	...	56.534	84.955	113.364	680.05	4.2	[89]

Empirical formula	Compound	T, °K	e^2Qq, MHz	η, %	References, notes
		^{51}V			
$C_{10}H_7O_3V$	π-$C_7H_7V(CO)_3$	Room	2.0	—	[90]. NMR l.
Cl_3OV	$VOCl_3$	167	5.4	⩽20	[91]. NMR cr. cf. Cl
F_3OV	VOF_3	301	8.9	⩽40	[91]. NMR cr.
O_2V	VO_2	Room	6.86	49	[92, 93]. NMR cr.
		^{93}Nb			
N_xNb	NbN_x				[94]. NMR cr.
	ε-phase	Room	10.6	...	
	δ-phase	»	4.3	...	
F_5Nb	NbF_5	77	114.27 115.77	16.8 12.4	*

Empirical formula	Structural formula	T, °K	ν, MHz	s/n	References, notes
	Group I element — chlorine bond				
Cl_3CsCu	$CsCuCl_3$	77	11.430 10.930	...	*[153]. At 300 °K $\eta_1 = 38.4\%$; $\eta_2 = 13.1\%$
		300	11.249 10.701		
Cl_3CuK	$KCuCl_3$	4.2	12.048 12.380 10.407	...	[112]. $\eta_1 = 22.5\%$; $\eta_2 = 51.3\%$; $\eta_3 = 91.6\%$

	Group II element — chlorine bond				
CH_3ClHg	$ClHgCH_3$	77	14.75	3	*
$C_6H_3Cl_3Hg$	2,6-$Cl_2C_6H_3HgCl$	77	17.964	. . .	* cf. C—Cl
$C_6H_4Cl_2Hg$	2-ClC_6H_4HgCl	77	17.328	. . .	cf. C—Cl
$Cl_2Hg + C_6H_4O_2$	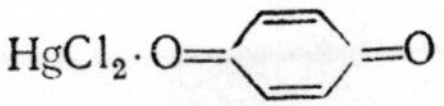	296	20.540	. . .	[113]
$Cl_2Hg + C_7H_8O_2$	$HgCl_2 \cdot CH_3C{=}CHCOCH{=}C(CH_3)O$ (ring closure C…O)	296	19.155	. . .	[113]
$Cl_2Hg + C_8H_8O$	$HgCl_2 \cdot C_6H_5COCH_3$	296	20.872 19.396	. . .	[113]
$Cl_2Hg + C_{13}H_{10}O$	$HgCl_2 \cdot C_6H_5COC_6H_5$	296	21.685 20.742	. . .	[113]
	Group III element — chlorine bond				
$C_2H_{11}B_{10}Cl$	$HCB_{10}H_9ClCH$ B-2-chloro-p-borane	77	23.938	. . .	*[113a]
Cl_3Tl	$TlCl_3$	300	14.812 14.756	. . .	[114]
Cl_3Y	YCl_3	4.2	4.212 4.273	2 1	[114]
Cl_3La	$LaCl_3$	4.2	4.167	. . .	[114]. $\eta = 50\%$
Cl_3Ce	$CeCl_3$	4.2	4.387	. . .	[114]
Cl_3Pr	$PrCl_3$	4.2	4.5667	. . .	[114]. $\eta = 58\%$
Cl_3Tb	$TbCl_3$	77	7.45 4.30	. . .	[115]
Cl_3Dy	$DyCl_3$	77 300	6.15 4.203	. . .	[114, 115]
Cl_3Ho	$HoCl_3$	77	4.3027 4.3647	2 1	[114]

Empirical formula	Structural formula	T, °K	ν, MHz	s/n	References, notes
Cl_3Er	$ErCl_3$	77	4.4495	2	[114]
			4.506	1	
Cl_3Yb	$YbCl_3$	77	4.7817	2	[114]
			4.8366	1	
Cl_3U	UCl_3	77	6.24	...	[115]
			4.51	...	[114]
	Group IV element — chlorine bond				
$C_3H_2Cl_5F_3Si$	$F_3CCH_2CCl_2SiCl_3$	77	20.096	...	*cf. C—Cl
			19.524		
			19.470		
$C_{12}H_{10}Cl_2FeGeO_4$	$[(C_5H_5)Fe(CO)_2]_2GeCl_2$	77	17.355	8	*
$CH_2Cl_4Sn + C_4H_{10}O$	$ClCH_2SnCl_3 \cdot O(C_2H_5)_2$	77	21.540	20	*cf. C—Cl
			21.400	10	
$CH_2Cl_4Sn + C_6H_6$	$ClCH_2SnCl_3 \cdot C_6H_6$	77	21.512	20	*cf. C—Cl
			21.392	10	
$C_8Cl_2Co_2O_8Sn$	$Cl_2Sn[Co(CO)_4]_2$				[97, 101]
	Phase I	298	17.66	...	
			17.14		
	Phase II	298	17.78	...	cf. Co
			17.08		
Cl_3CsSn	$CsSnCl_3$	77	10.844	3	*
			10.768	1.5	
$Cl_4Sn + C_3H_8O_2$	$SnCl_4 \cdot CH_2(OCH_3)_2$	77	20.720	3	*
			20.464	3	
			19.758	3	
			19,530	3	

$Cl_4Sn + 2C_3H_8O_2$	$SnCl_4 \cdot 2CH_2(OCH_3)_2$	77	19.734	4	*
			19.656	4	
			19.500	4	
			19.236	4	
$Cl_4Sn + 2H_2O$	$SnCl_4 \cdot 2H_2O$	77	21.439		[116]
$Cl_4Sn + 3H_2O$	$SnCl_4 \cdot 3H_2O$	77	20.384		[116]
$Cl_4Sn + 3D_2O$	$SnCl_4 \cdot 3D_2O$	77	20.438		[116]
$Cl_4Sn + 5H_2O$	$SnCl_4 \cdot 5H_2O$	293	18.054		[116]
			17.656		
$Cl_4Sn + 5D_2O$	$SnCl_4 \cdot 5D_2O$	293	18.102		[116]
			17.735		
Cl_6Cs_2Sn	Cs_2SnCl_6	300	16.06		[106]
	Group V element — chlorine bond				
$C_2H_5Cl_2OP$	$C_2H_5OPCl_2$				*
	Phase I	77	23.392	10	
			23.425	10	
	Phase II	77	23.536	2	
			23.640	8	
			23.672	4	
			23.072	2	
	Phase III	77	23.744	3	
			23.548	3	
			23.480	3	
			23.336	3	
	Phase IV	77	23.664	8	
			23.368	8	
$C_4H_7Cl_2O_2P$	$C_2H_5OCH{=}CHP(O)Cl_2$	77	26.56	2	*
			26.30	2	
			26.23	4	
$C_4H_{10}ClP$	$(C_2H_5)_2PCl$	77	25.49	2	*

Empirical formula	Structural formula	T, °K	ν, MHz	s/n	References, notes
$C_4H_{10}Cl_2NP$	$(C_2H_5)_2NPCl_2$	77	24.616	4	*
			24.640	4	
			22.960	4	
			22.944	4	
$C_6H_4Cl_2NO_3P$	$3\text{-}NO_2C_6H_4P(O)Cl_2$	77	26.935	10	*
$C_6H_4Cl_5P$	$4\text{-}ClC_6H_4PCl_4$	77	33.840	10	*[117]. cf. C—Cl
			33.420	10	
			25.535	8	
			25.060	8	
$C_6H_5Cl_4P$	$C_6H_5PCl_4$	77	33.744	10	*[117, 118]
			33.588	10	
			25.510	8	
			24.608	6	
$C_6H_5Cl_9P_2$	$[C_6H_5PCl_3]^+PCl_6^-$	77	30.894	2	*
			30.792	6	
			30.654	2	
			30.624	2	
			30.144	6	
			30.060	2	
			29.880	4	
$C_7H_7Cl_2OP$	$4\text{-}CH_3OC_6H_4PCl_2$	77	26.380	5	*
			25.710	5	
$C_7H_7Cl_2OPS$	$4\text{-}CH_3OC_6H_4P(S)Cl_2$	77	28.030	5	*
			27.530	5	
$C_7H_7Cl_2O_2P$	$4\text{-}CH_3OC_6H_4P(O)Cl_2$	77	26.66	10	*
			26.47	20	
			26.02	10	
$C_7H_7Cl_4OP$	$4\text{-}CH_3OC_6H_4PCl_4$	77	31.536	6	*[117]
			30.606	2	
$C_8H_6Cl_2NO_4P$	$3\text{-}NO_2C_6H_4OCH{=}CHP(O)Cl_2$	77	26.355	5	*

$C_8H_7Cl_2OP$	$C_6H_5CH{=}CHP(O)Cl_2$	77	26.650 26.325 26.220	2 3 3	*
$C_8H_7Cl_2P$	$C_6H_5CH{=}CHPCl_2$	77	25.875	3	*
$C_8H_{18}ClP$	$(tert\text{-}C_4H_9)_2PCl$	77	25.70	4	*
$C_{10}H_{11}Cl_2OP$	$(CH_3)_2C{=}C(C_6H_5)P(O)Cl_2$	77	26.125	3	*
$C_{12}H_{10}ClP$	$(C_6H_5)_2PCl$	77	26.32	2	*
$C_{12}H_{10}Cl_3P$	$(C_6H_5)_2PCl_3$	77	33.450 22.336	3 6	*[117]
$Cl_3OP + BBr_3$	$POCl_3 \cdot BBr_3$	77	31.135 31.104	2 4	*
$Cl_3P + BBr_3$	$PCl_3 \cdot BBr_3$	77	30.120	30	*
$Cl_3P + C_8H_5Cu$	$PCl_3 \cdot CuC{\equiv}CC_6H_5$	77	27.945	25	*
Cl_4PCu	$PCl_3 \cdot CuCl$	77	27.975	50	*
Cl_4NO_2PS	Cl_3PNSO_2Cl	77	30.135 30.251	10 10	[118]
$Cl_4N_3O_2PS_2$	$Cl_2PN[S(NO)Cl]_2$	294	28.660 29.836	. . .	[119]
	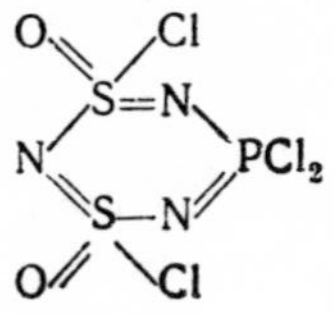	77	29.91	. . .	[118]. cf. S—Cl

Empirical formula	Structural formula	T, °K	ν, MHz	s/n	References, notes
Cl_5NOP_2	$Cl_3PNP(O)Cl_2$	77	30,742 30,680 30,403 30,360 30,034 29,966 29,768 29,734 29,534 29,422 27,219 26,845 26,775 26,692 26,555	...	*
$Cl_5P + C_{18}H_{15}OP$	$Cl_5P \cdot (C_6H_5)_3P(O)$	77	31,128 29,755 29,580	...	*
$AsCl_3 + {}^1/_2C_6H_6$	$2AsCl_3 \cdot C_6H_6$	77	25,634 25,560 25,406 25,261 24,765 24,378	...	[120]
$AsCl_3 + {}^1/_2C_8H_5Cu$	$AsCl_3 \cdot CuC{\equiv}CC_6H_5$	77	26,73	...	*
$AsCl_3 + {}^1/_2C_8H_{10}$	$2AsCl_3 \cdot 1,4\text{-}(CH_3)_2C_6H_4$	77	25,535 25,169 24,040	...	[120]
	$2AsCl_3 \cdot C_6H_5C_2H_5$	77	25,321 25,238 25,174 25,135	...	[120]

			25.059 24.309		
$AsCl_3 + C_8H_{10}$	$AsCl_3 \cdot C_6H_5C_2H_5$	77	25.220 25.102 24.928	...	[120]
$AsCl_3 + {}^1/_2C_{10}H_8$	$2AsCl_3 \cdot C_{10}H_8$	77	25.805 25.350 24.720	...	[120]
$AsCl_3 + {}^1/_2C_{13}H_{12}$	$2AsCl_3 \cdot (C_6H_5)_2CH_2$	77	25.388 24.858 24.701	...	[120]
CH_3AsCl_2	CH_3AsCl_2 Phase I	77	25.125 24.865	...	*
	Phase II	77	23.956 23.520	...	*[86]. cf. As
$C_2H_5AsCl_2$	$C_2H_5AsCl_2$	77	24.668 24.296	...	The same
$C_3H_7AsCl_2$	$(C_3H_7)AsCl_2$	77	24.328 24.440	...	»
$C_4H_9AsCl_2$	$(C_4H_9)AsCl_2$	77	24.600 25.060	...	»
$C_{12}H_{10}AsCl$	$(C_6H_5)_2AsCl$	77	24.312	...	»
$C_{18}H_{15}AsCl_2$	$(C_6H_5)_3AsCl_2$	77	16.15 16.36	...	»
$AgCl_6Cs_2Sb$	$Cs_2AgSbCl_6$	77	12.344	...	*
$C_6H_6Cl_5Sb$	(cis-ClCH=CH)$_3$SbCl$_2$	77	16.640 16.860	3 3	*cf. C—Cl, Sb
	(cis-ClCH=CH)·(trans-ClCH=CH)$_2$·SbCl$_2$	77	16.91	2	The same
$C_6H_8Cl_5Sb$	SbCl$_4$(3-Cl acac)	77	24.18 24.71 26.27 26.53	...	[121]. cf. C—Cl

Empirical formula	Structural formula	T, °K	ν, MHz	s/n	References, notes
$C_6H_9Cl_4O_2Sb$	$SbCl_4$ (acac)	77	24.28 25.79	...	[121]
$C_{12}H_{16}BCl_8O_4Sb$	B(3-Cl acac)$_2\cdot SbCl_6$	77	24.1—24.6	...	[121]. cf. C—Cl
Cl_6Cs_2LiSb	$Cs_2LiSbCl_6$	77	13.500	...	*
Cl_6Cs_2RbSb	$Cs_2RbSbCl_6$	77	13.495	...	*
$BiCl_6Cs_2K$	Cs_2KBiCl_6	77	13.314 12.770 12.526 12.433	...	*cf. Bi
$BiCl_6Cs_2Na$	$Cs_2NaBiCl_6$	77	13.330 12.734 12.518 12.432	...	*
$Cl_3V + 6H_2O$	$VCl_3\cdot 6H_2O$	297	8.412	...	[122]
Cl_6CsNb	$CsNbCl_6$	300	8.59 8.25	...	[106]
Cl_6CsTa	$CsTaCl_6$	300	9.12 8.80	...	[106]
	Group VI element — chlorine bond				
C_4H_7ClO	$(CH_3)_3COCl$	77	54.94 55.28	15 15	*
ClHO	HOCl	$e^2Qq = -121.7$ MHz			[123]. g
Cl_2O	Cl_2O	$e^2Qq = -141.06$ MHz			[124]. g
CH_2Cl_2S	ClH_2CSCl	77	38.9		*[155]. cf. Cl—C
CH_3ClS	CH_3SCl	$e^2Qq = -80.69$ MHz; $\eta = 7.8\%$			[125]. g
C_2H_3ClOS	$CH_3C(=O)SCl$	77	39.756 39.742	10 10	*[155]

C_2H_5ClS	C_2H_5SCl	77	36.978	20	*[155]
C_2H_6ClNS	$(CH_3)_2NSCl$	77	35.598	10	*[155]
			36.00	10	
C_3H_7ClS	C_3H_7SCl	77	37.562	30	*[155]
	iso-C_3H_7SCl	77	37.226	40	*[155]
C_4H_9ClS	C_4H_9SCl	77	36.925	10	*[155]
			37.268	10	
			37.751	10	
C_7H_7ClS	$C_6H_5CH_2SCl$	77	38.78	3	*[155]
			38.41	3	
$C_{10}H_6ClNO_4S$	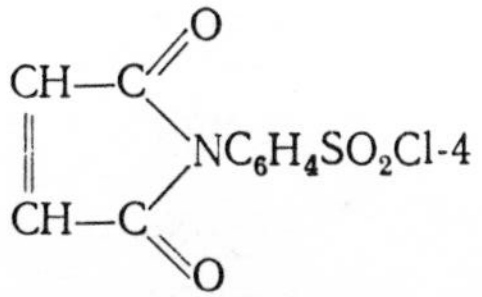	77	32.22	20	*
Cl_2S	SCl_2	77	40.250	10	*[155]
			39.088	10	
			39.018	10	
			39.011	10	
$Cl_3N_3O_3S_3$	α-$(NSOCl)_3$				[118, 119]
	Phase I	77	36.993	2	
			37.225	1	
	Phase II	294	36.134		
			36.445		
			37.550		
$Cl_4N_3O_2PS_2$	$Cl_2PN(NSOCl)_2$	294	34.521		[119]. cf. P—Cl
			35.472		
Cl_6Cs_2Te	Cs_2TeCl_6	77	15.625	30	*[106]
Cl_6Rb_2Te	Rb_2TeCl_6	77	15.150	15	*

Empirical formula	Structural formula	T, °K	ν, MHz	s/n	References, notes
$C_4H_{10}Cl_3CrN_2 + HCl$	$(Cren_2Cl_2)Cl \cdot HCl$	Room	$e^2Qq = 20.55$ MHz $\eta = 7.3\%$ $e^2Qq = 5.26$ MHz $\eta = 90.2\%$		[126]. NMR cr.
Cl_5Mo	$MoCl_5$	300	14.08	...	[106]
Cl_6Cs_2Mo	Cs_2MoCl_6	300	10.74	...	[106]
Cl_6K_3Mo	K_3MoCl_6	300	9.83 9.54	...	[106]
Cl_6CsW	$CsWCl_6$	300	11.75 11.60 11.28	...	[106]
Cl_6Cs_2W	Cs_2WCl_6	300	10.91	...	[106]
Cl_6KW	$KWCl_6$	300	11.46 11.34	...	[106]
Cl_6RbW	$RbWCl_6$	300	11.56 11.32	...	[106]
Cl_6Rb_2W	Rb_2WCl_6	300	10.58	...	[106]
	Group VII element — chlorine bond				
$Cl_2 + C_4H_8O_2$	$Cl_2 \cdot$ diox	77	53.257	...	[127]
$Cl_2 + C_4H_{10}O$	$Cl_2 \cdot O(C_2H_5)_2$	77	53.573	...	[127]
$Cl_2 + C_6H_5NO_2$	$Cl_2 \cdot C_6H_5NO_2$	77	54.510 53.675 53.050	...	[127]
$Cl_2 + C_6H_6$	$Cl_2 \cdot C_6H_6$	77	53.463	...	[127]
$Cl_2 + C_7H_8$	$Cl_2 \cdot C_6H_5CH_3$	77	53.730 53.595 53.338 53.128	...	[127]

$Cl_2 + C_8H_{10}$	$Cl_2 \cdot 1.2\text{-}(CH_3)_2C_6H_4$	77	53.863 53.530 53.400	. . .	[127]
	$Cl_2 \cdot 1.4\text{-}(CH_3)_2C_6H_4$	77	53.533	. . .	[127]
$C_6H_2Cl_5I$	2,4,6-$Cl_3C_6H_2ICl_2$	77	26.548	. . .	[128]
Cl_3KMn	$KMnCl_3$	77	10.26	. . .	[115]
Cl_5Re	$ReCl_5$	77	17.430 17.210	. . .	*
	Group VIII element — chlorine bond				
$C_4H_{10}Cl_3CoN_2 + HCl$	$(Coen_2Cl_2)Cl \cdot HCl$	Room	$e^2Qq = 31.91$ MHz $\eta = 20\%$ $e^2Qq = 5.04$ MHz $\eta = 90.4\%$		[126]. NMR cr.
Cl_3CsNi	$CsNiCl_3$	77	8.489	. . .	[153]. At 300 °K $\eta = 38.3\%$
$C_8H_{24}Cl_4N_2Ni$	$[N(CH_3)_4]_2NiCl_4$	273	8.85 9.05	. . .	[130]
$C_{12}H_{27}Cl_2NiP_2$	$[(C_4H_9)_3P]_2NiCl_2$	273	15.99	. . .	[130]
$C_{16}H_{40}Cl_4N_2Ni$	$[N(C_2H_5)_4]_2NiCl_4$	77	8.96 9.45	. . .	[130]
Cl_3Ru	$RuCl_3$	77	16.880 16.815	. . .	[115]
$C_4H_{10}Cl_3N_2Rh + HCl$	$(Rhen_2Cl_2)Cl \cdot HCl$	293	17.089	. . .	[131]
$C_{10}H_{12}Cl_4N_2Pd$	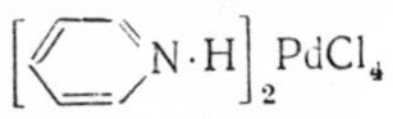	273	16.61 17.08	. . .	[130]
$C_{16}H_{40}Cl_4N_2Pd$	$[(C_2H_5)_4N^+]_2PdCl_4$	273	17.97	. . .	[130]
$C_{24}H_{54}Cl_4P_2Pd_2$	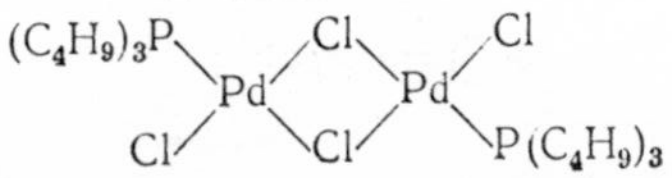	273	12.77 19.50	. . .	[130]

Empirical formula	Structural formula	T, °K	ν, MHz	s/n	References, notes
Cl_2Pd	α-$PdCl_2$	296	18.069	...	[132]. $\eta_{calc.}=15.7\%$
	β-$PdCl_2$	296	20.242	...	[132]
$Cl_2Pd + 2C_2H_6OS$	$PdCl_2 \cdot 2DMSO$	273	19.75	...	[130]
$Cl_2Pd + 2C_3H_5N$	$PdCl_2 \cdot 2C_2H_5CN$	273	20.30	...	[130, 133]
$Cl_2Pd + 2C_5H_5N$	$PdCl_2 \cdot 2Py$	273	17.72	...	[130, 133]
$Cl_2Pd + 2C_5H_{11}N$	$PdCl_2 \cdot 2$NH	273	16.11 16.31	...	[130, 133]
$Cl_2Pd + 2C_7H_5N$	$PdCl_2 \cdot 2C_6H_5CN$	273	20.58	...	[130, 133]
$Cl_2Pd + C_{10}H_8N_2$	$PdCl_2 \cdot 2,2'$-bipy	273	17.82 18.35	...	[130]
$Cl_2Pd + C_{12}H_{27}As$	$PdCl_2 \cdot 2As(C_4H_9)_3$	273	18.23 18.59	...	[130, 133]
$Cl_2Pd + C_{12}H_{27}P$	$PdCl_2 \cdot 2P(C_4H_9)_3$	273	18.37 18.50 18.58 18.63	...	[130, 133]
$C_2H_4Cl_3KPt + H_2O$	$K[\pi\text{-}CH_2{=}CH_2PtCl_3] \cdot H_2O$	77	20.420 20.520	1 2	*
$C_8H_{24}Cl_4N_2Pt$	$[N(CH_3)_4]_2PtCl_4$	273	18.38 19.16	...	[130]
$C_9H_{17}Cl_3N_2Pt$	$[N(CH_3)_4]C_5H_5NPtCl_3$	273	16.88 17.89 20.68	...	[130]
$C_{10}H_{12}Cl_4N_2Pt$	[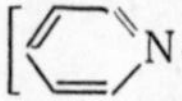N·H]$_2$ $PtCl_4$	273	18.41 18.60	...	[130]
Cl_2Pt	β-$PtCl_2$	296	23.615 23.035	...	[132]

$Cl_2Pt + 2C_2H_7N$	$PtCl_2 \cdot 2(CH_3)_2NH$-cis	273	17.21	. . .	[130]
	$PtCl_2 \cdot 2(CH_3)_2NH$-trans	273	18.16	. . .	[130, 133]
$Cl_2Pt + 2C_3H_5N$	$PtCl_2 \cdot 2C_2H_5CN$-cis	273	21.05 21.33	. . .	[130]
$Cl_2Pt + 2C_5H_5N$	$PtCl_2 \cdot 2C_5H_5N$-cis	273	17.70	. . .	[130]
	$PtCl_2 \cdot 2C_5H_5N$-trans	273	19.62	. . .	[130, 133]
$Cl_2Pt + 2C_6H_{15}P$	$PtCl_2 \cdot 2P(C_2H_5)_3$-trans	273	20.99	. . .	[130, 133]
$Cl_2Pt + 2C_9H_{21}P$	$PtCl_2 \cdot 2P(C_3H_7)_3$	273	15.46 22.36	. . .	[130]
$Cl_2Pt + 2C_{10}H_8N_2$	$PtCl_2 \cdot 2(2,2'$-bipy)	273	18.98	. . .	[130]
$Cl_2Pt + 2C_{10}H_{15}P$	$PtCl_2 \cdot 2P(C_2H_5)_2(C_6H_5)$-cis	273	17.82 17.99	. . .	[130]
$Cl_2Pt + 2C_{12}H_{27}P$	$PtCl_2 \cdot 2P(C_4H_9)_3$-cis	273	17.73 17.79 17.89 17.96	. . .	[130]
	$PtCl_2 \cdot 2P(C_4H_9)_3$-trans	273	20.90 21.04 21.08	1 2 1	[130, 133]
$Cl_2Pt + 2C_{14}H_{23}P$	$PtCl_2 \cdot 2P(C_4H_9)_2(C_6H_5)$-cis	273	18.33	. . .	[130]
	$PtCl_2 \cdot 2P(C_4H_9)_2 \cdot (C_6H_5)$-trans	273	21.32 21.64	. . .	[130]
$Cl_2Pt + 2H_3N$	$PtCl_2 \cdot 2NH_3$-trans	273	17.30	. . .	[130, 133]

C—Cl bond

The data were obtained at 77°K unless stated otherwise.

Empirical formula	Structural formula	ν, MHz	s/n	References, notes
CCl_3I	Cl_3CI	40.117 39.480	200 100	*
CCl_3HgI	Cl_3CHgI	39.16 39.12 38.30 38.25	1.5 1.5 3 3	*cf. I
$CHCl_3+{}^1/_2C_4H_8O_2$	$2CHCl_3 \cdot$ diox	37.741 38.008 38.070	6 6 6	[134]
CH_2ClS	ClH_2CSCl	35.0	2	*[155]. cf. Cl—S
$CH_2Cl_4Sn+C_4H_{10}O$	$ClCH_2SnCl_3 \cdot O(C_2H_5)_2$	37.404	30	*cf. Cl—Sn
$C_2Cl_3F_3$	CF_3CCl_3	40.53 40.65 40.77	2 5 4	[135]
$C_2H_2BrClHg$	trans-ClCH=CHHgBr	33.246	10	* cf. Br
	cis-ClCH=CHHgBr	33.078	3	* cf. Br
C_2H_2ClHgI	trans-ClCH=CHHgI	33.77	10	*
	cis-ClCH=CHHgI	32.95	10	*
C_2H_2ClF	CH_2=CFCl	$e^2Qq =$ $= -73.42$ MHz $\eta = 6.18\%$		[136]. g
$C_2H_2Cl_2Hg$	H_2C=ClCHgCl	34.13		*
$C_2H_2BrClHg$	H_2C=ClCHgBr	34.16		*
C_2H_2ClHgI	H_2C=ClCHgI	33.95		*
$C_2H_3ClF_2$	CH_3CF_2Cl	$e^2Qq =$ $= -72.63$ MHz $\eta = 1\%$		[137]
$C_3Cl_2F_2$	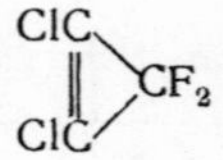	38.14 38.37 38.58 38.75	...	[138]
C_3Cl_3F	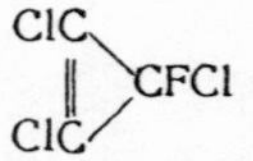	38.15 38.38 38.65 36.08 36.34	0.5 0.5 1 0.5 0.5	[138]
$C_3Cl_4F_2$	CCl_3CCl=CF_2 (glass)	38.8 36.8	10 3	*

Empirical formula	Structural formula	ν, MHz	s/n	References, notes
$C_3Cl_5F_3O$	$CFCl_2CF_2OCCl_3$	41.00 40.25 39.80 39.51 39.35	20 20 20 20 20	*
	$CF_2ClCFClOCCl_3$	40.1 38.05	8 5	*
C_3Cl_6	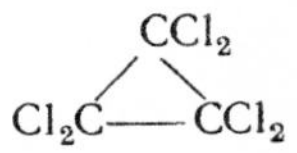CCl2 Cl2C — CCl2	40.124 40.313 40.439 40.572	5 5 30 10	*
C_3Cl_8	$CCl_3CCl_2CCl_3$	40.637 40.750 40.765 40.850 40.932 41.046 41.056 41.180	. . .	139]
$C_3H_2Cl_2O_2$	$CH_2(COCl)_2$	30.564 30.780 30.810 31.200	30 30 30 30	*[140]
$C_3H_2Cl_5F_3Si$	$CF_3CH_2CCl_2SiCl_3$	38.486 37.646	10 10	*
$C_3H_3Cl_2F_3$	$CF_3CH_2CHCl_2$	36.652 37.534	50 5	*
$C_3H_3Cl_2F_3O$	$CF_2ClCFClOCH_3$	37.6 35.65	5 5	*
$C_3H_3Cl_5$	$CCl_3CH_2CHCl_2$	38.976 38.689 38.514 36.977 36.893 36.680 36.309	30 30 30 20 20 20 20	*
C_3H_4ClF	cis-$CH_3CF{=}CHCl$	$e^2Qq =$ $= -73.49$ MHz $\eta = 8.44\%$		[141]. g
$C_3H_{11}ClB_{10}Br_2$	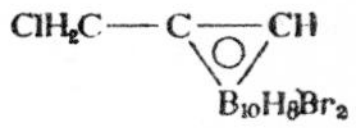ClH2C — C — CH B10H8Br2 C-chloromethyl-B-10,12-dibromo-o-borane	37.358	. . .	[154]

Empirical formula	Structural formula	ν, MHz	s/n	References, notes
$C_3H_{11}ClB_{10}I_2$	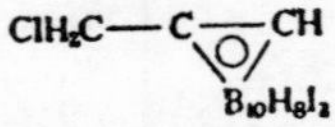 C-chloromethyl-B-10,12-diiodo-borane	36.848	...	[154]
$C_3H_{13}ClB_{10}S$	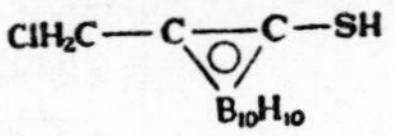	36.874	...	[154]
$C_4H_2Cl_2N_2$	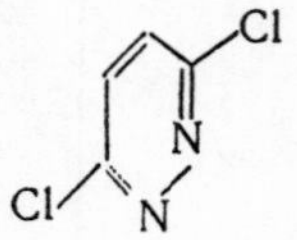	36.443 36.427 36.015 35.905	...	[149]
	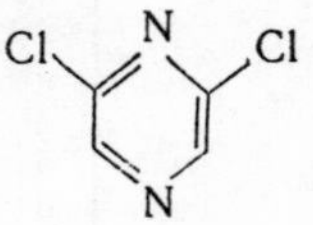	36.694 35.219	...	[149]
$C_4H_3ClN_2$	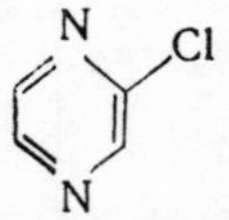	35,049	...	[149]
$C_4H_3Cl_2$	$CCl_2{=}CClCH{=}CH_2$	37.782 37.674 36.528 36.396	12 15 8 2	*
$C_4H_5Cl_3$	$ClCH_2C(CH_3)ClCOCl$ (glass)	36.8 35.4 31.5	5 5 3	*
$C_4H_5Cl_3OS$	$H_2C(SCl)CCl(CH_3)COCl$	36.5 31.7	6 4	*
$C_4H_7Cl_3$	$Cl_2CHCH_2CHClCH_3$	35.8 32.65	3 3	*
	$CCl_3CH_2CH_2CH_3$	38.0	10	*
$C_4H_7Cl_3$	$(ClCH_2)_2CClCH_3$	34.44 34.98	8 5	*
$C_4H_9Cl_2NO_4$	$ClCH_2COOH \cdot ClCH_2COONH_4$	35.123 33.710	...	[142]. $T = 20.1$ °K $T = 293$ °K
$C_4H_{13}ClB_{10}$		37.756 37.070 36.865 36.786	...	[154]

Empirical formula	Structural formula	ν, MHz	s/n	References, notes
$C_4H_{14}ClB_{10}ON$	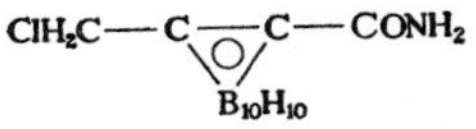	36.207	...	[154]
$C_4H_{15}ClB_{10}$	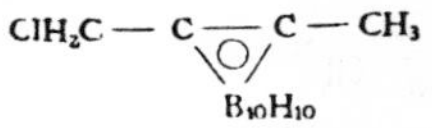	37.02	...	[154]
$C_4H_{15}ClB_{10}Hg$		35.565	...	[154]
$C_4H_{15}ClB_{10}S$	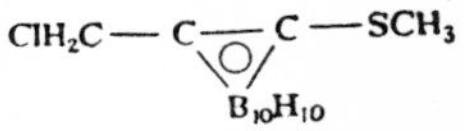	36.881	...	[154]
C_5Cl_{12}	$(CCl_3)_4C$	41.230 41.146 40.992 40.852 40.800 40.752 40.705	1 2 5 1 1 1 1	*
$C_5H_4ClN+ClI$	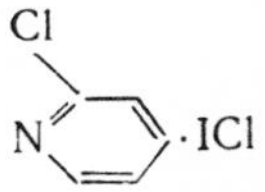	34.9	...	[105]. T=298 °K. cf. I
$C_5H_5ClN_2$	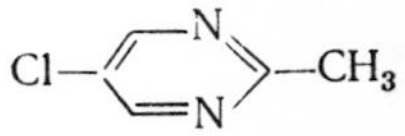	35.725	20	*
$C_5H_5ClN_2O$	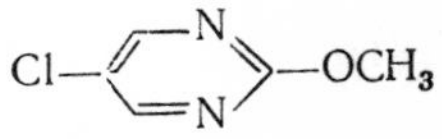	36.243	30	*
$C_5H_6Cl_2O_2$	$CH_2(CH_2COCl)_2$	28.770 29.340	10 10	*[140]
$C_5H_6Cl_4$	$Cl_2C(CH_2CH_2Cl)_2$	35.595 34.470	20 20	*
$C_5H_{11}ClNO_2Tl$	$(C_2H_5)_2TlCHClNO_2$	35.574 35.418	5 5	*
$C_5H_{15}ClB_{10}O_2$	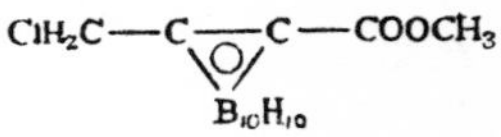	36.232	...	[154]

Empirical formula	Structural formula	ν, MHz	s/n	References, notes
$C_5H_{17}ClB_{10}$	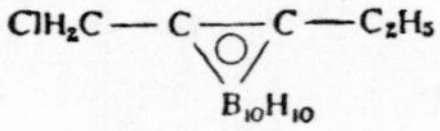	37.026	...	[154]
$C_6Cl_4O_2$	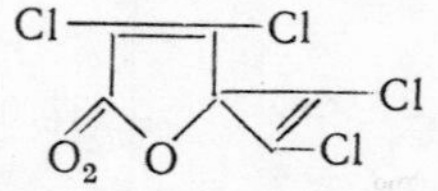	38.62 38.77 37.97 36.54	...	[138]
$C_6H_2BCl_3F_4N_2$	$[2,4,5\text{-}Cl_3C_6H_2N_2]BF_4$	38.859 38.775 38.572 37.980 37.943 37.660 37.274 37.244 37.191	...	[143]
	$[2,4,6\text{-}Cl_3C_6H_2N_2]BF_4$ Phase I from H_2O Phase II from CH_3CN	38.587 38.546 36.910 38.932 38.872 35.991	...	[143]
	$[3,4,5\text{-}Cl_3C_6H_2N_2]BF_4$	38.094 38.038 38.002	...	[143]
$C_6H_2BrCl_3$	$2,4,6\text{-}Cl_3C_6H_2Br$	36.594 36.038	5 3	[128]
$C_6H_2Cl_3I$	$2,4,6\text{-}Cl_3C_6H_2I$	37.034 36.356 35.807	2 2 2	[128]
$C_6H_2Cl_3NO_2$	$2,4,6\text{-}Cl_3C_6H_2NO_2$ unstable phase stable phase	 35.517 36.783 37.736 36.249	 8 4 4 2	[128]
$C_6H_2Cl_3NS_2$	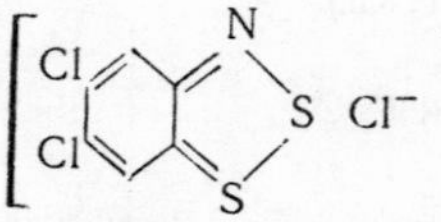	36.99	5	*
$C_6H_2Cl_5I$	$2,4,6\text{-}Cl_3C_6H_2ICl_2$	37.349 36.937 36.570	...	[128]
$C_6H_3BCl_2F_4N_2$	$[2,5\text{-}Cl_2C_6H_3N_2]BF_4$	37.996 37.219	...	[143]

Empirical formula	Structural formula	ν, MHz	s/n	References, notes
$C_6H_3BCl_2F_4N_2$	$[2,6\text{-}Cl_2C_6H_3N_2]BF_4$	38.104 37.779	...	[143]
	$[3,4\text{-}Cl_2C_6H_3N_2]BF_4$	37.680 36.616	...	[143]
	$[3,5\text{-}Cl_2C_6H_3N_2]BF_4$	37.612 37.371	...	[143]
$C_6H_3BrCl_2$	$3,4\text{-}Cl_2C_6H_3Br$	36.350 35.963	6 5	[128]
	$2,6\text{-}Cl_2C_6H_3Br$	36.084 36.096	...	*
	$3,5\text{-}Cl_2C_6H_3Br$	35.442 35.322 35.052	5 12 5	*Room temp.
$C_6H_3BrCl_2O$ + + $^1/_2C_6H_6Br$	$2(2,6\text{-}Cl_2\text{-}4\text{-}BrC_6H_2OH)\cdot$ $\cdot 4\text{-}BrC_6H_4NH_2$	35.280 35.484 36.132 36.180	10 10 10 10	*cf. Br
$C_6H_3BrCl_2O$ + + C_6H_6IN	$2(2,6\text{-}Cl_2\text{-}4\text{-}BrC_6H_2OH)\cdot$ $\cdot 4\text{-}IC_6H_4NH_2$	35.328 35.520 36.060	5 5 10	*cf. Br, I
$C_6H_3BrCl_2O + C_6H_7N$	$2,6\text{-}Cl_2\text{-}4\text{-}BrC_6H_2OH\cdot C_6H_5NH_2$	35.748 36.126	10 10	*cf. Br
$C_6H_3Br_2ClO + C_6H_7N$	$2,6\text{-}Br_2\text{-}4\text{-}ClC_6H_2OH\cdot C_6H_5NH_2$	35.676	20	*cf. Br
$C_6H_3ClN_2$	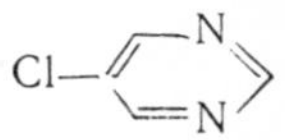	36.123	20	*
$C_6H_3Cl_2F$	$2,6\text{-}Cl_2C_6H_3F$	36.324	50	*
$C_6H_3Cl_2I$	$3,4\text{-}Cl_2C_6H_3I$	36.512 36.300 36.012 35.942	2 2 2 2	[128]
	$3,5\text{-}Cl_2C_6H_3I$	35.592	10	*
$C_6H_3Cl_2NO_2$	$3,5\text{-}Cl_2C_6H_3NO_2$	36.642 36.076	...	[143]
$C_6H_3Cl_3Hg$	$2,6\text{-}Cl_2C_6H_3HgCl$	34.440 34.824	40 40	*cf. Cl—Hg
$C_6H_3Cl_3O + C_6H_6IN$	$2,4,6\text{-}Cl_3C_6H_2OH\cdot 4\text{-}IC_6H_4NH_2$	35.538 35.748 35.784 36.132 36.156	20 10 10 10 10	*
$C_6H_3Cl_3O + C_6H_7N$	$2,4,6\text{-}Cl_3C_6H_2OH\cdot C_6H_5NH_2$	35.724 35.742 36.120	10 10 10	*

Empirical formula	Structural formula	ν, MHz	s/n	References, notes
$C_6H_4BClF_4N$	$[3\text{-}ClC_6H_4N_2]BF_4$	36.864	...	[143]
C_6H_4BrClO	$2\text{-}Cl\text{-}4\text{-}BrC_6H_3OH$	35.160 35.100 34.992	8 8 8	*
$C_6H_4BrCl_2N$	$3,5\text{-}Cl_2\text{-}4\text{-}BrC_6H_2NH_2$	36.183	7	*
$C_6H_4ClNO_3$	$2\text{-}Cl\text{-}6\text{-}NO_2C_6H_3OH$	36.747	20	*
$C_6H_4Cl_2Hg$	$2\text{-}ClC_6H_4HgCl$	33.984	10	*cf. Cl—Hg
$C_6H_4Cl_5P$	$4\text{-}ClC_6H_4PCl_4$	35.238	10	*[117]. cf. Cl—P
C_6H_5ClS	$4\text{-}ClC_6H_4SH$	34.368	3	*T=243 °K
$C_6H_6Br_2Cl_3Sb$	cis-$(ClCH{=}CH)_3SbBr_2$	34.710 34.520 34.446 34.326	4 4 8 8	*cf. Br. Sb
$C_6H_6Cl_3Sb$	trans-$(ClCH{=}CH)_3Sb$	33.024 32.940 32.748	5 5 5	*cf. Sb
$C_6H_6Cl_5Sb$	cis-$(ClCH{=}CH)_3SbCl_2$	34.746 34.248 34.206	7 5 5	*cf. Sb
$C_6H_8BClF_2$	$(3\text{-}Clacac)BF_2$	37.72	...	[121]
$C_6H_9ClN_3$	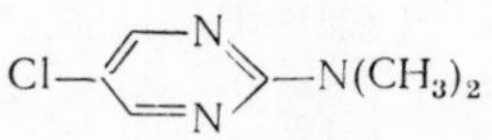	35.088	20	*
$C_6H_8Cl_5O_2Sb$	$(3\text{-}Clacac)SbCl_4$	38.22	...	[121]. cf. Cl—Sb
$C_6H_{12}Cl_2O_2$	$CHCl_2CH(OC_2H_5)_2$	36.379 36.631	15 5	*
$C_6H_{21}ClB_{10}Si$	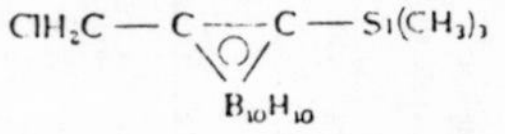	36.349	...	[154]
C_7Cl_8	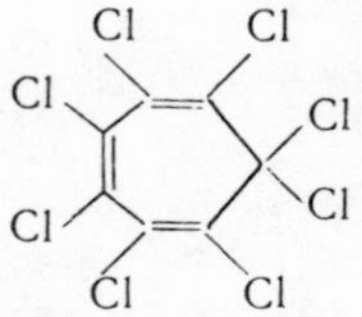	37.0 37.1 37.4 39.3 39.7	2 2 2 1 1	[129]
$C_7H_2Cl_3N$	$2,4,6\text{-}Cl_3C_6H_2CN$	37.647 37.052 36.442	...	[128]
$C_7H_3BrClF_3$	$2\text{-}Br\text{-}5\text{-}ClC_6H_3CF_3$	35.974 35.680	2 2	[128]

Empirical formula	Structural formula	ν, MHz	s/n	References, notes
$C_7H_3BrClF_3$	3-Br-4-$ClC_6H_3CF_3$	36.559	3	[128]
$C_7H_3ClF_3I$	2-I-5-$ClC_6H_3CF_3$	35.401	5	[128]
$C_7H_3Cl_2F_3$	2,5-$Cl_2C_6H_3CF_3$	36.788	3	[128]
		36.681	3	
		35.982	3	
		35.790	3	
	3,4-$Cl_2C_6H_3CF_3$	36.627	3	[128]
		36.277	3	
C_7H_4ClIO	3-IC_6H_4COCl	30.650	50	*
	4-IC_6H_4COCl	29.988	2	*
		30.372	2	
C_7H_4ClNS	4-ClC_6H_4SCN	34.638	20	*
$C_7H_4Cl_3NO$	2,4,6-$Cl_3C_6H_2CONH_2$			[128]
	Phase I	36.948	...	
		36.834		
		36.649		
		36.276		
		36.181		
		36.105		
		35.824		
		35.636		
		35.582		
	Phase II	36.472		
		35.894		
		35.765		
$C_7H_5ClF_3N$	2-Cl-5-$CF_3C_6H_3NH_2$	34.766	5	[128]
	4-Cl-2-$CF_3C_6H_3NH_2$	35.332	3	[128]
		35.262	3	
$C_7H_5ClN_2$	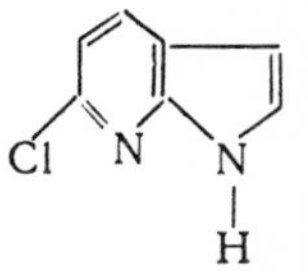	33.588	10	*
C_7H_5ClO	3-ClC_6H_4CHO	34.800	20	*
$C_7H_5ClO_2$	3-Cl-2-OHC_6H_3CHO	36.148	10	*
$C_7H_5Cl_3Hg$	$C_6H_5HgCCl_3$			*
	single crystal	34.452	15	
		36.761	15	
		36.824	15	
	polycrystal	36.198	5	
		36.276	5	
		36.300	5	
		36.600	5	
		36.660	5	
		37.116	5	

Empirical formula	Structural formula	ν, MHz	s/n	References, notes
$C_7H_5Cl_3O$	2,4,6-$Cl_3C_6H_2OCH_3$	35.709 36.323 36.767	. . .	[128, 144]
		At 292 °K $\eta_1 = 21\%$; $\eta_2 = 14\%$; $\eta_3 = 15\%$		
C_7H_6BrCl	2-Br-5-$ClC_6H_5CH_3$	34.806	2	[128]
C_7H_6ClI	2-I-5-$ClC_6H_3CH_3$	35.049 34.946	2 2	[128]
	3-I-4-$ClC_6H_3CH_3$	35.439	2	[128]
$C_7H_6ClNO_2$	4-Cl-3-$NO_2C_6H_3CH_3$	37.430	5	[128]
$C_7H_6Cl_2$	2,4-$Cl_2C_6H_3CH_3$	35.338 35.080	1,5 1,5	[128]
$C_7H_6Cl_3$	3,4-$Cl_2C_6H_3CH_3$	35.942 35.840	. . .	[128]
$C_7H_6Cl_3NO_2S_2$	$C_6H_5SO_2NHSCCl_3$	40.080 39.144 38.780	9 10 10	*
C_7H_7ClO	2-$ClC_6H_4OCH_3$	35.511	15	*
	2-CH_3-4-ClC_6H_3OH	34.818	15	*[145]
	2-Cl-4-$CH_3C_6H_3OH$	35.2	3	*
C_7H_7ClS	4-$ClC_6H_4SCH_3$	34.662	20	*
$C_7H_{75}ClS_2$	4-$ClC_6H_4SSCH_3$	34.890	20	*
C_7H_8ClN	2-CH_3-4-$ClC_6H_3NH_2$	34.134	3	[128]
C_7H_9Cl	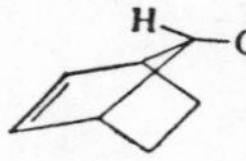	32.125 32.191	. . .	[151]
	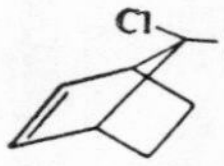	33.016 33.026	. . .	[151]
$C_7H_{10}Cl_2O_2$	$CH_2(CH_2CH_2COCl)_2$	29.170	50	*[140]
$C_7H_{13}Cl_3$	$CH_3CH_2CH_2CH(CH_3)CH_2CCl_3$	38.0	10	*
	$CH_3CHClCH_2CH(CH_3)\cdot CH_2CHCl_2$	35.5 32.55	10 10	*
$C_8H_2Cl_3F_3O_2$	2,4,6-$Cl_3C_6H_2OCOCF_3$	37.843 37.161 36.302	4 4 4	[128]

Empirical formula	Structural formula	ν, MHz	s/n	References, notes
$C_8H_2Cl_6O_2$	$2,4,6-Cl_3C_6H_2OCOCCl_3$	40.733	7	[128]
		40.457	7	
		40.161	7	
		37.244	6	
		37.126	5	
		35.729	5	
$C_8H_3Cl_2F_3O_2$	$2,4-Cl_2C_6H_3OCOCF_3$	37.012	2	[128]
		36.415	2	
$C_8H_3Cl_5O_2$	$2,4-Cl_2C_6H_3OCOCCl_3$	40.396	4	[128]
		40.179	4	
		39.930	4	
		36.493	4	
		35.338	4	
$C_8H_3Cl_5O_2$	$2,4,6-Cl_3C_6H_2OCOCHCl_2$	38.807	3	[128]
		38.743	3	
		38.254	3	
		38.105	3	
		37.205	3	
		37.127	2	
		36.972	3	
		36.914	3	
		36.461	3	
		36.366	3	
C_8H_4AgCl	$4-ClC_6H_4C{\equiv}CAg$	35.100	5	*
C_8H_4ClCu	$4-ClC_6H_4C{\equiv}CCu$	34.716	5	*
$C_8H_4Cl_4O_2$	$2,4-Cl_2C_6H_3OCOCHCl_2$	38.652	3	[128]
		38.527	3	
		36.655	3	
		35.695	4	
	$2,4,6-Cl_3C_6H_2OCOCH_2Cl$	37.325	4	[128]
		37.276	4	
		36.856	3	
		36.015	4	
$C_8H_4Cl_6$	$1,3-(CCl_3)_2C_6H_4$	38.472	20	*[146]
		38.582	20	
		38.694	20	
		38.772	10	
		38.916	10	
		38.988	10	
		39.094	10	
		39.468	10	
		38.488	10	
C_8H_5Cl	$4-ClC_6H_4C{\equiv}CH$	34.626	10	*

Empirical formula	Structural formula	ν, MHz	s/n	References, notes
$C_8H_5Cl_3O_2$	2,4,6-$Cl_3C_6H_2OCOCH_3$		...	[128]
	Phase I	36.787		
		36.748		
		36.648		
		36.636		
		36.150		
		35.650		
	Phase II	36.788		
		36.781		
		36.338		
$C_8H_5Cl_4O_2$	2,4-$Cl_2C_6H_3OCOCH_2Cl$	36.608	2	[128]
		36.561	2	
		35.213	3	
$C_8H_5Cl_5$	2-$CCl_3C_6H_4ClICl_2$	37.076	100	[147]
		37.302	100	
		38.976	140	
		39.281	140	
		39.879	140	
$C_8H_6Cl_2N_2O_4$	2,5-Cl_2-3,6-$(NO_2)_2C_6(CH_3)_2$-1,4	37.374	20	*
$C_8H_6Cl_2O_2$	4-$ClC_6H_4OCOCH_2Cl$			[128]
	Phase I	36.792	2	
		36.372	3	
		35.412	2	
		35.162	2	
	Phase II	36.396	4	
		35.848	4	
		35.017	4	
		34.733	4	
$C_8H_6Cl_2O_2$	2,4-$Cl_2C_6H_3OCOCH_3$	36.507	2	[128]
		36.467	3	
		35.400	2	
		35.314	2	
C_8H_7ClO	3-$CH_3C_6H_4COCl$	29.610	3	*
C_8H_7ClOS	4-$ClC_6H_4SCOCH_3$	34.968	40	*
$C_8H_7ClO_2$	2-OH-5-$ClC_6H_3COCH_3$	35.069	...	[128]
		34.961		
$C_8H_7ClO_2$	2-$ClC_6H_4OCOCH_3$	35.284	4	[128]
	4-$CH_3OC_6H_4COCl$	30.154	30	*
$C_8H_7Cl_3O$	2,4,6-$Cl_3C_6H_2OC_2H_5$	36.542	3	[128]
		36.158	4	
		35.736	4	
C_8H_9Cl	2,5-$(CH_3)_2C_6H_3Cl$	34.031	2	[128]
C_8H_9ClO	2,6-$(CH_3)_2$-4-ClC_6H_2OH	34.572	20	*
$C_8H_{10}ClNS$	4-$ClC_6H_4SN(CH_3)_2$	35.052	20	*

Empirical formula	Structural formula	ν, MHz	s/n	References, notes
$C_8H_{10}Cl_2O_2$	cis-1,4-$(ClOC)_2C_6H_{10}$-cyclo	29.00	30	*
»	trans-1,4-$(ClOC)_2C_6H_{10}$-cyclo	28.990	50	*
$C_8H_{12}Cl_2O_2$	$ClOC(CH_2)_6COCl$	29.085	30	*[140]
$C_8H_{14}Cl_2$	$(ClCH_2)_2C(CH_3)CH_2C(CH_3){=}CH_2$	33.924 33.504	10 10	*
$C_9H_3ClO_4$	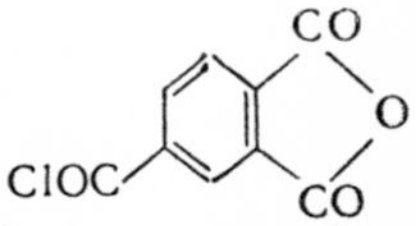	31.428 31.092	5 5	*
$C_9H_3Cl_3O_3$	1,3,5-$(COCl)_3C_6H_3$	31.158 31.188 30.888	10 10 5	*
$C_9H_3Cl_9$	1,3,5-$(CCl_3)_3C_6H_3$	39.023 39.088 39.117 39.238 39.402 39.644	. . .	[146]
$C_9H_5Cl_3$	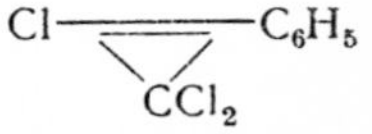	37.71 35.71 35.17	. . .	[138]
$C_9H_5Cl_3O$	$C_6H_5COCCl{=}CCl_2$	36.701 37.744 38.059	5 5 8	*
$C_9H_8Cl_2N_2O_2$	2,4-$(ClOCNH)_2C_6H_3CH_3$	33.17 32.94	3 4	*
$C_9H_{10}Cl_2O$	2,4-Cl_2-3,5-$(CH_3)_2\cdot C_6HOCH_3$	35.699 34.660	3 3	[128]
$C_9H_{11}ClO$	2,6-$(CH_3)_2$-4-$ClC_6H_2OCH_3$	34.392	15	*[128]
$C_9H_{14}Cl_2O_2$	$CH_2(CH_2CH_2CH_2COCl)_2$	29.290	5	*[140]
$C_{10}Cl_8$	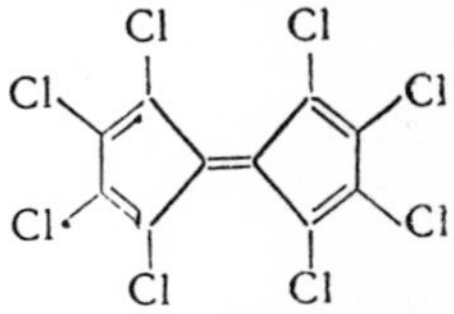	36.570 36.584 37.673 37.786	. . .	[148]
$C_{10}H_2Cl_4O_2$	1,2,4,5-$C_6H_2(COCl)_4$	31.220 31.500	20 20	*
$C_{10}H_7ClN_2$	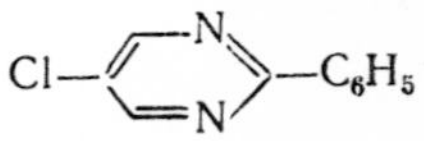	35.854	. . .	*

Empirical formula	Structural formula	ν, MHz	s/n	References, notes
$C_{10}H_8ClF_3O_2$	$3,5\text{-}(CH_3)_2\text{-}4\text{-}ClC_6H_2OCOCF_3$	34.348 34.322	1.7 2	[128]
$C_{10}H_8Cl_4O_2$	$3,5\text{-}(CH_3)_2\text{-}4\text{-}ClC_6H_2OCOCCl_3$	40.493 40.249 39.809 34.237	2 2 1.5 1.5	[128]
$C_{10}H_9ClN_2O_2$	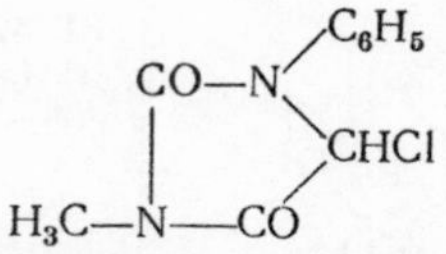	36.400	...	*
$C_{10}H_9Cl_3O_2$	$3,5\text{-}(CH_3)_2\text{-}4\text{-}ClC_6H_2OCOCHCl_2$	38.657 38.397 34.204	3 3 2	[128]
$C_{10}H_{10}Cl_2O_2$	$3,5\text{-}(CH_3)_2\text{-}4\text{-}ClC_6H_2OCOCH_2Cl$	36.579 34.540	1.5 2	[128]
$C_{10}H_{11}ClO_2$	$3,5\text{-}(CH_3)_2\text{-}4\text{-}ClC_6H_2OCOCH_3$	34.308	3	[128]
$C_{10}H_{11}ClO_4$	$3,4,5\text{-}(OCH_3)_3C_6H_2COCl$	30.120	10	*
$C_{10}H_{12}Cl_2O$	$2,4\text{-}Cl_2\text{-}3,5\text{-}(CH_3)_2C_6HOC_2H_5$	36.038 35.625 35.567 34.752 34.354 33.322	1.5 1.5 1.5 1.5 1.5 1.5	[128]
$C_{10}H_{13}ClO$	$3,5\text{-}(CH_3)_2\text{-}4\text{-}ClC_6H_2OC_2H_5$	33.768	2	[128]
$C_{10}H_{13}ClS$	$4\text{-}ClC_6H_4SC(CH_3)_3$	34.548 34.704 34.944	10 5 5	*
$C_{10}H_{19}B_{10}Cl$	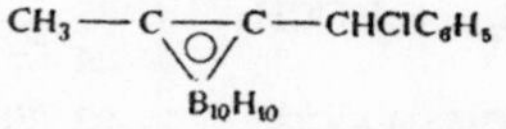	37.016	5	*
$C_{10}H_{19}Cl_3$	$CH_3CH_2CH_2CH(CH_3)CH_2CH\cdot$ $\cdot(CH_3)CH_2CCl_3$	38.0	5	*
$C_{11}H_{16}Cl_4O_4$	$CH_3OCOCH_2CHClCH_2—$ $CH(CH_2CCl_3)CH_2OCOCH_3$	38.35	2	*
$C_{12}H_7Cl_2FHgO$	$2,6\text{-}Cl_2\text{-}4\text{-}FC_6H_2OHgC_6H_5$	35.084 35.168	10 10	*
$C_{12}H_8BrClHgO$	$2\text{-}Cl\text{-}4\text{-}BrC_6H_3OHgC_6H_5$	33.336	10	*
$C_{12}H_8ClFHgO$	$2\text{-}Cl\text{-}4\text{-}FC_6H_3OHgC_6H_5$	32.928	7	*
$C_{12}H_8ClHgNO_3$	$2\text{-}Cl\text{-}4\text{-}NO_2C_6H_3OHgC_6H_5$	34.818	20	*

Empirical formula	Structural formula	ν, MHz	s/n	References, notes
$C_{12}H_8ClHgNO_3$	2-Cl-6-$NO_2C_6H_3OHgC_6H_5$	34.956	6	*
$C_{12}H_8Cl_2S_2$	4-$ClC_6H_4SSC_6H_4Cl$-4′	35.076	30	*
$C_{12}H_9ClF_6NP$	[(4-$ClC_6H_4CN)Fe(C_5H_5)]\cdot PF_6$	37.49	1.5	*
$C_{12}H_9ClHgO$	4-$ClC_6H_4OHgC_6H_5$	34.542	20	*
$C_{12}H_9ClS$	4-$ClC_6H_4SC_6H_5$	34.866	5	*
$C_{12}H_{12}ClF_6FeOP$	[(3-$ClC_6H_4OCH_3)Fe(C_5H_5)]\cdot PF_6$	37.66	2	*
	[(4-$ClC_6H_4OCH_3)Fe(C_5H_5)]\cdot PF_6$	37.16	2	*
$C_{12}H_{16}BCl_3O_4$	[(3-Clacac)$_2$B]ClO_4	37.77 38.02	. . .	[121]
$C_{12}H_{16}BCl_8Sb$	[(3-Clacac)$_2$B]$SbCl_6$	38.53	. . .	[121]. cf. Cl—Sb
$C_{12}H_{16}BeCl_2$	Be(3-Clacac)$_2$	35.47 36.18	. . .	[121]
$C_{13}H_7ClN_2O_6$	3,5-$(NO_2)_2C_6H_3COOC_6H_4Cl$-4′	35.098	3	*
$C_{13}H_8ClFO_2$	4-$ClC_6H_4COOC_6H_4F$-4′	35.208	15	*[150]
$C_{13}H_8ClNO_4$	4-$ClC_6H_4COOC_6H_4NO_2$-4′	35.244	20	*[150]
	4-$NO_2C_6H_4COOC_6H_4Cl$-4′	34.704	7	*[150]
$C_{13}H_8Cl_2O_2$	4-$ClC_6H_4COOC_6H_4Cl$-4′	35.028 34.790	3 3	*[150]
$C_{13}H_9ClHgO_2$	2-Cl-6-(CHO)$C_6H_3OHgC_6H_5$	33.090	4	*
$C_{13}H_9ClN_2$	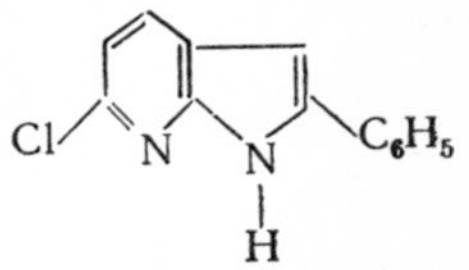	33.990 33.486	5 2	*
$C_{13}H_9ClO_2$	$C_6H_5COOC_6H_4Cl$-4′	34.872	4	*[150]
	4-$ClC_6H_4COOC_6H_5$	34.800	10	*[150]
$C_{13}H_{10}ClNO_2$	$C_6H_5NHCOOC_6H_4Cl$-2	35.700 35.898	20 20	*
	$C_6H_5NHCOOC_6H_4Cl$-4	34.854 35.110	15 15	*
$C_{13}H_{10}ClHgO$	2-Cl-4$CH_3C_6H_3OHgC_6H_5$	32.76	5	*
$C_{13}H_{11}ClS$	4-$ClC_6H_4SCH_2C_6H_5$	35.006	10	*
$C_{13}H_{12}ClF_6FeOP$	[(3-$ClC_6H_4COOCH_3)Fe(C_5H_5)]PF_6$	37.17	1.5	*

Empirical formula	Structural formula	ν, MHz	s/n	References, notes
$C_{14}H_4Cl_4O_2$	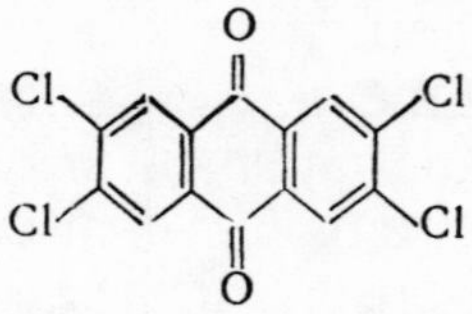	36.834 36.360	20 20	*
$C_{14}H_8Cl_2N_2O_2$	3-$ClOCC_6H_4N{=}NC_6H_4COCl$-3	30.696	3	*
$C_{14}H_9Cl$	4-$ClC_6H_4C{\equiv}CC_6H_5$	34.500	5	*
$C_{14}H_{10}Cl_4N_2S_2$	$C_6H_5NSCCl_2N(C_6H_5)SCCl_2$	39.074 39.032	20 20	*
$C_{14}H_{10}Cl_6HfO_4$	$(C_5H_5)_2Hf(COOCCl_3)_2$	40.047 39.650 39.501 39.228 39.368 38.794 38.542	2 2 2 4 8 1.5 2	*
$C_{14}H_{10}Cl_6OZr$	$(C_5H_5)_2Zr(COOCCl_3)_2$	39.480 39.130 37.982	...	*
$C_{14}H_{11}ClO$	$C_6H_5CH_2C_6H_4COCl$-4	30.480	15	*
$C_{14}H_{11}ClO_2$	4-$CH_3C_6H_4COOC_6H_4Cl$-4′	34.860	10	*[150]
$C_{14}H_{11}ClO_3$	4-$ClC_6H_4COOC_6H_4OCH_3$-4′	34.860	20	*[150]
	4-$CH_3OC_6H_4COOC_6H_4Cl$-4′	34.776	3	*[150]
$C_{14}H_{11}Cl_3NO_4S_2$	2-CH_3-5-$NO_2C_6H_3SO_2$— —$N(SCCl_3)C_6H_5$	39.746 39.382 39.256	5 5 5	*
$C_{15}H_8Cl_2O_3$	$(4\text{-}ClOCC_6H_4)_2CO$	30.762	30	*
$C_{15}H_8Cl_3NO_2$	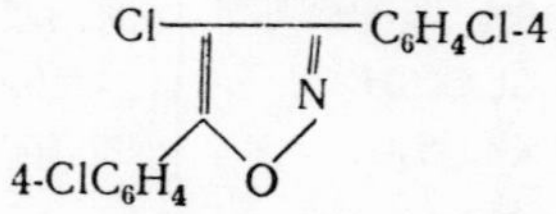	34.836 37.625 37.856 37.961	10 5 10 5	*
$C_{15}H_9BrClNO$	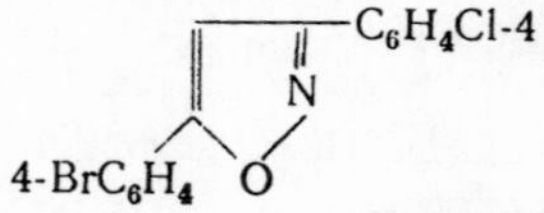	35.538	3	*
	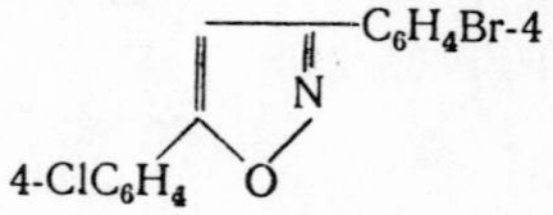	35.580	3	*

Empirical formula	Structural formula	ν, MHz	s/n	References, notes
$C_{15}H_9ClN_2O_3$	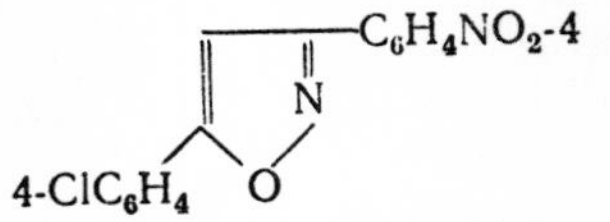	34.914	10	*
	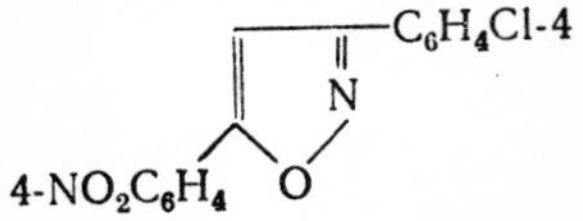	35.172	5	*
$C_{15}H_9Cl_2NO$	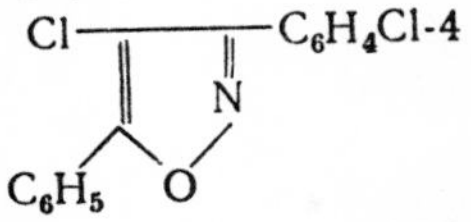	37.149 34.824	5 5	*
	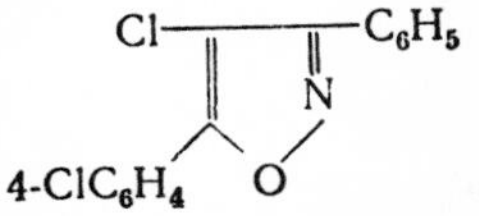	37.655 37.786 35.076 35.322	2 5 3 2	*
	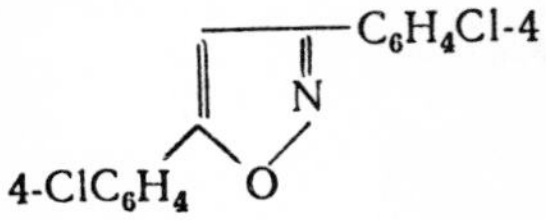	35.592	10	*
$C_{15}H_9Cl_3O_2$	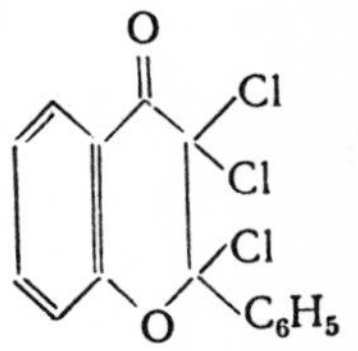	34.986 38.248 39.060	10 10 10	*
$C_{15}H_{10}ClNO$	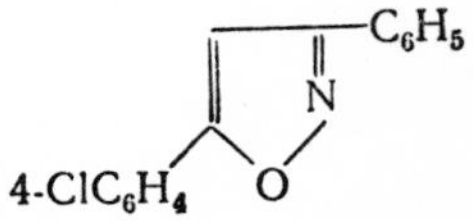	34.692		*
	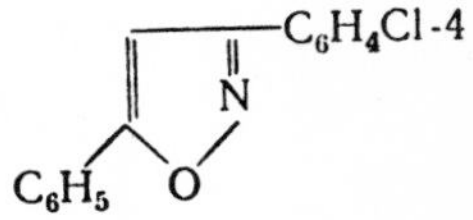	34.602	50	*
$C_{15}H_{10}Cl_2$	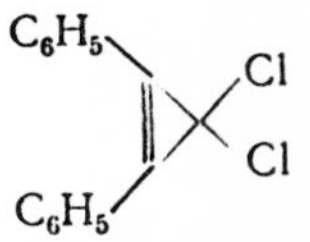	34.78 35.02	...	[138]
$C_{15}H_{10}Cl_2O_2$	$CH_2(C_6H_4COCl\text{-}4)_2$	29.994	3	*

Empirical formula	Structural formula	ν, MHz	s/n	References, notes
$C_{15}H_{10}Cl_3NO_2$	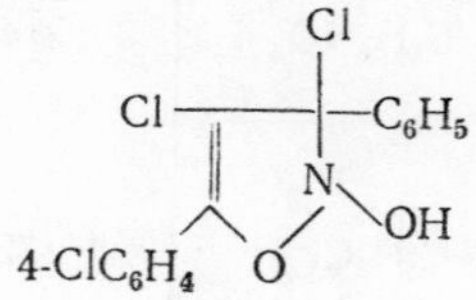	38.969 38.290 34.770	10 10 10	*
	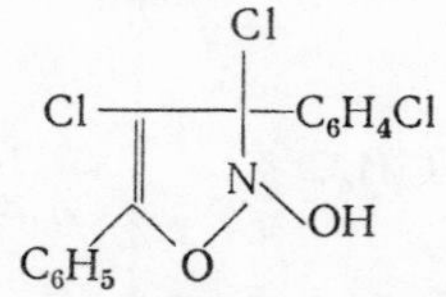	39.382 38.423 35.189	10 10 5	*
$C_{15}H_{11}ClO_3$	$4\text{-}ClC_6H_4COOC_6H_4COCH_3\text{-}4'$	35.382	15	*[150]
$C_{15}H_{11}Cl_2NO_2$	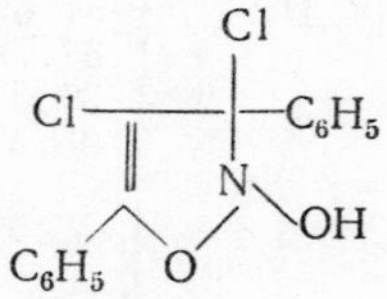	35.688	6	*
$C_{15}H_{13}ClN_2$	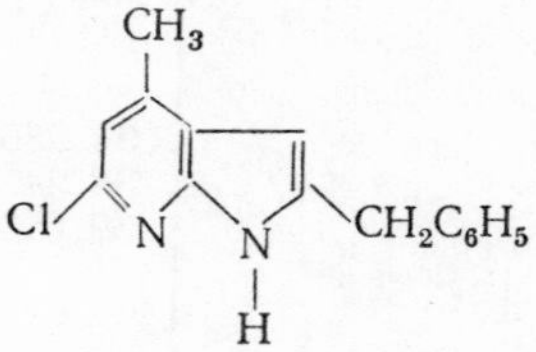	33.900	10	*
$C_{15}H_{13}ClN_2O$	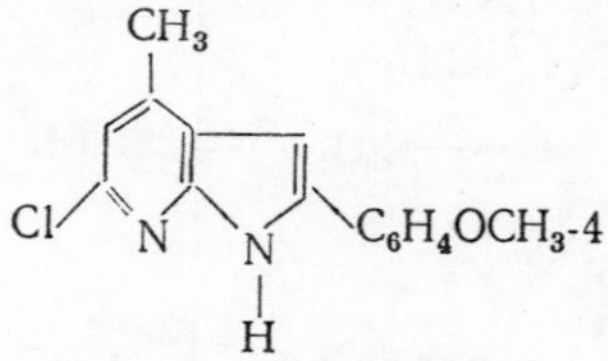	33.95	3	*
$C_{15}H_{14}Cl_2O_2$	$(3\text{-}Cl\text{-}4\text{-}HOC_6H_3)_2C(CH_3)_2$	35.634 34.866 34.686 34.596	10 10 10 10	*
$C_{18}H_{24}Cl_3Co$	$(3\text{-}Clacac)_3Co$	36.23	...	[123]
$C_{29}H_{22}Cl_2O_4$	$(CH_3)_2C(C_6H_3Cl\text{-}3\text{-}OCOC_6H_5\text{-}4)_2$	35.526 36.030	10 10	*

The data were obtained at 77°K unless stated otherwise.

Empirical formula	Structural formula	ν, MHz	s/n	References, notes
$Br_2Cd + 2C_5H_5N$	$CdBr_2 \cdot Py$	49.050	10	*
$Br_2Hg + C_4H_8O$	$HgBr_2 \cdot THF$	153.40 160.47	2 2	* Conversion
$Br_2Hg + C_4H_8O_2$	$HgBr_2 \cdot diox$	160.03 161.82		[102]. $T = 296$ °K
$C_2H_2BrClHg$	cis-ClCH=CHHgBr	130.50	~2	cf. Cl
$C_3H_4BrHgN_3O_9$	$(NO_2)_3CH_2CH_2HgBr$	136.50	5	*
C_7H_7BrHg	$C_6H_5CH_2HgBr$	121.46 120.22		*
BrInO	InOBr	85.65		*[49]. cf. In
$C_4H_{10}BrIn$	$(C_2H_5)_2InBr$	74.73		*[49]. cf. In
Br_4Th	$ThBr_4$	49.47 55.22 58.18 61.30 64.77	10 5 10 8 5	*
$Br_4Th + 10H_2O$	$ThBr_4 \cdot 10H_2O$	20.07	15	*
C_3H_5Br		$e^2Qq =$ $= 550.6$ MHz; $\eta < 4\%$		[103]. g
C_3H_6Br	trans-CH_3CH=CHBr	$e^2Qq =$ $= 557.7$ MHz; $\eta = 6\%$		[104]. g
$C_5H_3Br_2N + BrI$	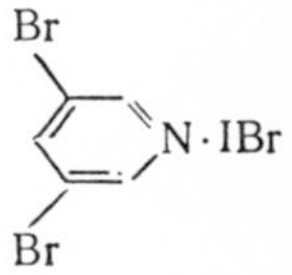	294.6 291.9	...	[105]. cf. I
$C_5H_3Br_2N + Br_2$	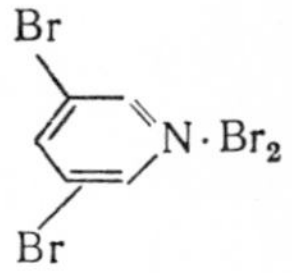	293.4 289.9	...	[105]. cf. Br
$C_5H_3Br_2N + ClHO_4$	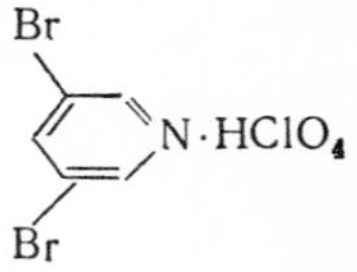	307.3 305.0	...	[105]

Empirical formula	Structural formula	ν, MHz	s/n	References, notes
$C_5H_3Br_2N + ClI$	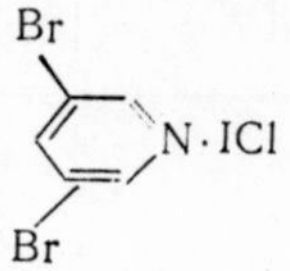	292.2 288.3	···	[105]. cf. I. $T=298$ °K.
$C_5H_4BrN + BrI$	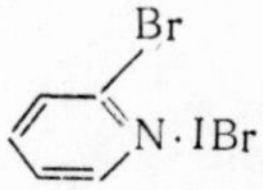	286.6	···	[105]. cf. I
$C_5H_4BrN + Br_2$	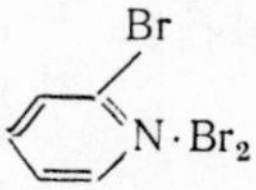	286.2	···	[105]. cf. Br
$C_5H_4BrN + ClHO_4$	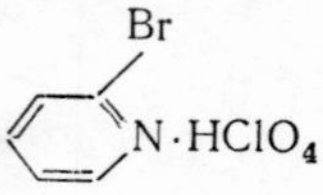	305.5	···	[105]
$C_5H_4BrN + ClI$	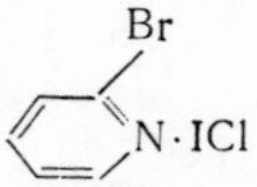	280.6	···	[105]. cf. I. $T=298$ °K
C_6BrCl_5	Cl_5C_6Br	307.8	2—3	*Conversion cf. Cl
$C_6H_3BrCl_2O +$ $+ {}^1/_2C_6H_6BrN$	2,6-Cl_2-4-BrC_6H_2OH· ·${}^1/_2$(4-$BrC_6H_4NH_2$)	286.40 281.21	5 5	*cf. Cl
$C_6H_3BrCl_2O +$ $+ {}^1/_2C_6H_6ClN$	2,6-Cl_2-4-BrC_6H_2OH· ·${}^1/_2$(4-$ClC_6H_4NH_2$)	286.60 280.90	5 5	*cf. Cl
$C_6H_3BrCl_2O +$ $+ {}^1/_2C_6H_6IN$	2,6-Cl_2-4-BrC_6H_2OH· ·${}^1/_2$(4-$IC_6H_4NH_2$)	286.30 281.30	5 5	*cf. Cl and I
$C_6H_3BrCl_2O +$ $+ C_6H_7N$	2,6-Cl_2-4-BrC_6H_2OH·$C_6H_5NH_2$	279.14	3	* cf. Cl
$C_6H_3Br_2ClO +$ $+ C_6H_6ClN$	2,6-Br_2-4-ClC_6H_2OH·4-$ClC_6H_4NH_2$	285.93 281.51	4 4	* cf. Cl
$C_6H_3Br_2ClO +$ $+ C_6H_7N$	2,6-Br_2-4-ClC_6H_2OH·$C_6H_5NH_2$	284.40 280.49	10 10	* cf. Cl
$C_6H_3Br_2FO$	2,6-Br_2-4-FC_6H_2OH	291.27 280.88	5 5	*Conversion
$C_6H_3Br_3O + C_6H_6NI$	2,4,6-$Br_3C_6H_2OH$·4-$IC_6H_4NH_2$	286.38 281.90 279.78	5 4 4	* cf. I

Empirical formula	Structural formula	ν, MHz	s/n	References, notes
$C_6H_3Br_3O + C_6H_7N$	2,4,6-$Br_3C_6H_2OH \cdot C_6H_5NH_2$	284.56 280.16 278.35	5 5 4	*
$C_6H_4Br_2O$	2,6-$Br_2C_6H_3OH$	284.00 277.00	2 2	*
$2C_6H_6BrN +$ $+ C_6H_3Cl_3O$	2·(4-$BrC_6H_4NH_2$)·2,4,6-$Cl_3C_6H_2OH$	272.43	2	* cf. Cl
$C_7H_6Br_2O$	2,6-Br_2-4-$CH_3C_6H_2OH$	283.47 281.93	3 3	*Conversion
»	2,6-$Br_2C_6H_3OCH_3$	281.70 280.60	5 5	*
$C_{12}H_9BrHgO$	2-$BrC_6H_4OHgC_6H_5$	255.69	2—3	*
Br_6Cs_2Sn	Cs_2SnBr_6	131.68	...	[106]. Conversion T=300 °K
CH_3Br_3Sn	H_3CSnBr_3	165.24 177.54	30 15	*
$C_2H_6Br_2Sn$	$(CH_3)_2SnBr_2$	131.65 133.08	100 90	*
C_3H_9BrSn	$(CH_3)_3SnBr$	111.92	100	*
Br_2Pb	$PbBr_2$ orthorhombic rhombic	15.955 33.240 24.688	30 50 20	*[107]
Br_3CsPb	$CsPbBr_3$ Phase I Phase II	61.570 41.629 70.43 67.16 38.42 37.11	... ...	*[107a]
$C_8H_{20}Br_3NPb$	$(C_2H_5)_4NPbBr_3$	43.344 43.064 41.090 40.222 37.314	...	*
$C_6H_7Br_3Ti$	$(CH_3C_5H_4)TiBr_3$	65.30 65.89 67.45	8 10 10	*
Br_7Cs_2CuSb	$Cs_2CuSbBr_7$	115.7	4	*

Empirical formula	Structural formula	ν, MHz	s/n	References, notes
$C_6H_6Br_2Cl_2Sb$	(cis-ClCH=CH)$_3$SbBr$_2$	140.64 143.93	5 5	* cf. Cl, Sb
$C_{18}H_{15}Br_2Sb$	$(C_6H_5)_3SbBr_2$	129.84 130.00 130.75 133.54	8 10 10 6	* cf. Sb
$Br_3V + 6H_2O$	$VBr_3 \cdot 6H_2O$	68.27	...	[108]. $T = 297$ °K
Br_6Rb_2Te	Rb_2TeBr_6	132.678	...	*
Br_2Cs_2W	Cs_2WBr_6	90.13	...	[106]. Conversion T=300 °K
Br_6W	WBr_6	86.33	...	The same
$Br_2 + C_5H_3Br_2N$	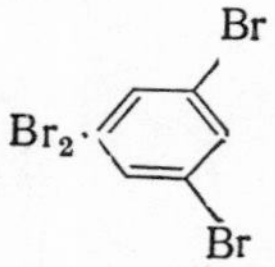	439.3 251.7	...	[105]. cf. Br—C
$Br_2 + C_5H_4BrN$	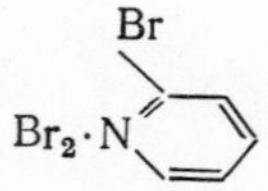	444.2	...	[105]. cf. Br—C
$Br_2 + C_5H_4ClN$	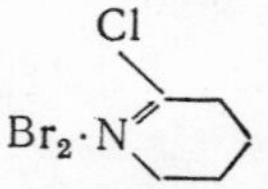	444.8	...	[105]
$C_7H_5BrReO_2$	$(CO)_2BrReC_5H_5$	196.09 199.08 201.74 205.24	4 4 4 4	*
$C_7H_5BrFeO_2$	$(CO)_2BrFeC_5H_5$	148.15	3	*
$C_9H_{15}B_{10}BrFeO_2$	$(CO_2)BrFeC_5H_4$—C—CH (bridged by O, $B_{10}H_9$)	140.88	5	*
Br_6Cs_2Os	Cs_2OsBr_6	139.85	...	[106]. Conversion T=300 °K
$Br_6H_8N_2Os$	$(NH_4)_2OsBr_6$	135.01	...	The same
Br_6K_2Os	K_2OsBr_6	133.81	...	»

Empirical formula	Structural formula	T, °K	$\nu_{1/2-3/2}$, MHz	s/n	$\nu_{3/2-5/2}$, MHz	s/n	e^2Qq, MHz	η, %	References, notes
			^{127}I						
AgI	β-AgI	298	—	—	—	—	7.91	...	[109]. NMR cr.
Ag_2HgI_4	Ag_2HgI_4	77	109.54	3	...	...	...	...	*
CCl_3HgI	Cl_3CHgI	77	155.67	3	...	...	...	...	* cf. Cl
C_2H_2ClHgI	trans-ClCH=CHHgI	77	131.27	10	262.60	10	875.2	0	* cf. Cl
	cis-ClCH=CHHgI	77	135.82	4	271.60	2	905.4	1	* cf. Cl
C_2H_3HgI	$H_2C{=}CHHgI$	77	126.00	...	...	...	...	...	*
C_4H_9HgI	C_4H_9HgI	77	124.17	...	...	...	...	...	*
$C_7H_6I_2O$	2,6-I_2-4-$H_3CC_6H_2OH$	77	294.30	20	...	...	...	...	*
			278.50	20	...	...	...	...	
$C_9H_{11}HgI$	1,3,5-$(CH_3)_3C_6H_2HgI$	77	145.17	4	290.09	4	967.1	2.5	*
Cs_2HgI_4	Cs_2HgI_4	77	97.82	3	195.44	20	651.6	2.8	*
			104.36	5	208.25	35	694.4	4.2	
			120.12	2	234.55	20	784.8	3.7	
$HgI_2 + C_4H_8O_2$	HgI_2·diox	296	172.77	...	...	...	...	...	[102]
$C_2H_{10}B_{10}I_2$		77	195.72	6	...	...	...	...	*
	B-10,12-diiodo-o-borane		196.14	6	...		...	...	
	$HCB_{10}H_8I_2CH$	77	204.22	7	...	...	...	...	*
	B-9,10-diiodo-m-borane		202.88	7	...		...	...	
			198.36	7	...		...	...	
			196.63	7	...		...	...	

Empirical formula	Structural formula	T, °K	$\nu_{1/2-3/2}$, MHz	s/n	$\nu_{3/2-5/2}$, MHz	s/n	e^2Qq, MHz	η, %	References, notes
$C_2H_{11}B_{10}I$	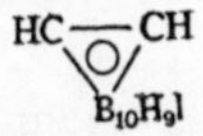 B-10(12)-iodo-o-borane	77	192.92	4	384.8	5	1283.2	4.5	*
	$HCB_{10}H_9ICH$ B-9(10)-iodo-m-borane	77	197.78	20	395.5	20	1318,4	1	*
	$HCB_{10}H_9ICH$ B-2-iodo-p-borane	77	204.76	20	405.4	20	1353.6	8.9	*
IInSe	InSeI	77	164.6 166.1 172.7 174.0 178.0	4 4 2 5 2	... ...	...	... ...	... ...	*
$C_2H_{10}B_{10}I_2$	$ICB_{10}H_{10}CI$ C,C-diiodo-p-borane	77	333.21	10	666.1	7	2220.6	1.9	*
$C_6H_3Cl_2I$	3,5-$Cl_2C_6H_3I$	77	274.40 278.00	21 7	... ...	...	... ...	... ...	*cf. Cl
$C_6H_3FI_2O$	2,6-I_2-4-FC_6H_2OH	77	295.35 290.85 290.10 288.35	3 3 3 3	... ...	...	... ...	... ...	*
C_6H_4FIO	2-I-4-FC_6H_3OH	77	285.75 288.30	40 20	564.8 571.0	20 10	1886.2 1906.3	9.6 8.7	*
C_6H_6IO	3-IC_6H_4OH	77	277.80	20	553.8	20	1846.9	5	*
C_6H_6IN + + $2C_6H_3BrCl_2O$	4-$IC_6H_4NH_2$· ·2(2,6-Cl_2-4-BrC_6H_2OH)	77	277.10	10	...	...	...	...	*cf. Cl, Br
$C_6H_6IN + C_6H_3Br_3C$	4-$IC_6H_4NH_2$·2,4,6-$Br_3C_6H_2OH$	77	271.96	5	...	...	...	...	*cf. Br
$C_6H_6IN + C_6H_3Cl_3O$	4-$IC_6H_4NH_2$·2,4,6-$Cl_3C_6H_2OH$	77	266.84	5	...	...	...	...	*cf. Cl

C_7H_7I	2-$IC_6H_4CH_3$	77	276.20	3	...	...	...	...	*
	3-$IC_6H_4CH_3$	77	276.75	10	553.2	10	1844.2	2	*
C_7H_7I	4-$IC_6H_4CH_3$	77	270.9	2	...	...	...	...	*
$C_8H_4MnIO_3$	$(CO)_3MnC_5H_4I$	77	305.55	10	610,6	5	2035,6	2.5	*cf. Mn
$C_{12}H_9HgIO$	2-$IC_6H_4OHgC_6H_5$	77	258.34	3	...	...	...	...	*
C_3H_9GeI	$(CH_3)_3GeI$	77	151.66	30	303,24	20	1010.9	1.2	*
$C_{12}H_{27}GeI$	$(C_4H_9)_3GeI$	77	154.92 157,04	10 10	... ...	...	... ...	... ...	*
CsI_3Pb	$CsPbI_3$	77	56,60	10	71,16	10	257.3	73.3	*[107a]
I_2Pb	PbI_2	77	17.50	5	34,53	10	116.6	0	*[107]
$I_6S_2Pb_5$	$Pb_5S_2I_6$	77	49,428 56,480 62,520	20 8 8	... ...	...	... ...	... ...	*
AsIS	AsSI	77	197.9	10	390.0	3	1303.0	10.7	* cf. As
AsISe	AsSeI	77	176,6	4—5					* cf. As
F_5I	IF_5	77	—	—	—	—	1073	...	[110]
F_7I	IF_7	77	—	—	—	—	—148	...	[110]
ClI + $C_5H_3Br_2N$	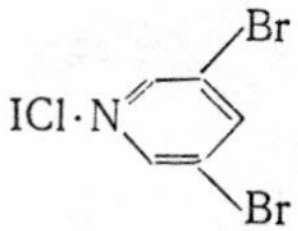	298	463.7	...	...	...	...	...	[105]. cf. Br
ClI + C_5H_4BrN	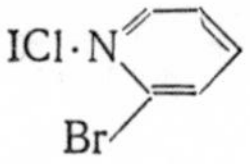	298	474,3	...	...	...	...	...	[105]. cf. Br
ClI + C_5H_4ClN	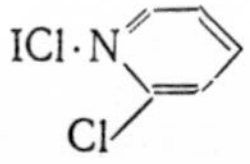	77 290	480.0 ...	...	... 940	... ...	... ...	... ...	[105]. cf. Cl

Empirical formula	Structural formula	T, °K	$\nu_{1/2-3/2}$, MHz	s/n	$\nu_{3/2-5/2}$, MHz	s/n	I^2Qq, MHz	η, %	References, notes
$ClI + C_5H_5N$	ICl·Py	298	450	...	...	...	...	...	[111]
		298	464.0	...	...	...	...	...	[105]
$ClIC_6H_7N$		298	462.5	...	...	...	...	...	[105]
$ClI + C_6H_{12}N_4$	$ICl \cdot N_4(CH_2)_6$	77	447.6	...	...	...	...	...	[105]
$BrI + C_5H_3Br_2N$	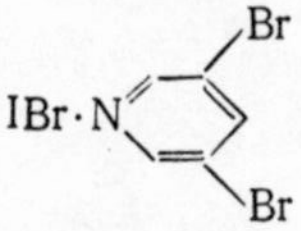	77	449.2	...	...	...	...	...	[105]. cf. Br
$BrI + C_5H_4BrN$	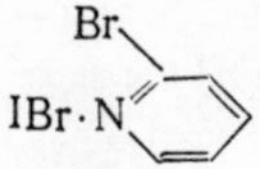	298	445.2	...	...	...	...	...	[105]. cf. Br
$BrI + C_5H_4ClN$	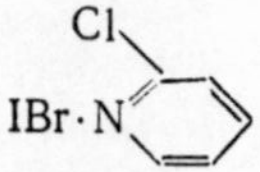	298	444.0	...	...	...	...	...	[105]
$BrI + C_5H_5N$	IBr·Py	298	445.7	...	...	...	...	...	[105]
$BrI + C_6H_7N$	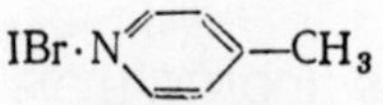	298	445.2	...	...	...	...	...	[105]
$BrI + C_6H_{12}N_4$	$IBr \cdot N_4(CH_2)_6$	298	437.4	...	...	...	...	...	[105]
$I_2 + C_3H_9N$	$I_2 \cdot N(CH_3)_3$	298	401.8	...	...	...	...	...	[105]
$I_2 + {}^1/_2C_5H_4N$	$2I_2 \cdot Py$	77	177.0	...	315.5	...	1091	28	[111]
			332.0	...	477.0	...	1713	58	
			376.0	...	...	...	2506	...	
$I_2 + C_6H_6$	$I_2 \cdot C_6H_6$	77	333.5	...	...	...	...	...	[111]

Empirical formula	Structural formula	T, °K	ν, MHz		e2Qq, MHz	η, %	References, notes
			1/2—3/2	3/2—5/2			
			^{55}Mn				
$C_7H_4MnNO_3$	$(C_4H_4N)Mn(CO)_3$	77 298	10.2287 10.1331	19.9355 19.6952	66.72 66.95	14.25 14.90	[95]
$C_{16}H_{26}MnO_2P$	$(C_5H_5)Mn(CO)_2 \cdot P(iso\text{-}C_3H_7)_3$	77	...	19.620	...	...	*
$C_{25}H_{20}MnO_2P$	$(C_5H_5)Mn(CO)_2 \cdot P(C_6H_5)_3$	77	...	18,78	...	...	*
$C_{25}H_{10}MnO_2Sb$	$(C_5H_5)Mn(CO)_2 \cdot Sb(C_6H_5)_3$	77	...	19.86	...	...	* cf. Sb
$C_{25}H_{20}MnO_5P$	$(C_5H_5)Mn(CO)_2 \cdot P(OC_6H_5)_3$	77	...	19,920	...	...	*
$C_{25}H_{38}MnO_2P$	$(C_5H_5)Mn(CO)_2 \cdot P(cyclo\text{-}C_6H_{11})_3$	77	10,27	19.50	65.4	18.6	*
$C_{28}H_{26}MnO_2P$	$(C_5H_5)Mn(CO)_2 \cdot P(CH_2C_6H_5)_3$	77	10.27	18.82	63.6	26.9	*
F_2Mn	MnF_2	Room	—	—	11,7	...	[96]. NMR cr.
Mn_2Sc	$ScMn_2$	4.2	—	—	21.3 10.58	0 71	[52]. NMR cr. cf. Sc

Empirical formula	Structural formula	T, °K	ν, MHz			e2Qq, MHz	η, %	References, notes
			1/2—3/2	3/2—5/2	5/2—7/2			
			^{59}Co					
$BrC_{12}Co_3O_{12}Sn$	$BrSn[Co(CO)_4]_3$	298	...	18.82	28,47	133.0	14	[97]
$Br_2C_8Co_2O_8Sn$	$Br_2Sn[Co(CO)_4]_2$	298	...	20.98 20.61	31.64 31,09	148.0 145.1	16 0	[97]
$Br_3C_4CoGeO_4$	$Br_3GeCo(CO)_4$	298	...	22,918	34,573	161,34	12	[98]
$Br_3C_4CoO_4Sn$	$Br_3SnCo(CO)_4$	298	11,65	22.79	34.26	160.0	11	[97, 98]
$CH_{12}BrCoN_4O_3$	$[Co(NH_3)_4](CO_3)Br$	Room	—	—	—	18.8	...	[99]. NMR cr.
$CH_{15}Cl_2CoN_6$	$[Co(NH_3)_5CN]Cl_2$	»	—	—	—	26,6	...	[99]. NMR cr.

Empirical formula	Structural formula	T, °K	ν, MHz $^1/_2$—$^3/_2$	ν, MHz $^3/_2$—$^5/_2$	ν, MHz $^5/_2$—$^7/_2$	e^2Qq, MHz	η, %	References, notes
$C_4Cl_3CoO_4Si$	$Cl_3SiCo(CO)_4$	298	9.825	18.500	27.957	130.67	13	[98]
		185	...	18.675	...	...	...	
		149	...	18.750	...	...	...	
		123	...	18.800	...	...	...	
$C_4Cl_3CoO_4Sn$	$Cl_3SnCo(CO)_4$	298	11.68	23.37	35.02	163.4	0	[97. 98]
$C_4CoGeI_3O_4$	$I_3GeCo(CO)_4$	298	...	22.694	34.078	159.08	4	[98]
$C_4CoI_3O_4Sn$	$I_3SnCo(CO)_4$	298	11.052	21.815	32.786	153.04	6	[98]
$C_4H_{10}CoN_5O_7$	trans-$[Co\ en_2(NO_2)_2]NO_3$	Room	—	—	—	13.2	...	[99]. NMR cr.
$C_4H_{22}B_{18}CoCs$	$Cs(1,2\text{-}B_9C_2H_{11})_2Co$	298	11.62	23.06	34.62	161.6	4.3	[100]
$C_7H_9CoO_4Sn$	$(CH_3)_3SnCo(CO)_4$	298	...	13.81	20.73	96.8	3	*[97]
			...	...	20.47	...	...	
$C_8Co_2HgO_8$	$Hg[Co(CO)_4]_2$	298	...	15.99	24.00	112.0	5	[97]
			...	15.81	23.75	110.8	0	
$C_{10}H_6CoO_8$	$(CH_3)_2[Co(CO)_4]_2$	298	8.15	15.26	22.65	105.7	0	[97]
$C_{10}H_{10}CoNO_3+H_2O$	$[Co(C_5H_5)]NO_3\cdot H_2O$	Room	...	...	...	165.6	...	[99]
$C_{12}H_5ClCo_2O_8Sn$	$C_6H_5SnCl[Co(CO)_4]_2$	298	...	18.356	27.500	128.4	5.1	[101]
			...	18.100	27.250	126.9	8.9	
$C_{13}H_3Co_3O_{12}Sn$	$CH_3Sn[Co(CO)_4]_3$	298	...	17.80	26.70	124.6	0	[97]
			...	17.56	26.47	122.5	0	
			...	17.22	25.77	120.2	0	
			...	16.97	25.57	119.3	10	
			...	16.66	25.34	118.6	18	
$C_{16}Co_4O_{16}Sn$	$Sn[Co(CO)_4]_4$	298	...	18.47	27.69	129.2	0	[97]
$C_{16}H_{10}BrCoGeO_4$	$(C_6H_5)_2BrGeCo(CO)_4$	77	...	18.150	27.315	127.65	9.2	[101]
$C_{20}H_{10}CoO_8Sn$	$(C_6H_5)_2Sn[Co(CO)_4]_2$	298	...	16.075	24.219	113.1	9.4	[101]
			...	16.050	24.089	112.7	6.3	
$C_{22}H_{15}CoO_4Pb$	$(C_6H_5)_3PbCo(CO)_4$	298	...	15.806	23.741	110.82	5	[98]
$C_{22}H_{15}CoO_4Si$	$(C_6H_5)SiCo(CO)_4$	298	...	14.434	21.662	101.09	0	[98]
$C_{22}H_{15}CoO_4Sn$	$(C_6H_5)_3SnCo(CO)_4$	77	...	15.035	22.55	105.38	2.3	*[98]
		298	...	14.852	22.304	104.11	5	
$C_{32}H_{12}CoN_8Na_4O_{12}S_4$	Tetrasodium salt of cobalt phthalocyanine sulfonic acid	77	...	—	—	168	...	[33]. EPR

BIBLIOGRAPHY

1. Pyykkö, P. and B. Pedersen. — Chem. Phys. Letters 2 (1968), 297.
2. Genin, G. J., D. E. O'Reilly, E. M. Peterson, and T. Tsang. — J. Chem. Phys. 48 (1968), 4525.
3. Ragle, J. L. and K. L. Sherk. — J. Chem. Phys. 50 (1959), 3553.
4. Caspary, W. J., F. Milett, M. Reichbach, and B. P. Dailey. — J. Chem. Phys. 51 (1969), 623.
4a. Zeiler, M. — Ber. Bunsengesellsch. phys. Chem. 69 (1965), 659.
5. Kukolich, S. G. — J. Chem. Phys. 51 (1969), 358.
5a. Kern, C. — J. Chem. Phys. 46 (1967), 4543.
6. Weiss, V. and W. Flygare. — J. Chem. Phys. 45 (1966), 3475.
7. Derbyshire, W., T. C. Gorvin, and D. A. Wagner. — Molec. Phys. 17 (1969), 401.
8. McCalley, R. C. and A. Kwiram. — Cited after No. 11.
9. Merchant, S. Z. and B. M. Fung. — J. Chem. Phys. 50 (1969), 2265.
10. Hilbers, C. W. and C. McLean. — Molec. Phys. 17 (1969), 433.
11. Olympia, P. L., I. Y. Wei, and B. M. Fung. — J. Chem. Phys. 51 (1969), 1610.
12. Hutton, G. and B. Pederson. — J. Phys. Chem. Soc. 30 (1969), 235.
13. Soda, G. and T. Chiba. — J. Phys. Soc. Japan 26 (1969), 249.
14. Soda, G. and T. Chiba. — J. Chem. Phys. 48 (1968), 4328.
15. Chiba, T. and G. Soda. — Bull. Chem. Soc. Japan 41 (1968), 1524.
16. Genin, D. J. and D. E. O'Reilley. — J. Chem. Phys. 50 (1969), 2842.
17. O'Reilly, D. E. — J. Chem. Phys. 52 (1970), 2407.
18. Hovi, V., U. Lähteenmäki, and R. Tuulensuu. — Phys. Rev. Lett. 29A (1969), 520.
19. Dmitrieva, L. V., Z. N. Zonn, and V. A. Ioffe. — Izv. AN SSSR, ser. inorg. mater. 5 (1969), 1269.
20. Soda, G. and T. Chiba. — J. Phys. Soc. Japan 26 (1969), 717.
21. Ossman, G. W. and J. W. McGrath. — J. Chem. Phys. 49 (1968), 783.
22. Peterson, G. E. and P. M. Bridenbaugh. — J. Chem. Phys. 51 (1969), 2610.
23. Cvikl, B. and J. W. McGrath. — J. Chem. Phys. 49 (1968), 3323.
24. Derbyshire, W. and J. P. Stuart. Magnetic Resonance and Relaxation, p. 403. Amsterdam, 1969.
25. Blinč, R., J. Stepisnik, and I. Zupančič. — Phys. Rev. 176 (1968), 732.
26. Soda, G. and T. Chiba. — J. Phys. Soc. Japan 26 (1969), 723.
27. Weiss, A. and W. Weyrich. — Z. Naturf. 24A (1969), 474.
28. Ketudat, S., I. Berthold, E. Paschalis, and A. Weiss. Magnetic Resonance and Relaxation, p. 1119. Amsterdam, 1967.
29. Blinč, C. R. and M. Mali. — Phys. Rev. 179 (1969), 552.
30. Gupta, L. C. and D. L. Setty Radhakrishna. — Chem. Phys. Letters 4 (1969), 263.
31. Matsumura, C. and D. R. Lide, Jr. — J. Chem. Phys. 50 (1969), 71.
32. Gupta, L. C., V. Rao, and Udaya Shaukar. — Phys. Rev. Lett. 28A (1968), 187.
33. Rollman and S. I. Chan. — J. Chem. Phys. 50 (1969), 3416.
34. Burkert, P. K. and H. P. Fritz. — Z. Naturf. ser. B, 24 (1969), 253.
35. Machmer, P. — J. Inorg. Nucl. Chem. 30 (1968), 2627.
36. Graybeal, J. and R. J. McKown. — J. Phys. Chem. 73 (1969), 3156.
37. Casabella, P. A. and T. Oja. — J. Chem. Phys. 50 (1969), 4814.
38. Farrar, T. C. and T. Tsang. — J. Res. Natn. Bur. Stand. A73 (1969), 195.
38a. Boden, N., H. Lutowsky, and J. Nansen. — J. Chem. Phys. 46 (1967), 2849.
39. Wiedemann, K. and J. Voitländer. — Z. Naturf. 24a (1969), 566.
40. Boesch, H. R., E. Brun, and B. Derighetti. — Helv. phys. Acta 41 (1968), 417.
41. Derighetti, B. and S. Ghose. — Phys. Rev. Lett. 28A (1969), 523.
42. Edmonds, D. T. and A. J. Lindop. — J. Appl. Phys. 39 (1968), 1008.
43. Jones, E. D. and V. H. Schmidt. — J. Appl. Phys. 40 (1969), 1406.
44. Brog, K. C., W. H. Jones, and C. M. Verber. — Phys. Rev. Lett. 20 (1966), 258.
45. Brinkmann, D., J. L. Staehli, and S. Ghose. — Helv. phys. Acta 42 (1969), 584.
46. Vinogradova, I. S. — Fiz. Tverd. Tela 10 (1968), 2204.
47. Vinogradova, I. S. and L. G. Falaleeva. — Izv. AN SSSR, phys. ser. 33 (1969), 254.
47a. Dewar, M. and D. Patterson. — Chem. Communs., p. 544, 1970.
48. Svergun, V. I., L. M. Bednova, O. Yu. Okhlobystin, and G. K. Semin. — Izv. AN SSSR, chem. ser., No. 6 (1970), 1449.
49. Svergun, V. I., P. Khagenmyuller, I. Port'e, A. F. Volkov, and G. K. Semin. — Izv. AN SSSR, chem. ser., p. 1676, 1970.

50. Svergun, V.I., Ya. Kh. Grinberg, V.M. Kuznets, and T.A. Babushkina. — Izv. AN SSSR, chem. ser., p. 1448, 1970.
51. Lovas, F.J. and T. Törring. — Z. Naturf. 24a (1969), 634.
52. Barnes, R.G. and B.K. Lunde. — J. Phys. Soc. Japan 28 (1970), 408.
53. Le Dang Khoi. — Phys. Rev. Lett. 28A (1969), 671.
54. Jones, E.D. — Phys. Rev. Lett. 27A (1968), 204.
55. Bray, P. — Phys. Rev. Lett. 27A (1968), 263.
56. Bray, P. — Phys. Rev. Lett. 28A (1968), 16.
57. Marino, R.A. and T. Oja. — Chem. Phys. Letters 4 (1970), 489.
58. McDonald, J.N., D. Taylor, J.K. Tyler, and J. Sheridan. — J. Molec. Spectrosc. 26 (1968), 285.
59. Pierce, L., D. Hayes, and J. Beecher. — J. Chem. Phys. 46 (1967), 4352.
60. Kemp, M.K. and W.H. Flygare. — J. Am. Chem. Soc. 90 (1968), 6267.
61. Coffey, D., C.O. Britt, and J.E. Boggs. — J. Chem. Phys. 49 (1968), 591.
62. Oja, T. — Phys. Rev. Lett. 30A (1969), 343.
63. Onda, S., R. Ikeda, D. Nakamura, and M. Kubo. — Bull. Chem. Soc. Japan 42 (1969), 1771.
64. Bodensch, H.K. and K. Morgenstern. — Z. Naturf. 25a (1970), 150.
65. Osokin, D.Ya., I.A. Safin, and I.A. Nuretdinov. — Doklady AN SSSR 190 (1970), 357.
66. Colligiani, A., R. Ambrosetti, and R. Angelone. — J. Chem. Phys. 52 (1970), 5022.
67. Rogowski, R.S. and R.H. Schwendeman. — J. Chem. Phys. 50 (1969), 397.
68. Anderson, L.O., M. Gourdij, L. Guibé, and W. Proctor. — Compt. rend. 267B (1968), 803.
69. Kintzinger, J.P. and J.M. Lehn. — Molec. Phys. 17 (1969), 135.
70. Wollrab, J.E. and V.W. Laurie. — J. Chem. Phys. 48 (1968), 5058.
71. Osokin, D.Ya., I.A. Safin, and I.A. Nuretdinov. — Doklady AN SSSR 186 (1969), 1128.
72. Okeda, R., D. Nakamura, and M. Kubo. — J. Phys. Chem. 72 (1968), 2982.
73. Moniz, W.B. and C.F. Poranski. — J. Phys. Chem. 73 (1969), 4145.
74. Wollrab, J.E., Le Roy, H. Scharpen, D.P. Ames, and J.A. Merritt. — J. Chem. Phys. 49 (1968), 2405.
75. Krause, L. and M. Whitehead. — J. Chem. Phys. 52 (1970), 2787.
76. Schempp, E. and P.J. Bray. — J. Chem. Phys. 49 (1968), 3450.
77. Ikeda, R., S. Onda, D. Nakamura, and M. Kubo. — J. Phys. Chem. 72 (1968), 2501.
78. Marino, R.A., L. Guibé, and P.J. Bray. — J. Chem. Phys. 49 (1968), 5104.
79. Yim, C.T., M.A. Whitehead, and D.H. Lo. — Can. J. Chem. 46 (1968), 3595.
80. Singh, K. and S. Singh. — Indian J. Phys. 41 (1968), 862.
81. Zussman, A. and S. Alexander. — J. Chem. Phys. 48 (1968), 3534.
82. Onda, S., R. Ikeda, D. Nakamura, and M. Kubo. — Bull. Chem. Soc. Japan 42 (1969), 2740.
83. Lide, D.R. — J. Chem. Phys. 38 (1963), 456.
84. Kirchhoff, W.H. and D.R. Lide. — J. Chem. Phys. 51 (1969), 467.
85. Kravchenko, E.A., S.A. Dembovskii, A.P. Chernov, and G.K. Semin. — Physica Status Solidi 31 (1969), K19.
86. Svergun, V.I., T.A. Babushkina, G.H. Shvedova, L.V. Kudryavtseva, and G.K. Semin. — Izv. AN SSSR, chem. ser., p. 179, 1970.
87. Nenasheva, S.N., I.N. Pen'kov, and I.A. Safin. — Doklady AN SSSR 183 (1968), 90.
88. Penkov, I.N., I.A. Safin, and S.N. Nenasheva. — Doklady AN SSSR 183 (1968), 689.
89. Petrov, L.N., I.A. Kyuntsel', and V.S. Grechishkin. — Vest. Leningr. State Univ., No. 1 (1969), 167.
90. Whitesides, G.M. and H.L. Mitchell. — J. Am. Chem. Soc. 91 (1969), 2245.
91. Allerhand, A. — J. Chem. Phys. 52 (1970), 2162.
92. Kasumoto, H., K. Narita, and E. Yamada. — J. Chem. Phys. 42 (1965), 1458.
93. Narita, K., J. Umeda, and H. Kasumoto. — J. Chem. Phys. 44 (1966), 2723.
94. Bennett, R.A., H.O. Hooper, and U. Roy. — J. Appl. Phys. 40 (1969), 2441.
95. Fayer, M.D. and C.B. Harris. — Inorg. Chem. 8 (1969), 2792
96. Yasuoka, H., N. Tin, V. Jaccarino, and H. Guggenheim. — Phys. Rev. 177 (1969), 667.
97. Spencer, D.D., J.L. Kirsch, and T.L. Brown. — Inorg. Chem. 9 (1970), 235.
98. Brown, T.L., P.A. Edwards, C.B. Harris, and J.L. Kirsch. — Inorg. Chem. 8 (1969), 763.
99. Spiess, H.W., H. Haas, and H. Hartmann. — J. Chem. Phys. 50 (1969), 3057.
100. Harris, C.B. — Inorg. Chem. 7 (1968), 1517.
101. Graybeal, J., S.D. Ing, and M.W. Hsu. — Inorg. Chem. 9 (1970), 678.
102. Brill, T.B. — J. Inorg. Nucl. Chem. 32 (1970), 1869.

103. Lam, F. M. K. and B. P. Dailey. — J. Chem. Phys. 49 (1968), 1588; 36 (1962), 2931.
104. Beaudet, R.A. — J. Chem. Phys. 50 (1969), 2002.
105. Bowmaker, G. A. and S. Hacobian. — Aust. J. Chem. 22 (1969), 2047.
106. Brown, L., W. G. McDugle Jr., and L. G. Kent. — J. Am. Chem. Soc. 92 (1970), 3745.
107. Volkov, A. F. and G. K. Semin. — Izv. AN SSSR, chem. ser., p. 1634, 1970.
107a. Volkov, A. F., Yu. N. Venevtsev, and G. K. Semin. — Physica Status Solidi 35 (1969) K167.
108. Podmore, L. P., P. W. Smith, and R. Stoessiger. — Chem. Communs., p. 221, 1970.
109. Segel, S. L., T. A. Walter, and G. T. Troup. — Physica Status Solidi 31 (1969), 43K.
110. Bukshpan, S., C. Goldstein, J. Soriano, and J. Shamir. — J. Chem. Phys. 51 (1969), 3976.
111. Thieme, H. — Diss. München, 1965.
112. Hynes, J. E., B. B. Garrett, and W. G. Moulton. — J. Chem. Phys. 52 (1970), 2671.
113. Brill, T. B. and Z. Z. Hugus. — Inorg. Chem. 9 (1970), 984.
113a. Stanko, V. I., E. V. Bryukhova, Yu. V. Gol'tyapin, and G. K. Semin. — Zh. Strukt. Khim. 10 (1969), 745.
114. Carlson, E. H. and H. S. Adams. — J. Chem. Phys. 51 (1969), 390.
115. Carlson, E. H. — Phys. Rev. Lett. 29A (1969), 696.
116. Negita, H., T. Okuda, and M. Mishima. — Bull. Chem. Soc. Japan 42 (1969), 2509.
117. Svergun, V. I., V. G. Rozanov, E.F. Grechkin, B.R. Timokhin, Yu. K. Maksyutin, and G. K. Semin. — Izv. AN SSSR, chem. ser., p. 1918, 1970.
118. Kaplansky, M., R. Clipsham, and M. A. Whitehead. — J. Chem. Soc. A, p. 584, 1969.
119. Clipsham, R., R. M. Hart, and M. A. Whitehead. — Inorg. Chem. 8 (1969), 2431.
120. Biedenkapp, D. and A. Weiss. — Z. Naturf. 23b (1968), 174.
121. Hammel, J. C., R. J. Lynch, J. A. S. Smith, A. Barabas, and E. Isfan. — J. Chem. Soc. A, p. 3000, 1969.
122. Podmore, L. P., P. W. Smith, and R. Stoessiger. — Chem. Communs., p. 221, 1970.
123. Lindsay, D. C., D. G. Lister, and D. J. Millen. — Chem. Communs., p. 950, 1969.
124. Jackson, R. H. and D. J. Millen. — Adv. Spectrosc. 3 (1962), 1157.
125. Guarnieri, A. — Z. Naturf. 23A (1968); 1867; 25A (1970), 18.
126. Hartmann, H., M. Fleissner, G. Gann, and H. Sillescu. — Theoret. Chim. Acta 3 (1965), 355.
127. Grechishkim, V. S., A. D. Gordeev, and Yu. A. Galishevskii. — Zh. Strukt. Khim. 10 (1969), 743.
128. Biedenkapp, D. and A. Weiss. — J. Chem. Phys. 49 (1968), 3933.
129. West, R. and K. Kusuda. — J. Am. Chem. Soc. 90 (1968), 7354.
130. Fryer, C. W. and J. A. S. Smith. — J. Chem. Soc. A, p. 1029, 1970.
131. Schreiner, A. F. and T. B. Brill. — Inorg. Nucl. Chem. Letters 6 (1970), 355.
132. Bronswyk, W. van and R. Nyholm. — J. Chem. Soc. A, p. 2084, 1968.
133. Fryer, C. W. and J. A. S. Smith. — J. Organometall. Chem. 18 (1969), 35.
134. Grechishkin, V. S. and I. A. Kyuntsel'. — Trudy ENI Permsk. State Univ. 12, No. 2 (1969), 11.
135. Barton, B. L. — J. Chem. Phys. 51 (1969), 4672.
136. Stone, R. G. and W. H. Flygare. — J. Molec. Spectrosc. 32 (1969), 233.
137. Graner, G. and C. Thomas. — J. Chem. Phys. 49 (1968), 4160.
138. Smith, R. M. and R. W. West. — Tetrahedron. Lett., p. 2141, 1969.
139. Zeil, W. and B. Haas. — Z. Naturf. 23A (1968), 1225.
140. Voronkov, M. G., S. A. Geller, I. N. Goncharov, L. I. Mironova, and V. P. Feshin. — Izv. AN LatvSSR, chem. ser., p. 250, 1969.
141. Stone, R. G., S. L. Srivastava, and W. H. Flygare. — J. Chem. Phys. 48 (1968), 1890.
142. Yamamoto, T., N. Nakamura, and H. Chihara. — J. Phys. Soc. Japan 25 (1968), 291.
143. Pies, W. and A. Weiss. — Z. Naturf. 24b (1969), 1268.
144. Penau, A. and L. Guibé. — Compt. rend. 266, B, No. 20 (1968), 1321.
145. Kantimati, B. — Curr. Sci. 38 (1969), 335.
146. Grechishkin, V. S., I. V. Izmest'ev, and G. B. Soifer. — Zh. Fiz. Khim. 43 (1969), 757.
147. Grechishkin, V. S., I. V. Izmest'ev, and G. B. Soifer. — Trudy ENI Permsk. State Univ. 12, No. 2 (1969), 3.
148. Agranat, J. D. Gill, M. Hayer, and R. M. J. Loewenstein. — J. Chem. Phys. 51 (1969), 2756.
149. Stidham, H. and H. Farrell. — J. Chem. Phys. 49 (1968), 2462.
150. Korshak, V. V., S. V. Vinogradova, V. A. Vasnev, E. V. Bryukhova, and G. K. Semin. — Izv. AN SSSR, chem. ser., p. 681, 1970.

151. Chihara, H., N. Nakamura, and T. Irie. — Bull. Chem. Soc. Japan 42 (1969), 3034.
152. Svergun, V. I., A. E. Borisov, N. V. Novikova, T. A. Babushkina, E. V. Bryukhova, and G. K. Semin. — Izv. AN SSSR, chem. ser., p. 484.
153. Hartmann, H., H. Haas, and H. Rinneberg. — J. Chem. Phys. 50 (1969), 3064.
154. Voronkov, M. G., V. P. Feshin, A. P. Snyakin, V. I. Kalinin, and L. I. Zakharkin. Chemistry of Heterocyclic Compounds, p. 565, 1970. (Russian)
155. Babushkina, T. A., V. S. Levin, M. I. Kalinkin, and G. K. Semin. — Izv. AN SSSR, chem. ser., p. 2340, 1969.

SUBJECT INDEX